ADVANCES IN

FOOD AND NUTRITION RESEARCH

VOLUME 51

ADVANCES IN

FOOD AND NUTRITION RESEARCH

VOLUME 51

Edited by

STEVE L. TAYLOR
Department of Food Science and Technology
University of Nebraska
Lincoln, Nebraska
USA

ELSEVIER

AMSTERDAM • BOSTON • HEIDELBERG • LONDON
NEW YORK • OXFORD • PARIS • SAN DIEGO
SAN FRANCISCO • SINGAPORE • SYDNEY • TOKYO
Academic Press is an imprint of Elsevier

Academic Press is an imprint of Elsevier
525 B Street, Suite 1900, San Diego, California 92101-4495, USA
84 Theobald's Road, London WC1X 8RR, UK

This book is printed on acid-free paper. 

For information on all Academic Press publications
visit our Web site at www.books.elsevier.com

ISBN-13: 978-0-12-016451-6
ISBN-10: 0-12-016451-5

PRINTED IN THE UNITED STATES OF AMERICA
06 07 08 09 9 8 7 6 5 4 3 2 1

CONTENTS

Food Components That Reduce Cholesterol Absorption

Timothy P. Carr and Elliot D. Jesch

Imaging Techniques for the Study of Food Microstructure: A Review

Pasquale M. Falcone, Antonietta Baiano, Amalia Conte, Lucia Mancini, Giuliana Tromba, Franco Zanini, and Matteo A. Del Nobile

Electrodialysis Applications in the Food Industry

Marcello Fidaleo and Mauro Moresi

CONTRIBUTORS TO VOLUME 51

Numbers in parentheses indicate the pages on which the authors' contributions begin.

Antonietta Baiano, *Department of Food Science, University of Foggia, Foggia, FG 71100, Italy (205)*

Timothy P. Carr, *Department of Nutrition and Health Sciences, University of Nebraska-Lincoln, Lincoln, Nebraska 68583 (165)*

Amalia Conte, *Department of Food Science, University of Foggia, Foggia, FG 71100, Italy (205)*

Matteo A. Del Nobile, *Department of Food Science, University of Foggia, Foggia, FG 71100, Italy (205)*

Pasquale M. Falcone, *Department of Food Science, University of Foggia, Foggia, FG 71100, Italy (205)*

Marcello Fidaleo, *Department of Food Science and Technology, University of Tuscia, Via San Camillo de Lellis, 01100 Viterbo, Italy (265)*

Clifford Hall III, *Department of Cereal and Food Sciences, North Dakota State University, Fargo, North Dakota 58105 (1)*

Elliot D. Jesch, *Department of Nutrition and Health Sciences, University of Nebraska-Lincoln, Lincoln, Nebraska 68583 (165)*

Lucia Mancini, *Sincrotrone Trieste S.C.p.A. in Area Science Park I, Basovizza, TS 34012, Italy (205)*

Mauro Moresi, *Department of Food Science and Technology, University of Tuscia, Via San Camillo de Lellis, 01100 Viterbo, Italy (265)*

A. V. Rao, *Department of Nutritional Sciences, Faculty of Medicine, University of Toronto, Toronto, Ontario, Canada (99)*

L. G. Rao, *Department of Medicine, Calcium Research Laboratory, St. Michael's Hospital, University of Toronto, Toronto, Ontario, Canada (99)*

M. R. Ray, *London Regional Cancer Program, London, Ontario, Canada (99)*

Giuliana Tromba, *Sincrotrone Trieste S.C.p.A. in Area Science Park I, Basovizza, TS 34012, Italy (205)*

Mehmet C. Tulbek, *Northern Crops Institute, North Dakota State University, Fargo, North Dakota 58105 (1)*

Yingying Xu, *Department of Cereal and Food Sciences, North Dakota State University, Fargo, North Dakota 58105 (1)*

Franco Zanini, *Sincrotrone Trieste S.C.p.A. in Area Science Park I, Basovizza, TS 34012, Italy (205)*

FLAXSEED

CLIFFORD HALL III,* MEHMET C. TULBEK,[†]
AND YINGYING XU*

*Department of Cereal and Food Sciences, North Dakota State University
Fargo, North Dakota 58105
[†]Northern Crops Institute, North Dakota State University
Fargo, North Dakota 58105

ADVANCES IN FOOD AND NUTRITION RESEARCH VOL 51

ISSN: 1043-4526
DOI: 10.1016/S1043-4526(06)51001-0

I. INTRODUCTION

Flaxseed or linseed (*Linum usitatissimum*) is an ancient crop that has been used for food and fiber. In North America, flaxseed is the preferred term for flax used in human consumption whereas Europeans use the term linseed for edible flax (Vaisey-Genser and Morris, 2003). Historical records indicate that flaxseed dates back to around 9000–8000 B.C. in Turkey (van Zeiste, 1972), Iran (Helbaek, 1969), Jordon (Hopf, 1983; Rollefson *et al.*, 1985), and Syria (Hillman, 1975; Hillman *et al.*, 1989). Although the evidence does not clearly show that flaxseed was cultivated, the seeds have been found alongside domesticated wheat and barley (Zohary and Hopf, 2000). Domestication of flaxseed is clearly evident around 7000–4500 B.C. (reviewed in Vaisey-Genser and Morris, 2003; Zohary and Hopf, 2000). The first probable use of flaxseed as food may have been as an ingredient in breads (Stitt, 1994) and as a laxative (Judd, 1995).

Flaxseed is grown in approximately 50 countries most of which are in theNorthern Hemisphere. In 2002, Canada was the largest producer of flaxseed accounting for approximately 33%, of the 2 million metric tons produced, followed by China (20%), the United States (16%), and India (11%) (Berglund, 2002). In general, the world supply of flaxseed has remained constant. Flaxseed acreage in the United States reached 516,000 harvested acres in 2004. Ninety-four percent of the flaxseed was grown in North Dakota followed by Montana (4%), and South Dakota (2%) (National Agricultural Statistics Service, 2005). Production in North Dakota totaled 10.5 million bushels and the yield per acre was 20.3 bushels (National Agricultural Statistics Service, 2005). The estimated harvest acreage for 2005 in North Dakota was approximately 955,000 acres (National Agricultural Statistics Service, 2006). Flaxseed acreage in the Canada totaled 1.9 million with a production of approximately 42 million bushels in 2004 (Agriculture and Agri-Food Canada, 2006).

History shows that flaxseed has been used as an ingredient in breakfast cereals and breads; however, since the 1990s, a number of products containing flaxseed have been developed primarily for the health food market. The renewed interest in flaxseed as a food source is due to findings that suggest that flaxseed can provide a variety of health benefits (Thompson and Cunnane, 2003). The components that contribute the health benefits include lignans (secoisolariciresinol diglucoside [SDG] being the predominant form), α-linolenic acid (ALA), and nonstarch polysaccharides (i.e., gum or fiber).

Flaxseed is an oilseed that contains roughly 38–45% oil. ALA, a polyunsaturated lipid, accounts for 52% of the fatty acids in the oil. Flaxseed is also a rich source of plant lignans (up to 13 mg/g flaxseed). The interest in ALA and lignans as food ingredients has opened opportunities for the utilization of flaxseed in foods. In contrast, the same level of interest has not been observed for other flaxseed components, such as protein and dietary fiber, which account for 20% and 28% of the flaxseed, respectively (Carter, 1993). This chapter will provide a general overview of flaxseed research completed over the past 50 years with the major focus being on data from 1990 to 2006. It will highlight the basic composition, health benefits, and finally the processing and application of flaxseed.

II. FLAXSEED COMPONENTS

A. FLAXSEED OIL

Flaxseed oil content falls in the range 38–45% oil depending on location, cultivar, and environmental conditions (Daun *et al.*, 2003; Oomah and Mazza, 1997). Kozlowska (1989) reported an average of 41.4% oil content for Polish cultivars. North Dakota flaxseed cultivars ranged from 31.9% to 37.8% oil (Hettiarachchy *et al.*, 1990). Wakjira *et al.* (2004) reported oil contents between 29.1% and 35.9% among flaxseed cultivars grown in Ethiopia.

In addition to oil content, fatty acid distribution in flaxseed can be affected by environmental conditions (Taylor and Morrice 1991). Growing conditions and variety can influence the unsaturated fatty acid content in flaxseed (Daun *et al.*, 2003). In contrast, the environment may also have an undesired impact on flaxseed composition. Early and late frosts, heat damage, and drought could have detrimental effects on flaxseed quality (Daun *et al.*, 2003). Significantly lower oil contents and a darken oil from frost-damaged immature seeds was reported by Gubbels *et al.* (1994). In addition, higher concentrations of palmitic (P), linoleic (La), and linolenic (Ln) acids and lower oleic (O) acid were observed in damaged seed compared to normal seeds.

Wanasundara *et al.* (1999) reported that neutral lipids (acylglycerols and fatty acids) constitute 96% of the total lipid in flaxseed, whereas polar lipids (glycolipids and phospholipids) account for 1.4%. Stenberg *et al.* (2005) observed similar findings except that less phospholipid was detected. Froment *et al.* (1999) discussed the effects of cultivar, location, and late harvest on phospholipid content. Neutral lipid fraction of flaxseed meal was 95–98% triacylglycerols (TAG) and thus accounts for the predominant lipid in flaxseed (Oomah *et al.*, 1996).

Ayorinde (2000) reported trilinolenate (sn-LnLnLn; 35%) as the most predominant TAG in flaxseed oil. In a study, Holcapek *et al.* (2003) observed comparable results (Table I). TAG and diacylglycerol (DAG) composition of

TABLE I

TRIACYLGLYCEROL DISTRIBUTION OF FLAXSEED[a,b]

Triacylglycerols (TG)	%	Triacylglycerols (TG)	%
LnLnLn	30.4	OLnP	3.1
LaLnLn	18.7	LnLaP	3.0
OLnLn	13.5	SLaLa	1.1
LnLnP	6.9	OLaLa	1.0
OLaLn	5.9	LaLaLa	0.9
LaLaLn	5.3	OLaO	0.8
OLnO	4.2	LaOP	0.6
SLnLn	4.1	PLnP	0.5

[a]Adapted from Holcapek *et al.* (2004).

[b]P, palmitic acid; S, stearic acid; O, oleic acid; La, linoleic acid; Ln, linolenic acid.

flaxseed oil was analyzed by high-performance liquid chromatography-mass spectrometry (HPLC-MS) with atmospheric pressure chemical ionization. The most abundant TAG was again sn-LnLnLn (30.4%) but was slightly lower than the observations of Ayorinde (2000) who used matrix-assisted laser desorption ionization–time of flight mass spectrometry (MALDI-TOF MS) as a means to detect the TAG species. Other flaxseed TAG included sn-LaLnLn, sn-OLnLn, sn-LnLnP, sn-LaLaLn, and sn-OLaLn. In total, 16 TAGs were detected, which is lower than the TAG totals of soybean, sunflower, almond, pistachio, hazelnut, poppy seed, palm, Brazil nut, macadamia, and rapeseed oil samples. The composition of the DAG was mainly LnLn in flaxseed oil, which exhibited the lowest equivalent carbon number (24) among oil samples.

Flaxseed has a high ALA content, generally constituting 50–62% of total fatty acids (Daun *et al.*, 2003). Dorrell (1970) reported that fatty acid distribution in flaxseed varied depending on the anatomical fractions. The hull is the main source of palmitic acid, but it has a relatively low oil content. Lower oleic and ALA and higher linoleic contents are present in the embryo compared to whole seed. Oomah and Mazza (1997) also observed higher levels of palmitic acid in the hull. However, the hull and whole seed gave similar ALA values, compared to dehulled seed.

Fatty acid composition of flaxseed was affected by cultivar (Froment *et al.*, 1998; Oomah and Mazza, 1997). DeClercq (2005) reported that average ALA content of Canadian flaxseed grown in 2004 was 61.9%. Similar high (56.5–61%) ALA levels were reported by Bozan and Temelli (2002) for Turkish flaxseed. Results were higher than the average values of flaxseed grown in Poland (57.1%) (Kozlowska, 1989), Ethiopia (51.9%) (Wakjira *et al.*, 2004), and North Dakota (48.4%) (Hettiarachchy *et al.*, 1990). In our laboratory, we observe ALA contents in organic flaxseed of approximately 50–52%. Linola, a low-ALA cultivar, developed for commercial vegetable oil market was reported to contain 3–4% ALA and 75% linoleic acid (LA) (Lukaszewicz *et al.*, 2004). Stenberg *et al.* (2005) reported that high-ALA flaxseed from Sweden contained 60.4% ALA, whereas high-LA flaxseed exhibited only 2.7% ALA. Furthermore, differences in oil content and fatty acid profile could be related to methods used to measure these components. Type of solvent, extraction time, and sample preparation could have a major impact on analytical results. Supercritical fluid CO_2 extraction gave a higher average ALA content (60.5%) compared to the soxhlet extraction method (56.7%) (Bozan and Temelli, 2002). Sikorska *et al.* (2005) discussed scanning fluorescence spectroscopy method as a means to detect adulteration in various oils including flaxseed oil.

B. PROTEIN

The protein content in flaxseed has been reported to be between 10.5% and 31% (Oomah and Mazza, 1993). The protein content of the most important flaxseed cultivars grown in different countries are presented in Table II. Protein values of flaxseed from Poland were generally well above 24%, whereas those from Canada were lower than 20% except for the cultivar AC Linora. The protein content of 24.1% for No. 1 Canadian Western flaxseed, from the 2001 harvest survey, was 1.7% higher than in 2000 and 1.6% higher than the 10-year mean of 22.5% (Canadian Grain Commission, 2001). Khategaon cultivar grown in India had a protein content of 21.9% (Madhusudhan and Singh, 1983). The protein contents of 11 flaxseed cultivars grown in North Dakota ranged from 26.9% to 31.6% (Hettiarachchy *et al.*, 1990). The USDA reports the protein content of flaxseed as 19.5% (USDA, 2005). Differences in protein can be attributed to both genetics and environment. However, a conversion factor of 6.5 (Oomah and Mazza, 1993) from nitrogen to protein was used in the calculation of protein content in Canada varieties, whereas a factor of 5.41 (Oomah and Mazza, 1998a) was also used in the calculation of flaxseed protein content. Thus, protein differences may be due to the conversion factor used in the determination of protein (Table II).

The protein contents of flaxseed fractionation from research work in different countries is presented in Table III. The proximate protein contents of dehulled and/or defatted flaxseed varied considerably, depending on cultivar, growth location, and seed processing. Oomah and Mazza (1997) reported that the hull fraction contains lower protein levels, and

TABLE II

PROTEIN CONTENT OF DIFFERENT FLAXSEED CULTIVARS[a]

Poland		Canada		USA	
Kozlowska, 1989		Oomah and Mazza, 1993		Hettiarachchy *et al.*, 1990	
Variety	Protein %	Variety	Protein[b] %	Variety	Protein %
Avangard	24.7	Linott	19.81	Clark	27.3
Reina	25.2	Noralta	19.82	Culbert	30.0
Viking	25.6	Dufferin	18.42	Dufferin	31.6
Ottawa	25.8	McGregor	19.07	Flor	31.0
Hera	27.4	NorLin	19.49	Linott	26.9
Zielona	27.9	NorMan	19.45	Linton	29.4
Bionda	28.0	Vimy	19.03	McGegor	27.1
LCSD200	28.2	Andro	19.89	Neche	29.5
Svapo	28.6	Somme	19.67	Norlin	28.1
–	–	Flanders	19.33	Norman	27.0
–	–	AC Linora	20.18	Verne	29.5

[a]Protein contents are calculated on dry basis.
[b]N × 6.5.

that dehulling increases protein content of flaxseed from 19.2% to 21.8%. Bhatty and Cherdklatgumchal (1990) reported a protein content of about 20% in flaxseed hull, which was identical to that reported in an earlier study by Peterson (1958). Bhatty and Cherdklatgumchal (1990) also reported that the mean protein contents for laboratory-prepared meals and commercial meals were statistically different. Researchers from Germany (Krause *et al.*, 2002) reported that flaxseed meal from dehulled seed had a protein content of 50%, which was close to the data reported by Indian researchers (Madhusudhan and Singh, 1983).

The protein content in various extracts obtained from different extraction methods is given in Table IV. Krause *et al.* (2002) extracted protein by isoelectric precipitation and micellization. The protein contents for isoelectric-precipitated protein isolate and micelle protein isolate were the same (Table IV). According to Dev and Quensel (1988), the protein contents of high-mucilage protein concentrate from seed (HMPC-S) and high-mucilage protein concentrate from expeller cake (HMPC-EC) were comparable (63.4% and 65.5%), while low-mucilage protein isolate (LMPI) contained much higher protein (86.6%). These results were consistent as Zhang (1994) also observed protein contents of 66.3% and 86.5% in the high-mucilage protein isolates HMPI and LMPI, respectively.

TABLE III

PROTEIN CONTENT OF FLAXSEED FRACTION[a]

Fractionation	Protein %	Cultivar	Reference	Country
Seed	19.2	NorMan, Linola™947, McGregor, NorLin, Omega, Flanders, and Vimy	Oomah and Mazza, 1997	Canada
Dehulled seed	21.8			
Hull	17.3			
Dehulled seed	23.9	NorMan	Lei *et al.*, 2003	
Meal	22.9			
Hull	20.3	–	Bhatty and Cherdklatgumchal, 1990	
Laboratory-prepared meal	43.9	Norlin, NorMan, and McGregor		
Commercial meal	34.7	–		
Dehulled meal	50.0	–	Krause *et al.*, 2002	Germany
Dehulled meal	48.9	Khategaon	Madhusudhan and Singh, 1983	India
Meal	49.0	Viking	Sammour, 1999	France

[a]Protein data are based on dry basis.

TABLE IV
PROTEIN CONTENT OF PROTEIN EXTRACT

Protein extract	Protein %	Reference	Country
Micelle protein isolate	93.0	Krause *et al.*, 2002	Germany
Isoelectric-precipitated protein isolate	89.0		
LMF[a]	56.4[b]	Dev and Quensel, 1988	
LMPC[a]	59.7[b]		
HMPC-S[a]	63.4[b]		
HMPC-EC[a]	65.5[b]		
LMPI[a]	86.6[b]		
HMPI	66.3	Zhang, 1994	China
LMPI	86.5		

[a]LMF, low mucilage flour; LMPC, low-mucilage protein concentrate; HMPC-S, high-mucilage protein concentrate from seed; HMPC-EC, high-mucilage protein concentrate from expeller cake; LMPI, low-mucilage protein isolate.
[b]N × 6.25.

Albumin- and globulin-type proteins are the major proteins in flaxseed. According to Madhusudhan and Singh (1983), flaxseed albumin comprised 20% of total meal protein. Marcone *et al.* (1998) reported that the globulin fraction makes up 73.4% of total protein, and the albumin constitutes about 26.6% of total protein. In contrast, Youle and Huang (1981) reported that 2S proteins accounted for 42% of the total seed proteins. Sammour (1999) also reported albumin accounted for 40.2% of the total protein content.

The amino acid profiles of flaxseed cultivars from different countries are given in Table V. Although there were variations among varieties, all flaxseeds had similar amino acid profiles. Flaxseed proteins are relatively high in arginine, aspartic acid, and glutamic acid, whereas lysine, methionine, and cystine were the limiting amino acid. Madhusudhan and Singh (1985a) investigated the amino acid composition of water-boiled flaxseed meal. Boiling appeared not to affect amino acids as the amino acid composition between the boiled and raw flaxseed meal were not significantly different. In contrast, germination of the seed significantly changed the amino acid content in the flaxseed. Wanasundara *et al.* (1999) reported that the total amino acid content of the flaxseed, after an 8-day germination, increased by about 15 times, with the greatest increase (i.e., 200 times) being observed in glutamine and leucine compared to the original seed. Wanasundara and Shahidi (1993) reported the amino acid profiles of solvent-extracted flaxseed meal and compared those to commercial meal. Commercial meal had slightly lower values for all amino acids than the laboratory-prepared meal (Table VI). They also noted that there were only minor changes in the amino acid content of flaxseed meal

TABLE V

AMINO ACID CONTENT (G/100 G PROTEIN) OF DIFFERENT FLAXSEED CULTIVARS

	Poland	Canada[a]			India
	Kozlowska, 1989	Oomah and Mazza, 1993			Madhusudhan and Singh, 1985a,b
Amino acid	LCSD 200	Norlin	Foster	Omega	Khategaon
Ala	5.40	4.4	4.7	4.5	4.3
Arg	9.75	9.2	10.0	9.4	11.5
Asp	10.40	9.3	10.0	9.7	11.2
Cys	–	1.1	1.8	1.1	–
Ser	–	4.5	4.7	4.6	5.1
Glu	22.50	19.6	20.0	19.7	19.8
Gly	6.41	5.8	5.9	5.8	4.8
His	1.42	2.2	2.1	2.3	2.5
Leu	–	5.8	6.0	5.9	5.8
Ile	3.53	4.0	4.1	4.0	4.6
Lys	1.80	4.0	4.0	3.9	4.1
Met	1.44	1.5	1.4	1.4	1.7
Phe	4.94	4.6	4.8	4.7	5.9
Pro	3.16	3.5	3.8	3.5	4.6
Thr	–	3.6	3.8	3.7	3.9
Tyr	1.53	2.3	2.4	2.3	3.3
Val	5.69	4.6	5.1	4.7	5.6

[a]Adapted from Oomah and Mazza (1993).

when changing from a nonpolar solvent to a solvent with increased polarity. Bhatty and Cherdklatgumchal (1990) also reported higher amino acid contents in laboratory-prepared meal than commercial meals, which is comparable to amino acid data of linseed meal reported previously by Sosulski and Sarwar (1973) and Madhusudhan and Singh (1985a).

The amino acid profiles were consistent among researchers with abundances in arginine, aspartic acid, and glutamic acid but deficient in sulfur-containing amino acids. Amino acid profiles in different flaxseed protein fractions are listed in Table VII. Sammour *et al.* (1994) and Madhusudhan and Singh (1985a) found the contents of glutamic acid and lysine were higher in albumin than in globulin, whereas, methionine was higher in globulin. Chung *et al.* (2004) also found similar amino acid profiles (Table VII) of the major protein fractions to previous researchers (Madhusudhan and Singh, 1985b; Marcone *et al.*, 1998). The data suggests that the amino acid composition of the protein fractions is less variable than total protein content of the

TABLE VI

AMINO ACID COMPOSITION OF LABORATORY-PREPARED AND COMMERCIAL FLAXSEED MEALS (G/100 G PROTEIN)

	Wanasundrara and Shahidi, 1994			Bhatty and Cherdklatgumchal, 1990	
Amino acid	Hexane extracted	MAW-HE[a]	Commercial meal	Laboratory prepared	Commercial meal
Ala	4.81	4.64	4.61	5.4	5.5
Arg	11.50	11.20	9.78	11.8	11.1
Asp	9.18	9.16	8.03	12.5	12.4
Cys	3.29	3.39	3.16	3.8	4.3
Glu	16.70	16.36	14.45	26.3	26.4
Gly	6.44	6.26	5.64	7.0	7.1
His	2.69	2.46	2.36	2.9	3.1
Ile	4.78	4.54	4.19	5.2	5.0
Leu	6.70	6.39	5.96	6.8	7.1
Lys	4.38	4.14	3.92	4.1	4.3
Met	1.45	1.41	1.24	2.2	2.5
Phe	5.13	4.91	4.63	5.3	5.3
Pro	3.64	3.65	3.32	5.2	5.5
Ser	4.94	4.99	4.48	5.8	5.9
Thr	3.40	3.33	3.00	4.9	5.1
Try	0.46	0.46	0.25	1.8	1.7
Tyr	2.21	2.12	1.98	2.9	3.1
Val	5.75	5.64	5.02	5.6	5.6

[a]MAW-HE, meal extracted with a multiple solvents (methanol-ammonia-water/hexane-extracted meal).

flaxseed. Furthermore, the country of origin has minimal impact on amino acid composition even if total protein did change.

C. CYANOGENIC GLYCOSIDES

Flaxseed contains cyanogenic glycosides, such as linamarin, linustatin, lotasutralin, and neolinustatin (Figure 1), which release toxic hydrogen cyanide upon hydrolysis. Cyanogenic glycosides content differs depending on location in the plant and stage of development (Niedzwiedz-Siegien, 1998).

Ten Canadian flaxseed cultivars were analyzed for total cyanide content (Chadha, 1995) and contents of individual cyanogenic glycosides (Oomah *et al.*, 1992). Chadha *et al.* (1995) determined cyanide content in 10 cultivars of flaxseed using an autohydrolysis method that required up to 5 hours of hydrolysis time. The maximum cyanide values were typically obtained

TABLE VII

AMINO ACID CONTENT (G/100 G PROTEIN) OF FLAXSEED PROTEIN FRACTIONS

	France		India		USA	Canada	
	Sammour *et al.*, 1994		Madhusudhan and Singh, 1985b		Marocone *et al.*, 1998	Chung *et al.*, 2004	
Amino acid	Globulin	Albumin	12 S protein	Albumin	11 S protein	Major fraction	Whole protein extract
Ala	3.8	3.4	4.8	1.9	5.5	5.7	6.9
Arg	11.5	11.1	12.5	13.1	12.6	11.9	8.4
Asp	10.1	8.7	11.3	5.5	12.4	12.3	10.3
Cys	1.1	0.36	–	3.5	0.9	0.6	1.4
Glu	17.1	27.3	19.8	35	24.3	21.8	21.5
Gly	9.8	8.1	4.8	8.3	5.4	5.6	10.9
His	1.9	2.1	2.5	1.6	2.6	2.5	1.8
Ile	4.1	3.6	4.6	2.8	5.6	4.6	4.2
Leu	6.6	5.9	5.8	5.4	5.9	5.8	5.9
Lys	4.0	5.0	3.1	4.9	3.1	3.2	3.4
Met	2.0	1.1	1.7	0.8	1.3	1.3	1.3
Phe	4.0	3.1	5.9	2.4	6.3	5.8	4.1
Pro	2.1	2.2	4.5	3.0	–	4.2	3.9
Ser	8.0	5.7	5.1	3.9	6.5	4.6	6.4
Thr	6.1	4.1	3.9	2.1	3.6	3.1	3.9
Try	–	–	1.3	2.0	–	–	–
Tyr	2.4	2.2	2.3	1.4	2.4	2.4	1.7
Val	4.1	4.1	5.6	2.6	5.1	4.7	4.7

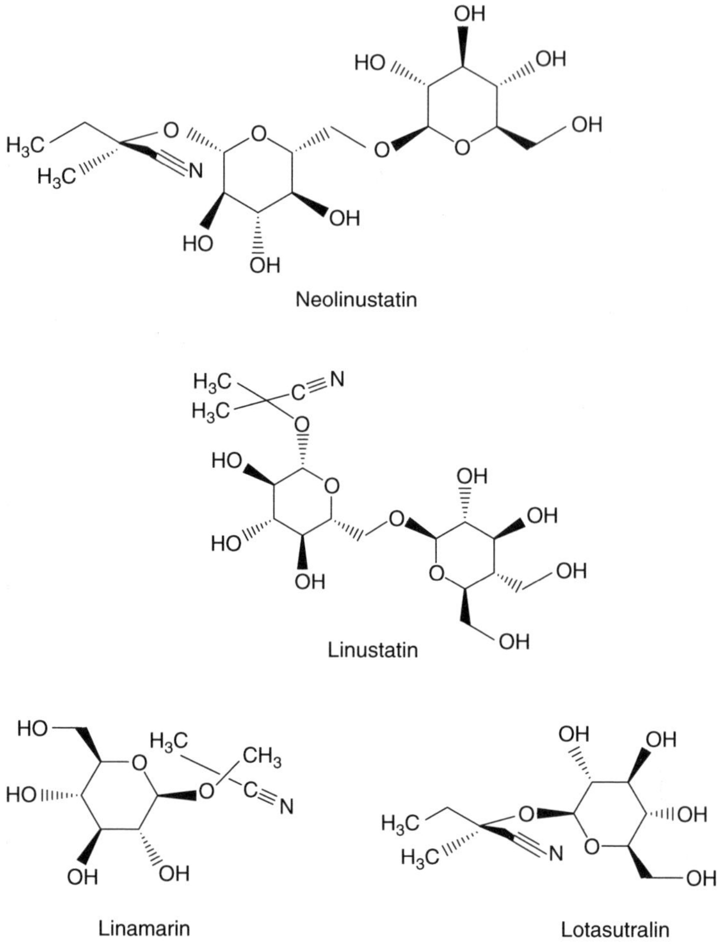

FIG. 1 Cyanogenic glycosides from flaxseed.

by 2–3 hours of hydrolysis (Table VIII). Oomah *et al.* (1992) reported that the amount of three cyanogenic glycosides present in 10 Canadian cultivars were significantly different, with linustatin and neolinustatin being the most abundant cyanide-containing compounds. Linamarin was reported to be present at very low levels (<32 mg/100 g seed) in 8 of the 10 cultivars (Table VIII). Oomah *et al.* (1992) concluded that the content of these three cyanogenic glycosides was dependent on cultivar, location, and the year of production of the seed, with cultivar being the most important factor.

TABLE VIII

CYANOGEN CONTENT (MG/100 G SEED) IN DIFFERENT FLAXSEED CULTIVARS

	Chadha *et al.*, 1995	Oomah *et al.*, 1992		
Cultivar	Cyanide	Linamarin	Linustatin	Neolinustatin
Ac Linora	14.5	19.8	269	122
Andro	19.6	16.7	342	203
Flanders	12.5	13.8	282	147
Linott	14.4	22.3	213	161
McDuff	15.4	–	–	–
McGregor	13.8	25.5	352	91
Noralta	–	20.3	271	163
Norlin	15.1	ND[a]	295	201
NorMan	12.4	ND[a]	231	135
Somme	16.6	27.5	322	149
Vimy	13.7	31.9	262	115

[a]ND, not detected.

Various procedures have been investigated to detect the cyanogenic glycoside via the release of hydrogen cyanide (HCN) (Amarowicz *et al.*, 1993; Bhatty, 1993; Kobaisy *et al.*, 1996; Kolodziejczyk and Fedec, 1995; Oomah *et al.*, 1992). Kobaisy *et al.* (1996) compared the accuracy of three methods, barbituric acid-pyridine, pyridine-pyrazolone, and HPLC, for the determination of cyanogens in flaxseed. They found that the total hydrogen cyanide values obtained by all three methods were not statistically different, although those obtained by the HPLC method were higher than those from the colorimetric methods. Their finding disagreed with those of Schilcher and Wilkens-Sauter (1986), who found that the HCN concentrations determined using HPLC were lower than those determined by a photometric method. Amarowicz *et al.* (1993) reported the chromatographic techniques for the preparation of linustatin and neolinustatin from flaxseed. The compounds isolated this way had a high purity (>99%) and could be used as standards for the analysis of cyanogenic glycosides by HPLC (Amarowicz *et al.*, 1993).

Various methods could be used to reduce the cyanogenic glycosides content in the seed. Wanasundara *et al.* (1993) investigated the use of solvents extraction as a means to reduce cyanogenic glycosides content of flaxseed meal. They found that extraction of flaxseed with alkanol-ammonia-water/hexane resulted in enhanced removal of cyanogenic compounds from 41 to

TABLE IX

HYDROGEN CYANIDE CONTENT (MG/100 G SEED) AFFECTED BY PROCESSING OF THE FLAXSEED

	Feng *et al.*, 2003		Yang *et al.*, 2004	
Processing method	HCN	Reduction (%)	HCN	Reduction (%)
Flaxseed raw	37.7		15.8	
Autoclaved	26.5	29.7	11.5	27
Microwave	6.35	83.2	2.8	82
Pelleted	9.88	73.8	–	–
Oven heated at 130°C, 10 minutes	3.16	16.2	–	–
Oven heated at 130°C, 20 minutes	2.91	22.8	–	–
Solvent extracted	–	–	1.7	89
Water boiled	–	–	ND[a]	100

[a]ND, not detected.

16 mg/100 g (total cyanogenic glycosides as HCN equivalents) compared to using hexane alone. Cyanogenic glycosides could be further removed by increasing solvent volumes, duration of the extraction period, and number of extraction stages. Yang *et al.* (2004) reported hydrogen cyanide reductions of 89%, 27%, 82%, and 100% using solvent extraction, autoclaving, microwave roasting, and water boiling, respectively (Table IX). These results were very close to those reported by Feng *et al.* (2003), who reported that autoclaving and microwave heating reduced hydrogen cyanide by 29.7% and 83.2%, respectively (Table IX). In their study, they also reported a 73.8% hydrogen cyanide reduction using a California pellet mill. In addition, hot air oven heating at 130°C for 10 and 20 minutes resulted in 16.2% and 22.8% reduction of hydrogen cyanide.

Research has also shown that commercial processing of flaxseed could affect the cyanogenic glycosides content in flaxseed (Oomah and Mazza, 1997, 1998a). These authors noted that dehulling increased linamarin content in the cotyledon or embryo fraction in all cultivars except Omega, where linamarin decreased by 50% to 23 mg/100 g seed (Oomah and Mazza, 1997). A reduction was also observed in linustatin and neolinustatin content. The mean values of seven cultivars show that linamarin increased from 19 mg/100 g seed to 67 mg/100 g seed, and linustatin and neolinustatin contents increased by 150%, after dehulling. Thus, total cyanogenic glycosides increased from 311 mg/100 g to 515 mg/100 g after dehulling (Oomah and Mazza, 1997). Commercial processing of flaxseed increased the cyanogenic glycosides from 309 to 476 mg/100 g (Oomah and Mazza, 1998a).

D. DIETARY FIBER (MUCILAGE OR GUM)

Flaxseed mucilage, associated with hull of flaxseed, is a gum-like material and composed of acidic and neutral polysaccharides. The neutral fraction of flaxseed mainly contains xylose (62.8%), whereas the acidic fraction of flaxseed is comprised mainly of rhamnose (54.5%), followed by galactose (23.4%) (Cui *et al.*, 1994a). A study by Warrand *et al.* (2005) found that the neutral monosaccharides were a mixture of three major families of polymers, arabinoxylans with a constant A/X ratio of 0.24, and various amount of galactose and fucose residues in the side chains. Acidic hydrolysis yields xylose, galactose, arabinose, rhamnose, galacturonic acid, fucose, and glucose (BeMiller, 1973; Erskine and Jones, 1957).

Extraction of flaxseed gum has been extensively investigated (Cui *et al.*, 1994b; Garden, 1993; Luo *et al.*, 2003; Oomah and Mazza, 2001). Luo *et al.* (2003) extracted the flaxseed gum by hot water (90–95°C) for 50 to 60 minutes and achieved the gum yield of 13–14%. Among the extract methods used to separate flaxseed gum, including centrifugation, steam heating, and vacuum filtration, the steam heating method was chosen for its efficient separation of gum from the rest of the seed (Luo *et al.*, 2003). Normal-pressure drying, reduced-pressure drying, freeze drying, microwave drying, and spray drying methods have all been tested to obtain a dry flaxseed gum. Spray drying was found to be the ideal method and was scaled up for industrial production (Luo *et al.*, 2003). Using response surface methodology, Cui *et al.* (1994b) determined that the optimum conditions for the gum extraction were a temperature of 85–90°C, a pH of 6.5 to 7.0 and water: seed ratio of 13:1. Response surface methodology was also used to optimize the spray drying to achieve maximum yield and functionality (rheological properties) of flaxseed gum (Oomah and Mazza, 2001). In another study, Garden (1993) extracted flaxseed gum from the outer coating of Neche flaxseed. In her research, a nonspecific protease was used to purify the flaxseed gum. Ultrafiltration with a hollow membrane cartridge was used to concentrate the mixture. Garden (1993) reported that the yield of crude flaxseed gum from flaxseed was 5.0% and that of purified gum was 4.5%.

Flaxseed mucilage composition varies due to the growing environment and cultivars. Oomah *et al.* (1995a) reported the variation (3.6–8%) in the content of water-soluble polysaccharides in flaxseeds from different geographical regions and cultivars. Glucose was reported to be the most abundant main monosaccharide in flaxseed gum with a mean value of 28.9% followed by xylose, galactose, rhamnose, and arabinose. However, xylose was reported to be present at the highest level (up to 40%) in the polysaccharides of seven Canadian cultivars extracted at 40°C (Chornick *et al.*, 2002). The xylose content was affected by ethanol, in that a 40% ethanol solution yielded higher mucilage than the 75% ethanol using precipitation as a means to concentrate the gum. In another study, Cui *et al.* (1996) studied yield and composition of flaxseed gums from six brown and six yellow cultivars. They noted that within yellow seeds, the minimum yield (5.2%) of gum was obtained from the APF 9006 cultivar and the maximum yield (6.5%) was from the Foster cultivar. They also noted that NorMan brown seed cultivar had the highest yield of 7.9%, whereas Royal has the lowest yield of 5.5%. Compared to gum yields (6.35%) from brown seeds, the yellow seed had a lower mean yield (5.85%) of gum among the 12 cultivars tested (i.e., six cultivars of each seed color). Xylose, galactose, rhamnose, and arabinose were the main components of both yellow seed and brown seed gums, with the content of 37.5, 19.2, 17.8, and 14.7% in yellow seeds, respectively, and 32.9, 21.6, 20.2, and 12.9% in brown seeds, respectively (Cui *et al.*, 1996). These results were consistent with data reported by Oomah *et al.* (1995a). However, the glucose contents (5.6% for yellow seed and 6.8% for brown seed) were much lower when compared to that of Oomah *et al.* (1995a), in which they reported the glucose (28.9%) to be the leading monosaccharide in flaxseed gum. A larger yield of mucilage was reported by Fedeniuk and Biliaderis (1994). They used higher extraction temperatures and Vega clay to purify the crude mucilage extract by reducing up to 80% of the protein that coextracted with the gum. More protein is typically extracted at higher temperature (80°C) than at lower 40–60°C temperatures.

Flaxseed gums extracted from different cultivars or by different methods exhibited various physiochemical properties (Cui and Mazza, 1996; Cui *et al.*, 1994a). Cui and Mazza (1996) tested for moisture, ash, mineral and nitrogen contents, amino acid composition, and intrinsic viscosity in both lab-prepared and commercial gums. Further characterization of the flaxseed gum was achieved by analysis of monosaccharide composition, galacturonic acid content, and ^{13}C NMR spectra. They found that intrinsic viscosity of flaxseed gum ranged from 434 to 658 ml/g, which was very different from gum Arabic (14.4 ml/g), guar gum (1135 ml/g), and xanthan gum (1355 ml/g). They concluded that among the gum extracted from the four cultivars, there were large variations in physicochemical properties. In another study, Cui *et al.* (1994a) noted that dialyzed flaxseed gums contain lower protein and higher carbohydrates and slightly lower mineral contents, with a similar monosaccharide ratio as crude flaxseed gums. These variations in flaxseed gums may be potentially suited for specific applications (Cui and Mazza, 1996). For additional review regarding applications of flaxseed gums, see the review of Chen *et al.* (2002).

Shan *et al.* (2000) reported that flaxseed gum had good foamability, stability, emulsibility, and salt resistance, and that flaxseed gum has the same rheology as non-Newtonian flow. The gum viscosity remains stabile over a broad pH 6 to12 (Shan *et al.*, 2000). In contrast, Mazza and Biliaderis (1989) reported that the viscosity of 0.05–0.5% (w/v) solutions exhibited Newtonian-like behavior at concentrations below 0.2% and shear thinning at concentrations above 0.2% (w/v) (Mazza and Biliaderis, 1989). Oomah and Mazza (1998b) found that the soluble carbohydrate content of flaxseed after commercial processing increased from 98.6 to 177.9 g/kg on a dry basis. Thus, variations in carbohydrate may account for different flow properties.

In addition to hot water extraction, chemical and enzymatic treatments were also investigated to remove the flaxseed mucilage. Wanasundara and Shahidi (1997) reported that soaking of seeds in sodium bicarbonate solution improved the removal of mucilage from seeds as compared with using water alone. Treatment with Celluclast® 1.5 L (45 mg protein/100 g for 3 hours or 22.5 mg protein/100 g for 6 hours) had a similar effect in reducing the mucilage content of the seed as soaking the seed in sodium bicarbonate solution (0.05 M, pH 8.16) for 12 hours. The effect of two enzymes, Pectinex Ultra SP and Viscozyme® had a similar effect in removal of mucilage as Celluclast® 1.5 L (Wanasundara and Shahidi, 1997).

E. POLYPHENOLS AND LIGNANS

Phenolic compounds are widely distributed in plants. In oilseeds, phenolic compounds occur as the hydroxylated derivatives of benzoic and cinnamic acids, coumarins, flavonoid compounds, and lignans (Ribereau-Gayon, 1972). Oomah *et al.* (1995b) reported the total phenolic acids (PA) in eight Canadian cultivars ranged from 790 to 1030 mg/100 g with esterified PA accounting for 48–66% of the total PAs, which was comparable to the 54% found by Varga and Diosady (1994). Dabrowski and Sosulski (1984) reported that flaxseed contained 811 mg/100 g phenolic compounds in defatted flour. In another study, Velioglu *et al.* (1998) reported the total phenolics contents of flaxseed and flaxseed gum prepared by different methods. They noted that gum prepared by different methods could retain total phenolics as high as 1422 mg/100 g and as low as 328 mg/100 g from flaxseeds with original total phenolics of 473 and 509 mg/100 g, respectively. A study in our laboratory (Hall and Shultz, 2001) measured the SDG and PAs, including ferulic, coumaric, caffeic, chlorogenic, gallic, protocatechuic, p-hydroxybenzoic, sinapic, and vanillic acids in defatted flaxseed and non-defatted flaxseed extracts (Table X). The total phenolic content in the defatted flaxseed extract and non-defatted flaxseed extract were 13,233, and 5420 mg/100 g extract, respectively.

The main lignan in flaxseed is SDG (Figure 2). Also present are a number of other lignans, that is, matairesinol (MAT), lariciresinol, hinokinin,

TABLE X
PHENOLIC COMPOUNDS CONTENTS (MG/100 G)[a]

Phenolic compounds	NDFE[b]	DFE[b]
Ferulic acid	161	313
Coumaric acid	87	130
Caffeic acid	4	15
Chlorogenic acid	720	1435
Gallic acid	29	17
Protocatechuic acid	7	7
p-Hydroxybenzoic acid	1719	6454
Sinapic acid	18	27
Vanillin	22	42
Total	2767	8440
SDG	2653	4793

[a]Adapted from Hall and Shultz (2001).
[b]NDFE, non-defatted flaxseed extract; DFE, defatted flaxseed extract.

arctigenin, divanillyl tetrahydrofuran nordihydroguaiaretic acid, isolariciresinol, and pinoresinol (Muir *et al.*, 2000). The main interest in these compounds is as precursor to mammalian lignans, which have been shown to have health-promoting activity (see Section III later).

Analytical methods of extraction and hydrolysis for quantifying lignan content in flaxseed has been extensively investigated (Charlet *et al.*, 2002; Degenhardt *et al.*, 2002; Eliasson *et al.*, 2003; Harris and Haggerty, 1993; Liggins *et al.*, 2000; Mazur *et al.*, 1996; Obermeyer *et al.*, 1995; Rickard *et al.*, 1996; Westcott and Muir, 1996). Liggins *et al.* (2000) reported lignan content of MAT, secoisolariciresinol (SECO), and shonanin in Cambridge and Argentinian linseed. The former had a higher SECO and shonanin content (1262 mg/100 g dried seed) than the latter (880 mg/100 g dried seed), and a lower MAT content (5.9 mg/100 g dried seed) compared to that of the latter (9.1 mg/100 g dried seed). However, lignan contents in flaxseed and its meal reported by several other researchers were slightly lower, ranging from 81 to 371 mg/100 g dried wt (Mazur *et al.*, 1996; Obermeyer *et al.*, 1995; Setchell *et al.*, 1999; Thompson *et al.*, 1991, 1997). Johnsson *et al.* (2000) reported SDG contents of defatted flaxseed flour and whole flaxseeds for cultivars grown in Sweden and Denmark. They noted that among the 14 cultivars from Sweden, cultivars Flanders and Mikael had the lowest and highest SDG contents in defatted flaxseed flour of 1170 and 2270 mg/100 g dry matter, respectively. The SDG content of defatted flaxseed flour in 15 cultivars grown in Denmark ranged from 1440

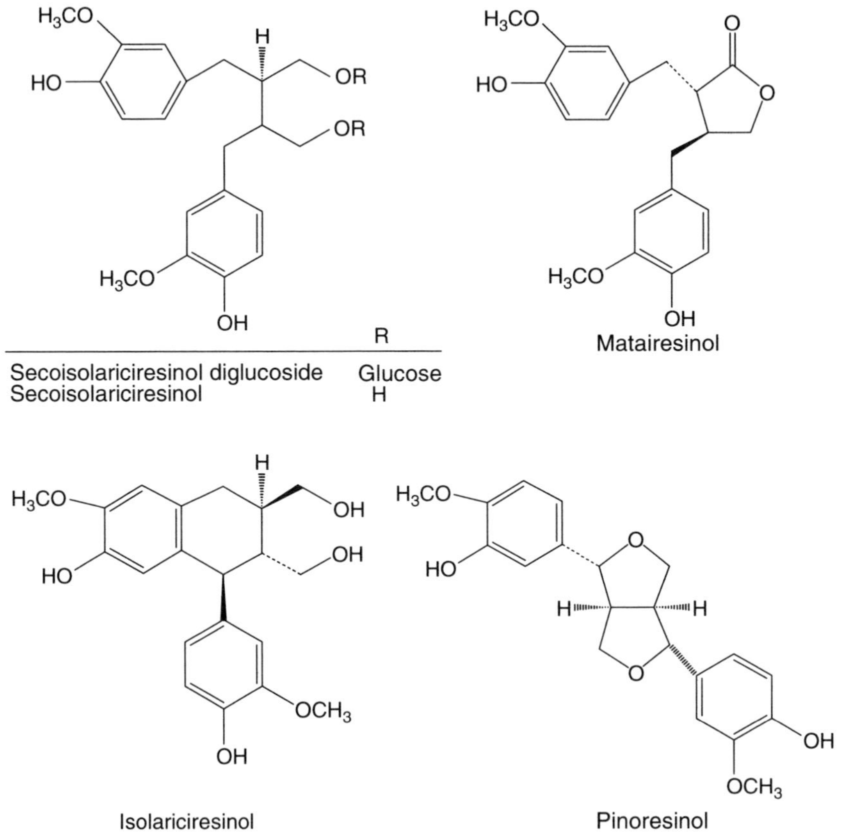

FIG. 2 Major lignans found in flaxseed.

to 2080 mg/100 g dry matter (Johnsson *et al.*, 2000). SDG contents in 27 Swedish cultivars were 1410 to 2590 mg/g dried seed (Eliasson *et al.*, 2003), which was slightly higher than the findings of Johnsson *et al.* (2000). Zimmermann *et al.* (2006) reported that flaxseed obtained from sites grown in Germany and Spain had lignan contents that were strongly correlated to cultivar and to a lesser extent the growing environment.

In addition to solvent extraction of lignans, a dry mechanical method for concentrating the lignan SDG was developed by Madhusudhan *et al.* (2000). As a result, the content of SDG increased from 1290 and 1430 mg/100 g in whole Neche and Omega seed, respectively, to 2760 and 2380 mg/100 g in the hull-rich fractions (Madhusudhan *et al.*, 2000).

F. OTHER COMPONENTS

Tocopherols consist of α, β, γ, and δ isomers and are effective antioxidants. Oomah *et al.* (1997a) observed that γ-tocopherol (9.04 mg/100 g seed) was the predominant isomer of Canadian flaxseed cultivars. Total tocopherol ($r = 0.42$) and γ-tocopherol ($r = 0.41$) values were correlated with seed oil content. Kamm *et al.* (2001) reported the distribution of tocopherols and tocotrienols in high and low linolenic flaxseed. Results were similar to the findings of Oomah *et al.* (1997a) in which γ-tocopherol content was greater (430–575 mg/kg oil) in high ALA flaxseeds, whereas low linolenic flaxseed exhibited lower (170 mg/kg oil) values. Bozan and Temelli (2002) compared tocopherol levels in oil extracted by supercritical CO_2 fluid and soxhlet (Table XI). Soxhlet-extracted oil had greater tocopherol levels (76.4 mg/100 g oil). These authors speculated that the temperature–pressure combination may have influenced the tocopherol extraction by supercritical CO_2 fluid.

Major flaxseed sterols are stigmasterol, camp sterol, and δ-5 avenasterol (Daun *et al.*, 2003). Obtusifoliol, gramisterol, and citrostadienol constituted 45%, 22%, and 12%, respectively, of the total 4α-monomethylsterol in flaxseed (Kamm *et al.*, 2001). Squalene content of flaxseed oil was reported as 4 mg/100 g oil, which was significantly lower than olive, corn, and rice bran oils. Squalene content is an intermediate compound of biosynthesis of plant sterols, which may have protective effects on lipid quality. Squalene could act as a peroxy radical scavengers in high polyunsaturated fatty acid oil (Dessi *et al.*, 2002).

Pretova and Vojtekova (1985) reported the presence of lutein, β-carotene, and violaxanthin in flaxseed. Carotenoids may serve as secondary antioxidants and scavenge singlet oxygen. In addition, carotenoids can function as chain-breaking antioxidants by trapping lipid-free radicals, in the absence of singlet oxygen (Belitz *et al.*, 2004). Daun *et al.* (2003) reported that Canadian flaxseed exhibited a range of 0–2 mg/kg chlorophyll which mostly disappeared during maturation.

TABLE XI

DISTRIBUTION OF TOCOPHEROLS AND TOCOTRIENOLS IN FLAXSEED (MG/100 MG OIL)[a]

Extraction method	α-Tocopherol	α-Tocotrienol	β- + γ-Tocopherol	δ-Tocopherol
Supercritical CO_2	0.21	0.67	53.74	0.95
Soxhlet	0.46	0.42	73.90	1.64

[a]Adapted from Bozan and Temelli (2002).

III. HEALTH BENEFITS

A. INTRODUCTION

1. α-Linolenic acid (ALA) and n-6 to n-3 fatty acid ratio

Flaxseed is one of the leading sources of ALA (Hauman, 1998), an omega-3 (n-3) fatty acid, which is essential for maintaining human health. The US Institute of Medicine, Food and Nutrition Board (2002) and Health Canada (Morris, 2003a) recommend ALA intakes of 1.1 and 1.6 g/day for women and men, respectively. Further recommendations of 1.3 and 1.4 g/day have been extended to lactating and pregnant women. Healthy individuals who have no signs of ALA deficiency are the basis for these recommendations. In infants, a 500 mg/day intake of long chain omega-3 has been recommended for proper neurological and cognitive development (Carver, 2003; Institute of Medicine, 2002). Other recommendations have been summarized by Nettleton (2003).

Nutrient deficiencies are difficult to research in humans because the test subject would have to be removed from that nutrient for an extended period of time, possibly to leading a state of starvation. Thus, clinical deficiencies of n-3 fatty acids have been reported by only two researchers. Both of these studies involved one patient each and different clinical symptoms were observed in the patient prior to ALA supplementation (Bjerve *et al.*, 1988; Holman *et al.*, 1982). Thus, our knowledge of symptoms of ALA deficiency is not complete. Furthermore, a complete removal of ALA may not be required for the manifestation of clinical symptoms related to an improper balance between ALA and other fatty acids in the diet. Many of these symptoms can be related to the overproduction of the proinflammatory eicosanoids, synthesized from fatty acids other than ALA, resulting in heart disease and cancer (Morris, 2003b).

Within the last decade, the health benefits of ALA have been documented in numerous studies and may be related to an improved n-6 to n-3 fatty acid intake. Nettleton (2003) summarized the recommendations of leading health organizations regarding the proper ratio of n-6 to n-3 fatty acid intake. Most organizations agree that a 5:1 to 10:1 n-6 to n-3 fatty acid ratio is preferred (Institute of Medicine, 2002; WHO/FAO, 2003). However, a typical western diet has an n-6 to n-3 fatty acid ratio well beyond 10:1; thus, flaxseed can be a valuable lipid source to improve the n-6 to n-3 fatty acid ratio due to the high n-3 content of flaxseed oil.

The recommendation for the n-6 to n-3 ratio stems from studies dating back to 1950 (Greenberg *et al.*, 1950). Studies show that ALA and LA (L, n-6) function synergistically at low concentration, but become competitive at higher intakes (Bourre *et al.*, 1990; Greenberg *et al.*, 1950). Furthermore, excessive ALA intake may not necessarily be better since high ALA can act as a substrate inhibitor in the conversion to long chain n-3 fatty acids (Vermunt *et al.*, 2000). Arguments against the necessity of ALA in the diet have focused on the lack of ALA conversion to long chain n-3. The conversion of ALA to eicosapentaenoic acid (EPA) and docosahexaenoic acid (DHA) in humans varies from 0.2% to 6% (Emken *et al.*, 1994; Pawlosky *et al.*, 2001); suggesting dietary factors may ultimately influence conversion. This was supported by several studies showing that dietary lipid influenced the conversion of ALA to EPA and DHA (Allman *et al.*, 1995; Emken *et al.*, 1999; Kestin *et al.*, 1990; Valsta *et al.*, 1996). Simopoulos (1999) recommends a LA to ALA ratio of 4:1 (e.g., 15 g LA: 3.7 g ALA) based on optimal conversion of ALA (11 g) to EPA (1 g). Thus, the proper balance between the n-6 and n-3 fatty acids in studies assessing conversion of ALA to long chain n-3 fatty acids EPA and DHA must be considered. Furthermore, iron, zinc, and possibly magnesium and vitamin B6 deficiencies could influence the conversion of ALA to long chain n-3 fatty acids (Cunnane and Yang, 1995; Cunnane *et al.*, 1987).

2. *Lignans*

Flaxseed is the richest source of plant lignans (Thompson *et al.*, 1991). SDG is the predominant lignan in flaxseed with minor amounts of pinoresinol and MAT (Meagher *et al.*, 1999; Thompson *et al.*, 1991). The lignans of flaxseed are phytoestrogens and serve as precursors in the production of mammalian lignans. Flaxseed lignans convert to the mammalian lignans enterolactone and enterodiol by intestinal flora (Adlercreutz *et al.*, 1982; Axelson and Setchell, 1981; Axelson *et al.*, 1982; Borriello *et al.*, 1985; Wang *et al.*, 2000).

A dose-dependent formation of mammalian lignans has been reported (Nesbitt *et al.*, 1999; Rickard *et al.*, 1996). In humans, a dose-dependent response was based on daily dietary intake of flaxseed between 5 and 25 grams as flaxseed or in the form of a baked product (Nesbitt *et al.*, 1999). Higher flaxseed intakes may not produce additional mammalian lignan concentrations based on the observation in rats that showed 4.4 μmol SDG/d did not produce higher mammalian lignan concentrations than the 2.2 μmol SDG/d treatment (Rickard *et al.*, 1996).

The concentration of the plasma lignans was also time dependent. The first day of the study showed that there was a continued increase in lignan concentrations in the plasma over a 24-hour period after the consumption of 25 grams of flaxseed. Whereas by the eighth day of the study, a high-level mammalian lignans was observed at the time of the flaxseed intake and that no significant change in plasma lignan was observed (Nesbitt *et al.*, 1999). The dose and time dependences are critical parameters that must be considered in clinical investigations and may account for differences observed by researchers.

The metabolic products of the mammalian lignans have not been fully elucidated. Jacobs and Metzler (1999) found that enterolactone and enterodiol converted into 12 and 6 metabolites, respectively, using an *in vitro* hepatic microsome assay. This research group (Jacobs *et al.*, 1999; Niemeyer *et al.*, 2000) also found these metabolites in the urine of humans fed flaxseed and in the urine and bile of rats given the pure mammalian lignans intraduodenally. The metabolites were oxidation products of the mammalian lignans. The most interesting metabolite was the addition of hydroxyl units to the aromatic rings to make dihydroxy benzene ring structures (Figure 3). Heinonen *et al.* (2001) also found similar metabolites from syringaresinol using a human fecal inoculum assay.

The presence of the oxidized metabolites is unique and may provide additional reasons for the health benefits of lignans. Classical antioxidant mechanisms show that the addition of an ortho hydroxyl group to a monophenol enhances the antioxidant activity of the original monophenol. Thus, some of the mammalian lignan metabolites may actually have greater or different activity than the parent lignan. Kitts *et al.* (1999) reported that enterolactone and enterodiol had greater antioxidant activity than the parent

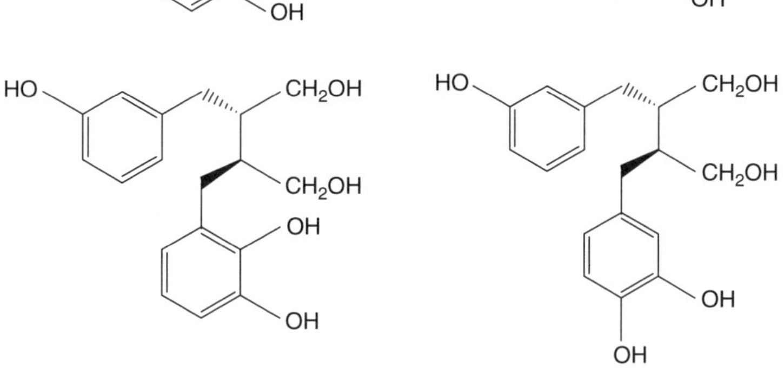

FIG. 3 Oxidized mammalian lignans observed by Jacobs *et al.* (1999) and Niemeyer *et al.* (2000) in rat and human urine.

lignan (SDG), suggesting that the metabolites might be the reason for the health benefits of the plant lignans.

The health benefits of flaxseed lignans is thought to be due to antioxidant activity, primarily as hydroxyl radical scavengers (Kitts *et al.*, 1999; Prasad, 1997a), and as estrogenic and antiestrogenic compounds due in part to the structural similarity to 17-β-estradiol (Adlercreutz *et al.*, 1992; Waters and Knowler, 1982). The behavior of the lignans depends on the biological levels of estradiol. At normal estradiol levels, the lignans act as estrogen antagonists, but in postmenopausal women (i.e., low estradiol levels) can act as weak estrogens (Hutchins and Slavin; 2003; Rickard and Thompson., 1997). Other activities related to estrogen include the *in vivo* synthesis of 2-hydroxy estrogen, a compound that may protect against cancer (Haggans *et al.*, 1999), and inhibit the binding of estrogen and testosterone to receptors on sex-binding globulin (Martin *et al.*, 1996). For further information regarding the estrogenic and antiestrogenic activity of flaxseed, see the review of Hutchins and Slavin (2003).

3. *Other components*

As previously mentioned, flaxseed is a rich source of dietary fiber (28%). Dietary fiber has been widely viewed as a component essential to lowering the risk of colon cancer. The flaxseed protein is similar to soy thus may be beneficial to health. Bhathena *et al.* (2002) first reported that flaxseed protein was effective in lowering plasma cholesterol and triacylglycerides (TAG) compared to soy and casein protein in male F344 and obese SHR/N-cp rats. The role of protein in disease prevention warrants further investigation. Components, such as PAs and flavonoids, may also contribute to the health benefits of flaxseed.

B. ROLE OF FLAXSEED IN CARDIOVASCULAR DISEASE PREVENTION

1. *Affect on biological lipid status*

a. Affect of ALA and flaxseed oil on lipid status. Kelley *et al.* (1993) reported that serum TAG, cholesterol, high-density lipoproteins (HDL), apoproteins A-I and B (markers for cardiovascular disease) of men fed a flaxseed oil (6.3% ALA) diet were not significantly different from the basal diet. However, an increase in ALA in serum and peripheral blood mononuclear cells (PBMNC) was found, as were EPA and DHA in the PBMNC. Men given a high ALA/low LA diet had significantly higher ALA in their plasma phospholipid, cholesteryl esters, and TAG and neutrophil phospholipids (Mantzioris *et al.*, 1994). The men on the diet containing flaxseed oil had higher (2.5 fold) EPA levels in their plasma lipids and neutrophil phospholipids (Mantzioris *et al.*, 1994). These results support the suggestion of Chan *et al.* (1993) that the low LA:ALA (n-6 to n-3) ratio in the diet was important for incorporation of long chain n-3 into platelets and plasma phospholipids. Ranhotra *et al.* (1992) noted that flaxseed oil or blends of flaxseed oil and sunflower oil promoted cholesterol reduction in hypercholesterolemic rats compared to diets formulated with hard fats. These authors suggested that a diet with the appropriate balance of n-6 and n-3 fatty acids was preferred over diets high in n-6 fatty acids.

Wilkenson *et al.* (2005) also found that dietary fatty acid intake could affect biological lipid status. ALA and EPA levels in the plasma erythrocytes increased by 225% and 150%, respectively, in subjects fed a high ALA diet (i.e., 0.5 to 1 LA:ALA ratio). A reduction in arachadonic acid was also observed in the patients fed the high ALA diet compared to the high LA (i.e., 27.9 to 1 LA:ALA ratio). A 12.4% reduction in total plasma cholesterol and a 10% reduction in HDL cholesterol were also observed in patients on the flaxseed oil diet (Wilkenson *et al.*, 2005). Hussein *et al.* (2005) also reported similar findings in 57 male subject fed high ALA diets (17 g/day). ALA and EPA contents increased by three- and twofold in erythrocyte phospholipids, respectively. A 50% increase in docosopentanoic acid (DPA) levels in erythrocyte phospholipids was also observed but increases in LA, AA, or DHA contents were not found. Furthermore, ^{13}C tracer studies supported the observations in that the AA formation was directly related to LA intake and inversely related to ALA intakes (Hussein *et al.*, 2005). Gerster (1998) found radioisotope-labeled ALA conversion to EPA and DPA was affected by diet. In diets high in saturated fats, the ALA conversion to EPA and DHA was 6% and 3.8%, respectively. A 40–50% reduction in ALA conversion was found for diets high in n-6 polyunsaturated fats compared to diets high in saturated lipids.

Clandinin *et al.* (1997) found that fish oil significantly lowered plasma TAG compared to flaxseed and olive oils. However, total cholesterol and low density lipoprotein (LDL) levels in humans were slightly lower, but not significantly, in flaxseed and olive oils diets compared to fish oil. In this study, a lack of a difference between the control diet (i.e., olive oil) and flaxseed diet on plasma lipids may have been due to the small number of subjects (i.e., 26). In a placebo-controlled, parallel study involving 150 hyperlipidemic subjects, significant reduction in fasting plasma TAG after 2 months in subjects given 1.7 g EPA+DHA per day was observed (Finnegan and Minihane, 2003). However, after 6 months on this diet, the significance in the reduction diminished. The 9.5 g ALA/day significantly increased the EPA concentrations in plasma phospholipids without significantly altering plasma TAG (Finnegan and Minihane, 2003).

b. Affect of flaxseed on lipid status. Weanling rats fed diets containing 20–40% flaxseed for 90 days had significantly lower total serum cholesterol and TAG levels than rats on flaxseed-free diets (Ratnayake *et al.*, 1992). Kritchevsky (1995) noted that rats fed a 20% flaxseed diet had a serum and liver cholesterol level that was 25% lower than rats fed a diet containing 10% flaxseed. The combination of defatted flaxseed meal and flaxseed oil caused a significant reduction in cholesterol levels in rats whereas the full fat flaxseed and flaxseed oil alone were not as effective (Ranhotra *et al.*, 1993).

Incorporating full fat flaxseed meal into the diet eliminated the adverse effect of hydrogenated soybean oil on serum cholesterol levels in hypercholesterolemic rats (Ranhotra *et al.*, 1993). In addition, full fat flaxseed meal enhanced the cholesterol-lowering effect of diets containing flaxseed oil. Cunnane *et al.* (1993) reported a 9% reduction in cholesterol (18% for LDL) in females fed 50 g of flaxseed per day. In addition, the n-3 fatty acid level increased in plasma and erythrocytes. No differences in plasma ALA were found in subjects fed 50 g flaxseed or 20 g flaxseed oil. Thus, when evaluating plasma ALA, the form in which ALA is consumed may not be as important as the level because 20 g flaxseed oil and 50 g flaxseed have equivalent ALA concentrations (12 g). However, a decreased postprandial glucose response was found in subjects fed 50 g flaxseed diet suggesting that soluble fiber has a positive impact on health and supports the consumption of flaxseed as a source of n-3 fatty acids. Jenkins *et al.* (1999) also noted a 5% and 8% reduction, respectively, in serum total and LDL levels in subjects fed partially defatted flaxseed. These researchers attributed the LDL reduction to the gum (i.e., soluble fiber) component of the flaxseed.

Arjmandi *et al.* (1998) reported a 6.9% reduction in cholesterol of postmenopausal women fed 38 g flaxseed. A 14.7% reduction in serum LDL was noted whereas serum HDL cholesterol and TAG were not affected. A 7.4% reduction in apoprotein A levels was also noted in the subjects. Identical reduction (7.5%) in apoprotein A levels was also noted in postmenopausal women on a 40-g/day flaxseed diet (Lucas *et al.*, 2002). Dodin *et al.* (2005) reported that significant reductions in serum total and HDL cholesterol of postmenopausal women on a 40-g/day flaxseed diet were observed in comparison to a wheat germ control diet. However, the authors considered these changes to be of little clinical importance. A significant observation was that flaxseed and wheat germ could reduce the severity of menopausal symptoms. Lemay *et al.*, (2002) also noted that a diet containing 40 g flaxseed/day improved menopausal symptoms in 25 hypercholesterolemic menopausal women. Glucose and insulin levels were lowered by the flaxseed intake; however, only small nonsignificant changes in cholesterol levels were observed.

An extensive study was completed on rats to determine the effects of flaxseed intake during pregnancy and throughout the lives of their offspring (Wiesenfeld *et al.*, 2003). The general finding of the study support others in that an increase in ALA and EPA and decrease in AA were observed in the rats and their offspring fed flaxseed. Pregnant rats fed 20% and 40% flaxseed diets had significantly lower plasma LA compared to the rats on the control diet. A dose-dependent increase in plasma ALA were observed in all flaxseed treatments (20% and 40% flaxseed, 13% and 26% low-fat flaxseed meal) compared to the plasma of rats fed a nonflaxseed diet. Plasma AA contents also were significantly lower than the control for rats on all flaxseed diets except 20% flaxseed (Wiesenfeld *et al.*, 2003). Only the 20% and 40% diets were sufficient to promote a significant increase in plasma EPA. Similar trends in the results were observed in the offspring (i.e., F_1 rats). The F_1 rats on the 40% flaxseed diets had significantly lower total cholesterol after 90 days from weaning. As in other studies, HDL cholesterol accounted for most of the reduction. Based on their finding, Wiesenfeld *et al.* (2003) concluded that high flaxseed (40%) and flaxseed meal (26%) diets were safe based on the lack of observed deleterious effects in pregnant and F_1 rats.

2. *Affect on inflammatory markers and atherosclerosis*

Atherosclerosis is a disease that is characterized by deposition and accumulation of cholesterol and other blood lipids in blood vessel walls. Factors affecting the development atherosclerosis include high cholesterol (i.e., hypercholesterolemia), inflammatory compounds, such as eicosanoids and cytokines, and reactive oxygen species (Prasad, 1997b; Ross, 1999). Tumor necrosis factor (TNF), interleukin 1-β (IL-1 β), and platelet-activating factor (PAF) have been shown to promote the production of oxygen-free radicals by polymorphonuclear leucocytes (PMNLs) (Braquet *et al.*, 1989; Paubert-Braquet *et al.*, 1988; Stewart *et al.*, 1990). Reactive oxygen species can be synthesized during the metabolism of arachidonic acid to prostaglandins (Panganamala *et al.*, 1976) and leukotrienes (Murota *et al.*, 1990). Thus, control of these factors would result in a reduction of atherosclerosis and cardiovascular disease.

Caughey *et al.* (1996) reported that flaxseed oil inclusion into the diet of healthy volunteers resulted in 30% reduction in the production of the cytokines TNF-α and IL-1 β. These cytokines were inversely related to EPA levels in mononuclear cells. An intake of 1.8 g EPA and DHA per day and 9.0 g ALA per day over 4 weeks resulted in a 20%, 26%, and 36% reduction in IL-1 β, prostaglandin E_2 (PGE_2), and Thromboxane B_2 (TXB_2), respectively (Mantzioris *et al.*, 2000). As in the previous study, EPA levels in mononuclear cells were inversely related to the inflammatory markers. Kew *et al.* (2003) reported a reduction of PGE_2 production in spleen mononuclear cells when EPA and DHA were located at the *sn*-1(3) position of the TAG. In contrast, EPA and DHA incorporation into the spleen mononuclear cells phospholipids was not dependent on the location of the fatty acid on the dietary TAG.

Allman *et al.* (1995) noted that platelet EPA levels were more than double for individuals fed flaxseed oil compared to sunflower oil group. Platelet EPA:arachidonic acid ratio (i.e., marker for thromboxane production and platelet aggregation potential) increased in the flaxseed group, thus a protective effect against cardiovascular disease, over LA-rich oils, would be expected. Their findings support the decreased platelet aggregation observed in hyperlipidemic subjects fed flaxseed (Bierenbaum *et al.*, 1993).

Zhao *et al.* (2004) reported that a high-LA (12.6% and 3.6% energy from LA and ALA, respectively) or -ALA (10.5% and 5.6% energy from LA and ALA, respectively) diet was significantly better than an average American (7.7% and 0.8% energy from LA and ALA, respectively) diet in controlling inflammatory markers in 23 subjects. Subjects on the high-ALA diet had lower serum LA and AA but higher serum ALA, EPA, and DPA than the high-LA diet. Serum C-reactive protein was also improved in subjects on the high-LA and -ALA diets compared to the average American diet. However, the reduction in C-reactive protein was only significant in the high-ALA diet (Zhao *et al.*, 2004). Vascular cell adhesion molecule-1 and E-selectin were lower in subjects on the ALA diet when compared to the LA diet. These parameters were inversely related to serum EPA and to a lesser extent DPA. The conclusion of the study was that the reduction of CVD by ALA occurs through several mechanisms based on the findings that the high ALA affected both lipids/lipoproteins and C-reactive protein/cell adhesion molecular parameters (Zhao *et al.*, 2004).

Prasad (1997b) found that a 46% reduction in hypercholesterolemic atherosclerosis (HCA) resulted from a daily flaxseed intake of 7.5 g/kg body weight in New Zealand white rabbits. This reduction was remarkable considering the fact that a lowering of serum cholesterol was not achieved during the 8-week feeding study. No atherosclerotic plaques developed in the rabbits on the cholesterol-free diets. Furthermore, flaxseed diets were able to decrease the oxygen-free radicals production by PMNLs. Subsequent work by this author assessed the role of ALA and lignan in the control of HCA (Prasad, 2005; Prasad *et al.*, 1998).

Rabbits fed a low-ALA flaxseed diet (7.5 g/kg body weight/day) plus 1% cholesterol had higher serum TAG and very low density lipoprotein (VLDL) cholesterol than the control group on a 1% cholesterol diet. However, total cholesterol and LDL cholesterol at week 4 were lower by 14% and 17%, respectively, in the flaxseed plus 1% cholesterol diet compared to the 1% cholesterol diet. Further lowering (31% and 32%, respectively) of total cholesterol and LDL cholesterol were observed by week 8; addition of flaxseed to the diet was favorable (Prasad *et al.*, 1998). A 69% reduction in atherosclerotic plaques formation was observed in rabbits fed flaxseed plus 1% cholesterol diet compared to the 1% cholesterol diet. These authors suggested that the ALA was not responsible for the prevention of atherosclerosis. Prasad (2005) suggested that the lignan may act to prevent oxygen radical production by PMNLs, thus effectively reducing atherosclerosis. Data to support this hypothesis was based on several observations.

Rabbits fed a 0.5% cholesterol diet had atherosclerotic plaques on over 50% of the aorta surface whereas the 0.5% cholesterol plus 40 mg lignan complex/kg body/per day reduced atherosclerosis by 34.4% (Prasad, 2005). Furthermore, the added lignan complex lowered the total cholesterol, LDL cholesterol, serum, and aortic malondialdehyde by 20%, 14%, 35%, and 58%, respectively. Unlike previous studies, the lignan enhanced HDL cholesterol by 25% and 30% in normocholesterolemic and hypercholesterolemic rabbits, respectively. Slightly higher (33% and 35%) reductions in total cholesterol and LDL cholesterol, respectively, were observed in rabbits on a 1% cholesterol diet containing 15 mg SDG/kg/day (Prasad, 1999). A 73% reduction in atherosclerosis was observed in the rabbits given the SDG diets compared to the 1% cholesterol diets.

Similar trends were observed in ovariectomized Syrian hamsters (Lucas *et al.*, 2004). Flaxseed diets of 7.5%, 15%, or 22.5% for 120 days were able to prevent the rise in plasma cholesterol that was observed in the ovariectomized hamsters. Nearly 61% of the aorta contained fatty streaks in the ovariectomized hamsters, whereas the ovariectomized hamsters on the 7.5% and 22.5% flaxseed diets had aortic fatty streaks over 7.5% and 7.2% of the aorta, respectively (Lucas *et al.*, 2004). This difference was nearly an 88% reduction in atherosclerosis compared to ovariectomized hamsters.

3. *Significant clinical and epidemiological studies*

Although a large number of clinical studies on small groups of patients have been completed, only a small number of studies have evaluated large populations regarding potential benefits of flaxseed. Mozaffarian *et al.* (2005) completed a 14-year follow-up involving 45,722 men in the Health Professionals Follow-up Study. The first major conclusion from the study was that both plant and seafood n-3 fatty acids could reduce the risk of coronary heart disease. Furthermore, the association between ALA and reduced risk of coronary heart disease was enhanced in subjects consuming low levels of

EPA and DHA. In fact, these researchers observed a 47% and 58% lower risk of coronary heart disease and nonfatal myocardial infarctions, respectively, for each 1 g increase in ALA in subjects consuming low levels of EPA and DHA (Mozaffarian *et al.*, 2005). In contrast to low-EPA and -DHA consumption, ALA intake was not associated with reduced coronary heart disease in men with high- ($\geq$250 mg/d) EPA and DHA intake. This observation may be due to the way in which ALA is metabolized and that the already high EPA and DHA diminished the conversion of ALA to other anti-inflammatory lipids.

A second conclusion was that n-6 content in the diet did not appear to influence the benefit of ALA (Mozaffarian *et al.*, 2005), which also was supported from earlier studies (Djoussé *et al.*, 2001; Hu *et al.*, 1999). That the n-6 levels did not significantly alter ALA efficacy may be attributed to a number of factors such as LA intakes by the test subjects may not reflect the general population (Mozaffarian *et al.*, 2005). However, increasing concentrations of ALA reduced the risk of coronary heart disease (Djoussé *et al.*, 2001; Hu *et al.*, 1999). In the Family Heart Study (Djoussé *et al.*, 2001), higher ALA intakes were always associated with reduced prevalence odds ratio of coronary heart disease, even at high-LA intakes. These researchers also concluded that the ALA and LA acted synergistically. In contrast, a higher ALA and lower LA (i.e., approximately 5:1 LA:ALA) intake resulted in decreased platelet reactivity in Moselle farmers, suggesting a lower risk of myocardial infarctions (Renaud *et al.*, 1986). Zhao *et al.* (2004) also reported higher total n-3 fatty acids as serum LA:ALA ratios decreased. Thus, further research is needed to clarify discrepancies in experimental and clinical data.

In the Lyon Diet Heart Study, survivors of myocardial infarctions on a Mediterranean style diet rich in ALA were less likey to experience a second episode (de Lorgeril *et al.*, 1994). These researchers reported that nonfatal myocardial infarctions and total death in subjects on the experimental and control diets were reduced by 75% and 70%, respectively. Fewer nonfatal myocardial infarctions also were reported in subjects consuming an Indo-Mediterranean diet rich in ALA (Singh *et al.*, 2002). The Nurse's Health Study (Albert *et al.*, 2004) also showed that diets rich in ALA reduce the risk of dying from coronary heart disease. The study was a 16-year follow-up involving approximately 76,000 women. Women on the highest ALA (1.5 g/day) diet had a 21% and 46%, respectively, lower risk of dying from coronary heart disease or sudden cardiac death compared to women on a 0.7 g ALA/day diet.

The concentration of ALA in adipose tissue was also associated with a reduction in nonfatal acute myocardial infarctions (Baylin *et al.*, 2003; Guallar *et al.*, 1999). However, only the relationship in the study by Baylin *et al.* (2003) was sufficiently strong. In this study, 964 survivors of myocardial infarction in

Costa Rica were the test subjects. The results showed that the subjects with the highest adipose tissue ALA had a lower risk of additional myocardial infarctions.

The National Heart, Lung, and Blood Institute Family Heart Study showed that higher dietary ALA intake was associated with lower prevalence of carotid plaques (Djoussé *et al.*, 2003a). One interesting observation in the Djoussé *et al.* (2003a) investigation was that ALA was inversely related to the thickening of internal and bifurcation segments of the carotid arteries whereas LA and long chain n-3 fatty acids were not related to carotid artery disease. In a cross-sectional designed study involving approximately 4400 men and women between the ages of 24 and 93, ALA consumption was inversely related to plasma TAG concentrations (Djoussé *et al.*, 2003b). The men and women consuming the highest ALA had 26% and 14.6%, respectively, lower plasma TAG than the subject with the lowest ALA intake. This observation was in contrast to other reports that suggested ALA consumption increases plasma TAG (Bemelmans *et al.*, 2002; Layne *et al.*, 1996; McManus *et al.*, 1996). Djoussé *et al.* (2003a) suggested that higher ALA intakes might have been the reason for the discrepancies between studies.

The Multiple Risk Factor Intervention Trial (Dolecek, 1992) included over 12,000 men over an 8-year period. The results showed that higher ALA intakes were associated with lower risks of death due to coronary heart disease and cardiovascular disease. Furthermore, a 28% reduction in risk of stroke was associated with a 0.06% increase in the ALA content of serum phospholipids (Simon *et al.*, 1995). Other studies have since supported the association between ALA and reduction in stroke risk (Leng *et al.*, 1999; Vartiainen *et al.*, 1994). Vartiainen *et al.* (1994) followed a Finnish population of approximately 28,000 men and women over 20 years and found that a 60% reduction in mortality from stroke was associated with increased ALA consumption. In a study involving approximately 1,100 subjects, individuals suffering a stroke had significantly lower ALA concentrations in the red blood cell (Leng *et al.*, 1999).

Inflammation is an important factor in the development of cardiovascular disease. Most clinical studies involving inflammation parameters have been relatively small. The Nurses Health Study involving 727 women was the largest study designed to determine the effects of n-3 fatty acids on biomarkers of inflammation and endothelium activation (Lopez-Garcia *et al.*, 2004). They found an inverse association between ALA intake and plasma concentrations of C-reactive protein (a marker for inflammation), Interlukin-6, and E-selectin. Bemelmans *et al.* (2004) also found an inverse association between C-reactive protein and ALA intake in a randomized, double-blind placebo-controlled study involving 103 hypercholesterolemic subjects.

In general, ALA intake was associated with reduced risks of cardiovascular disease in many of the large population studies despite inconsistencies in serum lipid data. Only one study (Zutphen Elderly Study) did not report a benefit of ALA in reducing cardiovascular disease (Oomen *et al.*, 2001). However, this study relied on dietary intake of ALA from margarine and meat thus raising questions regarding the outcome of the study.

The benefits of flaxseed and flaxseed oil have been demonstrated in a number of animal and human studies. The resulting biological benefits may be related to the potential of ALA to block the formation and release of proinflammatory eicosanoids and cytokines, reducing apolipoprotein B formation, blocking platelet activation factor, and improving blood vessel flexibility (Morris, 2003b).

C. ROLE OF FLAXSEED IN CANCER PREVENTION

1. *Introduction*

Cancer is a complex disease characterized by several stages: initiation, promotion, proliferation (i.e., rapid growth), and metastasis (spread). Because of the complexity of this disease, developing treatments to control or eliminate the disease is difficult. Furthermore, not all cancers are alike, thus treatments may work against one type of cancer but not others. Treatment is complicated because many factors, such as inflammatory compounds, hormones, nutrients, and exposure to toxins, all contribute to the growth and spread of the disease. This section of the chapter will highlight only a few studies involving the role of flaxseed in cancer prevention. For more in-depth information, please see the reviews of Bougnoux and Chajés (2003), Saarinen *et al.* (2003), and Thompson (2003)

2. *Breast cancer*

a. General studies using flaxseed. Early cancer studies used Sprague-Dawley (SD) rats injected with the carcinogen dimethylbenzanthracene (DMBA) as a means to assess the anticarcinogenic activity of flaxseed. These rats were generally fed 5% or 10% flaxseed or defatted flaxseed prior to preinitiation (i.e., before injection of DMBA) or during the early promotion stage of carcinogenesis. Reduction in tumor cell proliferation (Serraino and Thompson, 1991, 1992), reduction in mammalian tumor size and number (Thompson *et al.*, 1996a,b), and having a positive affect in controlling the initiation and promotional stages of mammalian cancers (Serraino and Thompson, 1992) were observed in these early investigations.

The susceptibility of a women to breast cancer is believed to start very early in life (Russo *et al.*, 2001). Maternal and prepuberty dietary intakes and hormonal exposure play a significant role in the development to breast cancer (Hilakivi-Clarke *et al.*, 2001; Russo and Russo, 1995). Terminal end buds (TEBs) formation is a critical stage in the development of the mammary glands. During puberty or exposure to estrogen, the TEBs mature into alveolar buds and lobules (i.e., differentiated TEBs). Prior to this time, TEBs remain undifferentiated. Due to the high number of proliferative epithelial cells, undifferentiated TEBs are very susceptible to carcinogens (Russo and Russo, 1978). Furthermore, a correlation exists between the number of undifferentiated TEBs and cancer risk (Russo and Russo, 1978). Thus, interventions early in life may reduce cancer risks.

Tou and Thompson (1999) observed endocrine changes in virgin female rat offspring of dams fed either 5% or 10% flaxseed or the equivalent amount of SDG in 5% flaxseed (SDG-5f) diets. Mammary structure changes, primarily the density of TEBs, were observed in 50-day-old rats exposed to the 5% flaxseed diet during gestation and lactation and over a lifetime exposure. However, no changes in TEBs were observed in these rats fed a 5% flaxseed diet after weaning. Alveolar bud density increased in the 50-day-old rats fed the 10% flaxseed diet whereas TEB density decreased. The 50-day-old rats exposed to the SDG-5f diet during gestation and lactation had fewer TEBs and Alveolar buds. The major conclusion from this study was that the lower TEB formation in rats exposed to low flaxseed levels was due to delayed puberty and reduced number of estrous cycles (Tou and Thompson, 1999). In contrast, the lower TEB formation and higher alveolar bud density in rats exposed to 10% flaxseed levels was due to an earlier onset of puberty and increased number of estrous cycles. The final conclusion from this study was that mammalian cancers could be due to the reduction in TEBs (Tou and Thompson, 1999).

Further support for early exposure to flaxseed has recently been reported (Chen *et al.*, 2003). In this study, dams were fed either a basal diet or one containing 10% flaxseed or the equivalent amount of SDG in 10% flaxseed (SDG-10f) during the lactation period (i.e., up to 21 days). At no time was the rat offspring (i.e., pups) exposed directly to flaxseed during the suckling/lactation period. The pups were then given DMBA and 21 weeks later sacrificed. The tumor incidence, tumor load, mean tumor size, and tumor number were all significantly lower by approximately 31%, 51%, 44%, and 47%, respectively, in the 10% flaxseed group and 42%, 63%, 68%, and 45%, respectively, in the SDG-10f group compared to the control (Chen *et al.*, 2003). The observed improvements in tumorigenesis were due to mammary gland differentiation for rats exposed to flaxseed or SDG (Tan *et al.*, 2004). Near the end of the study (i.e., day 49–51 postnatal), lower numbers of

TEBs were observed and expression of epidermal growth receptor and estrogen receptors (ER) decreased in rats exposed to both the flaxseed- and SDG-fortified diets.

Downregulation of expression of epidermal growth receptor, insulin-like growth factor I, and vascular endothelial growth factor in mice injected with human breast cancer has been reported (Chen *et al.*, 2002; Dabrosin *et al.*, 2002). In both these reports, MDA MB 435 (cells that are known to metastasize) human breast cancer cells were injected into athymic mice and subjected to the basal diet or 10% flaxseed diet at week 8. Tumor growth rates and total incidence of metastasis were reduced significantly during the 7-week feeding (Chen *et al.*, 2002). The suppression of angiogenesis may account for the slow tumor growth rate and lower incidence of metastasis in mice on the 10% flaxseed diet (Dabrosin *et al.*, 2002).

In contrast to MDA MB 435 cells, MCF-7 human breast cancer cells lines are ER-positive (i.e., estrogen dependent). Chen *et al.* (2004) assessed the role of flaxseed on the inhibition of ER-positive cancers using MCF-7 human breast cancer cells lines injected into mice. In this study, a 10% flaxseed diet was fed to mice with low- (0.035 nmol/L) and high- (0.3 nmol/L) blood estrogen levels. A 74% reduction in pretreatment tumor size was observed in the low-estrogen mice. In contrast, the tamoxifen group initially reduced tumor size but in the end, no tumor size reduction occurred (Chen *et al.*, 2004). The combination of flaxseed and tamoxifen produced a 53% reduction in tumor size in mice as compared to the tamoxifen-treated mice. In mice with high blood estrogen, 22%, 41%, and 50% reductions in tumor size were observed in mice on the flaxseed, tamoxifen, and flaxseed plus tamoxifen diets, respectively, compared to the estrogen positive control. Increased apoptosis and reduced tumor proliferation were the reasons for the tumor size reductions (Chen *et al.*, 2004).

In human studies, flaxseed consumption has had positive impacts on carcinogenesis and parallel observations from animal studies. A double-blind, placebo-controlled, prospective clinical trial involving 39 women, newly diagnosed with breast tumors, was completed to evaluate the effect of flaxseed consumption on tumor growth (Thompson *et al.*, 2000). Subjects given flaxseed (25 g/day) diets had reduced tumor cell proliferation and c-erB-2 expression, and an increased apoptosis index compared with women who ate whole-wheat muffins.

Haggans *et al.* (1999) reported that 28 postmenopausal women on diets supplemented with 10 g/day of ground flaxseed had significantly higher urinary 2-hydroxyestrone (2-OHE1) excretion and ratio of 2-OHE1:16α-OHE1 (16α-hydroxyestrone). A high level of 16α-OHE1is associated with increased cell proliferation and thus is a measure of possible cancer risk (Gupta *et al.*, 1998). Brooks *et al.* (2004) also reported significant higher concentrations of 2-OHE1 in urine of women fed a diet containing 25 g flaxseed/day than in the control and soy flour–fortified diets. In this study, 2-OHE1 and the ratio of 2-OHE1:16α-OHE1 increased by 103% and 98%, respectively, compared to the levels found by Haggans *et al.* (1999) supporting a dose-dependent response between the levels tested. Higher 2-OHE1 excretion and ratio of 2-OHE1:16α-OHE1 were also reported by Slavin *et al.* (2002). Furthermore, 10 g flaxseed/day decreased estradiol levels and estrone-sulfate concentrations.

Thompson *et al.* (2005) reported that tumor cell proliferation and c-erbB2 expression decreased by 34% and 71%, respectively, whereas cell apoptosis increased (31%) in menopausal women fed a diet containing 25 g of flaxseed. Changes in c-erbB2 score and apoptosis index were correlated with total flaxseed intake whereas cell proliferation, as measured by Ki-67 labeling index, was not. A main conclusion from these studies was that the lignans and, to a lesser extent, ALA were responsible for the anticarcinogenic activity.

b. Studies involving lignans. Lignans could be a significant part of a treatment regimen for cancer based on the large number of small-scale studies. Thompson *et al.* (1996a) reported a 46% reduction in the number of tumors per rat fed a diet that contained purified SDG at 1.5 mg/day (equivalent to a 5% flaxseed/day intake) compared to the basal diet. The feeding started 1 week after DMBA induction. A second study involving the same level of SDG showed that established tumor volumes were reduced by 54% and that new tumor numbers were 50% lower than the control and other treatments (Thompson *et al.*, 1996b). In this particular study, the 2.5% and 5% flaxseed diets also had positive effects on tumor reduction and growth. In contrast, 0.7 or 1.5 mg SDG/day dietary intakes did not significantly reduce tumor size in SD rats injected with N-methyl-N-nitrosourea (Rickard *et al.*, 1999). However, tumor size was lowest in rats fed the 5% diet and all treatments reduced tumor multiplicity and grade. The anticancer effect of flaxseed against N-methyl-N-nitrosourea-promoted cancers was thought to be due partly to the reduction of insulin-like growth factor I (Rickard *et al.*, 1999).

Mousavi and Adlercreutz (1992) found that enterolactone stimulated the growth of MCF-7 cells *in vitro* at low concentrations (0.5 and 10 μM), but at concentrations above 10 μM an inhibitory activity was observed. Estrogen stimulated the growth of MCF-7 cells, similar to 1 μM enterolactone, at concentrations of 1 nM. However, the combination of estrogen (1 nM) and enterolactone (1 μM) inhibited cell growth. These authors suggested that enterolactone might have inhibited aromatase and 17 β-hydroxysteroid dehydrogenase, enzymes important in estrogen production. Brooks and Thompson (2005) reported that 10 μM concentrations of enterolactone and enterodiol significantly inhibited aromatase activity in MCF-7 cells by 37% and 81%, respectively. Enterolactone concentration of 20 μM also significantly inhibited (25%) aromatase activity; however, 1 and 50 μM had no inhibitory activity. In contrast, all concentrations of enterodiol inhibited the aromatase activity (Brooks and Thompson, 2005). The greatest 17 β-hydroxysteroid dehydrogenase inhibitions (84 and 59%) were found at the 50 μM concentrations of enterolactone and genistein, respectively. A significant reduction in androstenedione-stimulated cell proliferation was also observed in enterolactone-treated MCF-7 cells. The enterodiol treatment did not reduce androstenedione-stimulated cell proliferation and in some instances (10 and 50 μM) increased cell proliferation (Brooks and Thompson, 2005). Saarinen *et al.* (2002) reported that enterolactone inhibited the growth of DMBA-inducted cancers in rats. To achieve this inhibition, a dose of 10 mg/kg body weight was required, which resulted in a plasma level of 0.4 μM. Although weak, aromatase activity inhibition was observed.

In a study by Chen and Thompson (2003), the effects of enterolactone, enterodiol, tamoxifen, or combinations of these were evaluated in estrogen receptor negative human breast cancer cells MDA-MB-435 and MDA-MB-231. At physiologically relevant concentrations, all three chemicals were effective against steps (i.e., cancer cell adhesion, invasion, migration) in the metastasis process. However, the effects were concentration dependent. Enterolactone inhibited MDA-MB-435 cell adhesion at 1 μM concentration but not at 5 μM, whereas cell adhesion was not inhibited at 1 μM concentration of enterodiol. However, the 5 μM concentration of enterodiol was effective against cell adhesion. In the MDA-MB-231 cells, enterolactone (5 μM) but not enterodiol inhibited cell adhesion. Tamoxifen prevented cell adhesion in both cell lines (Chen and Thompson, 2003). Combinations of enterolactone, enterodiol, and tamoxifen, at 1 μM concentration each, had the greatest reduction (36%) in cell adhesion. Cell invasion in general was inhibited by the lignans and tamoxifen; however, the 5 μM concentration was most effective in both cell lines (Chen and Thompson, 2003). In contrast, Magee *et al.* (2004) found that enterolactone was not effective against cell migrations at concentrations of 2.5, 10, or 50 μM. Furthermore, SDG and enterodiol were only effective against cell migrations at the 50 μM concentration. Differences in the results may be due to the methodology used by the investigators. Enterolactone and enterodiol both inhibited cell migrations at all concentrations (0.1, 1, and 10 μM) tested whereas tamoxifen was ineffective against cell migration (Chen and Thompson, 2003). The authors concluded that combination of these three chemicals could be effective against breast cancer.

A number of factors may contribute to the various anticancer activity of flaxseed (Thompson *et al.*, 2005). The behavior of the lignans depends on the biological levels of estradiol. At normal estradiol levels, the lignans act as estrogen antagonists but in postmenopausal women (i.e., low estradiol levels) can act as weak estrogens (Hutchins and Slavin, 2003; Rickard and Thompson., 1997). Tumor size reductions were observed in MCF-7 breast tumors in ovariectomized mice fed a diet containing flaxseed or flaxseed lignan; however, tumors were larger in rats fed diets containing soy products (Powers *et al.*, 2004). These authors suggested that the tumor growth and MCF-7 cell proliferation in mice fed soy-based diets was due to an estrogenic effect.

The presence of flaxseed lignans in MCF-7 tumors and the observed lignan binding to ER suggests that lignan function may be ER mediated (Adlercreutz *et al.*, 1992; Saarinen *et al.*, 2000). Although the lignans have been shown to be protective against breast cancer, minor structural alterations may influence overall activity (Saarinen *et al.*, 2005). Thus, many of the aforementioned benefits might be the results of specific structural features needed for lignans to bind to ER.

Flaxseed was among the best food sources in the prevention of *in vivo* spontaneous chromosomal damage in mice (Trentin *et al.*, 2004). The exact reason for the chromosomal damage prevention was not identified; however, the mechanism may be related to the antioxidant function of flaxseed components. Lignans have antioxidant activity and thus may contribute to the anticancer activity of flaxseed (Kangas *et al.*, 2002; Kitts *et al.*, 1999; Prasad 1997a, Yuan *et al.*, 1999).

c. Studies involving ALA. Few studies involving the role of ALA in breast cancer have been reported. Most of these studies used flaxseed oil and not ALA directly. Thompson (2003) reported a summary of these studies and in general a positive benefit has been associated with flaxseed oil intake and tumor prevention. A 10% flaxseed oil diet reduced tumor growth and metastasis incidents. The presence of ALA in breast adipose tissue was inversely related to breast cancer risk (Klein *et al.*, 2000; Maillard *et al.*, 2002). However, ALA content in other biomarkers did not provide the same results (Bougnoux and Chajès, 2003). Rao *et al.* (2000) reported that the number of tumors and tumor growth in general was influenced by the ratio of n-6 to n-3 fatty acids. The closer to a ratio of one gave the greatest benefit. This suggests the role of ALA in breast cancer has not been completely elucidated.

3. *Other cancer models*

a. Prostate cancer. Tumor cell proliferation decreased and apoptosis increased in TRAMP (i.e., transgenic adenocarcinoma mouse prostate) mice fed a 5% flaxseed diet for 30 weeks (Lin *et al.*, 2002). These authors reported that no differences in cell proliferation or apoptosis were observed at the 20-week evaluation. Byland *et al.* (2000) also reported that tumor numbers and growth decreased and an increase in apoptosis in nude mice model. These authors concluded that enterolactone was not related to the antitumor effect.

In contrast, enterodiol and enterolactone were believed to be partly responsible for the growth inhibition of three human prostate cancer cell lines (Lin *et al.*, 2001). Morton *et al.* (1997) reported that higher enterolactone levels in prostatic fluid were associated with populations with a low risk of prostate cancer. In a small clinical study, prostate cancer cell proliferation decreased and apoptosis increased in men fed 30 g of flaxseed per day (Demark-Wahnefried *et al.*, 2001). A significant factor which may have influenced this study was that the subjects were on a low-fat diet. A subsequent study by the authors further supported the role of flaxseed in combination with a low-fat diet as a means to control prostate growth (Demark-Wahnefried *et al.*, 2004). In this study, prostate-specific antigen level and cell proliferation both decreased from baseline after only 6 months on the dietary regime.

Cameron *et al.* (1989) reported that flax oil prevented tumor formation in mice whereas mice on the corn oil and safflower oil diets had the greatest number of tumors. A diet consisting of 10% flax oil was sufficient to reduce tumor growth and metastasis in mice compared with corn or fish oil diets (Fritsche and Johnston, 1990). Although flaxseed and flaxseed oil and lignans have been shown to be beneficial, additional studies are needed to identify the mechanisms by which flaxseed or its components function to reduce prostate cancer.

b. Colon and skin cancer. Although not extensively evaluated, flaxseed has been shown to inhibit colon and skin cancers in cell cultures and in animal studies as reviewed by Thompson (2003) and Morris (2003a). Oikarinen *et al.* (2005) reported that flaxseed oil may be responsible for preventing colon carcinogenesis in multiple intestinal neoplasia (Min) mice. Dwivedi *et al.* (2005) also supported this finding that flaxseed oil prevented colon tumor development in rats. These authors used an azoxymethane-induced colon tumor model and reported that corn oil, high in omega-6, did not prevent tumor development. Danbara *et al.* (2005) reported that a 10 mg/kg dose of enterolactone, by subcutaneous injection three times per week, reduced the expression of colo 201 human colon cancer cell in athymic mice. Using various testing protocols, Danbara *et al.* (2005) concluded that the tumor suppression was due to apoptosis and decreased cell proliferation. In general, flaxseed may be a valuable tool in the fight against various cancers. Further research is needed in clinical settings to support the role of flaxseed in cancer prevention in human populations.

4. *Epidemiological studies*

a. Breast cancer. In general, the consumption of flaxseed or lignans has been inversely associated with breast cancer. Ingram *et al.* (1997) reported a significant reduction in breast cancer risk among women consuming high levels of the enterolactone. The reduction in breast cancer risk was further supported by the increased excretion of lignan (Dai *et al.*, 2003). Linseisen *et al.* (2004) reported that enterodiol and enterolactone were inversely related to breast cancer risks in premenopausal German women. Women with palpable cysts and high-serum enterolactone concentrations had lower risks of developing breast cancer than women with palpable cysts with a low enterolactone intake (Boccardo *et al.*, 2004).

Low plasma concentration of enterolactone was associated with an increased breast cancer risk in 248 breast cancer cases selected from three population-based cohorts in Sweden (Hulten *et al.*, 2002). Tonkelaar *et al.* (2001) reported that high urinary enterolactone was weakly associated with increased breast cancer risks in postmenopausal women. Kilkkinen *et al.* (2004) concluded that high serum enterolactone was not associated with a reduced risk of breast cancer, as serum enterolactone did not differ between the case and control groups. The results may be due to the population of women tested in the study. Using data collected from a case-controlled study nested within the New York University Women's Health prospective cohort study, Zeleniuch-Jacquotte *et al.* (2004) compared serum enterolactone and breast cancer risk. They found higher enterolactone concentrations in the case group compared to the control group for the premenopausal women. Howevere, no significant differences in serum enterolactone concentrations were found between the control and case group for the postmenopausal women. These findings led the authors (Zeleniuch-Jacquotte *et al.*, 2004) to conclude that circulating enterolactone did not provide protection against breast cancers.

McCann *et al.* (2004) reported that enterolactone intakes were not associated with breast cancer risk in postmenopausal women. However, the highest quartile of dietary lignan intake was strongly associated with reduced cancer risks in premenopausal women. The different observations (McCann *et al.*, 2004; Zeleniuch-Jacquotte *et al.*, 2004) regarding the premenopausal women might suggest that biomarkers beyond serum enterolactone concentrations may be required to assess the relationship between enterolactone and breast cancer risks. Furthermore, Olsen *et al.* (2004) reported that breast cancer risk increased as plasma enterolactone decreased. However, this relationship was statistically stronger for estrogen receptor negative breast cancer and only a weak association between plasma enterolactone and lower breast cancer risk in estrogen receptor positive cases. Thus, the contradictory reports may be due to differences in the type of breast cancer (i.e., estrogen receptor positive or negative). An inverse association between high serum enterolactone and lower risk of breast cancer was observed in Finnish women (Pietinen *et al.*, 2001). This effect was observed in both pre- and postmenopausal women and contradicts other reports. Again, this result shows the high variability in epidemiological studies and that additional research may be needed to clarify the role of enterolactone and breast cancer risk.

In contrast to lignans, few studies have been reported regarding the relationship between ALA and breast cancer. In a meta-analysis, Saadatian-Elahi *et al.* (2004) reported a significant protective effect of total n-3 fatty acids and breast cancer risk. In this assessment, three cohort and seven case-control studies were reviewed. The case-control studies revealed an inverse association between ALA and breast cancer risk. High dietary intakes of ALA were correlated with a reduced breast cancer risk (Franceschi *et al.*, 1996). This study involved 2569 women with breast cancer and the result was supported by a cohort study conducted in the Netherlands (Voorrips *et al.*, 2002). A significant inverse association was found between ALA content in breast adipose tissue and breast cancer risk (Klein *et al.*, 2000; Maillard *et al.*, 2002). Furthermore, ALA to LA ratio close to one was also significantly associated with lower breast cancer risk (Maillard *et al.*, 2002).

Minimal to no association between breast cancer risk and subcutaneous ALA has been reported (London *et al.*, 1993; Simonsen *et al.*, 1998). In contrast, an increased risk of breast cancer has been associated with high ALA intakes (De Stéfani *et al.*, 1998). The link between ALA and breast cancer is weak in this study because most of the ALA was obtained from red meat. Thus, differentiating the effects from red meat and ALA intakes on breast cancer is difficult.

b. Prostate cancer. In the majority of the epidemiological studies, the authors concluded that prostate cancer was associated with ALA intake. However, inconsistencies in the reported data suggest that ALA may not be the component responsible for the observed risk. The influence of total dietary fat intake and caloric intake has been associated with increased risk of prostate cancer (Denis *et al.*, 1999; Giles and Ireland, 1997). Thus, many of the studies that draw conclusions on the association between ALA and prostate cancer may not be real.

Giovannacci *et al.* (1993) concluded from prospective data in the Health Professionals follow-up study that prostate cancer risk was associated with ALA intake. In contrast, Schuurman *et al.* (1999) reported a nonsignificant inverse relationship between ALA intake and prostate risk from data obtained in a cohort study completed in the Netherlands. In the Physicians'

Health Study (Gann *et al.*, 1994) and the Health Professionals Follow-up Study (Giovannacci *et al.*, 1993) eating red meat also emerged as a risk factor for prostate cancer. Leitzmann *et al.* (2004) reported that the ALA from both plant and animal sources was "suggestively" associated with an increased risk of advanced prostate cancer. However, total prostate cancer risk or early stage prostate cancer was not associated with ALA intake.

To complicate further the relationship between ALA and prostate cancer risk, Mannisto *et al.* (2003) reported that prostate cancer risk and ALA decrease when smokers were included in the model. This suggests that factors other than diet need to be considered in the models for determining causative relationships. Case-controlled studies also provide conflicting conclusions regarding the role of ALA in prostate cancer. For a complete review of these studies, see Bougnoux and Chajès (2003).

Kilkkinen *et al.* (2003a,b) reported that serum enterolactone was not associated with an increased prostate cancer risk. However, they also concluded that serum enterolactone concentration did not provide a protective effect against prostate cancer in male smokers. Statin *et al.* (2002) also concluded that high serum enterolactone was not associated with a lower risk of prostate cancer in a nested case-controlled study. The conflicting results between clinical and epidemiological studies suggest that the role of ALA and lignans in prostate cancer are far from being conclusive and that additional research in this arena is needed.

D. ROLE OF FLAXSEED IN DIABETES PREVENTION

Low glycemic index foods containing soluble fiber not only prevent certain metabolic ramifications of insulin resistance, but also reduce insulin resistance (Reaven *et al.*, 1993). Soluble fiber and other components of flaxseed fractions could potentially affect insulin secretion and its mechanisms of action in maintaining plasma glucose homeostasis. Flaxseed was shown to reduce the postprandial blood glucose response in humans (Cunnane *et al.*, 1993; Jenkins *et al.*, 1999). A consumption of 50 g/day ground flaxseed by young females over a 4-week period caused a reduction in blood glucose levels (Cunnane *et al.*, 1993). Similar findings were observed in postmenopausal women fed a 40 g/day flaxseed fortification diet (Lemay *et al.*, 2002). Bread containing 25% flaxseed gave a glycemic response that was 28% lower than the control (no flaxseed) bread (Jenkins *et al.*, 1999).

Prasad *et al.* (2000) reported that rats fed 22 mg SDG/kg and treated with the diabetes-promoting chemical streptozotocin had 75% lower incidence of type-1 diabetes than the streptozotocin-treated control group. However, the serum glucose of the SDG plus streptozotocin-treated rats had significantly higher serum glucose levels than streptozotocin-treated control group.

In contrast, the rats treated with SDG and streptozotocin that did not develop diabetes had lower serum glucose than the streptozotocin-treated control group. Rats fed no streptozotocin (control) or 22 mg SDG/kg body weight without streptozotocin did not develop type-1 diabetes and had serum glucose that where not significantly different. Similar results were observed in the formation of malondialdehyde, an oxidative stress indicator (Prasad *et al.*, 2000). In another type-1 model using Bio-Breed diabetic-prone rats, 71% of the rats fed 22 mg SDG/kg body weight did not show signs of diabetes whereas only 27% of the Bio-Breed diabetic prone rats had no signs of diabetes (Prasad, 2000a). Furthermore, pancreatic and serum malondialdehyde contents were lower in the rats treated with SDG whereas pancreatic antioxidant reserves were higher in these same rats.

The addition of 40 mg SDG/kg body weight reduced the incidence of type-2 diabetes in Zucker diabetic fatty (ZDF) rats by 80% (Prasad, 2001). In addition, blood glucose, glycated hemoglobin, and serum malondialdehyde were all lower in the rats treated with SDG at 72 days of age. In contrast, blood glucose levels were not significantly different between the SDG-treated rats and the control at 42 days. Oxidative stress plays a significant role in the etiology of type-1 and type-2 diabetes (Seghrouchni *et al.*, 2002; Shinomiya *et al.*, 2002). Kaneto *et al.* (1999) reported that oxidative stress can promote the overexpression of cyclin-dependent kinase inhibitor p21 mRNA resulting in reduced insulin production. Thus, SDG may indirectly affect this inhibitor by acting as an antioxidant (Prasad, 1997a, 2000b). Furthermore, diabetics have higher phosphoenolpyruvate carboxykinase, a liver enzyme important for gluconeogenesis, levels that result in higher glucose production (Consoli *et al.*, 1989). Prasad (2002) found that SDG inhibited phosphoenolpyruvate carboxykinase thus indicating another means by which SDG can act as an antidiabetic agent.

In contrast to ground flaxseed and SDG, ALA did not regulate glucose levels in elderly (Ezaki *et al.*, 1992) and type-2 diabetic (McManus *et al.*, 1996) populations fed 3 g/day and 35 mg/kg body weight, respectively. Kleeman *et al.* (1998) also reported that flaxseed oil had no effect on insulitis in diabetic-prone rats. In contrast to the antioxidant functions of SDG, *in vivo* oxidation of ALA may be in part one reason for the lack of activity of ALA in diabetes.

E. SAFETY OF FLAXSEED

1. Antinutrients

Flaxseed has several compounds that may negatively influence health and well-being. In some cases, the negative impact might simply be an assumption based on literature reports of similar compounds from other foods. The two components that have been questioned most frequently are the cyanogenic glycosides and linatine, an antipyridoxine factor.

Cyanogenic glycosides are not exclusive to flaxseed. These compounds can be found in a number of plants including apples, cherries, and cassava. Many of the health concerns regarding cyanogenic glycosides stem from studies showing that cassava was toxic to animals and humans (McMahon *et al.*, 1995). However, cassava contains significantly more cyanogenic glycosides than flaxseed. Furthermore, the release of hydrogen cyanide from flaxseed would be minimal and below the toxic, lethal dose. At the recommend daily intake of about 1–2 tablespoons, approximately 5–10 mg of hydrogen cyanide is released from flaxseed, which is well below the estimated acute toxic dose for an adult of 50–60 mg inorganic cyanide and below the 30–100 mg/day humans can detoxify (Roseling 1994). Daun *et al.* (2003) reported that a person would have to consume 8 cups (i.e., 1 kg) of ground flaxseed to achieve acute cyanide toxicity.

In addition to cyanogenic glycosides, trypsin inhibitor, linatine and phytic acid are other antinutrients contained in flaxseed. Trypsin inhibitor activity (TIA) in flaxseed was low when compared to those in soybean and canola seeds. Bhatty (1993) reported laboratory-prepared flaxseed meals containing 42–51 units of TIA, which was slightly higher than 10–30 units observed by Madhusudhan and Singh (1983) and commercially obtained flaxseed meal (14–37 units). The contents of phytic acid were significantly different among cultivars. AC Linora has a lowest phytic acid content of 2280 mg/100 g and low ALA yellow-seeded cultivar Linola 947 has the highest content (3250 mg/100 g seed) among the eight cultivars reported (Oomah *et al.*, 1996).

Kratzer (1946) reported that pyridoxine supplementation in chicks on diets containing a linseed meal was necessary to counteract the vitamin B_6 deficiency. Klosterman *et al.* (1967) identified the antipyridoxine factor linatine. Although linatine is a problem in chicks, flaxseed has not been associated with a vitamin B_6 deficiency in humans. In fact, no affect on serum pyridoxine levels in subjects consuming 45 grams of flaxseed per day over 5 weeks was observed (Dieken, 1992).

2. *Role of flaxseed in reproduction*

Extensive investigations into the effects of flaxseed consumption, by mothers during gestation and lactation, on the offspring have been reported. Various effects in the offspring of rats fed flaxseed at levels between 5% and 40% have been observed (Collins *et al.*, 2003; Flynn *et al.*, 2003; Orcheson *et al.*, 1998; Tou and Thompson, 1999; Tou *et al.*, 1998, 1999; Ward *et al.*, 2000, 2001a).

Flynn *et al.* (2003) reported that flaxseed (20% or 40%) or flaxseed meal (13% or 26%) diets did not alter the toxicological properties of rat serum in a way that would disrupt organogenesis. No significant embryotoxicity was observed in rat embryos cultured in serum obtained from pregnant rats fed flaxseed diets. Although the growth of the embryo was affected, no dose-dependent response was observed thus the findings might be a chance occurrence unrelated to flaxseed treatment (Flynn *et al.*, 2003). These authors concluded that the exposure serum obtained from pregnant rats fed flaxseed diets was not teratogenic to embryos.

Flaxseed consumption did not affect the pregnancy of the rats (Collins *et al.*, 2003; Tou *et al.*, 1998). Flaxseed (20% and 40%) and flaxseed meal (13% and 26%) diets did not affect fertility, litter size, or survival of offspring (Collins *et al.*, 2003). However, length of gestation, anogenital distance and index, onset of puberty, and estrous cycles were affected by flaxseed consumption.

Lower offspring birth weights were observed in rats fed a 10% flaxseed diet (Tou *et al.*, 1998). In contrast, the higher flaxseed (20% and 40%) and flaxseed meal (13% and 26%) diets did not significantly affect birth weight (Collins *et al.*, 2003). With the exception of the 40% flaxseed diet, female rats gain more weight on the diets containing 20% flaxseed and 13% and 26% flaxseed meal than the control rats from weaning to maturity (Collins *et al.*, 2003).

Female rats on a 5% flaxseed diet had delayed onset of puberty whereas earlier puberty onset was observed in female rats on a 10% flaxseed diet (Tou *et al.*, 1998, 1999). In both studies, the average weight of the female rats at puberty was lower than the control group. Higher flaxseed (20% and 40%) and flaxseed meal (13% and 26%) diets did not affect the onset of puberty or weight of the female rats at puberty (Collins *et al.*, 2003). Lengthened estrous cycle of the female rats was reported in all studies and supports the anti-estrogenic activity of flaxseed (Collins *et al.*, 2003; Orcheson *et al.*, 1998; Tou *et al.*, 1998, 1999). Orcheson *et al.* (1998) reported that a 10% flaxseed diet produced irregular estrous cycles similar to the addition of 3 mg SDG/kg fortified diet.

Ward *et al.* (2000) reported that terminal mammary end buds decreased and alveolar buds increased in rat offspring exposed to 10% flaxseed or the equivalent SDG diet during pregnancy and lactation. However, exposure to the same flaxseed or SDG levels only during lactation did not affect reproductive organs (Ward *et al.*, 2001a). Lina *et al.* (2005) reported a decrease in ovary weights in rats fed the lignan 7-hydroxymatairesinol. These authors suggested a 160 mg/kg body weight/day no adverse effect level, which is similar to the 177.2 mg SDG/kg body weight/day reported by Ward *et al.* (2000). Furthermore, no adverse effect level is approximately 160-fold greater than the proposed daily lignan intake in humans (Lina *et al.*, 2005).

In male offspring, reduced postnatal weight gains were observed in the lower flaxseed diets (Tou *et al.*, 1998) whereas high flaxseed diets did not affect postnatal weight gain, although the 40% flaxseed diet did have lower weight gain than the control (Collins *et al.*, 2003). At puberty, the rats on the flaxseed or flaxseed meal were heavier than the rats on the control diets. The onset of puberty was delayed in rats fed the 20% flaxseed and 13% and 26% flaxseed meal diets, but not in the 40% flaxseed diet. Collins *et al.* (2003) suggested that hormonal effects on the male reproductive system might be the cause of the observed delay in puberty.

Tou *et al.* (1999) reported that serum testosterone and estradiol levels were higher in rats fed a 10% flaxseed diet compared to the control. In addition, this diet produced higher relative accessory sex gland and prostate weight (Tou *et al.*, 1999). In contrast, lifetime exposure to 5% flaxseed diets reduced prostate weights. Sprando *et al.* (2000a) also found significant reductions in prostate weight in offspring from dams fed flaxseed (20%) or flaxseed meal (13% or 26%) diets. Testes weight was not significantly altered in offspring of dams fed flaxseed (20% or 40%) or flaxseed meal (13% or 26%) diets (Sprando *et al.*, 2000b) during suckling. Ward *et al.* (2001a) also reported no adverse effects on testes and prostate weights of offspring from dams fed a flaxseed (10%) diet during lactation. The absolute volume in the seminiferous tubules was significantly lower in offspring exposed to 20% or 40% flaxseed diets. However, this difference was not considered biologically relevant because the seminiferous tubule diameter, an indicator of spermatogenic activity, was not affected (Sprando *et al.*, 2000b). This combined with the observation that homogenization-resistant spermatid counts and daily sperm production rates were not affected indicated further that spermatogenesis was not affected by the high flaxseed or meal diets (Sprando *et al.*, 2000a,b). Furthermore, testis structure was not significantly affected by exposure to high flaxseed or meal diets (Sprando *et al.*, 2000b).

The development bone is a hormone dependent process. Thus, flaxseed and in particular lignans could influence bone development. Ward *et al.* (2001b) found that rats exposed to 88 or 177.3 mg SDG/kg body weight/day had higher bone strength than the basal diet at 50 days postnatal. However, the significant higher bone strengths were lost at day 132 postnatal. The bone mineral content were generally not significantly different between the basal and treatment groups except at postnatal day 132 where the higher lignan treatment had lower bone mineral content than the basal diet. In contrast, bone strength at postnatal day 50 were significantly lower in male rats fed 10% flaxseed diets compared to the basal diet (Ward *et al.*, 2001c). However, by postnatal day 132 no differences in bone strength, bone mineral density, or bone mineral content were observed. The authors concluded that the exposure to SDG did not have a negative effect on bone strength (Ward *et al.*, 2001b,c).

Babu *et al.* (2003) reported that a significant reduction in concanavalin A spleen lymphocytes proliferation in pregnant rats fed a 40% flaxseed diet. Spleen lymphocytes proliferation was significantly lower in a 90-day-old offspring on the 40% flaxseed diet and exposed to phytohemagglutinin. In both cases, interleukin-2 formation was not affected by flaxseed intake. The levels of LA and AA in serum and tissues combined with relatively constant interleukin-2 concentrations lead to the conclusion that flaxseed could be used in treating autoimmune diseases (Babu *et al.*, 2003).

3. Metabolism of flaxseed lignan

Rickard and Thompson (1998) followed the metabolism of ^{3}H-SDG in rats. They found that a chronic (10 day, 1.5 mg/day) exposure to SDG could increase lignan levels up to threefold compared to acute (1 day) exposure. In a follow-up study involving feeding of ^{3}H-SDG to rats, Rickard and Thompson (2000) reported that blood radioactivity peaked at 9 hours for both the chronic and acute groups. Furthermore, only slight reductions in blood radioactivity were noted after 24 hours in both treatment groups. Regardless of the treatment group, 75–80% of urine radioactivity was due to enterodiol, enterolactone and SECO. This combined with blood radioactivity profiles suggests that lignans are metabolized in a similar manner between the chronic and acute treatment groups.

Humans appear to be able to effectively handle single dose of lignans. Nesbitt *et al.* (1999) observed that mammalian lignan contents in urine started to increase 9 hours after ingesting 5, 15, or 25 grams of ground flaxseed. The excreted lignan content was dose dependent but remained higher than baseline values for up to 24 hours. The pharmacokinetics of ingested SDG has been reported in humans (Kuijsten *et al.*, 2005). The maximum plasma levels of enterodiol and enterolactone were reached 14.8 and 19.7 hours after ingestion of SDG, respectively. The mean elimination half-life was 4.4 and 12.6 hours for enterodiol and enterolactone, respectively. Mean residence times of 20.6 and 35.8 hours for enterodiol and enterolactone were observed, respectively (Kuijsten *et al.*, 2005). In all pharmacokinetic measurements, the time to reach maximum mammalian lignan concentration in plasma and urine was longer in men. Given the vast number of research reports on health benefits combined with positive safety results, flaxseed may be an effective food ingredient that could enhance health status in humans.

IV. FLAXSEED QUALITY AND END USE FUNCTIONALITY

A. INTRODUCTION

Flaxseed is utilized as a main food ingredient in order to enhance functional foods (Oomah and Mazza 1999). Whole or ground flaxseed can be used in various food products, such as bread (Carter, 1993; Garden, 1993), pasta (Lee *et al.*, 2003; 2004a; Manthey *et al.*, 2000, 2002a), candy, chocolate bar, chocolate (Kozlowska, 1989), muffin, bagel, bun, cereals, salad toppings (Carter, 1993), corn snack (Ahmed, 1999), cake (Lee *et al.*, 2004b), tortilla (Ghosh *et al.*, 2004), ice cream (Hall and Schwarz, 2002), yogurt (Hall *et al.*, 2004), or can be consumed as a roasted snack (Kozlowska, 1989; Schorno *et al.*, 2003; Tulbek *et al.*, 2004). Flaxseed is also processed for oil, which is a major product for the organic food industry (Wiesenborn *et al.*, 2004). Oomah (2001) and Hall (2002) reported flaxseed as a functional food source. In this section, end product use of flaxseed will be discussed.

B. SENSORY PROPERTIES OF FLAXSEED

In general, flaxseed has a pleasant nutty flavor (Carter, 1996). However, sensory characteristics have not been fully evaluated and consumer preference of flaxseed is largely dependent on seed quality. Appearance, color, and flavor attributes could be variable depending on cultivar and growing conditions. Carter (1996) indicated that ground flaxseed samples from different cultivars had acceptable quality. However, the Omega cultivar exhibited the highest sensory score whereas McGregor was statistically rated the lowest (Figure 4). A composite sample was tasted first and accepted as a reference

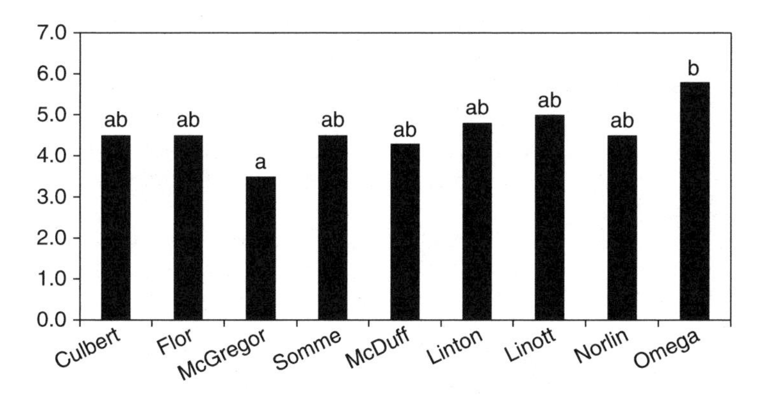

FIG. 4 Sensory scores of flaxseed cultivars observed by Carter (1996).

TABLE XII
PHENOLIC ACID CONTENTS (G/KG) OF WHOLE AND DEHULLED FLAXSEED[a,b]

	Total phenolic acid	
	Whole	Dehulled
Flanders	16.85b	21.03a
Linola™947	5.37e	9.75d
McGregor	6.99d	19.78ab
NorLin	8.05c	16.53c
NorMan	16.85b	16.71c
Omega	16.91b	18.61b
Vimy	18.15a	18.85b

[a]Adapted from Oomah and Mazza (1997).
[b]Values not sharing a common letter are significantly different ($p \leq 0.05$).

with a rating of 5. Linott and Linton exhibited similar flavor profile, slightly lower than Omega. The panelists found McGregor to have a slightly higher bitterness than other cultivars, whereas Omega was slightly milder.

Grinding can generate flavor volatiles and enhance the accessibility of the cyanogenic glycosides and phenolic components, which may then be more easily detected by sensory panelists. In addition, components, such as PAs and cyanogenic glycoside, may contribute to bitterness (Chiwona-Karltun *et al.*, 2004). Oomah and Mazza (1997) analyzed the phenolic and cyanogenic compounds of flaxseed cultivars. Results indicated that dehulled McGregor cultivar yielded the highest linamarin and linustatin content (Table VIII). In contrast, Flanders and Omega cultivars had the lowest linamarin content and total cyanogenic glycoside levels. Cultivars showed statistical differences ($p \leq 0.05$) in terms of PA contents in McGregor, NorLin and Linola™947 cultivars were the lowest, whereas Omega, Flanders, NorMan, and Vimy had higher values (Table XII). Linola™947 had the lowest PA content among the dehulled samples, whereas Flanders the highest. Based on this data, the sensory scores reported by Carter (1996) may be due to the lower cyanogenic glycoside present in the Omega cultivar and not as much on the PA content.

C. FLAXSEED PROCESSING

1. Milling and fractionation

Schorno *et al.* (2004) compared milled flaxseed production using several mills which included roller, burr, hammer, and centrifugal cutting mills. The burr mill was found to be the least efficient for flaxseed milling. The roller mill caused particle adherence to rolls due to the high-surface lipid content. High-surface lipid content, which could trigger lipid oxidation, increased due to the higher feed values. The centrifugal cutting mill and hammer mill showed greater potential for flaxseed milling.

Flaxseed has two flattened cotyledons, which constitute the greater portion of the embryo. The embryo is surrounded by a seed coat that consists of a hull and an adherent layer of endosperm. Hull and cotyledon can be utilized separately as a functional ingredient (Oomah and Mazza 1997; Wiesenborn *et al.* 2002). Hence dehulling and milling studies have been carried out for flaxseed utilization purposes.

Oomah and Mazza (1997) reported that the dehulling process significantly decreased water absorption capacity and viscosity of flaxseed. Dehulled Omega showed the lowest viscosity among the cultivars, whereas Linola™847 and NorMan had the highest. NorMan gave the least reduction in terms of viscosity. Oomah and Mazza (1998b) noted that microwave drying was a valuable pretreatment for abrasive dehulling applications. Higher hull yields resulted from the microwave pretreated seeds compared to untreated seeds.

Tostenson *et al.* (2000) reported the use of a pearler for fractionation of the flaxseed. The data indicated higher hull fraction yields compared to abrasive dehulling. Madhusudhan *et al.* (2000) found that centrifugal cutting mill could be successfully applied to fractionate hull and germ fractions. Results suggested a negative correlation between the oil and SDG contents. Hull fractions could be utilized as a good source of SDG, whereas dehulled seed could be processed for flaxseed oil. Wiesenborn *et al.* (2003) discussed continuous abrasive milling and results showed that the process efficiency was affected by cultivar, moisture content, and feed rate. SDG content of fractions was negatively related to oil content, which was consistent with the results of Madhusudhan *et al.* (2000). Wiesenborn *et al.* (2003) noted that the SDG content was an indicator of purity of the embryo and hull fractions, which was related to cultivar and growing location and conditions.

2. *Storage stability of milled flaxseed*

White and Jayas (1991) observed a twofold increase in free fatty acids (FFA), which was correlated to discolored seed content. This observation also triggered a rapid loss of germination capacity. Significant color changes (from red–brown to dark brown) in flaxseed were observed at various storage times when the seed was stored at 40–50°C at a relative humidity level of 35% and above. Seed storage conditions can affect seed color, which might affect flaxseed end use. Dark flaxseed percentage in a seed lot was related to poor lipid stability in milled flaxseed (Pizzey and Luba, 2002). The authors investigated lipid oxidation in milled flaxseed samples containing

2.7% and 25% dark flaxseed. The milled flaxseed sample with 25% dark seed produced higher FFA and peroxide values, suggesting that presence of dark seed could cause undesirable quality of the milled flaxseed if not removed prior to milling. Results were consistent with the conclusions of White and Jayas (1991) in terms of seed color. Pizzey and Luba (2002) demonstrated the fact that that Canadian and US number 1 grade flaxseed could contain high levels of dark seed depending on the season. The role of lipase and other enzymes in flaxseed stability is not well understood.

Wanasundara *et al.* (1999) showed that flaxseed lipase had an activity of 160 units/g prior to germination and that after germination activity increased to as high as 354 units/g. High FFA content in immature seed reported by Malcolmson *et al.* (2000) indicates that lipase activity might be greater in immature seed than in mature seed. Information concerning lipase activity in immature flaxseed was not found. However, Daun (1993) presented data showing that frost-damaged immature canola had higher FFA content than normal canola. Though this information is not for flaxseed, the same trend may be observed for flaxseed.

Whole flaxseed remains oxidatively stable for many years; however, high moisture conditions during storage can trigger enzymatic-promoted oxidation. Enzymes, such as lipoxygenase (LOX), might promote sufficient oxidation to affect flaxseed quality. Thus, knowing the enzyme level might provide flaxseed handlers with a strategy to prevent enzymatic-promoted oxidation. LOX is soluble in the cytoplasm of cells and has been found in chloroplasts (Vick and Zimmerman, 1987) and lipid bodies of oilseeds (Feussner *et al.*, 1995). LOX requires substrate with the pentadiene system; thus only the unsaturated fatty acids, linoleic, and linolenic acids, in flaxseed can function as LOX substrate. LOX has a preference for FFA substrates more than TAG (Hamilton, 1994). Flaxseed LOX produces hydroperoxides at the C-13 position (80%) and C-9 position (20%) in linoleic substrates (Zimmerman and Vick, 1970). With linolenic acid, 88% of the C-13 hydroperoxide and 12% of the C-9 hydroperoxide is formed (Zimmerman and Vick, 1970).

LOX in young and developing plant tissues generally have greater activity than in mature tissues (Siedow 1991; Zhuang *et al.* 1992). Thus, flaxseed with high LOX activity may be indicative of either germinated and or immature seeds. LOX content is affected by cultivar × location × year interaction but quantitative differences in LOX content and LOX activity are primarily due to cultivar (Oomah *et al.*, 1997b). Flaxseed was reported to contain from 1.63 to 5.98 g/kg of LOX (Oomah *et al.*, 1997b) with Linola cultivar having the highest content. Linola is very high in LA and low in linolenic acid, and that LA is the preferred flaxseed LOX substrate (Zimmerman and Vick, 1970). This suggests that linolenic acid content is less important than LA content in regards to LOX content in flaxseed. Investigations by Oomah *et al.* (1997b) showed that LOX content varied significantly with cultivar; however, the relationship between LOX and flaxseed stability was not determined. The data suggests that the fatty acids in the TAG are an important factor in lipoxygenase-promoted oxidation of flaxseed. Additional research to correlated lipoxygenase and oxidative stability is needed.

Thermal and oxidative stability of whole and milled flaxseed and extracted flaxseed oil at 178°C has been reported (Chen *et al.*, 1994). Researchers showed that milled flaxseed exhibited the highest level of oxygen consumption followed by flaxseed lipid extract, whereas whole seed showed little change over a period of 90 minutes at 178°C. The ALA content decreased significantly from 3.8% to 3.3% for milled flaxseed and flaxseed lipid extract, respectively. Results indicated that oxidative susceptibility of milled flaxseed was related to particle size. Chen *et al.* (1994) reported that oxygen consumption was highest in the coarse fraction (>950 micron), followed by the fine fraction (<500 micron), and the granular (intermediate) size fractions (500–710 and 710–850 micron) were the most stable as measured by the lower oxygen consumption. High oxygen consumption by large-sized particles was suggested to be due to ample air space between flaxseed particles. Thus, greater oxygen diffusion into the pile of coarse-milled flaxseed may be more probable than for the fine-milled flaxseed. Surface area and tight packing of fine-milled flaxseed particles might have contributed to the lower oxidation rate observed in contrast to coarse-milled flaxseed (Chen *et al.*, 1994). We have observed similar observations in pasta made from different sized flaxseed particles, whereby pasta made with large flaxseed particles oxidized more readily than pasta made with small flaxseed particles.

Malcolmson *et al.* (2000) reported the storage stability of milled flaxseed stored at ambient temperatures. The result showed that peroxide value and conjugated diene did not significantly change during storage (128 day). Linott cultivar exhibited a significant increase in terms of FFA (0.3% to 1.58%). About 5% of the Linott seed was slightly discolored indicating the seed lacked full maturity. Malcolmson *et al.* (2000) speculated that high FFA content in milled flaxseed sample was due to the presence of immature seed in the sample. Chlorophyll and moisture content could also be significant parameters, which are likely to be high in immature seed and may trigger lipid oxidation in milled flaxseed.

In the follow-up report, Przybylski and Daun (2001) tested different lots of milled flaxseed for oxidation. These samples were stored in loosely closed plastic bags protected from light at ambient temperatures for up to 20 months. The authors observed that one sample stored for 11 months had higher FFA content (9.70%) and had higher off-flavor characteristic than samples that were stored for 0 and 20 months. Peroxide values for samples varied between 2.4–3.4 meq/kg and flaxseed stored for 11 months actually had the lowest value in contrast to FFA. Researchers suggested the presence of sufficient moisture or damaged seed might have promoted lipolytic activity during storage.

In a study conducted in the author's research lab, storage stability of flaxseed stored under various conditions was tested. Ground flaxseed packed in plastic or aluminum bags were stored for 4 weeks at 30°C, then for 3 weeks at 60°C in a temperature-controlled chamber under light. Peroxide values of all samples increased from 0.8 to 1.3 meq/kg oil in flaxseed stored in aluminum foil bags at 60°C at the end of 7 weeks of storage, which showed protective effect on flaxseed oxidation against light exposure and high temperature. Peroxide values of the flaxseed stored in plastic bags averaged 1.5 meq/kg oil at the end of 4 weeks of storage, but then increased to 5.0 meq/kg oil after three additional weeks of storage at 60°C. In terms of GC-MS analysis, flaxseed stored in plastic bags exhibited significantly higher propanol, pentanol, and hexanol secondary oxidation products, whereas these same volatiles were observed to be at lower concentrations in flaxseed stored in aluminum bags. In contrast, butanol, the primary volatile, was higher in the samples stored in aluminum bags compared to the samples stored in the plastic bags. Alkoxy radicals can react with unsaturated lipid to form stable and innocuous alcohols (Frankel, 2005). However in this case, hydroperoxides produced by lipoxygenase might go through an isomerization and rearrangement to produce aldehydes, short chain alcohols, vinyl ethers, or oxo acids (Frankel, 2005). We mainly observed the oxidation end product volatiles of LA during storage. We speculate that lipoxygenase, which might promote enzymatic oxidation (Frankel, 2005) may have been triggered during the milling process thus promoting the formation of LA primary oxidation and eventually to the secondary products. One particular compound, 2-pentyl furan (butter, green bean flavor), was detected in flaxseed stored in plastic bags, which is a main secondary end product volatile of LA (Frankel, 2005). High temperature storage (60°C) increased significantly the headspace volatile concentration in flaxseed; however, peroxide values did not increase to a great extent at the end of 7-week storage. This indicated that possible decomposition of the hydroperoxides into secondary oxidation products occurred. Overall the aluminum bags provided better shelf life stability for ground flaxseed during elevated storage.

D. FLAXSEED OIL

1. *Extraction of flaxseed*

Edible flaxseed oil is largely sold in organic food markets, thus mechanical pressing is preferred by the food industry due to the absence of alternative and convenient solvents (Wiesenborn *et al.*, 2004; Zheng *et al.*, 2003). Zheng *et al.* (2002) reported that pressing dehulled flaxseed yielded twice the oil output compared to whole flaxseed. Wiesenborn *et al.* (2004) compared the processing of whole flaxseed and dehulled seed. Oil yield was significantly higher in whole seed, possibly indicating that higher oil temperature (66°C) promoted the oil extraction compared to lower temperature (33°C) expelling. Results were consistent with the previous results indicating the aid of fiber during pressing (Zheng *et al.*, 2003). Nevertheless higher oil temperatures of whole flaxseed during cold pressing could be a concern for the industry, because autooxidation may be triggered. Zheng *et al.* (2005) discussed the specific mechanical energy during screw pressing of whole and dehulled seed. Results indicated that a reduction in moisture content of flaxseed (from 12.6% to 6.3%) triggered friction among the seeds in turn, causing a significant increase in specific mechanical energy and heat. Data were consistent with the previous study and suggested that pretreatment of the flaxseed prior to pressing help to reduce exposure of the oil to high temperatures and preventing possible off flavors.

In order to improve oxidative stability, supercritical CO_2 fluid extraction could be used for flaxseed oil extraction (Bozan and Temelli, 2002). This method demonstrated higher ALA content compared to soxhlet extraction. In contrast, tocopherol content was lower (Table XI). Temperature and pressure profiles did not alter the fatty acid profile and exhibited similar results at 50 and 70°C and pressures of 35 and 55Mpa. Stability of oil was not reported, and since limited knowledge is available further research is required with regards to shelf life and comparisons to cold-pressing operation.

2. *Flaxseed oil stability*

Hadley (1996) reported the use of flaxseed oil in frying applications. At higher temperatures (177–191°C) lipid oxidation rate was rapid and ALA content was significantly reduced. Furthermore, a fishy flavor in the oil developed during frying. Although not evaluated, 1-penten-3-one may have formed during this process causing the fishy flavor. Van Ruth *et al.* (2001) noted the use of soybean extracts in flaxseed oil as an antioxidant source. Extracts of soybean seeds reduced the formation of primary oxidation products up to 30%, and secondary lipid oxidation products up to 99%. Various antioxidants could be applied in flaxseed oil (Rudnik *et al.*, 2001). Ascorbyl palmitate, citric acid, ascorbic acid, ethoxylated glycol, and α-tocopherol blend exhibited stronger antioxidant capacity compared to butylated hydroxyanisole.

Lukaszewicz *et al.* (2004) reported the oxidative properties of oil extracted from various flax cultivars. Linola cultivar (high LA), with the lowest content of linolenic acid, exhibited the highest conjugated diene values when heated

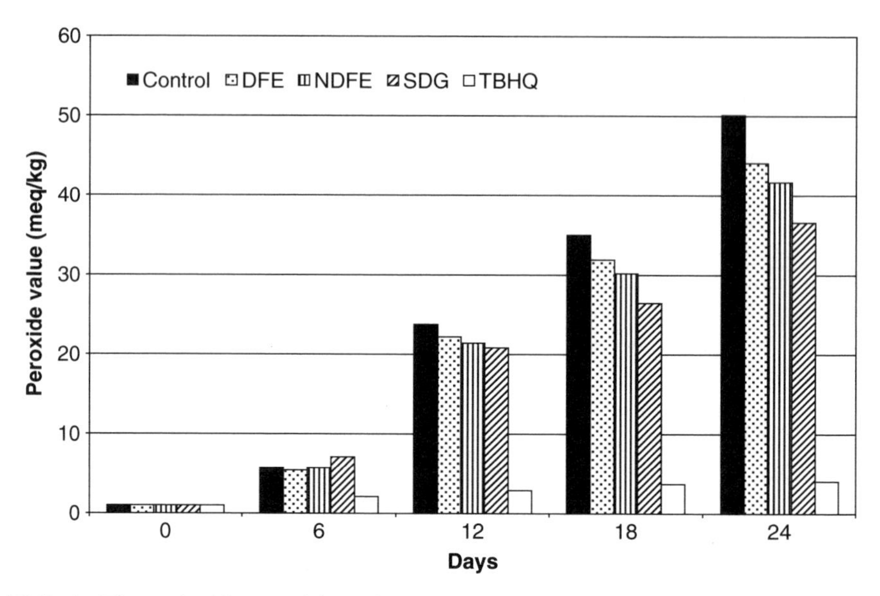

FIG. 5 The antioxidant activity of non-defatted flaxseed extract (NDFE), defatted flaxseed extract (DFE), SDG, and tertiary-butylhydroquinone (TBHQ) in stripped corn oil observed by Hall and Shultz (2001).

for 40 minutes at 140°C. Abby (high linolenic acid), which is a British flaxseed cultivar, gave the lowest conjugated diene and thiobarbutiric acid values. No relationship was demonstrated between oxidation attributes of cultivars and ALA content whereas ALA was negatively correlated with other fatty acids. The observations by Lukaszewicz *et al.* (2004) suggest factors other than oil may be responsible for the observed oxidation behavior of the different cultivars. In the author's laboratory (Hall and Shultz, 2001), antioxidant and prooxidant activities of flaxseed phenolics and SDG were tested. SDG significantly ($p < 0.05$) decreased lipid oxidation rate in corn oil, compared to ground defatted and nondefatted flaxseed extracts (Figure 5). At the 18- and 24-day evaluations, the SDG-treated oil samples had significantly ($p < 0.05$) lower peroxide values than the control oil. However, no differences in peroxide values of the oil treated with the flaxseed extracts and SDG were observed (Figure 5). In the second part of the study, SDG was compared to tocopherol and phenolic compounds. The corn oil treated with SDG had significantly ($p < 0.05$) lower peroxide values at the 11-day storage evaluation than the control (Figure 6). Although not significant, the peroxide values of the combined PA and SDG were lower than the control. The tocopherol treatment (220 ppm) was prooxidant, but

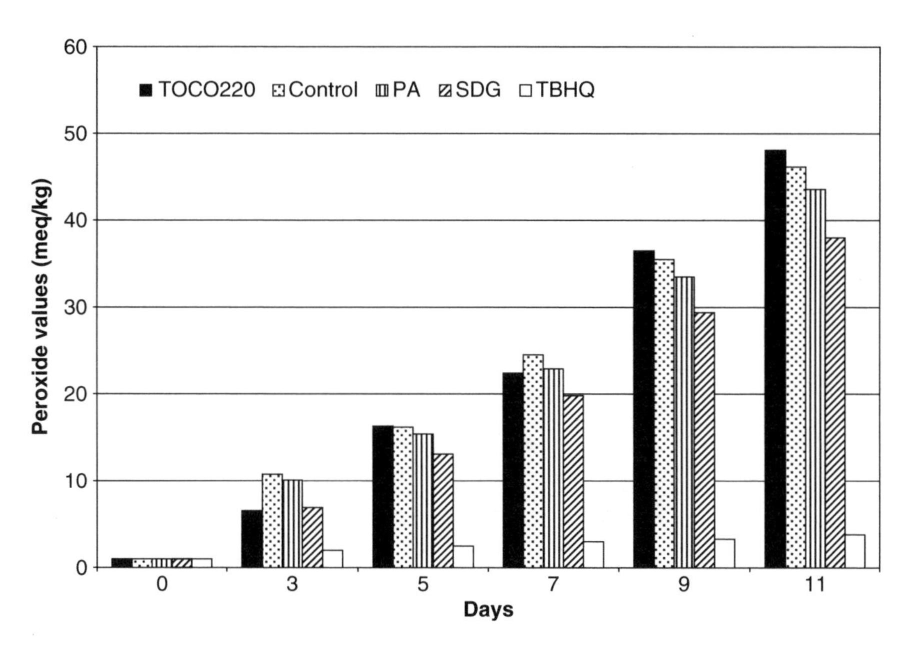

FIG. 6 The antioxidant activity of SDG, phenolic acid (PA), tocopherol (TOCO = 220 ppm), and tertiary-butylhydroquinone (TBHQ) in stripped corn oil observed by Hall and Shultz (2001).

combined with SDG and PA the prooxidant activity of tocopherol was eliminated (Figure 7). The results from our work are consistent with previous findings of Shahidi *et al.* (1995) who also observed antioxidant activity of flaxseed extracts. Wiesenborn *et al.* (2005) reported a relationship between headspace volatile analysis and sensory properties of cold-pressed flaxseed oil. Results indicated that significant differences between samples for painty-bitter flavors and overall quality. Thus, solid phase microextraction analysis could be a means of determining flaxseed oil quality since head space volatiles correlated with descriptive sensory profile.

E. FLAXSEED GUM

Mason and Hall (1948) noted the potential use of flaxseed gum in soft drinks, candy, processed cheese, jellies, and fruit juice. Garden (1993) reported that flaxseed gum significantly improved bread quality and shelf life, and suggested the use of the gum fraction as a food ingredient in food products. Chemical, physical, and functional properties of flaxseed gum have been documented (Chornick *et al.*, 2002; Cui *et al.*, 1994b, 1994c; Mazza and

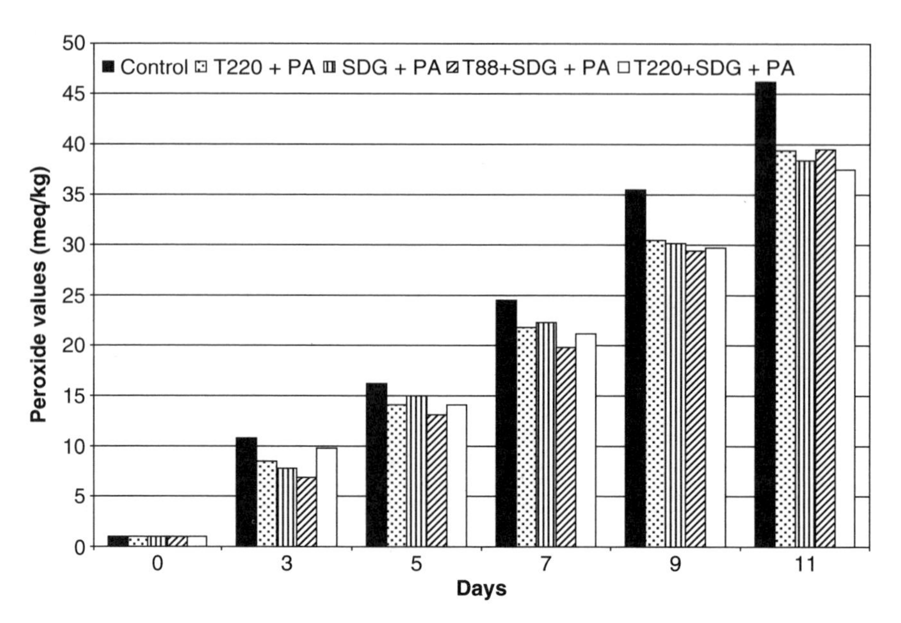

FIG. 7 The antioxidant activity of SDG, phenolic acid (PA), and tocopherol (TOCO = 88 or 220 ppm), alone or in combinations, in stripped corn oil observed by Hall and Shultz (2001).

Biliaderis, 1989; Oomah and Mazza, 1998b). Flaxseed gum exhibited good foam stability at a level of 1% and maximum viscosity at pH 6.0–8.0 (Mazza and Biliaderis, 1989). Oomah and Mazza (1998b) reported that lipid removal significantly increased apparent viscosity values of flaxseed gum. Furthermore, viscosity of seed, cake, and flake samples was significantly related to protein (r = 0.97) and carbohydrate (r = 0.91) fractions, which were related to mucilage fraction of the seed.

Cui *et al.* (1994c) reported genotypic differences of flaxseed gum in terms of chemical and rheological properties. Yellow flaxseed cultivars showed stronger dynamic rheological properties in aqueous solutions, compared to brown seeds. It was observed that neutral polysaccharide fraction of flaxseed gum caused weak gel forming. NorMan cultivar exhibited the highest properties among the brown cultivars. Moore *et al.* (1996) reported that apparent viscosity of flaxseed gum was stable at a wide pH range (4.5–7.0).

Cui and Mazza (1996) reported that neutral polysaccharides have a larger molecular size than acidic polysaccharides and flaxseed gum showed superior moisture retention properties compared to carboxymethylcellulose, Arabic, guar, and xanthan gums. Acidic polysaccharides show more shear thinning properties than neutral polysaccharides. The pH of a 0.5% flaxseed gum dispersion was observed to be 6.4 (Huang *et al.*, 2001). The emulsion made from 0.5% flaxseed gum lacked storage stability as microbial spoilage occurred before 30 days (Huang *et al.*, 2001). We have observed similar results with aqueous extracts of flaxseed gums where the gum solution becomes moldy within 30 days. Thus, the gum should be dried as quickly as possible once extracted from the flaxseed to prevent molding of the extract.

Flaxseed had the highest protein content of various gums tested by Huang *et al.* (2001); however, the stabilizing effect of the flaxseed gum on the emulsion was less than fenugreek, yellow mustard, gum Arabic, and methylcellulose. Apparent viscosity value of flaxseed gum was the third highest following xanthan and locust bean gum. Huang *et al.* (2001) attributed the results mainly to protein content, since protein content is significantly related to emulsion and foam stability of colloidal solutions.

Chornick *et al.* (2002) tested the mucilage content among the Canadian flaxseed cultivars. Results showed diversity in terms of molecular weight, structural conformation, and proportion of neutral to acidic polysaccharides. Xylose content exhibited significant correlation with viscosity values. Mucilage viscosity significantly increased with higher levels of neutral polysaccharide content, which was consistent with the results of Cui and Mazza (1996). In contrast, Warrand *et al.* (2005a) reported a reduction in flaxseed cake mucilage viscosity compared to whole seed mucilage. Steady shear rheological flow properties indicated a very low viscosity level which might have been due to oil extraction conditions (high temperature and pressure). Thus, oil extraction technique could be a major factor that impacts the mucilage and gum properties. In a follow-up study, Warrand *et al.* (2005b) investigated the neutral polysaccharides of flaxseed mucilage after fractionation by size-exclusion chromatography. The presence of arabinoxylans (1,4 linked β-D-xylans) with a constant Arabinose/Xylose ratio of 0.24 was observed with variable galactose and fucose residues connected to the side chains. These authors speculated that the association between polymers could be the reason for the observed rheological properties.

There has been an interest in utilizing flaxseed pectin as an encapsulation agent for shark liver oil (Diaz-Rojas *et al.*, 2004). Flaxseed pectin did not exhibit acceptable results; however, researchers suggested the use of the pectin fraction be combined with alginate coated by chitosan. In a study reported by Qin *et al.* (2005), flaxseed gum stabilized the cloudy appearance of carrot juice and reduced creaming. Thus, suggesting that the soluble gum could be used as dispersing and stabilizing agents in beverages. Limited knowledge is available regarding flaxseed gum and mucilage in food model systems, thus further research is required to understand the basic interactions between flaxseed gum and food components.

F. ROASTED FLAXSEED

Kozlowska (1989) discussed the use of roasted flaxseed in bread and confectionery products. Schorno *et al.* (2003) and Tulbek *et al.* (2004) reported roasted flaxseed quality. Higher roasting temperature and longer roasting time conditions indicated greater moisture loss and flaxseed exhibited significantly lower moisture contents (0.1–0.3%), after roasting samples for 24 minutes at 160 and 180°C. The recorded moisture loss might include volatiles that are lost under the conditions of the test; however, the major loss should be considered to be moisture. When the roasting temperature was increased to 180°C, significant changes were observed in color properties. Roasted flaxseed brightness significantly ($p \leq 0.05$) decreased at 160 and 180°C; however, no color differences were observed at 140°C. Samples roasted at 160 and 180°C were higher in redness and lower in yellowness than control and samples roasted at 140°C. Extractable lipid increased with roasting temperature but not with duration. Observations were consistent with the research of Yoshida and Takagi (1997), who found that roasting time and temperature increased lipid extraction in sesame seed.

Jung *et al.* (1999) extracted pyrazines from the oil extracted from roasted red pepper seeds and reported that the oil had a pleasant nutty and peanut butter-like aroma. Similar aroma was noted in roasted flaxseed oils after soxhlet extraction. GC-MS analysis indicated methyl pyrazine, 2,5-dimethyl pyrazine, trimethyl pyrazine, and 3-ethyl-2,5-dimethyl pyrazine as the most abundant Maillard end products in roasted flaxseed products (Figure 8). Pyrazines contributed to the roasted flavor of roasted products. The threshold levels of pyrazines are very low, thus low pyrazine contents could significantly affect the flavor profile of roasted products (Belitz *et al.*, 2004). According to the sensory ranking test, the top three roasted flaxseed treatments were 160°C for 16 and 24 minutes, and 180°C for 8 minutes. These treatments were screened for further sensory evaluation. Consumer preference tests indicated that roasted flaxseed had better flavor, texture, and overall acceptability than the nonroasted control sample. Roasting significantly improved the aroma and flavor properties of flaxseed. We believe that the presence of pyrazines in the roasted flaxseed affected consumer preference. Kozlowska (1989) reported similar results and the pleasant taste of roasted flaxseed. Principal component analysis of sensory results was used to show that overall acceptability was significantly related to flavor and texture, whereas appearance was not determined to be a consumer preference factor in the roasted flaxseed. The cracking or splitting of some of the flaxseed during roasting may be the cause for the lack of correlation between preference and appearance. Although minimal, the scanning electron micrograph (SEM) illustrates a split seed that occurs during roasting (Figure 9).

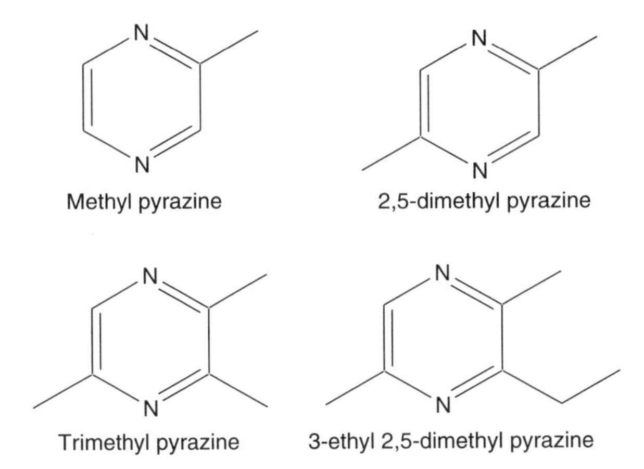

FIG. 8 Flavor compounds in roasted flaxseed observed by Meyers *et al.* (2004).

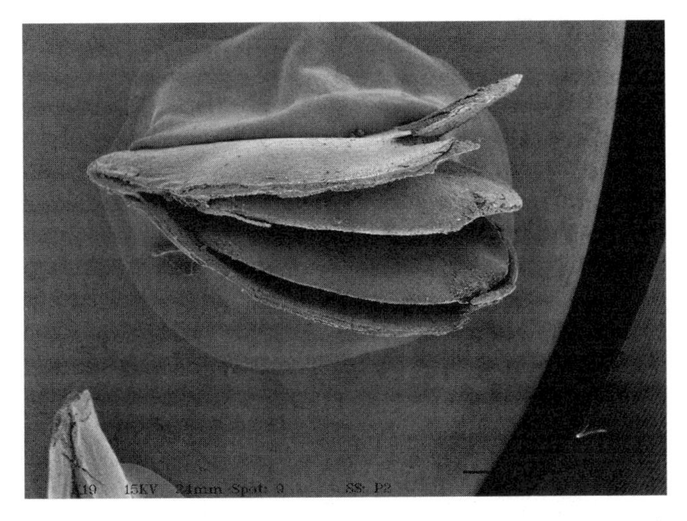

FIG. 9 Scanning electron micrograph of roasted flaxseed (160°C, 4 minutes, ×19) (Tulbek *et al.,* 2004).

Pearson correlation coefficients showed a relationship between color and extractable lipid. Lipid extraction was negatively correlated with brightness ($r = -0.68, p \leq 0.01$), and yellowness ($r = -0.67, p \leq 0.01$), whereas positively correlated with redness ($r = 0.40, p \leq 0.01$) color values. Thus, the extraction of Maillard reaction byproducts during oil extraction may account for the color of the roasted flaxseed oil, and thereby explain the increase in extractable lipid observed. Peroxide values of flaxseed decreased significantly with roasting. Roasted flaxseed showed significantly ($p \leq 0.05$) lower peroxide values (0–0.2 meq/kg) when compared to the unroasted sample (1.8 meq/kg). In contrast, Yoshida and Takagi (1997) reported that peroxide values of roasted sesame seed increased with roasting temperature and time. FFA content of roasted flaxseed decreased with roasting applications; however, flaxseed roasted for a shorter time (4–8 minutes) had lower FFA contents than samples roasted longer (16–24 minutes). Shimoda *et al.* (1997) reported that the consumption of FFA during flavor compound formation can occur, but at high temperature and long-roasting times FFA formation occurs faster than the consumption of FFA in flavor formation. Thus, this may explain the increased FFA we observed in the flaxseed as roasting time increased. FFA content was significantly correlated ($r = 0.54$, $p \leq 0.01$) with peroxide value, indicating the corresponding decrease observed in both parameters. However, as FFA content increased at longer roasting times, peroxide values did not respond similarly. This might be due to a breakdown of the hydroperoxides and also the protective effects of endogenous antioxidants such as SDG, tocopherols, or various pyrazines.

Fatty acid compositions of roasted flaxseed samples are presented in Table XIII. The distributions of fatty acid showed statistical variations among the treatments within similar ranges. α-Linolenic, oleic, and LA were the predominant fatty acids in roasted flaxseed respectively. Hettiarachchy *et al.* (1990) reported the fatty acid composition of 11 flaxseed cultivars grown in North Dakota in 1989 to be 4.6–6.3 palmitic, 3.3–6.1 stearic, 19.3–29.4 oleic, 14.0–18.2 linoleic, and 44.6–51.5% ALA. The fatty acid composition of all samples reported in this study fall within the ranges reported by Hettiarachchy *et al.* (1990). Results were furthermore consistent with the literature findings, as fatty acid profile of roasted flaxseed was stable during and after roasting process (Kozlowska, 1989). ALA slightly decreased at longer roasting time, whereas no significant differences were detected in relation to roasting temperature. Chen *et al.* (1994) heated flaxseed samples at 178°C for 90 minutes and found that no significant difference was observed in fatty acid composition when compared to the control. When the same flaxseed was ground prior to heating, a significant decrease was observed in ALA and a significant increase observed in oleic acid (Chen *et al.*, 1994). Oleic acid content was negatively correlated with brightness ($r = -0.49$, $p \leq 0.01$) and yellowness ($r = -0.44$, $p \leq 0.01$) color values, whereas ALA was positively correlated ($r = 0.40$, $p \leq 0.05$; $r = 0.40$, $p \leq 0.05$ respectively). These findings might indicate the role of triacylglycerols or diglycerols on the development of color and flavor compounds

TABLE XIII

FATTY ACID DISTRIBUTION OF LIPIDS EXTRACTED FROM ROASTED FLAXSEED[a,b]

Roasting temperature (°C)	Roasting time (minutes)	16:0 (%)	18:0 (%)	18:1 (%)	18:2 (%)	18:3 (%)
	Unroasted	5.49bc	4.55c	27.15e	16.23d	46.28ab
140	4	5.54abc	4.51c	27.00e	16.33bcd	46.24abc
	8	5.49bc	4.32d	27.16cde	16.56a	46.05bc
	16	5.46bc	4.27de	27.05e	16.48ab	46.20abc
	24	5.58abc	4.54c	27.20cde	16.56a	45.74cd
160	4	5.27c	4.66b	26.98e	16.25cd	46.06bc
	8	5.44bc	4.49c	27.05e	16.28cd	46.27ab
	16	5.85a	4.77a	27.89a	16.24cd	44.79e
	24	5.65ab	4.48c	27.16cde	16.28cd	46.19abc
180	4	5.45bc	4.20e	26.93e	16.40abc	46.67a
	8	5.69ab	4.31d	27.36bcd	16.40abcd	46.05bc
	16	5.56abc	4.50c	27.44bc	16.46ab	45.45d
	24	5.69ab	4.58bc	27.48b	16.38bcd	45.51d
P		0.05	0.0001	0.0001	0.0001	0.0001

[a]Adapted from Tulbek *et al.*, 2004.
[b]Values not sharing a common letter are significantly different ($p \leq 0.05$).

during flaxseed roasting. In addition, microbial load on flaxseed was significantly reduced during the high temperature processing. The impact was greatest at high temperature long time combinations. Kozlowska (1989) observed that the roasting process promoted the decomposition of cyanogenic glycosides from 16.32 to 0.82 mg/kg. Thus roasting process could be means of improving the microbial quality of flaxseed and removal of undesired compounds, such as cyanogenic glycosides and vitamin antagonists, in flaxseed. However, only the extreme conditions were detrimental to the SDG content (Figure 10).

Meyers *et al.* (2004) and Tulbek *et al.* (2004) reported the storage stability of roasted flaxseed. Roasted flaxseed was stored for 16 weeks in paper bags at 25 and 30°C. Roasting process significantly decreased water activity (a_w) of flaxseed ($a_w < 0.1$). Very low water activity has been associated with rapid lipid oxidation (Fennema, 1996). Furthermore, cracked seeds expose cotyledon to air and consequently we anticipated that oxidation could be a setback during ambient storage. The storage of samples exhibited an increase in peroxide values (Figure 11) and a decrease in FFA. Elevation in oxidation rate supported our theory that, lower a_w levels enhanced oxidation. Week 16 samples gave painty and fishy aroma and flavor at 30°C. Peroxide values,

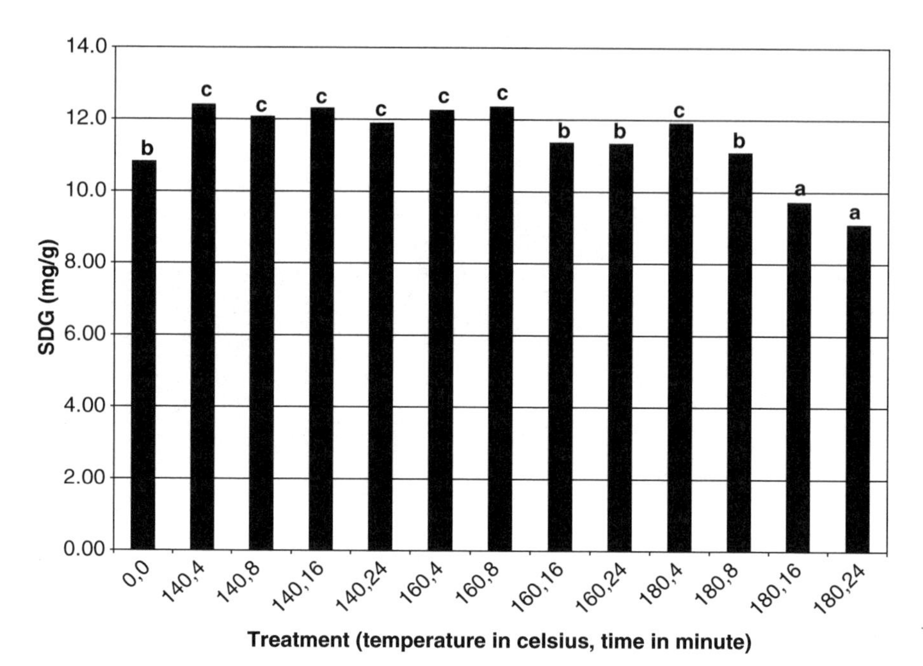

FIG. 10 Concentration of SDG (mg/g) in roasted flaxseed observed by Tulbek *et al.* (2004). Letters above bars represent significant ($p < 0.05$) differences.

propanal and hexanal contents significantly ($p < 0.05$) increased with storage time, indicating a decrease in stability of roasted flaxseed with paper bags. Data was not observed to be similar to Kozlowska (1989); however, water activity and harsh storage conditions were in a range that would accelerate the oxidation. Results indicated that storage in a paper bag and ambient temperature could not be appropriate for roasted flaxseed storage. Additional research is in progress to identify proper storage conditions.

G. BAKING APPLICATIONS

Garden (1993) reported the physical properties of wheat flour with added ground flaxseed. Dough strength was stable up to 5% ground flaxseed (Table XIV). Farinograph absorption significantly ($p \leq 0.05$) increased and dough strength remained stable up to 5% ground flaxseed addition. Peak time, an indirect indicator of dough mixing time, increased with flaxseed addition, which could lengthen baking time. However, a general weakening of farinogram curve was observed due to the decrease in stability over 5% level. Baking experiments indicated a reduction in terms of specific

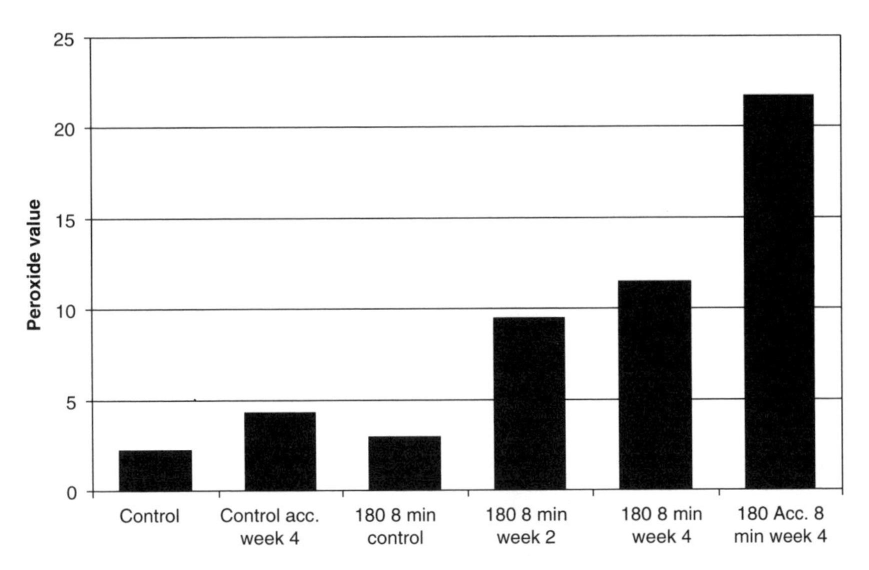

FIG. 11 Peroxide values (meq/kg) of roasted flaxseed (180°C, 8 minutes) stored at room temperature or 30°C (Accelerated = Acc) as observed by Tulbek *et al.* (2004).

volume and oven spring values with 5% ground flaxseed. The presence of TAG and flaxseed protein was suggested to cause the detrimental effects on bread properties. Ground flaxseed significantly increased bread firmness and reduced crust and crumb color quality due to the Maillards reactions possibly related to flaxseed protein and phenolic compounds (Garden, 1993). In contrast, the author observed superior results with flaxseed gum. Every 0.5% flaxseed gum addition increased water absorption of a hard spring wheat flour by 3%. Farinograph peak time slightly increased with flaxseed gum addition, whereas stability values decreased. Extensigraph analysis showed that extensibility, maximum resistance, and area values were significantly reduced with flaxseed gum, which was consistent with farinograph analysis. Nevertheless, flaxseed gum addition increased loaf volume and enhanced shelf stability of bread. Oven spring and specific volume values significantly increased at 1% flaxseed gum addition. Furthermore, crumb and crust color quality was similar to control, indicating no additional Maillard reactions. Carbohydrate–protein interactions in dough system could improve bread quality and explain some of the observations upon addition of flaxseed gum to dough.

Muir and Westcott (2000) investigated the stability of SDG in baked products. SDG content of flaxseed bread was stable during baking process

TABLE XIV
EFFECTS OF GROUND FLAXSEED ON FARINOGRAPH PROPERTIES OF LEN HARD RED SPRING WHEAT FLOUR[a,b]

Sample	Absorption (%)	Peak time (minutes)	Stability (minutes)
Control	60.5e	8.3d	21.8a
Control + 0.5% GF	61.5d	10.8c	22.0a
Control + 2% GF	62.0c	12.4b	22.0a
Control + 5% GF	63.0b	13.4a	20.8a
Control + 10% GF	65.0a	13.8a	15.4b

[a]Adapted from Garden (1993).
[b]Values not sharing a common letter are significantly different ($p \leq 0.05$).

and consistent with the observed range of SDG levels of flaxseed cultivars (Westcott and Muir, 1996). Muir and Westcott (2000) noted that recovery from the bread was only 73–75% of the theoretical yield of SDG. In their finding, the SDG in flaxseed bread was in a range of 34 to 136 mg/100 g, and that of other baked products including bagel, cookie, and muffin ranged from 60 to 120 mg/100 g. Nesbitt and Thompson (1997) also investigated lignan contents of homemade and commercial products containing flaxseed, which included bread, pancake, muffin, pizza dough, and breakfast cereal. The lignan content ranged from 1.1 to 32.4 μmol/100 g (0.33 to 9.72 mg/100 g by a conversion factor of 0.3, calculated from an average molecular weight of 300 of metabolites enterodiol and enterolactone) in the products due to the different flaxseed levels in the formulas and the different retaining ratio in the processing of different types of product (Nesbitt and Thompson, 1997). Gluten formation in bread or other bakery items could entrap flaxseed and interfere with SDG recovery, which might possibly lead to a low SDG theoretical yield.

Flaxseed powder and oat bran were utilized as a fat replacer in cake without any detrimental effects (Lee *et al.*, 2004b). Flaxseed significantly decreased viscosity at 20%; however, an increase in cake volume was observed, potentially due to the high lipid content which served as a shortening in the presence of nonstarch polysaccharides. However, more data are required to explain the observed phenomenon. Flaxseed powder addition significantly ($p \leq 0.05$) darkened crumb color and increased yellowness values. Results indicated that flaxseed was an acceptable additive when used with Nutrim oat bran.

Ghosh *et al.* (2004) reported the effects of ground flaxseed on wheat flour tortilla quality. Ground flaxseed addition significantly increased water absorption and decreased dough strength. However, the presence of nonstarch polysaccharides in flaxseed significantly increased ($p \leq 0.05$) rapid viscoanalyzer values. Cold paste viscosity values were observed to be higher than the control. Flaxseed addition significantly decreased brightness values due to the presence of specks. Tortilla tensile properties of fresh tortillas containing 15% or more ground flaxseed were significantly different than the control and lower flaxseed treatments (5% and 10%). However, 10-day storage results did not show statistically significant differences in tensile properties. Headspace volatile analysis indicated that flaxseed in the tortilla significantly increased propanal, the major secondary volatile of ALA. Tortillas are typically baked for 1–2 minutes in an air impingement oven. Surface area of the product and lower a_w on the surface of tortilla could be the reason for the increased lipid oxidation. However, proper storage and packaging greatly diminished oxidation (unpublished data).

H. EXTRUDED PRODUCTS

Ahmed (1999) reported that flaxseed flour addition significantly reduced expansion ratio, and increased bulk density and breaking strength results, indicating a denser product. Sensory evaluation scores of the extruded flaxseed-corn snack product were lower than control. The water absorption index and water solubility index of extruded samples significantly decreased in extruded samples, but increased in raw samples. Flaxseed significantly ($p \leq 0.05$) reduced brightness and increased redness. Results were consistent with the observations of Garden (1993). Fiber fraction of flaxseed could be effective in extrusion properties due to the extensive hydrogen bonding and linear structures of flaxseed gum.

Manthey *et al.* (2000) investigated the effects of ground flaxseed in spaghetti. Dough strength significantly decreased, with small particles having the most detrimental effect on dough strength. However, medium and coarse fractions resulted in spaghetti that was too brittle whereas the fine particle size flaxseed gave acceptable spaghetti quality. Manthey *et al.* (2002a) noted that flaxseed macaroni was stable during processing and storage. FFA content was reduced due to hydration and extrusion processes. Hydrated premix showed higher conjugated diene values compared to dry premix. Manthey *et al.* (2002b) reported results consistent with the previous study. Extractable lipid and FFA content decreased due to processing. Conjugated diene results indicated that TAG and ALA were stable during processing and cooking steps (Table XV).

Lee *et al.* (2003) tested the effects of boiling, refrigeration, and microwave heating on ground flaxseed macaroni quality. Boiled macaroni and boiled-refrigerated-microwave-heated macaroni exhibited similar appearance attributes. Cooked firmness values were highest with boiled macaroni,

TABLE XV

CONJUGATED DIENE CONTENT OF LIPID (%, DB[a]) EXTRACTED FROM SEMOLINA-FLAXSEED, UNCOOKED AND COOKED SPAGHETTI[b,c]

		Uncooked		Cooked	
Mixture	Premix[d]	LTDC[d]	HTDC[d]	LTDC	HTDC
Semolina	0.05a	0.22c	0.22c	0.18b	0.18b
Semolina + 15% WGF[e]	0.03a	0.07a	0.07a	0.01a	0.00a
Semolina + 15% SF[e]	0.03a	0.10ab	0.13b	0.01a	0.03a

[a]DB, dry basis.
[b]Adapted from Manthey *et al.* (2002b).
[c]Values not sharing a common letter are significantly different ($p \leq 0.05$).
[d]Premix, dry mixture of semolina and ground flaxseed prior to pasta processing; LTDC, low temperature-drying cycle; HTDC, high temperature-drying cycle.
[e]WGF, whole ground flaxseed; SF, sieved flaxseed.

intermediate with refrigerated macaroni, and lowest with microwave-heated macaroni. Yoshida *et al.* (1990) noted a significant increase in conjugated diene values (from 0.38% to 0.78%) in microwave-heated (20 minutes) flaxseed oil. However, flaxseed macaroni was observed to be stable and ALA remained unchanged. In terms of lipid stability, results were consistent with previous results and lipid oxidation was not observed with boiling (6 minutes), refrigeration (72 hours), and microwave (20 s) treatments. Lee *et al.* (2004a) reported the stability of ground whole flaxseed and flaxseed hull in pasta. Hydration step exhibited the greatest detrimental impact on content and stability of extractable lipid in macaroni supplemented with flaxseed. Hydration decreased the amount of ALA and FFA, whereas an increase in the conjugated diene content in lipid fraction was observed. Semolina-flaxseed premix was hydrated for 10 minutes to 30% moisture prior to extrusion. The extrusion and drying processes did not affect lipid extraction or stability (Lee *et al.*, 2004a).

Hall *et al.* (2005) reported the shelf life of flaxseed macaroni in terms of SDG and lipid stability. Processing and drying methods did not affect lipid oxidation as much as the pretreatment of the flaxseed with steam or addition of the hull component. Pasta made with hull and steam-treated flaxseed had higher oxidation than pasta made with ground flaxseed. Propanal, pentane, hexanal, 2t- and 3c-hexenal, heptadienal, octanal, and nonanal volatiles were detected. These volatiles observed are similar to those reported by Malcolmson *et al.* (2000) and Jelen *et al.* (2000). Ultrahigh and high temperature-drying cycles showed slightly better oxidative stability than low temperature-drying process. However, similar results were observed in the oxidation data for all drying applications. Pentane was observed to be the most abundant volatile of all treatments. Propanal significantly increased by week 32 in hull and steam-treated flaxseed, indicating a potential degradation of ALA. Thus, these pretreatments were detrimental to the oxidative stability of the flaxseed macaroni and were not recommended as possible alternatives to whole ground flaxseed (Hall *et al.*, 2005). Low temperature-dried pasta generally had higher levels of volatiles compared to high temperature- and ultrahigh temperature-dried macaroni. Samples did show early signs of oxidation but no detectable off-aroma was found by week 32. Nevertheless shelf life of pasta is limited as observed by high hexanal levels in pasta stored over 1 year. Volatile concentrations were similar to those reported by Malcolmson *et al.* (2000) hence macaroni containing ground flaxseed would be expected to have similar sensory properties. Minimal lipid oxidation could be due to the possible protection by native proteins of the flaxseed via an encapsulation of the oil body, formation of gluten during dough extrusion entrapping the ground flaxseed, the presence of a vacuum during extrusion, and the presence of antioxidants in durum wheat and flaxseed. In terms of SDG stability, low SDG recovery without a protease pretreatment of the samples supported the flaxseed-gluten entrapment hypothesized by Muir and Westcott (2000) in breads. No degradation in SDG was observed. Results indicated that macaroni containing ground flaxseed and dried at high temperature had very good lipid and SGD stability, which could be used as a means to enhance dietary ALA and SDG consumption.

Sinha *et al.* (2004) reported the effects of various levels of flaxseed (0–20%) on extrusion properties and cooking quality of fresh pasta. Appearance and cooking quality of fresh pasta made with flaxseed was superior at lower absorption level (29%), but brightness and yellowness scores were lower whereas redness score increased in fresh pasta. Flaxseed flour decreased energy requirement to extrude dough, by decreasing gluten strength. Cooked firmness values varied significantly ($p \leq 0.05$) (Figure 12). Control sample yielded the highest firmness and statistically differed from 5% and 10% flaxseed fortified fresh pasta ($p \leq 0.05$), whereas 15% and 20% flaxseed fortified fresh pasta gave the lowest firmness scores.

I. DAIRY PRODUCTS

Flaxseed oil has been proposed to be a valuable ingredient for ice cream products (Hall and Schwarz, 2002). Flaxseed oil replaced between 10% and 25% of the milk fat in ice cream formulas has been investigated. The 25% flaxseed product exhibited an oil-like mouth feel; however, the presence of

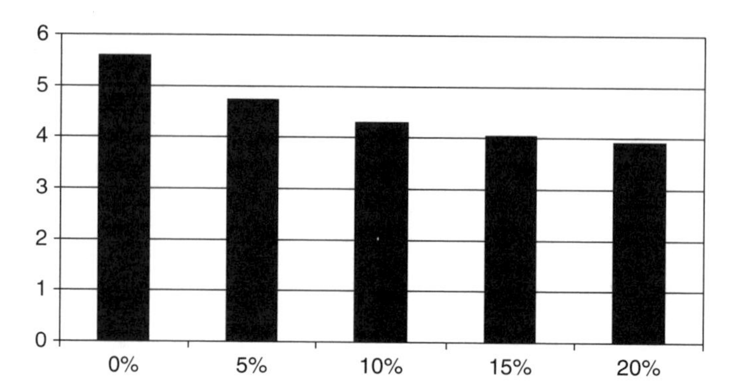

FIG. 12 Effect of flaxseed concentration on fresh pasta cooked firmness (determined by Texture Analyzer TA-XT2, g-cm) observed by Sinha *et al.* (2004).

the oil in product could not be detected by 60% of the panelists using an informal sensory evaluation. A trained sensory panel showed that 15% of the milk fat could be replaced in a vanilla ice cream without being detected. The melt time of flaxseed ice cream was not significantly different compared to control. The 25% flaxseed ice cream product gave a thin consistency compared to 10% product, whereas 10% flaxseed ice cream showed similar properties with control. Flaxseed oil addition significantly improved the fatty acid profile of frozen dessert (Table XVI). No ALA was detected in conventional chocolate ice cream, whereas ALA constituted 20.4% of the fatty acids in the lipid fraction of the flaxseed oil ice cream. Moreover, the ratio of n-3 to n-6 fatty acids was approximately 2.5 to 1 for flaxseed oil ice cream. Overall saturated fatty acid composition decreased from 64.6% to 42%. The ratio of saturated to unsaturated fatty acids was roughly 2.3 to 1 and 0.77 to 1 for the chocolate- and flaxseed oil-containing products, respectively.

Hall *et al.* (2004) reported the stability of lignan in yogurt. In addition, flaxseed extract addition did not have a negative impact on the fermentation. At 700 ppm of the flaxseed extract, lactic acid was observed to be higher than control, whereas the 7000 ppm flaxseed extract addition resulted in lactic acid levels similar to the control yogurt. In contrast, lactic acid bacteria counts in yogurts were lower than the control for the product containing 700 ppm but higher in the product containing 7000 ppm. Acetaldehyde content, the characteristic volatile of yogurt was not significantly influenced by flaxseed extract addition. Lactic acid content, pH, and acetaldehyde content were not significantly affected with regards to end product quality, whereas a reduction in SDG content was observed due to the possible interactions between SDG and

TABLE XVI
FATTY ACID PROFILE OF A COMMERCIAL CHOCOLATE ICE CREAM AND A CHOCOLATE FROZEN DESSERT CONTAINING 25% FLAXSEED OIL[a]

	% Fatty acids[b]	
	Flaxseed oil-fortified ice cream	Standard ice cream
C4–C12 fatty acids	5.2	8.8
Myristic 14:0	6.6	10.9
Palmitic 16:0	20.7	31.5
Stearic 18:0	9.5	13.4
Oleic 18:1	25.7	25.0
Linoleic 18:2	8.3	3.5
Linolenic 18:3	20.4	0.0

[a]Adapted from Hall and Schwarz (2002).
[b]Sums of the fatty acids do not equal 100% due to the fact that odd and branched chain fatty acids present in dairy are not included in the table.

yogurt proteins. SDG recovery results were consistent with the findings of Muir and Westcott (2000) and Hall *et al.* (2005).

J. ANTIFUNGAL PROPERTIES

Antifungal properties of flaxseed were tested in the author's research lab. Milled hull, embryo, or whole flaxseeds of Omega were milled and incorporated into fresh pasta at up to15% and stored in plastic bags under room temperature. Mold growth was monitored on a daily basis. Fresh pasta with 15% Omega flaxseed or fraction was mold-free after 20 days, whereas mold growth was present after 4 days in the control. Defatted flaxseed meal was also tested for their effect on the mold growth of fresh pasta. At 9% of defatted meal level, mold appeared on the same day as compared with control. In contrast, at the 15% level, there was no mold growth after 20 days. We first thought that the variance in mold growth on different fresh pasta samples was due to difference in water activity in different samples. Results from water activity showed that each sample had a water activity of approximately 0.97. These results suggested that components in flaxseed caused the observed phenomenon.

In a subsequent evaluation, a noodle formula was tested with 15% ground flaxseed addition. Noodle sheets were cut into circles and put into sterilized petri dish. The spot inoculation screening method using *Penicillium chrysogenum, Fusarium graminearum*, and *Aspergillus flavus* was conducted to test the antifungal activity of the whole ground flaxseeds. Three spots for each microorganism were inoculated on the noodle sheets and incubated for 5 days. After which time the colony size of the microganism growth was measured. Noodle sheets incorporated with both yellow and brown seeds had less or no mold growth after 6 days of incubation, which suggested that flaxseed exhibited antifungal activity in fresh noodle. Several lots of flaxseed purchased from different suppliers were used in the study. No significant ($p > 0.05$) impact on antifungal activity was observed between flaxseed obtained from various suppliers. The antifungal activities of the yellow and brown flaxseed also were not significantly different. In contrast, microorganisms showed significantly different responses to the flaxseed treatments. Further investigation has shown that the protein and polyphenolic fractions have antimicrobial activity (Xu *et al.*, 2006).

K. VALUE-ADDED ANIMAL PRODUCTS

Flaxseed as an animal feed has been limited until recently, although the benefits of feeding flaxseed to animals have been observed for nearly 100 years. Many of these observations have become folklore such as shinner coats and improved animal health. Quantifying the benefit to animal health is needed. However, some studies have shown general health improvement along with enhancement to animal production and end product quality. For a review of the health and production issues see the extension publications of Maddock *et al.* (2005), Novak and Scheideler (1998), and Puthpongsiriporn and Scheideler (2001).

1. *Poultry*

The most recent interest in flaxseed as a feed has focused on enhancing the ALA and other long chain fatty acid contents in eggs, meat, and milk. Scheideler and Froning (1996) found that ALA was the major omega-3 in egg yolks. However, significant amounts of longer-chain omega-3 fatty acids, such as EPA, DPA, and DHA, were incorporated into the egg yolk phospholipid fraction. Furthermore, a linear increase in ALA content in the yolk was observed with increasing (i.e., 5%, 10%, and 15%) dietary flaxseed (Scheideler and Froning, 1996). Ahn *et al.* (1995) reported that incorporation of ALA (3%) and tocopherol (120 IU/kg) into chicken diets enhanced omega-3 content in eggs by 6.5%. Over 70% of the omega-3 lipid in the egg was ALA. The remaining omega-3 lipids were DHA and EPA, which accounted for approximately 23% and 7%, respectively. Leeson and Caston (2004) reported that dehulling of flaxseed improved ALA deposition in eggs. However, a nonsignificant reduction in EPA and DHA was observed in eggs from hens fed a dehulled flaxseed diet compared to the whole flaxseed diet.

Although minor differences were reported, sensory scores of eggs from the chickens fed the control (tallow) diet were more favorable than the eggs from the ALA diet. Scheideler *et al.* (1997) reported that the overall acceptance of fresh eggs from hens fed a flaxseed diet were not significantly different from control eggs. However, slightly lower scores in appearance and flavor were observed in eggs obtained from hens fed 10% to 15% ground flaxseed compared to whole flaxseed. Parpinello *et al.* (2006) also reported that overall acceptance of the eggs from hens fed a 2% flaxseed diet was not significantly different from the control. In boiled egg evaluations, the overall acceptability of eggs from flaxseed fed hens was lower than the acceptability of control eggs (Leeson *et al.*, 1998). However, no difference in acceptability was detected between eggs from hens fed a 10% or 20% flaxseed. Leeson *et al.* (1998) also reported that vitamin E fortification in the hen's diet did not improve the sensory characteristics of the eggs. In fact, the combined high flaxseed and high vitamin E diet produced the lowest quality eggs. The appearance of the eggs from hens fed the golden (Omega) flaxseed variety was preferred by panelists over eggs of hens fed brown (Neche) flaxseed (Scheideler *et al.*, 1997). From these reports, a number of factors, such as flaxseed variety and level and egg preparation method, may be responsible for reported differences in sensory quality.

In addition of changes in egg composition, feeding flaxseed to chickens alters the meat omega-3 content. Ajuyah *et al.* (1993) reported significantly higher omega-3 lipids in both dark and white meat from broiler chickens fed flaxseed or flaxseed combined with an antioxidant. The white meat from the chicken fed the flaxseed diets with mixed tocopherol and mixed tocopherol plus canthaxanthin had significantly more omega-3 fatty acids compared to the flaxseed alone or the control diet. In contrast to the white meat, the dark meat had higher ALA levels and less EPA and DHA. However, the overall omega-3 content was higher in the dark meat due to the higher ALA content (Ajuyah *et al.*, 1993). Gonzalez-Esquerra and Leeson (2000) also reported that ALA was deposited in the dark meat broiler chickens feed flaxseed for 1 week prior to slaughter. They also found that more long chain omega-3 fatty acids were deposited in the white or breast meat. These authors reported that the sensory characteristics of the breast meat were not affected by chicken diet. However, only diets containing menhaden oil (7.5 g/kg) or flaxseed (100 g/kg) plus 0.75 g/kg menhaden oil negatively impact the sensory quality of the thigh meat (Gonzalez-Esquerra and Leeson, 2000). Loopez-Ferrer *et al.* (1999) reported that replacement of fish oil with flaxseed oil (8.2% in the diet) in broiler chicken diets increased meat ALA content. However, a decrease in long chain omega-3 content was also observed. Feeding of the flaxseed diet throughout a 5-week period resulted in unacceptable sensory quality. In contrast, feeding the 8.2% flaxseed diet for 1 or 2 weeks prior to slaughter resulted in meat with higher sensory scores compared to the fish oil diet (Loopez-Ferrer *et al.*, 1999).

In several studies, the deposition of the omega-3 in the TAG or phospholipids has been identified. ALA tends to deposit in the TAG fraction whereas long chain omega-3 fatty acids deposit in the phospholipids (Ajuyah *et al.*, 1993; Jiang *et al.,* 1991; Scheideler and Froning, 1996). The data reporting the deposition of omega-3 fatty acid in poultry eggs and meat is clear; however, sensory evaluation of the products give mixed results. For an extensive review of omega-3-enriched poultry meat and eggs see Elswyk (1997), González-Esquerra and Leeson (2001), and Scheideler (2003).

The enhancement of omega-3 in eggs has been used to create a new egg market. The "Omega eggs" contain 350 mg omega-3 fatty acids per egg and have lower amounts of saturated fatty acids and cholesterol (Scheideler and Lewis, 1997). The "Omega Egg" was developed at the University of Nebraska by feeding chickens a special diet containing flaxseed (Scheideler, 1998, 1999). In a small feeding trial, enriched omega-3 eggs, from hens fed flaxseed, were reported to enhance omega-3 fatty acids in blood platelet phospholipid male participants (Ferrier *et al.*, 1995). The data suggests that eggs from flaxseed fed hens could be a way to reduce the dietary omega-6 to omega-3 ratio in humans.

2. *Beef cattle*

Several researchers have looked at the use of flaxseed in beef cattle diets as a means to enhance omega-3 composition in meat. The conversion or deposition of omega-3 into milk and meat of cattle is far less efficient compared to poultry. Ruminal biohydrogenation of the unsaturated fatty acids is thought to be the main reason for the poor deposition omega-3 fatty acids in the meat. However, feeding flaxseed did enhance ALA content in beef when compared to barley- and corn-based diets (Maddock *et al.*, 2003) and Holstein steers (Drouillard *et al.* 2002, 2004). The level of long-chained omega-3 also increased in meat of cattle fed a flaxseed diet. Maddock *et al.* (2005) reported that ALA, EPA, DPA, and DHA levels were higher in the phospholipid fraction of cattle fed 8% flaxseed diet compared to the control. Drouillard *et al.* (2002, 2004) also reported enhanced EPA and DHA in cattle fed a 5% flaxseed diet. In contrast to the observed omega-3 enhancement, conflicting sensory characteristics of omega-3-enhanced beef have been reported (Drouillard *et al.* 2004; Maddock *et al.*, 2003, 2004).

Maddock *et al.* (2003) reported that marbling score were improved for when flaxseed was incorporated into the cattle diet. This observation was in agreement with that of Drouillard *et al.* (2002). The tenderness was lower for boneless strip loins from steers fed 6% flaxseed compared to those fed 3% flaxseed. Juiciness was rated higher for boneless strip loins from cattle fed a corn diets. All diets that included flaxseed resulted in statistically similar juiciness score. In contrast, flavor of the boneless strip loins was not affected by the addition of flaxseed to cattle diet (Maddock *et al.*, 2003). Drouillard *et al.* (2004) did not find significant differences in the tenderness, juiciness, or flavor of steaks from animals fed flaxseed or a control diets.

Further investigation by Maddock *et al.* (2004) showed that tenderness scores were best, but not significant, for steaks obtained from heifers that were on a diet that contained 8% rolled flaxseed. In contrast, the Warner-Bratzler shear force values for tenderness did show significantly improved tenderness values compared to the control corn diet. Furthermore, steaks obtained from heifers fed an 8% rolled or ground flaxseed diet were significantly more tender than those steaks obtained from heifers fed an 8% whole flaxseed diet (Maddock *et al.*, 2004). As in their previous study, juciness scores of the steaks from the flaxseed fed heifers were rated lower than the steaks from cattle fed the control diet. However, the differences in juiciness of steaks obtained from heifer on the rolled flaxseed diet and the control diet did not appear to be significant (Maddock *et al.*, 2004). Thus, the data suggests that preparation of the flaxseed did have some benefit to finished product quality.

3. *Dairy cattle*

The information obtained from studies with dairy cattle is mixed with regards to the influence of flaxseed in milk production and composition. However, the general trend is that flaxseed in the cattle diet does influence the composition of milk. The method of feeding flaxseed may be responsible for the sometime contradictory observations regarding milk production. Kennelly and Khorasani, (1992) reported that milk production and fat content were not affected by the incorporation of rolled flaxseed up to 15% in the diets of dairy cattle. Goodridge *et al.* (2001) and Ward *et al.* (2002) also reported that milk production and fat content were not affected by flaxseed diets. Mustafa *et al.* (2003) and Gonthier *et al.* (2005) also reported that milk yield was not significantly different between cattle fed the control diet or diets containing microionized or untreated flaxseed. However, these authors did report lower milk fat content in a diet containing ground flaxseed. Petit *et al.* (2001) also reported a lower fat content in milk and lower milk production from cattle fed formaldehyde-treated flaxseed. In contrast, Petit (2002) found that a whole flaxseed-containing diet resulted in higher milk production than the control diet containing Megalac® (rumin inert fat). However, the milk fat content was lower in the milk obtained from the cattle on the flaxseed diet compared to the Megalac® diet. When considering the fat production on a kilogram per day basis, no significant differences were observed in the milk fat production (Petit, 2002; Ward *et al.*, 2002). All research agreed that omega-3 content of milk was enhanced by the addition of flaxseed in the diet of cattle.

Kennelly and Khorasani (1992), Khorasani and Kennelly (1994), and Ward *et al.*, (2002) reported that protein content was lower in milk from cattle fed a diet containing flaxseed. In contrast, Goodridge *et al.* (2001), Petit (2002), and Petit *et al.* (2001) reported an increased protein composition of milk obtained from cattle fed flaxseed-fortified diets. Mustafa *et al.* (2003) reported similar protein contents for milk obtained from cattle on the control or flaxseed diets. Studies have been completed to evaluate the methods to reduce the rate of biohydrogenation of ALA and crude protein degradability in dairy cattle and model *in vitro* systems (Gonthier *et al.*, 2004a,b; Loor *et al.*, 2005; Petit *et al.*, 2002). Thus, the conflicting observations regarding production, and the fat and protein contents is likely due to the method of flaxseed incorporation into cattle diets and how the cattle assimilate the flaxseed. In any case, omega-3-enhanced milk can be achieved by incorporating flaxseed into the diets of dairy cattle.

4. *Swine*

Romans *et al.* (1995a,b) reported that incorporation of up to15% flaxseed into the diets of swine did not affect carcass traits. However, the amount of omega-3 incorporated into pork lipids varied depending on the carcass tissue. The ALA and EPA contents in the outer backfat were affected by flaxseed incorporation into the swine diet whereas no change in DHA was observed. No significant differences were observed in the ALA or EPA content of the outer backfat of swine fed a 10% or 15% flaxseed diet (Romans *et al.*, 1995a). In contrast, significantly more ALA and EPA were found in the middle/inner backfat for swine fed 15% flaxseed. The addition of 5% and 10% flaxseed in the swine diet also resulted in higher ALA and EPA contents in the backfat compared to the control. Thacker *et al.* (2004) also observed increased ALA content in backfat obtained from swine fed a 30% Linpro (i.e., extruded flaxseed and pea product) compared to the control diet containg soy meal. Significant increases in backfat ALA and EPA contents were also observed as the length of exposure to the flaxseed diets increased to 28 days (Romans *et al.*, 1995b). However, longer feeding trials showed reduced omega-3 contents in adipose tissue (Fontanillas *et al.*, 1998; Riley *et al.*, 2000).

Signficant increases in ALA and EPA were also observed in the belly and bacon of the swine. Belly and bacon from swine fed the 15% flaxseed diets had the highest ALA and EPA contents (Romans *et al.*, 1995a). Signficant increases in belly and bacon ALA and EPA contents were also observed as the length of exposure to the flaxseed diets increased to 28 days (Romans *et al.*, 1995b). Frying of the bacon resulted in a significant increase in EPA and DHA contents whereas ALA was not affected compared to uncooked bacon. The content of ALA was higher in the neutral lipid fraction compared to the polar lipid fraction. In contrast, the polar lipids contained more EPA (Romans *et al.*, 1995b). The increased EPA and DHA in the fried bacon may have been due to the incorporation of these fatty acids into the phospholipid (i.e., polar lipids) fraction and thus less impacted by the heating process. Microwaving proved to be detrimental to the omega-3 fatty acids, as signficiant reductions in their contents were observed (Romans *et al.*, 1995a,b).

The incorporation of omega-3 fatty acids into the loin was not as efficient compared to the other tissues. Feeding a diet greater than 10% flaxseed did not enhance loin ALA or EPA contents and only minimal changes were observed after 21 days of feeding (Romans *et al.*, 1995a,b). Kouba *et al.* (2003) also found increased ALA and EPA contents in the loin of the pig fed the flaxseed diet compared to the control. However, only minor changes were observed in fatty acid contents in lipids in swines fed flaxseed up to 60 days. A significant reduction in ALA and EPA contents were observed in pigs fed flaxseed for 100 days compared to the 60-day feeding.

Sensory panelists were able to identifiy the bacon from the animals fed 10% and 15% flaxseed. Romans *et al.* (1995b) reported that flavor intensity of bacon was greater for bacon obtained from animals fed flaxseed diets between 14 and 28 days compared to the control diet. As the length of the feeding trial increased, so did the fishy defect as reported by the sensory panel. In contrast, panelists were not able to identify the loin from the swine fed flaxseed up to 15% (Romans *et al.*, 1995a) or differentiate the eating quality of loin in swine fed up to 10% flaxseed (Matthews *et al.*, 2000). Riley *et al.* (2000) also observed that feeding flaxseed for 24 days resulted in loin steaks that were juicer and had greater tenderness than the loin steaks from the swine fed the control diet. No differences in taste, tenderness, and juiciness were observed between the loins from pigs fed diets containing 0.4, 0.7, or 1.0% ALA (van Oeckel *et al.*, 1996).

Feeding of animals in general can enhance the omega-3 levels in lipids and is well documented. In contrast, few studies have documented the stability of the lipids in stored meats or in processed meat products. Additional researcher is needed to evaluate the sensory characteristics of stored meats and of processed meat products containing enhanced omega-3 lipids.

V. CONCLUSION

Flaxseed is an oilseed that has had a long history dating back nearly 12,000 years. However, many people are still unaware of the potential health benefits of flaxseed and potential food applications. One cereal company (US Mills) has had flaxseed as a component in their Uncle Sams cereal for nearly 100 years. More recently, other cereal, baking, and pasta companies have incorporated flaxseed into their formulation. Many of these products are promoted as high fiber, but also have a nutrient content claim for the omega-3 fatty acid. Currently, flaxseed does not have a health claim or qualified health claim. The intent of this chapter was to provide a foundation for individuals not aware of the benefits of flaxseed and potential food applications. Although the body of evidence is growing in support of flaxseed consumption, many more studies are needed to resolve the conflicting reports regarding the health benefits, in particular the role of ALA in prostate cancer and cancer in general. Also, a daily recommendation for flaxseed has to be made by authoritative organizations. The general recommendation has been 1–3 tablespoons per day for ground flaxseed or 1 tablespoon for flaxseed oil (Morris, 2003a). These recommendations are based on data reported in the health benefits section of this chapter.

REFERENCES

Adlercreutz, H., Fotsis, T., Heikkinen, R., Dwyer, J., Woods, M., Goldin, B., and Borbac, S. 1982. Excretion of the lignans enterolactone, enterodiol and equol in omnivorous and vegetarian women and in women with breast cancer. *Lancet* **2**, 1295–1299.

Adlercreutz, H., Mousavi, Y., Clark, J., Hockerstedt, K., Hämälainen, E., Wähälä, K., Mäkelä, T., and Hase, T. 1992. Dietary phytoestrogens and cancer: *In vitro* and *in vivo* studies. *J. Steroid Biochem. Mol. Biol.* **41**, 331–337.

Agriculture and Agri-Food Canada. 2006. Canada: Grains and Oilseeds Outlook. http:/www.rayglen.com/pdf/jun2006_e.pdf (accessed May 29, 2006).

Ahmed, Z.S. 1999. Physico-chemical, structural and sensory quality of corn based flax-snack. *Nahrung* **4**, 253–258.

Ahn, D., Sunwoo, H., Wolfe, F., and Sim, J. 1995. Effects of dietary α-linolenic acid and strain of hen on the fatty acid composition, storage stability, and flavor characteristics of chicken eggs. *Poult. Sci.* **74**, 1540–1547.

Ajuyah, A.O., Ahn, D.U., Hardin, R.T., and Sim, J.S. 1993. Dietary antioxidants and storage affect chemical characteristics of w-3 fatty acid enriched broiler chicken meat. *J. Food Sci.* **58**, 43–46.

Albert, C.M., Kyungwon, O.H., and Whang, W. 2004. American Heart Association Annual 2004 Scientific Sessions. Abstract 3604. Presented Nov. 8, 2004. New Orleans, LA.

Allman, M.A., Pena, M.M., and Pang, D. 1995. Supplementation with flaxseed oil versus sunflower seed oil in healthy young men consuming a low fat diet: Effects on platelet composition and function. *Eur. J. Clin. Nutr.* **49**, 169–178.

Amarowicz, R., Chong, X., and Shahidi, F. 1993. Chromatographic techniques for preparation of linustatin and neolinustatin from flaxseed: Standards for glycoside anlyses. *Food Chem.* **48**, 99–101.

Arjmandi, B.H., Khan, D.A., Juma, S., Drum, M.L., Venkatesh, S., Sohn, E., Wei, L., and Derman, R. 1998. Whole flaxseed consumption lowers serum LDL-cholesterol and lipoprotein(a) concentrations in postmenopausal women. *Nutr. Res.* **18**, 1203–1214.

Axelson, M. and Setchell, K.D.R. 1981. The excretion of lignans in rats as evidence for an intestinal bacterial source for this new group of compounds. *FEBS Lett.* **123**, 337–343.

Axelson, M., Sjövall, J., Gustafsson, B.E., and Setchell, K.D.R. 1982. Origin of lignans in mammals and identification of a precursor from plants. *Nature* **298**, 659–660.

Ayorinde, F.O. 2000. Determination of the molecular distribution of triacylglycerol oils using matrix-assisted laser desorption/ionization time-of-flight mass spectrometry. *Lipid Technol.* **12**, 41–44.

Babu, U.S., Wiesenfeld, P.W., Collins, T.F.X., Sprando, R., Flynn, T.J., Black, T., Olejnik, N., and Raybourne, R.B. 2003. *Food Cheml. Toxicol.* **41**, 905–915.

Baylin, A., Kabagambe, E.K., and Ascherio, A. 2003. Adipose tissue alpha-linolenic acid and non-fatal acute myocardial infarction in Costa Rica. *Circulation* **107**, 1586–1591.

Belitz, H.D., Grosch, W., and Schieberle, P. 2004. "Food Chemistry", 3rd revised Ed. Springer-Verlag, Berlin, Germany.

Bemelmans, W.J.E., Broer, J., and Feskens, E.J.M. 2002. Effect of an increased intake of alpha-linolenic acid and group nutritional education on cardiovascular risk factors: The Mediterranean alpha-linolenic enriched Groningen dietary intervention (MARGIN) study. *Am. J. Clin. Nutr.* **75**, 221–227.

Bemelmans, W.J.E., Lefrandt, J.D., and Feskens, E.J.M. 2004. Increased alpha-linolenic acid intake lowers C-reactive protein, but has no effect on markers of atherosclerosis. *Eur. J. Clin. Nutr.* **58**, 1083–1089.

BeMiller, J.N. 1973. Quince seed, psyllium seed, flax seed and okra gums. *In* "Industrial Gumes" (R.L. Whistler and J.N. BeMiller, eds), 2nd Ed. Academic Press, New York.

Berglund, D. 2002. Flax: New uses and demands. *In* "Trends in New Crops and New Uses" (J. Janick and A. Whipkey, eds), pp. 258–360. ASHS Press, Alexandria, VA.

Bhathena, S.J., Ali, A.A., Mohamed, A.I., Hansen, C.T., and Velasquez, M.T. 2002. Differntial effects of dietary flaxseed protein and soy protein on plasma triglyceride and uric acid levels in animal models. *J. Nutr. Biochem.* **13**, 684–689.

Bhatty, R.S. 1993. Further compositional analyses of flax: Mucilage, trypsin inhibitors and hydrocyanic acid. *J. Am. Oil Chem. Soc.* **70**, 899–904.

Bhatty, R.S. and Cherdklatgumchal, P. 1990. Compositional analysis of laboratory-prepared and commercial samples of linseed meal and of hull isolated from flax. *J. Am. Oil Chem. Soc.* **67**, 79–84.

Bierenbaum, M.L., Reichstein, R., and Watkins, T.R. 1993. Reducing atherogenic risk in hyperlipemic humans with flax seed supplementation: A preliminary report. *J. Am. Coll. Nutr.* **12**, 501–504.

Bjerve, K.S., Thoresen, L., and Borsting, S. 1988. Linseed and cod liver oil induce rapid growth in a 7–year-old girl with n-3 fatty acid deficiency. *J. Parent. Enter. Nutr.* **12**, 521–525.

Boccardo, F., Lunardi, G., Guglielmini, P., Parodi, M., Murialdo, R., Schettini, G., and Rubagotti, A. 2004. Serum enterolactone levels and the risk of breast cancer in women with palpable cysts. *Eur. J. Cancer* **40**, 84–89.

Borriello, S.P., Setchell, K.D.R., Axelson, M., and Lawson, A.M. 1985. Production and metabolism of lignans by the human faecal flora. *J. Appl. Bacteriol.* **58**, 37–43.

Bougnoux, P. and Chajes, V. 2003. α-Linolenic acid and heart disease. *In* "Flaxseed in Human Nutrition" (S.C. Cunnane and L.U. Thompson, eds), 2nd Ed., pp. 232–241. AOCS Press, Champaign, IL.

Bourre, J.M., Piciotti, M., Dumont, O., and Durand, G. 1990. Dietary linoleic acid and polyunsaturated fatty acids in rat brain and other organs: Minimal requirements of linoleic acid. *Lipids* **25**, 465–472.

Bozan, B. and Temelli, F. 2002. Supercritical extraction of flaxseed oil. *J. Am. Oil Chem. Soc.* **79**, 231–235.

Braquet, P., Hosfard, D., Braquet, M., Bourgain, R., and Bussolino, F. 1989. Role of cytokines and platelet-activating factor in microvascular immune injury. *Int. Arch. Allerg. Appl. Immunol.* **88**, 88.

Brooks, J.D. and Thompson, L.U. 2005. Mammalian lignans and genistein decrease the activities of aromatase and 17β-hydroxysteroid dehydrogenase in MCF-7 cells. *J. Steroid Biochem. Mol. Biol.* **94**, 461–467.

Brooks, J.D., Ward, W.E., Lewis, J.E., Hilditch, J., Nickell, L., Wong, E., and Thompson, L.U. 2004. Supplementation with flaxseed alters estrogen metabolism in postmenopausal women to a greater extent than does supplementation with an equal amount of soy. *Am. J. Clin. Nutr.* **79**, 318–325.

Byland, A., Zhang, J.X., Bergh, A., Damber, J.E., Widmark, A., Johansson, A., Adlercreutz, H., Aman, P., Shepherd, M.J., and Hallmans, G. 2000. Rrye bran and soy protein delay growth and increase apoptosis of human LNCaP prostate adenocarcinoma in nude mice. *Prostate* **42**, 304–314.

Cameron, E., Bland, J., and Marcuson, R. 1989. Divergent effects of omega-6 and omega-3 fatty acids on mammary tumor development in C3H/Heston mice treated with DMBA. *Nutr. Res.* **9**, 383–393.

Canadian Grain Commission. 2001. Quality of western Canadian flaxseed. available at http://www.grainscanada.gc.ca/Quality/Flax/2001/flax01hs06–e.htm accessed on April (2005).

Carter, J. 1993. Potential of flaxseed and flaxseed oil in baked goods and other products in human nutrition. *Cer. Foods World* **38**, 753–759.

Carter, J. 1996. Sensory evaluation of flaxseed of different varieties. *In* "Proceedings of the 56th Flax Institute of the United States", pp. 201–203. Fargo, ND.

Carver, J.D. 2003. Advances in nutritional modifications of infant formulas. *Am. J. Clin. Nutr.* **77**, 1550S–1554S.

Caughey, G., Mantzioris, E., Gibson, R., Cleland, L., and James, M. 1996. The effect on human tumor necrosis factor α and interleukin 1 β production of diets enriched with n-3 fatty acids from vegetable oils or fish oil. *Am. J. Clin. Nutr.* **63**, 116–122.

Chadha, R.K., Lawrence, J.F., and Ratnayake, W.M.N. 1995. Ion chromatographic determination of cyanide released from flaxseed under autohydrolysis conditions. *Food Add. Cont.* **12**, 527–533.

Chan, J.K., McDonald, B.E., Gerrad, J.M., Bruce, V.M., Weaver, B.J., and Holub, B.J. 1993. Effect of dietary alpha-linolenic acid and its ratio to linoleic acid on platelet and plasma fatty acids and thrombogenesis. *Lipids* **28**, 811–817.

Charlet, S., Bensaddek, L., Raynaud, D., Gillet, F., Mesnard, F., and Fliniaux, M.A. 2002. An HPLC method for the quantification of anhydrosecoisolariciresinol. Application to the evaluation of flax lignan content. *Plant Physil. Biochem.* **40**, 225–229.

Chen, H., Xu, S., and Wang, Z. 2002. Research progress and application of flaxseed gum. *Food and Ferm. Ind.* **28**, 64–68.

Chen, J. and Thompson, L.U. 2003. Lignans and tamoxifen, alone or in combination, reduce human breast cancer cell adhesion, invasion and migration *in vitro*. *Breast Cancer Res. Treat.* **80**, 163–170.

Chen, J., Tan, K.P., Ward, W.E., and Thompson, L.U. 2003. Exposure to flaxseed or its purified lignan during suckling inhibits chemically induced rat mammary tumorigenesis. *Exp. Biol. Med.* **228**, 951–958.

Chen, J., Hui, E., Ip, T., and Thompson, L.U. 2004. Dietary flaxseed enhances the inhibitory effect of tamoxifen on the growth of estrogen-dependent human breast cancer (MCF-7) in nude mice. *Clin. Cancer Res.* **10**, 7703–7711.

Chen, Z.Y., Ratnayake, W.M.N., and Cunnane, S.C. 1994. Oxidative stability of flaxseed lipids during baking. *J. Am. Oil Chem. Soc.* **71**, 629–632.

Chiwona-Karltun, L., Brimer, L., Saka, J.D.K., Mhone, A., Mkumbira, J., Johansson, L., Bokanga, M., Mahungu, N.M., and Rosling, H. 2004. Bitter taste in cassava roots correlates with cyanogenic glucoside levels. *J. Sci. Food Agric.* **84**, 581–590.

Chornick, T., Malcolmson, L., Izydorczyk, M., Duguid, S., and Taylor, C. 2002. Effect of cultivar on the physicochemical properties of flaxseed mucilage. *In* "Proceeding of 59th Flax Institute of the United States", pp. 7–13. Fargo, ND.

Chung, M.W.Y., Lei, B., and Li-Chan, E.C.Y. 2004. Isolation and structural characterization of the major protein fraction from NorMan flaxseed (*Linum usitativissimum* L). *Food Chem.* **90**, 271–279.

Clandinin, M.T., Foxwell, A., and Goh, Y.K. 1997. Omega-3 fatty acid intake results in a relationship between the fatty acid composition of LDL cholesterol ester and LDL cholesterol content in humans. *Biochim. Biophys. Acta* **1346**, 247–252.

Collins, T.F.X., Sprando, R.L., Black, T.N., Olejnik, N., Wiesenfeld, P.W., Babu, U.S., Bryant, M., Flynn, T.J., and Ruggles, D.I. 2003. Effects of flaxseed and defatted flaxseed meal on reproduction and development in rats. *Food Chem. Toxicol.* **41**, 819–834.

Consoli, A., Nurjhan, N., Capani, F., and Gerich, J. 1989. Predominant role of gluconeo-genesis in increased hepatic glucose production in NIDDM. *Diabetes* **38**, 550–557.

Cui, W. and Mazza, G. 1996. Physicochemical characteristics of flaxseed gum. *Food Res. Int.* **29**, 397–402.

Cui, W., Mazza, G., and Biliaderis, C.G. 1994a. Chemical structure, molecular size distributions and rheological properties of flaxseed gum. *J. Agric. Food Chem.* **42**, 1891–1895.

Cui, W., Mazza, G., Oomah, B.D., and Biliaderis, C.G. 1994b. Optimization of an aqueous extraction process for flaxseed gum by response surface methodology. *Lebensm, Wiss. U. Technol.* **27**, 363–369.

Cui, W., Kenaschuk, E., and Mazza, G. 1994c. Flaxseed gum: Genotype, chemical structure and rheological properties. *In* "Proceedings of the 55th Flax Institute of the United States", pp. 166–177. Fargo, ND.

Cui, W., Kenaschuk, E., and Mazza, G. 1996. Influence of genotype on chemical composition and rheological properties of flaxseed gums. *Food Hydrocolloids* **10**, 221–227.

Cunnane, S.C. and Yang, J. 1995. Zinc deficiency impairs whole body accumulation of polyunsaturates and increases the utilization of [1–14C]-linoleate for *de novo* lipid synthesis in pregnant rats. *Can. J. Physiol. Pharmacol.* **73**, 1246–1252.

Cunnane, S.C., McAdoo, K.R., and Horrobin, D.F. 1987. Horrobin, iron intake influences essential fatty acids and lipid composition of rat plasma and erythrocyts. *J. Nutr.* **117**, 1514–1519.

Cunnane, S.C., Ganguli, S., and Menard, C. 1993. High α-linolenic acid flaxseed (*Linum usitatissimum*): Some nutritional properties in humans. *Br. J. Nutr.* **69**, 443–453.

Dabrosin, C., Chen, J., Wang, L., and Thompson, L.U. 2002. Flaxseed inhibits metastasis and decreases extracellular vascular endothelial growth factor in human breast cancer xenografts. *Cancer Lett.* **185**, 31–37.

Dabrowski, K.J. and Sosulski, F.W. 1984. Composition of free and hydrolysable phenolic acids in defatted flours of ten oilseeds. *J. Agric. Food Chem.* **32**, 128–130.

Dai, Q., Fanke, A.A., Yu, H., Shu, X., Jin, F., Hebert, J.R., Custer, L.J., Gao, Y., and Zheng, W. 2003. Urinary phytoestrogen excretion and breast cancer risk: Evaluating potential effect modifiers endogenous estrogens and anthropometrics. *Cancer Epid. Bio. Prev.* **12**, 497–502.

Danbara, N., Yuri, T., Tsujita-Kyutoku, M., Tsukamoto, R., Uehara, N., and Tsubura, A. 2005. Enterolactone induces apoptosis and inhibits growth of colo 201 human colon cancer cells both *in vitro* and *in vivo*. *Anticancer Res.* **25**, 2269–2276.

Daun, J.K. 1993. Oilseeds processing. "Grain Processing and Technology", Vol. 2, pp. 883–936. Canadian International Grains Institute, Manitoba, Canada.

Daun, J.K., Barthet, J.V., Chornick, T.L., and Duguid, S. 2003. Structure, composition, and variety development of flaxseed. *In* "Flaxseed in Human Nutrition" (S.C. Cunnane and L.U. Thompson, eds), 2nd Ed., pp. 1–40. AOCS Press, Champaign, IL.

de Lorgeril, M., Renaud, S., and Mamelle, N. 1994. Mediterranean alpha-linolenic acid-rich diet in secondary prevention of coronary heart disease. *Lancet* **343**, 1454–1459.

DeClercq, D.R. 2005. Quality of western Canadian flaxseed 2004: Canadian Grain Commission. Canadian Grain Commision, Winnipeg, MB. http://www.grainscanada.gc.ca/Quality/Flax/flaxmenu-e.htm

Degenhardt, A., Habben, S., and Winterhalter, P. 2002. Isolation of the lignan secoisolariciresinol diglucoside from flaxseed (*Linum ustitatissimum* L.) by high-speed counter-current chromatography. *J. Chromatogr.* **943**, 299–302.

Demark-Wahnefried, W., Price, D.T., Polascik, T.J., Robertson, C.N., Anderson, E.E., Paulson, D. F., Walther, P.J., Gannon, M., and Vollmer, R.T. 2001. Pilot study of dietary fat restriction and flaxseed supplementation in men with prostate cancer before surgery: Exploring the effects on hormonal levels, prostate-specific antigen, and histopathologic features. *Urology* **58**, 47–52.

Demark-Wahnefried, W., Robertson, C.N., Walther, P.J., Polascik, T.J., Paulson, D.F., and Vollmer, R.T. 2004. Pilot study to explore effects of low-fat, flaxseed-supplemented diet on proliferation of benign prostatic epithelium and prostate-specific antigen. *Urology* **63**, 900–904.

den Tonkelaar, I., Keinan-Boker, L., Veer, P.V., Arts, C.J.M., Adlercreutz, H., Thijssen, J.H.H., and Peeters, P.H.M. 2001. Urinary phytoestrogens and postmenopausal breast cancer risk. *Cancer Epid. Bio. Prev.* **10**, 223–228.

Denis, L., Morton, M.S., and Griffiths, K. 1999. Diet and its preventive role in prostatic disease. *Eur. Urol.* **35**, 377–387.

Dessi, M.A., Deiana, M., Dat, B.W., Rosa, A., Banni, S., and Corongiu, F.P. 2002. Oxidative stability of polyunsaturated fatty acids: Effect of squalene. *Eur. J. Lipid. Sci. Technol.* **104**, 506–512.

Dev, D.K. and Quensel, E. 1988. Preparation and functional properties of linseed protein products containing differing levels of mucilage. *J. Food Sci.* **53**, 1834–1837, 1857.

Diaz-Rojas, E.I., Pacheco-Aguilar, R., Lizardi, J., Arguelles-Monal, W., Valdez, M.A., Rinaudo, M., and Goycoolea, F.M. 2004. Linseed pectin: Gelling properties and performance as an encapsulation matrix for shark liver oil. *Food Hydrocolloids* **18**, 293–304.

Dieken, H.A. 1992. Use of flaxseed as a source of omega-3 fatty acids in human nutrition. *In* "The 54th Proceeding of Flax Institute of United States", pp. 1–4.

Djoussé, L., Pankow, J.S., and Eckfeldt, J.H. 2001. Relation between dietary linolenic acid and coronary artery disease in the National Heart, Lung, and Blood Institute Family Heart Study. *Am. J. Clin. Nutr.* **74**, 612–619.

Djoussé, L., Folsom, A.R., Province, M.A, Hunt, S., and Ellison, R.C. 2003a. Dietary linolenic acid and carotid atherosclerosis: The national heart, lung, and blood family heart study. *Am. J. Clin. Nutr.* **77**, 819–825.

Djoussé, L., Hunt, S.C., Arnett, D.K., Province, M.A., Eckfeldt, J.H., and Ellison, R.C. 2003b. Dietary linolenic acid is inversely associated with plasma triacylglycerol: The national heart, lung, and blood institute family heart study. *Am. J. Clin. Nutr.* **78**, 1098–1102.

Dodin, S., Lemay, A., Jacques, H., Légaré, F., Forest, J.C., and Masse, B. 2005. The effects of flaxseed dietary supplement on lipid profile, bone mineral density, and symptoms in menopausal women: A randomized, double-blinid, wheat germ placebo-controlled clinical trial. *J. Clin. Endocrinol. Metab.* **90**, 1390–1397.

Dolecek, T.A. 1992. Epidemiological evidence of relationships between dietary polyunsaturated fatty acids and mortality in the Multiple Risk Factor Intervention Trial. *Proc. Soc. Exp. Biol. Med.* **200**, 177–182.

Dorrell, D.G. 1970. Distribution of fatty acids within the seed of flax. *Canadian J. Plant Sci.* **50**, 71–75.

Drouillard, J.S., Good, E.J., Gordon, C.M., Kessen, T.J., Sulpizio, M.J., Montgomery, S.P., and Sindt, J.J. 2002. Flaxseed and flaxseed products for cattle: Effects on health, growth performance, carcass quality and sensory attributes. *In* "Proceedings of the 59th Flax Institute of the United States", pp. 72–87. Fargo, ND.

Drouillard, J.S., Seyfert, M.A., Good, E.J., Loe, E.R., Depenbusch, B., and Daubert., R. 2004. Flaxseed for finishing beef cattle: Effects on animal performance, carcass quality and meat composition. *In* "Proceedings of the 60th Flax Institute of the United States", pp. 108–117. Fargo, ND.

Dwivedi, C., Natarajan, K., and Matthees, D. 2005. Chemopreventive effects of dietary flaxseed oil on colon tumor development. *Nutr. Cancer.* **51**, 52–58.

Eliasson, C., Kamal-Eldin, A., Andersson, R., and Aman, P. 2003. High-performance liquid chromatographic analysis of secoisolariciresinol diglucoside and hydroxycinnamic acid glucosides in flaxseed by alkaline extraction. *J. Chromatogr.* **1012**, 151–159.

Elswyk, M. 1997. Comparison of n-3 fatty acid sources in laying hen rations for improvement of whole egg nutritional quality: A review. *Br. J. Nutr.* **78**(Suppl. 1), S61–S69.

Emken, E.A., Adlof, R.O., and Gulley, R.M. 1994. Dietary linoleic acid influences desaturation and acylation of deuterium-labeled linoleic and linolenic acids in young adult males. *Biochim. Biophys. Acta* **1213**, 277–288.

Emken, E.A., Adlof, R.O., Duval, S.M., and Nelson, G.J. 1999. Effect of dietary docosahexaenoic acid on desaturation and uptake *in vivo* of isotope-labeled oleic, linoleic, and linolenic acids by male subjects. *Lipids* **34**, 785–791.

Erskine, A.J. and Jones, J.K.N. 1957. The structures of linseed mucilage. Part I. *Can. J. Chem.* **7**, 301–312.

Ezaki, O., Tsuji, E., Momomura, K., Kasuga, M., and Itakura, H. 1992. Effects of fish and safflower oil feeding on subcellular glucose transporter distributions in rat adipocytes. *Am. J. Physiol.* **263**, E94–E101.

Fedeniuk, R.W. and Biliaderis, C.G. 1994. Composition and physicochemical properties of linseed (*Linum Usitatissimum* L.) mucilage. *J. Agric. Food Chem.* **42**, 240–247.

Feng, D., Shen, Y., and Chavez, E.R. 2003. Effectiveness of different processing methods in reducing hydrogen cyanide content of flaxseed. *J. Sci. Food Agric.* **83**, 836–841.

Fennema, O.R. 1996. Water and ice. *In* "Food Chemistry" (O.R. Fennema, ed.), 3rd Ed., pp. 17–94. Marcel Dekker, Inc, New York, USA.

Ferrier, L., Caston, L., Leeson, S., Squires, J., Weaver, B., and Holub, B. 1995. α-Linolenic acid- and docosahexaenoic acid-enriched eggs from hens fed flaxseed: Influence on blood lipids and platelet phospholipid fatty acids in humans. *Am J.Clin Nutr.* **62**, 81–86.

Feussner, I., Wasternack, C., Kindl, K., and Kuhn, H. 1995. Lipoxygenase-catalyzed oxygenation of storage lipids is implicated in lipid mobilization during germination. *Proc. Natl. Acad. Sci. USA* **92**, 11849–11855.

Finnegan, Y.E. and Minihane, A.M. 2003. Plant- and marine-derived n-3 polyunsaturated faty acids have differential effects on fasting and postprandial blood lipid concentrations and on the susceptibility of LDL to oxidative modification in moderately hyperlipidemic subjects. *Am. J. Clin. Nutr.* **77**, 783–795.

Flynn, T.J., Collins, T.F.X., Sprando, R.L., Black, T.N., Ruggles, D.I., Wiesenfeld, P.W., and Babu, U.S. 2003. Developmental effects of serum from flaxseed-fed rats on cultured rat embryos. *Food Chem Toxicol.* **41**, 835–840.

Fontanillas, R., Barroeta, A., Baucells, M.D., and Guardiola, F. 1998. Backfat fatty acid evolution in swine fed diets high in either *cis*-monounsaturated, *trans*, or (*n-3*)fats. *J. Anim Sci.* **76**, 1045–1055.

Franceschi, S., Favero, A., Decarli, A., Negri, E., La Vecchia, C., Ferraroni, M., Russo, A., Salvini, S., Amadori, D., Conti, E., Montella, M., and Giacosa, A. 1996. Intake of macronutrients and risk of breast cancer. *Lancet* **347**, 1351–1356.

Frankel, E.N. 2005. "Lipid Oxidation" (E.N. Frankel, ed.), 2nd Ed. The Oily Press, PJ Barnes Associates, Bridgewater, England.

Fritsche, K.L. and Johnston, P.V. 1990. Effect of dietary α-linolenic acid on growth, metastasis, fatty acid profile and prostaglandin production of two murine mammary adenocarcinomas. *J. Nutr.* **120**, 1601–1609.

Froment, M.A., Smith, J.M., Turley, D., Booth, E.J., and Kightley, S.P.J. 1998. Fatty acid profiles in the seed oil of linseed and fibre flax cultivars (*Linum Usitatissimum*) grown in England and Scotland. *Tests Agrochem. Cult.* **19**, 60–61.

Froment, M.A., Smith, J., and Freeman, K. 1999. Influence of environmental and agronomic factors contributing to increased levels of phopholipids in oil from UK linseed *Linum usitatissimum*. *Ind. Crops Prod.* **10**, 201–207.

Gann, P.H., Hennekens, C.H., and Sacks, F.M. 1994. Prospective study of plasma fatty acids and risk of prostate cancer. *J. Natl. Cancer Inst.* **86**, 281–286.

Garden, J. 1993. Flaxseed gum: Extraction, characterization, and functionality. Doctoral Dissertation of North Dakota State University, Fargo, ND, 157p.

Gerster, H. 1998. Can adults adequately convert α-linolenic acid (18:3n-3) to eicosapentaenoic acid (20:5n-3) and docosahexaenoic acid (22:6n-3). *Int. J. Vitm. Nutr. Res.* **68**, 159–173.

Ghosh, P., Chakraborty, M., Sorenson, B., Berzonsky, W.A., and Hall, C., III 2004. Effects of ground flaxseed on the processing and quality parameters of wheat flour tortilla. Presented at AACC Annual Meeting Conference, San Diego, CA.

Giles, G. and Ireland, P. 1997. Diet, nutrition and prostate cancer. *Int. J. Cancer* **10** (Suppl.), 13–17.

Giovannacci, E., Rimm, E.B., Colditz, G.A., Stampfer, M.J., Ascherio, A., Chute, C.C., and Willett, W.C. 1993. A prospective study of dietary fat and risk of prostate cancer. *J. Natl. Cancer Inst.* **85**, 1571–1579.

Gonthier, C., Mustafa, A.F., Berthiaume, R., Petit, H.V., and Ouellet, D.R. 2004a. Feeding micro-ionized and extruded flaxseed to dairy cows: Effects on digestion and ruminal biohydrogenation of long-chain fatty acids. *Can. J. Anim. Sci.* **84**, 705–711.

Gonthier, C., Mustafa, A.F., Berthiaume, R., Petit, H.V., Martineau, R., and Ouellet, D.R. 2004b. Effects of feeding micronized and extruded flaxseed on ruminal fermentation and nutrient utilization by dairy cows. *J. Dairy Sci.* **87**, 1854–1863.

Gonthier, C., Mustafa, A.F., Ouellet, D.R., Chouinard, P.Y., Berthiaume, R., and Petit, H.V. 2005. Feeding microionized and extruded flaxseed to dairy cows: Effects on blood parameters and milk fatty acid composition. *J. Dairy Sci.* **88**, 748–756.

González-Esquerra, R. and Leeson, S. 2000. Effects of menhaden oil and flaxseed in broiler diets on sensory quality and lipid composition of poultry meat. *British Poultry Sci.* **41**, 481–488.

González-Esquerra, R. and Leeson, S. 2001. Alternatives for enrichment of eggs and chicken meat with omega-3 fatty acids. *Can. J. Anim. Sci.* **81**, 295–305.

Goodridge, J., Ingalls, J., and Crow, G. 2001. Transfer of omega-3 linolenic acid and linoleic acid to milk fat from flaxseed or linola protected with formaldehyde. *Can. J. Anim. Sci.* **81**, 525–532.

Greenberg, S.M., Calbert, C.E., Savage, E.E., and Deuel, H.J., Jr. 1950. The effect of fat level of the diet on general nutrition. *J. Nutr.* **41**, 473–486.

Guallar, E., Aro, A., and Jimenez, F.J. 1999. Omega-3 fatty acids in adipose tissue and risk of myocardial infarction: The Euramic study. *Arterioscler Thromb. Vasc. Biol.* **19**(4), 1111–1118.

Gubbels, G.H., Bonner, D.M., and Kenaschuk, E.O. 1994. Effect of frost injury on quality of flaxseed. *Can. J. Plant Sci.* **74**, 331–333.

Gupta, M., McDougal, A., and Safe, S. 1998. Estrogenic and antiestrogenic activities of 16a- and 2–hydroxy metabolites of 17b-estradiol in MCF-7 and T47D human breast cancer cells. *J. Steroid. Biochem. Mol. Biol.* **67**, 413–419.

Hadley, M. 1996. Stability of flaxseed oil used in cooking/stir frying. *In* "Proceedings of the 56th Flax Institute of the United States", pp. 55–61. Fargo, ND.

Haggans, C.J., Hutchins, A.M., and Olson, B.A. 1999. Effect of flaxseed consumption on urinary estrogen metabolites in postmenopausal women. *Nutr. Cancer* **33**, 188–195.

Hall, C., III 2002. Flaxseed as a functional food. *In* "Proceedings of the 59th Flax Institute of the United States", pp. 1–6. Fargo, ND.

Hall, C., III and Shultz, K. 2001. Phenolic antioxidant interactions. *In* "Abstracts of the 92nd American Oil Chemists Society Annual Meeting and Expo". p. S88.

Hall, C., III and Schwarz, J. 2002. Functionality of flaxseeds in frozen desserts-preliminary report. *In* "Proceedings of the 59th Flax Institute of the United States", pp. 21–24. Fargo, ND.

Hall, C., III., Niehaus, M., Wolf-Hall, C., Bauer, A., Auch, C., and Stoerzinger, C. 2004. Flaxseed in yogurt: Lignan stability and effect on processing. *In* "Proceedings of the 60th Flax Institute of the United States", pp. 1–7. Fargo, ND.

Hall, C.A., III., Manthey, F.A., Lee, R.E., and Niehaus, M. 2005. Stability of α-Linolenic acid and secoisolariciresinol diglucoside in flaxseed fortified macaroni. *J. Food Sci.* **70**, 483–489.

Hamilton, R. 1994. The chemistry of rancidity in foods. *In* "Rancidity in Foods" (J.C. Allen and R.J. Hamilton, eds), pp. 1–21. Blackie Academic, London.

Harris, R.K. and Haggerty, W.J. 1993. Assays for potentially anticarcinogenic phytochemicals in flaxseed. *Cereal Foods World* **38**, 147–151.

Hauman, B.F. 1998. Alternative source for n-3 fatty acids. *Inform.* **9**, 1108–1109, 1112, 1115–1116, 1118–1119.

Heinonen, S., Nurmi, T., Liukkonen, K., Poutanen, K., Wahala, K., Deyama, T., Nishibe, S., and Aldercreutz, K. 2001. *In vitro* metabolism of plant lignans, new precursor of mammalian lignans enterolactone and enterodiol. *J. Agric. Food Chem.* **49**, 3178–3186.

Helbaek, H. 1969. Plant collecting, dry-farming and irrigation agriculture in prehistoric Deh Luran. *In* "Prehistory and Human Ecology of the Deh Lurah Plain, Memoirs of the Museum of Anthropology" (F. Hole, K.V. Flannery, and J.F. Neely, eds), pp. 386–426. University of Michiagan, Ann Arbor, MI.

Hettiarachchy, N., Hareland, G., Ostenson, A., and Balder-Shank, G. 1990. Composition of eleven flaxseed varieties grown in North Dakota. *In* "Proceedings of the 53rd Flax Institute of the United States", pp. 36–40. Fargo, ND.

Hilakivi-Clarke, L., Cho, E., deAssis, S., Olivo, S., Ealley, E., Bouker, K.B., Welch, J.N., Khan, G., Clarke, R., and Cabances, A. 2001. Maternal and prepubertal diet, mammary development and breast cancer risk.l. *J. Nutr.* **131**, 154S–157S.

Hillman, G. 1975. The plant remains from Tell Abu Hureyra: A preliminary report. *Proc. Prehist. Soc.* **41**, 70–73.

Hillman, G.C., Colledge, S.M., and Harris, D.R. 1989. Plant-food economy during the Epipalaeolithic period at Tell Abu Hureyra, Syria: Dietary diversity, seasonality, and modes of exploitation. *In* "Foraging and Farming: The Evolution of Plant Exploitation" (D.R. Harris and G.H. Hillman, eds), pp. 240–268. Unwin & Hyman, London.

Holcapek, M., Jandera, P., Zderadicka, P., and Hruba, L. 2003. Characterization of triacylglycerol and diacylglycerol composition of plant oils using high-performance liquid chromatography-atmospheric pressure chemical ionization mass spectrometry. *J. Chromatogr.* **1010**, 195–215.

Holman, R.T., Johnson, S.B., and Hatch, T.F. 1982. A case of human linolenic acid deficiency involving neurological abnormalities. *Am. J. Clin. Nutr.* **35**, 617–623.

Hopf, M. 1983. Jericho plant remains. *In* "Excavatioins at Jericho" (K.M. Kenyon and T.A. Holland, eds), Vol. 5, pp. 576–621. British School of Archaeology in Jerusalem, London.

Hu, F.B., Stampfer, M.J., and Manson, J.E. 1999. Dietary intake of α-linolenic acid and risk of fatal ischemic heart disease among women. *Am. J. Clin. Nutr.* **69**, 890–897.

Huang, X., Kakuda, Y., and Cui, W. 2001. Hydrocolloids in emulsions: Particle size distribution and interfacial activity. *Food Hydrocolloids* **15**, 533–542.

Hulten, K., Winkvist, A., and Lenner, P. 2002. An incident case-referent study onplasma enterolactone and breast cancer risk. *Eur. J. Nutr.* **41**(4), 168–176.

Hussein, N., Ah-Sing, E., Wilkinson, P., Leach, C., Griffin, B.A., and Millward, D.J. 2005. Long-chain conversion of [^{13}C] linoleic acid and α-linolenic acid in response to marked changes in their dietary intake in men. *J. Lipid Res.* **46**, 269–280.

Hutchins, A.M. and Slavin, J.L. 2003. Effects of flaxseed on sex hormone metabolism. *In* "Flaxseed in Human Nutrition" (L.U. Thompson and S.C. Cunnane, eds), 2nd Ed., pp. 126–149. AOCS Press, Champaign, IL.

Ingram, D., Sanders, K., Kolybaba, M., and Lopez, D. 1997. Case-control study of phytoestrogens and breast cancer. *Lancet* **350**, 990–994.

Institute of Medicine. Food and Nutrition Board 2002. "Dietary Reference Intakes for Energy, Carbohydrate, Fiber, Fat, Fatty Acids, Cholesterol, Protein, and Amino Acids", pp. 8–97. The National Academies Press, Washington, DC.

Jacobs, E. and Metzler, M. 1999. Oxidative metabolism of the mammalian lignans enterolactone and enterodiol by rat, pig, and human liver microsomes. *J. Agric. Food Chem.* **47**, 1071–1077.

Jacobs, E., Kulling, S.E., and Metzler, M. 1999. Novel metabolites of the mammalian lignans enterolactone and enterodiol in human urine. *J. Steroid Biochem. Mol. Biol.* **68**, 211–218.

Jelen, H.H., Obuchowska, M., Zawirska-Wojtasiak, R., and Wasowicz, E. 2000. Headspace solid-phase microextraction use for the characterization of volatile compounds in vegetable oils of different sensory quality. *J. Agric. Food Chem.* **48**, 2360–2367.

Jenkins, D.J.A., Kendall, C.W.C., and Vidgen, E. 1999. Health aspects of partially defatted flaxseed, including effects on serum lipids, oxidative measures, and *ex vivo* androgen and progestin activity: A controlled crossover trial. *Am. J. Clin. Nutr.* **69**, 395–402.

Jiang, Z., Ahn, D., and Sim, J. 1991. Effects of feeding flax and two types of sunflower seeds on fatty acid compositions of yolk lipid classes. *Poultry Sci.* **70**, 2467–2475.

Johnsson, P., Kamal-Eldin, A., Lundgren, L. N., and Aman, P. 2000. HPLC method for analysis of secoisolariciresinol diglucoside in flaxseeds. *J. Agric. Food Chem.* **48**, 5216–5219.

Judd, A. 1995. Flax-some historical considerations. *In* "Flaxseed in Human Nutrition" (L.U. Thompson and S.C. Cunnane, eds), 2nd Ed., pp. 1–10. AOCS Press, Champaign, IL.

Jung, M.Y., Bock, J.B., Baik, S.O., Lee, J.H., and Lee, T.K. 1999. Effects of roasting on pyrazine contents and oxidative stability of red pepper seed oil prior to its extraction. *J. Agric. Food Chem.* **47**, 1700–1704.

Kamm, W., Dionisi, F., Hischenhuber, C., and Engel, K.H. 2001. Authenticity assessment of fats and oils. *Food Rev. Int.* **17**, 249–290.

Kaneto, H., Kajimoto, Y., Fujitani, Y., Matsuoka, T., Sakamoto, K., Matsuhisa, M., Yamasaki, Y., and Hori, M. 1999. Oxidative stress induces p21 expression in pancreatic islet cell: Possible implication in beta-cell dysfunction. *Diabetologia* **42**, 1093–1097.

Kangas, L., Saarinen, N., and Mutanen, M. 2002. Antioxidant and antitumor effects of hydroxymatairesinol (HM-3000, HMR), a lignan isolated from the knots of spruce. *Eur. J. Cancer Prev.* **11**, S48–S57.

Kelley, D.S., Nelson, G.J., Love, J.E., Branch, L.B., Taylor, P.C., Schmidt, P.C., Mackey, B.E., and Iacono, J.M. 1993. Dietary alpha-linolenic acid alters tissue fatty acid composition, but not blood lipids, lipoproteins or coagulation status in humans. *Lipids* **28**, 533–537.

Kennelly, J.J. and Khorasani, G.R. 1992. Influence of flaxseed feeding on fatty acid composition of cows' milk. *In* "Proceedings of the 54th Flaxseed Institute of the United States", pp. 99–105. Fargo, ND.

Kestin, M., Clifton, P., Belling, G.B., and Nestel, P. 1990. n-3 Fatty acids of marine origin lower systolic blood pressure and triglycerides but raise LDL cholesterol compared with n-3 and n-6 fatty acids from plants. *Am. J. Clin. Nutr.* **51**, 1028–1034.

Kew, S., Wells, S., Thies, F., McNeill, G.P., Quinlan, P.T., and Clark, G.T. 2003. The effect of eicosapentaenoic acid on rat lymphocyte profileration depends upon its position in dietary triacylglycerols. *J. Nutr.* **133**, 4230–4238.

Khorasani, G. and Kennelly, J. 1994. Influence of flaxseed on the nutritional quality of milk. *In* "Proceedings of the 55th Flax Institute Conference" (J.F. Carter, ed.), pp. 127–134. North Dakota State University, Fargo, ND.

Kilkkinen, A., Valsta, L.M., Virtamo, J., Stumpf, K., Adlercreutz, H., and Pietinen, P. 2003a. Intake of lignans is associated with serum enterolactone concentration in Finnish men and women. *J. Nutr.* **133**, 1830–1833.

Kilkkinen, A., Virtamo, J., Virtanen, M.J., Adlercreutz, H., Albanes, D., and Pietinen, P. 2003b. Serum enterolactone concentration is not associated with prostate cancer risk in a nested case control study. *Cancer Epid. Bio. Prev.* **12**, 1209–1212.

Kilkkinen, A., Virtamo, J., Vartiainen, E., Sankila, R., Virtanen, M.J., Adlercreutz, H., and Pietinen, P. 2004. Serum enterolactone concentration is not associated with breast cancer risk in a nested case-control study. *Int. J. Cancer* **108**, 277–280.

Kitts, D.D., Yuan, Y.V., Wijewickreme, A.N., and Thompson, L.U. 1999. Antioxidant activity of the flaxseed lignan secoisolariciresinol diglycoside and its mammalian lignan metabolites enterodiol and enterolactone. *Mol. Cell. Biochem.* **202**, 91–100.

Kleeman, R., Scott, F.W., Worz-Pagenstert, U., Ratnayake, W.M.N., and Kolb, H. 1998. Impact of dietary fat on Thl/Th2 cytokine gene expression in the pancreas and gut of diabetes-prone BB rats. *J. Autoimmun.* **11**, 97–103.

Klein, V., Chajès, V., and Germain, E. 2000. Low alpha-linolenic acid content of adipose breast tissue is associated with an increased risk of breast cancer. *Eur. J. Cancer* **36**, 335.

Klosterman, H.J., Lamoureux, G.L., and Parsons, J.L. 1967. Isolation, characterization, and synthesis of linatine. A vitamin B_6 antagonist from flaxseed (*Linum ustitatissimum*). *Biochemistry* **6**, 170–177.

Kobaisy, M., Oomah, B.D., and Mazza, G. 1996. Determination of cyanogenic glycosides in flaxseed by barbituric acid-pyridine, pyridine-pyrazolone, and high-performance liquid chromatography methods. *J. Agric. Food Chem.* **44**, 3178–3181.

Kolodziejczyk, P. and Fedec, P. 1995. Processing flaxseed for human consumption. *In* "Flaxseed in Human Nutrition" (L.U. Thompson and S.C. Cunnane, eds), 2nd Ed., pp. 194–218. AOCS Press, Champaign, IL.

Kozlowska, J. 1989. The use of flax seed for food purposes. *In* "Flax in Europe Production and Processing. Proceedings of the European Regional Workshop on Flax" (R. Kozlowska, ed.), Poznan, Poland.

Kouba, M., Enser, M., Whittington, F.M., Nute, G.R., and Wood, J.D. 2003. Effect of a high-linolenic acid diet on lipogenic enzyme activities, fatty acid composition and meat quality in the growing pig. *J. Anim. Sci.* **81**, 1967–1979.

Kratzer, F.H. 1946. The treatment of linseed meal to improve its feeding value for chicks. *Poult. Sci.* **25**, 541–542.

Krause, J., Schultz, M., and Dudek, S. 2002. Effect of extraction conditions on composition, surface activity and rheological properties of protein isolates from flaxseed (*Linum usitativissimum* L). *J. Sci. Food Agric.* **82**, 970–976.

Kritchevsky, D. 1995. Fiber effects on hyperlipidemia. *In* "Flaxseed in Human Nutrition" (S.C. Cunnane and L.U. Thompson, eds), 1st Ed., pp. 174–184. AOCS Press, Champaign, IL.

Kuijsten, A., Arts, I.C.W., Vree, T.B., and Hollman, P.C.H. 2005. Pharmacolinetics of enterolignans in healthy men and women consuming a single dose of secoisolariciresinol diglucoside. *J. Nutr.* **135**, 795–801.

Layne, K.S., Goh, Y.K., and Jumpsen, J.A. 1996. Normal subjects consuming physiological levels of 18:3(n-3) and 20:5(n-3) from flaxseed or fish oils have characteristic differences in plasma lipid and lipoprotein fatty acid levels. *J. Nutr.* **126**, 2130–2140.

Lee, R.E., Manthey, F.A., and Hall, C.A., III 2003. Effects of boiling, refrigerating, and microwave heating on cooked quality and stability of lipids in macaroni containing ground flaxseed. *Cereal Chem.* **80**, 570–574.

Lee, R.E., Manthey, F.A., and Hall, C.A., III 2004a. Content and stability of hexane extractable lipid at various steps of producing macaroni containing ground flaxseed. *J. Food Proc. Preser.* **28**, 133–144.

Lee, S., Inglett, G.E., and Carriere, C.J. 2004b. Effect of nutrim oat bran and flaxseed on rheological properties of cakes. *Cereal Chem.* **81**, 637–642.

Leeson, S. and Caston, L.J. 2004. Feeding value of dehulled flaxseed. *Can. J. Anim. Sci.* **84**, 545–547.

Leeson, S., Caston, L.J., and Maclaurin, T. 1998. Organoleptic evaluation of eggs produced by laying hens fed diets containing graded levels of flaxseed and vitamin E. *Poult. Sci.* **77**, 1436–1440.

Leitzmann, M.F., Stampfer, M.J., and Michaud, D.S. 2004. Dietary intake of n-3 and n-6 fatty acids and the risk of prostate cancer. *Am. J. Clin. Nutr.* **80**, 204–216.

Lemay, A., Dodin, S., and Kadri, N. 2002. Flaxseed dietary supplement versus hormone replacement therapy in hypercholesterolemic menopausal women. *Obstet. Gynecol.* **100**, 495–504.

Leng, G.C., Taylor, G.S., and Lee, A.J. 1999. Essential fatty acids and cardiovascular disease: The Edinburgh artery study. *Vas. Med.* **4**, 219–226.

Liggins, J., Grimwood, R., and Bingham, S.A. 2000. Extraction and quantification of lignan phytoestrogens in food and human samples. *Anal. Biochem.* **287**, 102–109.

Lin, X., Switzer, B.R., and Demark-Wahnefried, W. 2001. Effect of mammalian lignans on the growth of prostate cancer cell lines. *Anticancer Res.* **21**, 3995–4000.

Lin, X., Gingrich, J.R., Bao, W., Li, J., Haroon, Z.A., and Demark-Wahnefried, W. 2002. Effect of flaxseed supplementation on prostatic carcinoma in transgenic mice. *Urology* **60**, 919–924.

Lina, B., Korte, H., Nyman, L., and Unkila, M. 2005. A thirteen week dietary toxicity study with 7–hydroxsymatairesinol potassium acetate (HMR lignan) in rats. *Regul. Toxicol. Pharmacol.* **41**, 28–38.

Linseisen, J., Piller, R., Hermann, S., and Chang-Claude, J. 2004. Dietary phytoestrogen intake and premenopausal breast cancer risk in a German case-control study. *Int. J. Cancer* **110**, 284–290.

London, S.J., Sacks, F.M., Stampfer, I.C., Henderson, I.C., Maclure, H., Tomita, A., Wood, W.C., Remine, S., Robert, N.J., and Dmochowski, J.R. 1993. Fatty acid composition of the subcutaneous adipose tissue and risk of proliferative benign breast disease and breast cancer. *J. Natl. Cancer Inst.* **85**, 785–793.

Loopez-Ferrer, S., Baucells, M., Barroeta, A., and Grashorn, M.A. 1999. n-3 enrichment of chicken meat using fish oil: Alternative substitution with rapeseed and linseed oils. *Poultry Sci.* **78**, 356–365.

Loor, J.J., Ueda, K., Ferlay, A., Chilliard, Y., and Doreau, M. 2005. Intestinal flow and digestibility of trans fatty acids and conjugated linoleic acids (CLA) in dairy cows fed a high-concentrate diet supplemented with fish oil, linseed oil, or sunflower oil. *Anim. Feed Sci. Tech.* **119**, 203–225.

Lopez-Garcia, E., Schulze, M.B., and Manson, J.A.E. 2004. Consumption of (n-3) fatty acids is related to plasma biomarkers of inflammation and endothelial activation in women. *J. Nutr.* **134**, 1806–1811.

Lucas, E.A., Wild, R.D., and Hammond, L.J. 2002. Flaxseed improves lipid profile without altering biomarkers of bone metabolism in postmenopausal women. *J. Clin. Endocrinol. Metab.* **87**, 1527–1532.

Lucas, E.A., Lightfood, S.A., Hammond, L.J., Devareddy, L., Khalil, D.A., Daggy, B.P., Smith, B.J., Westcott, N., Mocanu, V., Soung, D.Y., and Arhnabdi, B.H. 2004. Flaxseed reduces plasma cholesterol and atherosclerotic lesion formation in ovarietomized golden Syrian hamsters. *Atherosclerosis* **173**, 223–229.

Lukaszewicz, M., Szopa, J., and Krasowska, A. 2004. Susceptibility of lipids from different flax cultivars to peroxidation and its lowering by added antioxidants. *Food Chem.* **88**, 225–231.
Luo, T., Tang, Y., Shan, Q., Lu, Y., and Pan, Y. 2003. Research into a new process for extraction of flaxseed gum. *Min. Metal.* **12**, 92–95.
Maddock, T., Anderson, V., Maddock, R., Bauer, M., and Lardy, G. 2004. Effect of processing flax in feedlot diets on beef heifer performance, carcass composition, and trained sensory panel evaluations. In 2004 Beef Research Report http://www.ag.ndsu.nodak.edu/carringt/livestock/Beef%20Report%2004/Effect%20of%20Processing%20Flax%20in%20Feedlot%20Diets%20on%20Beef%20Heifer%20Performance.htm.
Maddock, T., Anderson, V., and Lardy, G. 2005. Using flax in livestock diets. North Dakota Experiment Station Publication AS-1283, North Dakota State University. Fargo, ND. (http://www.ext.nodak.edu/extpubs/ansci/livestoc/as1283w.htm).
Maddock, T.D., Anderson, V.L., Berg, P.T., Maddock, R.J., and Marchello, M.J. 2003. Influence of level of flaxseed addition and time fed flaxseed on carcass characteristics, sensory panel evaluation and fatty acid content of fresh beef. Proc. 56th Reciprocal Meats Conference, American Meat Science Association, Columbia, MO. http://www.ag.ndsu.nodak.edu/carringt/livestock/Beef%20Report%2003/Flaxseed%20Addition.htm.
Madhusudhan, B., Wiesenborn, D., Schwarz, J., Tostenson, K., and Gillespie, J. 2000. A dry mechanical method for concentrating the lignan secoisolariciresinol diglucoside in flaxseed. *Leb. Wiss. Tech.* **33**, 268–275.
Madusudhan, K.T. and Singh, N. 1983. Studies on linseed proteins. *J. Agric. Food Chem.* **31**, 959–963.
Madusudhan, K.T. and Singh, N. 1985a. Effect of detoxification treatment on the physicochemical properties of linseed proteins. *J. Agric. Food Chem.* **33**, 1219–1222.
Madusudhan, K.T. and Singh, N. 1985b. Isolation and characterization of the major fraction (12S) of linseed proteins. *J. Agric. Food chem.* **33**, 673–677.
Magee, P.J., McGlynn, H., and Rowland, I.R. 2004. Differential effects of isoflavones and lignans on invasiveness of MDA-MB-231 breast cancer cell *in vitro*. *Cancer Lett.* **208**, 35–41.
Maillard, V., Bougnoux, P., and Ferrari, P. 2002. N-3 and n-6 fatty acids in breast adipose tissue and relative risk of breast cancer in a case-control study in Tours, France. *Int. J. Cancer.* **98**, 78–83.
Malcolmson, L.J., Przybylski, R., and Daun, J.K. 2000. Storage stability of milled flaxseed. *J. Am. Oil. Chem. Soc.* **77**, 235–238.
Mannisto, S., Pietinen, P., and Virtanen, M.J. 2003. Fatty acids and risk of prostate cancer in a nested case-control study in male smokers. *Cancer Epidmol. Biomarkers. Prev.* **12**, 1422–1428.
Manthey, F.A., Lee, R.E., and Kegode, R.K. 2000. Quality of spaghetti containing ground flaxseed. *In* "Proceedings of the 58th Flax Institute of the United States", pp. 92–99. Fargo, ND.
Manthey, F.A., Lee, R.E., and Hall, C.A., III 2002a. Stability of alpha-linolenic acid in macaroni containing ground flaxseed. *In* "Proceedings of the 59th Flax Institute of the United States", pp. 14–20. Fargo, ND.
Manthey, F.A., Lee, RE., and Hall, C.A., III 2002b. Processing and cooking effects on lipid content and stability of α-linolenic acid in spaghetti containing ground flaxseed. *J. Agric. Food Chem.* **50**, 1668–1671.
Mantzioris, E., James, M.J., Gibson, R.A., and Cleland, L.G. 1994. Dietary substitution with an alpha-linolenic acid-rich vegetable oil increases eicosapentaenoic acid concentrations in tissues. *Am. J. Clin. Nutr.* **59**, 1304–1309.
Mantzioris, E., Cleland, L.G., Gibson, R.A., Neumann, M.A., Demasi, M., and James, M.J. 2000. Biochemical effects of a diet containing foods enriched with n-3 fatty acids. *Am. J. Clin. Nutr.* **72**, 42–48.
Marcone, M.F., Kakuda, Y., and Yada, R.Y. 1998. Salt-soluble seed globulins of dicotyleonous and monocotyledonous plants. II. Structural characterization. *Food Chem.* **63**, 265–274.

Martin, M.E., Haourigui, M., and Pelissero, C. 1996. Interactions between phytoestrogens and human sex steroid binding protein. *Life Sci.* **58**, 429–436.

Mason, C.T. and Hall, L.A. 1948. New edible colloidal gum from linseed oil meal cake. *Food Ingred.* **20**, 382–383.

Matthews, K.R., Homer, D.B., Thies, F., and Calder, P.C. 2000. Effect of whole linseed (*Linum usitatissimum*) in the diet of finishing pigs on growth performance and on the quality and fatty acid composition of various tissues. *Br. J. Nutr.* **83**, 637–643.

Mazur, W.M., Fotsis, T., Wahala, K., Ojala, S., Sakakka, A., and Adlercreutz, H. 1996. Isotope-dilution gas-chromotographic mass-spectrometric method for the determination of isoflavonoids, coumestrol and lignans in food samples. *Anal. Biochem.* **233**, 169–180.

Mazza, G. and Biliaderis, C.G. 1989. Functional proerties of flax seed mucilage. *J. Food Sci.* **54**, 1302–1305.

McCann, S.E., Muti, P., Vito, D., Edge, S.B., Trevisan, M., and Freudenheim, J.L. 2004. Dietary lignan intakes and risk of pre- and postmenopausal breast cancer. *Int. J. Cancer.* **111**, 440–443.

McMahon, J.M., White, W.L.B., and Sayre, R.T. 1995. Cyanogenesis in cassava (Manihot esculenta Crantz). *J. Exp. Botany.* **46**, 731–741.

McManus, R.M., Jumpson, J., Finegood, D.T., Clandinin, M.T., and Ryan, E.A. 1996. A comparison of the effects of n-3 fatty acids from linseed oil and fish oil in well-controlled type 2 diabetes. *Diabetes Care* **19**, 463–467.

Meagher, L.P., Beecher, G.R., Flanagan, V.P., and Li, B.W. 1999. Isolation and characterization of the lignans, isolariciresinol and pinoresinol, in flaxseed meal. *J. Agric. Food Chem.* **47**, 3173–3180.

Meyers, S., Tulbek, M.C., and Hall, C., III 2004. "Storage stability of roasted flaxseed. Presented at IFT Annual Meeting". Las Vegas, NE.

Moore, W., Garden-Robinson, J., and Nelson, R. 1996. Characterization of flaxseed gum in food model systems. *In* "Proceedings of the 56th Flax Institute of the United States", pp. 86–95. Fargo, ND.

Morris, D.M. 2003a. "Flax, a Health and Nutrition Primer", pp. 9–19. Flax Council of Canada, Winnipeg, Manitoba.

Morris, D.H. 2003b. Methodologic challenges in designing clinical studies to measure differences in the bioequivalence of n-3 fatty acids. *Mol. Cell. Biochem.* **246**, 83–90.

Morton, M.S., Chan, P.S.F., Cheng, C., Blacklock, N., Matos-Ferreira, A., Abranches-Montero, L., Correia, R., Lloyd, S., and Griffiths, K. 1997. Lignans and isoflavonoids in plasma and prostatic fluid in men: Samples from Portugal, Hong Kong, and the United Kingdom. *Prostate* **32**, 122–128.

Mousavi, Y. and Adlercreutz, H. 1992. Enterolactone and estradiol inhibit each other's proliferative effect on MCF-7 breast cancer cells in culture. *J. Steroid Biochem. Mol. Biol.* **41**, 615–619.

Mozaffarian, D., Ascherio, A., Hu, F.B., Stampfer, M.J., Willett, W.C., Siscovick, D.S., and Rimm, E.B. 2005. Interplay between different polyunsaturated fatty acids and risk of coronary heart disease in men. *Circulation* **11**, 151–164.

Muir, A., Westcott, N., Ballantyne, K., and Northrup, S. 2000. Flax lignans-recent developments I the analysis of lignans I plant and animal tissues. *In* "Proceedings of the 54th Flaxseed Institute of the United States", pp. 23–31. Fargo, ND.

Muir, A.D. and Westcott, N.D. 2000. Quantitation of the lignan secoisolariciresinol diglucoside in baked goods containing flax seed or flax meal. *J. Agric. Food Chem.* **48**, 4048–4052.

Murota, S.I., Morita, I., and Suda, N. 1990. The control of vascular endothelial cell injury. *Ann NY Acad. Sci.* **598**, 182.

Mustafa, A.F., Chouinard, P.Y., and Christensen, D.A. 2003. Effects of feeding micronised flaxseed on yield and composition of milk from Holstein cows. *J. Sci. Food. Agr.* **83**, 920–926.

National Agricultural Statistics Service. 2005. Flaxseed: Area planted, harvested, yield, and production by State and United States, 2002–2004. *In* "Crop Production: 2004 Summary", p. 37. United States Department of Agriculture. http://usda.mannlib.cornell.edu/reports/nassr/field/pcp-bban/cropan05.pdf (accessed May 31, 2005).

National Agricultural Statistics Service. 2006. Flaxseed: Area planted, harvested, yield, and production by State and United States, 2003–2005. *In* "Crop Production: 2005 Summary", p. 39. United States Department of Agriculture. http://usda.mannlib.cornell.edu/reports/nassr/field/pcp-bban/cropan06.pdf (accessed January 12, 2006).

Nesbitt, P.D. and Thompson, L.U. 1997. Lignans in homemade commercial products containing flaxseed. *Nutr Cancer* **29**, 222–227.

Nesbitt, P.D., Lam, Y., and Thompson, L.U. 1999. Human metabolism of mammalian lignan precursors in raw and processed flaxseed. *Am. J. Clin. Nutr.* **69**, 549–555.

Nettleton, J. 2003. Collected Recommendations for LC-PUFA Intake. PUFA Newsletterhttp://www.fatsoflife.com/article.asp?i=l&id=142 (accessed May, 2005).

Niedzwiedz-Siegien, I. 1998. Cyanogenic glucosides in Linum usitatissimum. *Phytochemistry* **49**, 59–63.

Niemeyer, H.B., Honig, D., Lange-Bohmer, A., Jacobs, E., Kulling, S.E., and Metzler, M. 2000. Oxidative metabolites of the mammalian lignans enterodiol and enterolactone in rat bile and urine. *J. Agric. Food Chem.* **48**, 2910–2919.

Novak, C. and Scheideler, S. 1998. The effect of calcium and/or vitamin D_3 supplementation of flax-based diets on production parameters and egg composition. *In* "The 1997–1998 Nebraska Poultry Report". University of Nebraska Cooperative Extension MP 70 (http://ianrpubs.unl.edu/poultry/mp70/mp70-10.htm).

Obermeyer, W.R., Musser, S.M., Betz, J.M., Casey, R.E., Pohland, A.E., and Page, S.W. 1995. Chemical studies of phytoestrogens and related compounds in dietary supplements: Flax and chaparral. *Proc. Soc. Exp. Biol. Med.* **208**, 6–12.

Oikarinen, S, Pajari, A., Salminen, I., Heinonen, S.M., Adlercreutz, H., and Mutanen, M. 2005. Effects of a flaxseed mixture and plant oils rich in alpha-linolenic acid on the adenoma formation in multiple intestinal neoplasia (Min) mice. *Br. J. Nutr.* **94**, 510–518.

Olsen, A., Knudsen, K.E.B., Thomsen, B.L., Loft, S., Stripp, C., Overvad, K., Moller, S., and Tjonneland, A. 2004. Plasma enterolactone and breast cancer incidence by estrogen receptor status. *Cancer Epidemiol. Biomarkers Prev.* **13**, 2084–2089.

Oomah, B.D. and Mazza, G. 1993. Flaxseed proteins—a review. *Food Chem.* **48**, 109–114.

Oomah, B.D. and Mazza, G. 1997. Effect of dehulling on chemical composition and physical properties of flaxseed. *Lebensm. Wiss. U. Technol.* **30**, 135–140.

Oomah, B.D. and Mazza, G. 1998a. Compositional changes during commercial processing of flaxseed. *Ind. Crops Prod.* **9**, 29–37.

Oomah, B.D. and Mazza, G. 1998b. Fractionation of flaxseed with a batch dehuller. *Ind. Crops Prod.* **9**, 19–27.

Oomah, B.D. and Mazza, G. 1999. Health benefits of phytochemicals from selected Canadian crops. *Tr. Food Sci. Technol.* **10**, 193–198.

Oomah, B.D. and Mazza, G. 2001. Optimization of a spray drying process for flaxseed gum. *Intl. J. Food Sci. Technol.* **36**, 135–143.

Oomah, B.D., Mazza, G., and Kenaschuk, E.O. 1992. Cyanogenic compounds in flaxseed. *J. Agric. Food Chem.* **40**, 1346–1348.

Oomah, B.D., Kenaschuk, E.O., and Mazza, G. 1995a. Phenolic acids in flaxseed. *J. Agric. Food Chem.* **43**, 2016–2019.

Oomah, B.D., Kenaschuk, E.O., Cui, W., and Mazza, G. 1995b. Variation in the composition of water-soluble polysaccharides in flaxseed. *J. Agric. Food Chem.* **43**, 1484–1488.

Oomah, B.D., Mazza, G., and Przybylski, R. 1996. Comparison of flaxseed meal lipids extracted with different solvents. *Leb. Wiss. Tech.* **29**, 654–658.

Oomah, B.D., Kenaschuk, E.O., and Mazza, G. 1997a. Tocopherols in flaxseed. *J. Agric. Food Chem.* **45**, 2076–2080.

Oomah, B.D., Kenaschuk, E.O., and Mazza, G. 1997b. Lipoxygenase enzyme in flaxseed. *J. Agric. Food Chem.* **45**, 2426–2430.

Oomah, D. 2001. Flaxseed as a functional food source. *J. Sci. Food. Agric.* **81**, 889–904.

Oomen, C.M., Ocké, M.C., and Feskens, E.J.M. 2001. α-Linolenic acid intake is not beneficially associated with 10-y risk of coronary artery disease incidence: The Zutphen elderly study. *Am. J. Clin. Nutr.* **74**, 457–463.

Orcheson, L.J., Rickard, S.E., Seidl, M.M., and Thompson, L.U. 1998. Flaxseed and its mammalian lignan precursor cause a lengthening or cessation of estrous cycling in rats. *Cancer Lett.* **125**, 69–76.

Panganamala, R.V., Sharma, H.M., Heikki, J.C., Geer, J.C., and Cornwell, D.G. 1976. Role of hydroxyle radical scavenger dimethylsulfoxide, alcohols and methional in the inhibition of prostaglandin biosynthesis. *Prostaglandins* **11**, 599.

Parpinello, G., Meluzzi, A., Sirri, F., Tallarico, N., and Versari, A. 2006. Sensory evaluation of egg products and eggs laid from hens fed diets with different fatty acid composition and supplemented with antioxidants. *Food Res. Int.* **39**, 47–52.

Paubert-Braquet, M., Longchampt, M.O., Koltz, P., and Gutlbaud, J. 1988. Tumor necrosis factor (TNF) primes human neutrophil (PMN) platelet-activating factor (PAF)-induced superoxide generation. Consequences in promoting PMn-mediated endothelial cell (EC) damages (abstract). *Prostaglandins* **35**, 803.

Pawlosky, R.J., Hibbeln, J.R., Novotny, J.A., and Salem, N., Jr. 2001. Physiological compartmental analysis of α-linolenic acid metabolism in adult humans. *J. Lipid Res.* **42**, 1257–1265.

Peterson, S.W. 1958. Linseed oil meal. *In* "Processed Plant Protein Foodstuffs" (A.M. Altschul, ed.), pp. 593–617. Academic Press Inc., New York.

Petit, H.V. 2002. Digestion, milk production, milk composition, and blood composition of dairy cows fed whole flaxseed. *J. Dairy Sci.* **85**, 1482–1490.

Petit, H.V., Dewhurst, R.J., Proulx, J.G., Khalid, M., Haresign, W., and Twagiramungu, H. 2001. Milk production, milk composition, and reproductive function of dairy cows fed different fats. *Can. J. Anim. Sci.* **81**, 263–271.

Petit, H.V., Tremblay, G.F., Tremblay, E., and Nadeau, P. 2002. Ruminal biohydrogenation of fatty acids, protein degradability, and dry matter digestibility of flaxseed treated with different sugar and heat combinations. *Can. J. Anim. Sci.* **82**, 241–250.

Pietinen, P., Stumpf, K., Mannisto, S., Kataja, V., Uusitupa, M., and Adlercreutz, H. 2001. Serum enterolactone and risk of breast cancer: A case-control study in eastern Finland. *Cancer Epidemiol. Biomarkers. Prev.* **10**, 339–344.

Pizzey, G.R. and Luba, T. 2002. Effect of seed selection and processing on stability of milled flaxseed. *In* "The 93rd AOCS Annual Meeting and Expo Book of Abstracts", p. S143. Published by the American Oil Chemist Society, Montreal, Canada.

Power, K.A., Saarinen, N.M., Chen, J., and Thompson, L.U. 2004. Lignans (enterolactone and enterodiol) negate the proliferative effect of sioflavone (genistein) on MCF-7 breast cancer cells *in vitro* and *in vivo* (abstract). *Proc. AACR* **45**, 878.

Prasad, K. 1997a. Hydroxyl radical-scavenging property of secoisolariciresinol diglucoside (SDG) isolated from flax-seed. *Mol. Cell. Biochem.* **168**, 117–123.

Prasad, K. 1997b. Dietary flax seed in prevention of hypercholesterolemic atherosclerosis. *Atherosclerosis* **132**, 69–76.

Prasad, K. 1999. Reduction of serum cholesterol and hypercholesterolemic atherosclerosis in rabbits by secoisolariciresinol diglucoside isolated from flaxseed. *Circulation* **99**, 1355–1362.

Prasad, K. 2000a. Oxidative stress as a mechanism of diabetes in diabetic BB prone rats: Effects of secoisolariciresinol diglucoside (SDG). *Mol. Cell. Biochem.* **209**, 89–96.

Prasad, K. 2000b. Antioxidant activity of secoisolariciresinol diglucoside-derived metabolites, secoisolariciresinol, enterodiol, and enterolactone. *Intl. J. Angiol.* **9**, 220–225.

Prasad, K. 2001. Secoisolariciresinol diglucoside from flaxseed delays the development of type 2 diabetes in Zucker rat. *J. Lab. Clin. Med.* **138**, 32–39.

Prasad, K. 2002. Suppression of phosphoenolpyruvate carboxykinase gene expression by secoisolariciresinol diglucoside (SDG), a new antidiabetic agent. *Intl. J. Angiol.* **11**, 107–109.

Prasad, K. 2005. Hypocholesterolemic and antiatherosclerotic effect of flax lignan complex isolated from flaxseed. *Atherosclerosis* **179**, 269–275.

Prasad, K., Mantha, S.V., Muir, A.D., and Westcott, N.D. 1998. Reduction of hypercholesterolemic atherosclerosis by CDC-flaxseed with very low alpha-linolenic acid. *Atherosclerosis* **136**, 367–375.

Prasad, K., Mantha, S.V., Muir, A.D., and Westcott, N.D. 2000. Protective effect of secoisolariciresinol diglucoside against streptozotocin-induced diabetes and its mechanism. *Mol. Cell. Biochem.* **206**, 141–150.

Pretova, A. and Vojtekova, M. 1985. Chlorophylls and carotenoids in flax embryos during embryogenesis. *Photosynthetica* **19**, 194–197.

Przybylski, R. and Daun, J.K. 2001. Additional data on the storage stability of milled flaxseed. *J. Am. Oil Chem. Soc.* **78**, 105–106.

Puthpongsiriporn, U. and Scheideler, S. 2001. Ratios of linoleic to linolenic acid and immune function in pullets. *In* "The 2000–2001 Nebraska Poultry Report", pp. 5–7. University of Nebraska Cooperative Extension MP75.

Qin, L., Xu, S., and Zhang, W. 2005. Effect of enzymatic hydrolysis on the yield of cloudy carrot juice and the effects of hydrocolloids on color and cloud stability during ambient storage. *J. Sci. Food Ag.* **85**, 505–512.

Ranhotra, G.S., Gelroth, J.A., and Glaser, B.K. 1992. Lipidemic response to rats fed flaxseed or sunflower oils. *Cereal. Chem.* **69**(6), 623–625.

Ranhotra, G.S., Gelroth, J.A., Glaser, B.K., and Potnis, P.S. 1993. Lipidemic responses in rats fed flaxseed oil and meal. *Cereal. Chem.* **70**(3), 364–366.

Rao, G.N., Ney, E., and Herbert, R.A. 2000. Effect of Melatonin and linolenic acid on mammary cancer in transgenic mice with c-neu breast cancer oncogene. *Breast Cancer Res. Treat.* **64**, 287–296.

Ratnayake, W.M.N., Behrens, W.A., Fischer, P.W.F., L'Abbe, M.R., Mongeau, R., and Beare-Rogers, J.L 1992. Chemical and nutritional studies of flaxseed (variety Linott) in rats. *J. Nutr. Biochem.* **3**(5), 232–240.

Reaven, G.M., Brand, R.J., Chen, Y.D.I., Mathur, A.K., and Goldfine, I. 1993. Insulin resistance and insulin secretion are determinants of oral glucose tolerance in normal individuals. *Diabetes* **42** (9), 1324–1332.

Renaud, S., Godsey, P., and Dumont, E. 1986. Influence of diet modification on platelet function and composition in Moselle farmers. *Am. J. Clin. Nutr.* **43**, 136–150.

Ribereau-Gayon, P. 1972. "Plant Phenolics". Edinburg, Oliver and Boyd.

Rickard, S. and Thompson, L. 1998. Chronic exposure to secoisolariciresinol diglycoside alters lignan disposition in rats. *J. Nutr.* **128**, 615–623.

Rickard, S. and Thompson, L. 2000. Urinary composition and postprandial blood changes in ^{3}H-secoisolariciresinol diglycoside (SDG) metabolites in rats do not differ between acute and chronic SDG treatments. *J. Nutr.* **130**, 2299–2305.

Rickard, S.E. and Thompson, L.U. 1997. Phytoestrogens and lignans: Effects on reproduction and chronic disease. *In* "Antinutrients and Phytochemicals in Foods" (F. Shahidi, ed.), pp. 273–293. Oxford University Press, New York.

Rickard, S.E., Orcheson, L.J., and Seidl, M.M. 1996. Dose-dependent production of mammalian lignans in rats and *in vitro* from the purified precursor secoisolariciresinol diglycoside in flaxseed. *J. Nutr.* **126**, 2012–2019.

Rickard, S.E., Yuan, Y.V., Chen, J., and Thompson, L.U. 1999. Dose effects of flaxseed and its lignan on N-methyl-N-nitrosourea-induced mammary tumorigenesis in rats. *Nutr. Cancer* **35**, 50–57.

Riley, P.A., Enser, M., Nute, G.R., and Wood, J.D. 2000. Effects of dietary linseed on nutritional values and other quality aspects of pig muscle and adipose tissue. *Anim. Sci.* **71**, 483–500.

Rollefson, G.O., Simmons, A.H., Donaldson, M.L., Gillespie, W., Kafafi, Z., Kohler-Rollefson, I.U., McAdam, E., Ralston, S.L., and Tubb, M.K. 1985. Excavation at the pre-pottery Neolithic B village of 'Ain Ghazal (Jordan), (1983). *Mitteilungen der Deuschen Orient-Gesellschaft zu Berlin* **117**, 69–116.

Romans, J.R., Johnson, R.C., Wulf, D.M., Libal, G.W., and Costello, W.J. 1995a. Effects of ground flaxseed in swine diets on pig performance and on physical and sensory characteristics and omega-3 fatty acid content of pork. I. Dietary level of flaxseed. *J. Anim. Sci.* **73**, 1982–1986.

Romans, J.R., Wulf, D.M., Johnson, R.C., Libal, G.W., and Costello, W.J. 1995b. Effects of ground flaxseed in swine diets on pig performance and on physical and sensory characteristics and omega-3 fatty acid content of pork. II. Duration of 15% dietary flaxseed. *J. Anim. Sci.* **73**, 1987–1999.

Roseling, H. 1994. Measuring effects in humans of dietary cyanide exposure to sublethal cyanogens from Cassava in Africa. *Acta Hort.* **375**, 271–283.

Ross, R. 1999. Atherosclerosis—an inflammatory disease. *N. Engl. J. Med.* **340**, 115–126.

Rudnik, E., Szczucinska, A., Gwardiak, H., Szulc, A., and Winiarska, A. 2001. Comparative studies of oxidative stability of linseed oil. *Thermo. Acta* **370**, 135–140.

Russo, J. and Russo, I.H. 1978. DNA labeling index and structure of the rat mammary gland as determinant of its susceptibility to carcinogenesis. *J. Natl. Cancer Inst.* **61**, 1451–1459.

Russo, J. and Russo, I.H. 1995. Hormonally induced differentiation: A novel approach to breast cancer prevention. *J. Cell Biochem.* **22**, 58–64.

Russo, J., Lynch, H., and Russo, I.H. 2001. Mammary gland architecture as a determing factor in the susceptibility of the human breast to cancer. *Breast J.* **7**, 278–291.

Saadatian-Elahi, M., Norat, T., and Goudable, J. 2004. Biomarkers of dietary fatty acid intake and the risk of breast cancer: A meta-analysis. *Int. J. Cancer* **111**, 584–591.

Saarinen, N.M., Warri, A., and Makela, S.I. 2000. Hydroxymatairesinol, a novel enterolactone precursor with antitumor properties from coniferous tree (*Picea abies*). *Nutr. Cancer* **36**, 207–216.

Saarinen, N.M., Huovinen, R., Varri, A., Makela, S.I., Valentin-Blasini, L., Sjoholm, R., Ammala, J., Lehtila, R., Eckerman, C., Collan, Y.U., and Santii, R.S. 2002. Enterolactone inhibits the growth of 7, 12-dimethylbenz (a) anthracene-induced mammary carcinomas in the rat. *Mol. Cancer Ther.* **1**, 869–876.

Saarinen, N.M., Makela, S., and Satti, R. 2003. Mechanism of anticancer effects of lignans with a special emphasis on breast cancer. *In* "Flaxseed in Human Nutrition" (S.C. Cunnane and L.U. Thompson, eds), 2nd Ed., pp. 223–228. AOCS Press, Champaign, IL.

Saarinen, N.M., Penttinen, P.E., Smeds, A.I., Hurmerinta, T.T., and Makela, S.I. 2005. Structural determinatnts of plant lignans for growth of mammary tumors and hormonal responses *in vivo*. *J. Steroid Biochem. Mol. Biol.* **93**, 209–219.

Sammour, R.H. 1999. Proteins of linseed (*Linum usitativissimum* L.), extraction and characterization by electrophoresis. *Bot. Bull. Acad. Sin.* **40**, 121–126.

Sammour, R.H., El-Shourbagy, M.N., Abo-Shady, A.M., and Abasary, A.M. 1994. The seed proteins of linseed (*Linum usitativissimum* L.). *Bot. Bull. Acad. Sin.* **35**, 171–177.

Scheideler, S. 1998. Omega-3 fatty acid enriched eggs and method for producing such eggs. International Patent WO 98/47389.

Scheideler, S. 1999. Feed to produce Omega-3 fatty acid enriched eggs and method for producing such eggs. U.S. Patent no. 5897890.

Scheideler, S. 2003. Flaxseed in poultry diets: Meat and eggs. *In* "Flaxseed in Human Nutrition" (L.U. Thompson and S.C. Cunnane, eds), 2nd Ed., pp. 423–428. AOCS Press, Champaign, IL.

Scheideler, S. and Froning, G. 1996. The combined influence of dietary flaxseed variety, level, form, and storage conditions on egg production and composition among vitamin E-supplemented hens. *Poult. Sci.* **75**, 1221–1226.

Scheideler, S. and Lewis, N. 1997. Omega eggs: A dietary source of N-3 fatty acids. Neb. Facts. NF97-354.

Scheideler, S.E., Froning, G., and Cuppett, S. 1997. Studies of consumer acceptance of high omega-3 fatty acid-enriched eggs. *J. Applied Poultry Res.* **6**, 137–146.

Schilcher, H. and Wilkens-Sauter, M. 1986. Quantitative determination of cyanogenic glycosides in *Linum usitatissimum* using HPLC. *Fette, Seife, Anstrichm.* **8**, 287–290.

Schorno, A., Tulbek, M.C., Hall, C., III, and Manthey, F. 2003.Evaluation of physical and chemical properties of roasted flaxseed. Presented at IFT Annual Meeting. Chicago, IL.

Schorno, A., Manthey, F., Wiesenborn, D., Hall, C., III., and Hammond, J. 2004. Flaxseed milling. *In* "Proceedings of the 60th Flax Institute of the United States", pp. 15–23. Fargo, ND.

Schuurman, A.G., van den Brandt, P.A., and Dorant, E. 1999. Association of energy and fat intake with prostate carcinoma risk: Results from the Netherlands cohort study. *Cancer* **86**, 1019–1027.

Seghrouchni, I., Drai, J., Bannier, E., Riviere, J., Calmard, P., Garcia, I., Orgiazzi, J., and Revol, A. 2002. Oxidative stress parameters in type I, type II and insulin-treated type 2 diabetes mellitus: Insulin treatment efficiency. *Clin. Chim. Acta* **321**, 89–96.

Serraino, M. and Thompson, L.U. 1991. The effect of flaxseed supplementation on early risk markers for mammary carcinogenesis. *Cancer Lett.* **60**, 135–142.

Serraino, M. and Thompson, L.U. 1992. The effect of flaxseed supplementation on the initiation and promotional stages of mammary tumorigenesis. *Nutr. Cancer* **17**, 153–159.

Setchell, K.D.R., Childress, C., Zimmer-Nechemias, L., and Cai, J. 1999. Method for measurement of dietary secoisolariciresinol using HPLC with multichannel electrochemical detection. *J. Medicinal Food* **2**, 193–198.

Shahidi, F., Wanasundra, U., and Amarowicz, R. 1995. Isolation and partial characterization of oilseed phenolics and evaluation of their antioxidant activity. *In* "Food Flavors: Generation, Analysis and Process Infleunce" (G. Charalambous, ed.), Elsevier Science.

Shan, Q., Tang, Y., and Luo, T. 2000. Study on properties of flaxseed gum. *Min. Metal.* **9**, 87–90.

Shimoda, M., Nakada, Y., Nakashima, M., and Osajima, Y. 1997. Quantitative comparison of volatile flavor compounds in deep-roasted and light-roasted sesame seed oil. *J. Agric. Food Chem.* **45**, 3193–3196.

Shinomiya, K., Fukunaga, M., Kiyomoto, H., Mizushige, K., Tsuji, T., Noma, T., Ohmori, K., Kohno, M., and Senda, S. 2002. A role of oxidative stress-generated eicosanoid in the progression of arteriosclerosis in type 2 diabetes mellitus model rats. *Hypertens. Res.* **25**, 91–98.

Siedow, J.N. 1991. Plant lipoxygenase: Structure and function. *Annual Rev. Plant Physiol. Plant Mol. Biol.* **42**, 145–149.

Sikorska, E., Gorecki, T., Khmenlinskii, I.V., Sikorski, M., and Koziol, J. 2005. Classification of edible oils using synchronous scanning fluorescence spectroscopy. *Food Chem.* **89**, 217–225.

Simon, J.A., Fong, J., Bernert, J.T., Jr., and Browner, W.S. 1995. Serum fatty acids and the risk of stroke. *Stroke* **26**, 778–782.

Simonsen, N., Van't Veer, N., Strain, J.J., Martin-Moreno, J.M., Huttunen, J.K., Navajas, J.F., Martin, B.C., Thamm, M., Kardinaal, A.R., Kok, F.J., and Kohlmeier, L. 1998. Adipose tissue omega-3 and omega-6 fatty acid content and breast cancer in the EURAMIC study. *Am. J. Epidemiol.* **147**, 342–352.

Simopoulos, A.P. 1999. Essential fatty acids in health and chronic disease. *Am. J. Clin. Nutr.* **70**, 560S–569S.

Singh, R.B., Dubnov, G., and Niaz, M.A. 2002. Effect of an Indo- Mediterranean diet on progression of coronary artery disease in high risk patients (Indo-Mediterranean Diet Heart Study): A randomized single-blind trial. *Lancet* **360**, 1455–1461.

Sinha, S., Yalla, S., and Manthey, F. 2004. Extrusion properties and cooking quality of fresh pasta containing ground flaxseed. *In* "Proceedings of the 60th Flax Institute of the United States", pp. 24–30. Fargo, ND.

Slavin, J., Hutchins, A., and Haggans, C. 2002. Can flaxseed protect against hormonally dependent cancer? *In* "Proceeding of 59th Flax Institute of United States", p. 52. Fargo, ND.

Sosulski, F.W. and Sarwar, G. 1973. Amino acid composition of oilseed meals and protein isolates. *Can. Inst. Food Sci. Tech. J.* **6**, 1–5.

Sprando, R.L., Collins, T.F.X., Black, T.N., Olejnik, N., Rorie, J.I., Scott, M., Wiesenfeld, P., Babu, U.S., and O'Donnell, M. 2000a. The effect of maternal exposure to flaxseed on spermatogenesis in F_1 generation rats. *Food Chem. Toxicol.* **38**, 325–334.

Sprando, R.L., Collins, T.F.X., Wiesenfeld, P., Babu, U.S., Rees, C., Black, T., Olejnik, N., and Rorie, J. 2000b. Testing the potential of flaxseed to affect spermatogenesisd: Morphometry. *Food Chem. Toxicol.* **38**, 887–892.

Statin, P., Adlercreutz, H., Tenkanen, L., Jellum, E., Lumme, S., Hallmans, G., Harvei, S., Teppo, L., Stumpf, K., Luostarinew, T., Lehtinen, M., Dillner, M., *et al.* 2002. Circulating enterolactone and prostate cancer risk: A Nordic nested case-control study. *Int. J. Cancer* **99**, 124–129.

Stenberg, C., Svensson, M., and Johansson, M. 2005. A study of the drying of linseed oils with different fatty acid patterns using RTIR-spectroscopy and chemiluminescence (CL). *Ind. Crops Prod.* **21**, 263–272.

Stewart, A.G., Dubbin, P.N., Harris, T., and Dusting, G.J. 1990. Platelet-activating factor may act as a second messenger in the release of eicosanoids and superoxide anions from leucocytes and endothelial cell. *Proc. Natl. Acad. Sci. USA* **87**, 3215.

Stitt, P.A. 1994. History of flax: 9000 years ago to (1986). *In* "55th Flax Institute of the United States", pp. 152–153. Fargo, ND.

Tan, K.P., Chen, J., Ward, W.E., and Thompson, L.U. 2004. Mammary gland morphogenesis is enhanced by exposure to flaxseed or its major lignan during suckling in rats. *Exp. Biol. Med.* **229**, 147–157.

Taylor, B.R. and Morrice, L.A.F. 1991. Effects of husbandry practices on the seed yield and oil content of linseed in Northern Scotland. *J. Sci. Food Agric.* **57**, 189–198.

Thacker, P.A., Racz, V.J., and Soita, H.W. 2004. Performance and carcass characteristics of growing-finishing pigs fed barley-based diets supplemented with Linpro (extruded whole flaxseed and peas) or soybean meal. *Can. J. Anim. Sci.* **84**, 681–688.

Thompson, L.U. 2003. Flaxseed, lignans, and cancer. *In* "Flaxseed in Human Nutrition" (L.U. Thompson and S.C. Cunnane, eds), 2nd Ed., pp. 194–218. AOCS Press, Champaign, IL.

Thompson, L.U., Robb, P., Serraino, M., and Cheung, F. 1991. Mammalian lignan production from various foods. *Nutr. Cancer* **16**, 43–52.

Thompson, L.U., Rickard, S.E., Orcheson, L.J., and Seidl, M.M. 1996a. Flaxseed and its lignan and oil components reduce mammary tumor growth at a late stage of carcinogenesis. *Carcinogenesis* **17**, 1373–1376.

Thompson, L.U., Seidl, M., Rickard, S., Orcheson, L., and Fong, H. 1996b. Antitumorigenic effect of a mammalian lignan precursor from flaxseed. *Nutr. Cancer* **26**, 159–165.

Thompson, L.U., Rickard, S.E., and Cheung, F. 1997. Variability in anticancer lignan levels in flaxseed. *Nutr. Cancer* **27**, 26–30.

Thompson, L.U., Li, T., Chen, J., and Goss, P.E. 2000. Biological effects of dietary flaxseed in patients with breast cancer (abstract). *Breast Cancer Res. Treat.* **64**, p. 50.

Thompson, L.U., Chen, J.M., Li, T., Strasser-Weippl, K., and Goss, P.E. 2005. Dietary flaxseed alters tumor biological markers in postmenopausal breast cancer. *Clin. Cancer Res.* **11**, 3828–3835.

Tostenson, K., Wiesenborn, D., Zhang, X., Kangas, N., and Schwarz, J. 2000. Evaluation of continuous process for mechanical flaxseed fractionation. *In* "Proceedings of the 58th Flax Institute of the United States", pp. 17–22. Fargo, ND.

Tou, J.C. and Thompson, L.U. 1999. Exposure to flaxseed or its lignan component during different developmental stages influences rat mammary gland structures. *Carcinogenesis* **20**, 1831–1835.

Tou, J.C.L., Chen, J., and Thompson, L.U. 1998. Flaxseed and its lignan precursor, secoisolariciesinol diglycoside, affect pregnancy outcome and reproductive development in rats. *J. Nutr.* **128**, 1861–1868.

Tou, J.C.L., Chen, J., and Thompson, L.U. 1999. Dose, timing, and duration of flaxseed exposure affect reproductive indices and sex hormone levels in rats. *J. Toxicol. Environ. Health* **56**(Part A), 555–570.

Trentin, G.A., Moody, J., Torous, D.K., Thompson, L.U., and Heddle, J.A. 2004. The influence of dietary flaxseed and other grains, fruits and vegetables on the frequency of spontaneous chromosomal damage in mice. *Mutat. Res.* **551**, 213–222.

Tulbek, M.C., Schorno, A., Meyers, S., Hall, C., III, and Manthey, F. 2004. Chemical and sensory analysis and shelf life stability of roasted flaxseed. *In* "Proceedings of the 60th Flax Institute of the United States", pp. 31–36. Fargo, ND.

USDA 2005. Nutrient Data Laboratory United States Department of Agriculture. available at http://www.nal.usda.gov/fnic/foodcomp/cgi-bin/list_nut_edit.placcessed on April, 2005.

Vaisey-Genser, M. and Morris, D. 2003. Introduction: History of the cultivation and uses of flaxseed. *In* "Flax: The Genus Linum" (A. Muir and N. Westcott, eds), pp. 1–21. Taylor and Francis, Ltd, London, England.

Valsta, L.M., Salminen, I., Aro, A., and Mutanen, M. 1996. Alpha-linolenic acid in rapeseed oil partly compensates for the effect of fish restriction on plasma long chain n-3 fatty acids. *Eur. J. Clin. Nutr.* **50**, 229–235.

Van Oeckel, M.J., Casteels, M., Warnants, N., Van Damme, L., and Boucque, V. 1996. Omega-3 fatty acids in pig nutrition: Impications for the intrinsic and sensory quality of the meat. *Meat Sci.* **44**, 55–63.

Van Ruth, S.M., Shaker, E.S., and Morrissey, P.A. 2001. Influence of methanolic extracts of soybean seeds and soybean oil on lipid oxidation in linseed oil. *Food Chem.* **75**, 177–184.

Van Zeiste, W. 1972. Palaeobotanical results in the 1970 seasons at Cayonu, Turkey. *Helinium* **12**, 3–19 (Cited by Zohary and Hopf, 1993).

Varga, T.K. and Diosady, L.L. 1994. Simultaneous extraction of oil and antinutritional compounds from flaxseed. *J. Am. Oil Chem. Soc.* **71**, 603–607.

Vartiainen, E., Puska, P., Pekkanen, J., Toumilehto, J., and Jousilahti, P. 1994. Changes in risk factors explain changes in mortality from ischemic heart disease in Finland. *Br. Med. J.* **309**, 23–27.

Velioglu, Y.S., Mazza, G., Gao, L., and Oomah, B.D. 1998. Antioxidant activity and total phenolics in selected fruits, vegetables, and grain products. *J. Agric. Food Chem.* **46**, 4113–4117.

Vermunt, S.H.F., Mensink, R.P., and Simonis, M.M.G. 2000. Effects of dietary alpha-linolenic acid on the conversion and oxidation of 13C-alphalinolenic acid. *Lipids* **35**, 137–142.

Vick, B.A. and Zimmerman, D.C. 1987. Oxidative systems for modification of fatty acids: The lipoxygenase pathway. *In* "The Biochemistry of Plants" (P.K. Stumpf and E.E. Conn, eds), Vol. 9, p. 53. Academic Press, Orlando, FL.

Voorrips, L., Brants, H., Kardinaal, A., Hiddink, G., and van den Brandt, P. 2002. Intake of conjugated linoleic acid, fat, and other fatty acids in relation to postmenopausal breast cancer: The Netherlands cohort study on diet and cancer. *Am. J. Clin. Nutr.* **76**, 873–882.

Wakjira, A., Labuschagne, M.T., and Hugo, A. 2004. Variability in oil content and fatty acid composition of Ethiopian and introduced cultivars of linseed. *J. Sci. Food Agric.* **84**, 601–607.

Wanasundara, J. and Shahidi, F. 1994. Functional properties and amino acid composition of solvent-extracted flaxseed meals. *Food Chem.* **49**, 45–51.

Wanasundara, P.K.J.D. and Shahidi, F. 1993. Functional properties and amino acid composition of solvent-extracted flaxseed meals. *Food Chemistry* **49**, 45–51.

Wanasundara, P.K.J.P.D. and Shahidi, F. 1997. Removal of flaxseed mucilage by chemical and enzymatic treatments. *Food Chem.* **59**, 47–55.

Wanasundara, P.K.J.P.D., Amarowicz, R., Kara, M.T., and Shahidi, F. 1993. Removal of cyanogenic glycosides of flaxseed meal. *Food Chem.* **48**, 263–266.

Wanasundara, P.K.J.P.D., Wanasundara, U.N., and Shahidi, F. 1999. Changes in flax (*Linum usitatissimum L.*) seed lipids during germination. *J. Am. Oil Chem. Soc.* **76**, 41–48.

Wang, L.Q., Meselhy, M.R., Li, Y., Qin, G.W., and Hattori, M. 2000. Human intestinal bacteria capable of transforming secoisolariciresinol diglucoside to mammalian lignans, enterodiol and enterolactone. *Chem. Pharm. Bull.* **48**, 1606–1610.

Ward, W.E., Jiang, F.O., and Thompson, L.U. 2000. Exposure to flaxseed or purified lignan during lactation influences rat mammary gland structures. *Nutr. Cancer* **37**, 187–192.

Ward, W.E., Chen, J., and Thompson, L.U. 2001a. Exposure to flaxseed or its purified lignan during suckling only or continuously does not alter reproductive indices in male and female offspring. *J. Toxicol. Environ. Health* **64**(Part A), 567–577.

Ward, W.E., Yuan, Y.V., Cheung, A.M., and Thompson, L.U. 2001b. Exposure to purified lignan from flaxseed (Linum usitatissimum) alters bone development in female rats. *Br. J. Nutr.* **86**, 499–505.

Ward, W.E., Yuan, Y.V., Cheung, A.M., and Thompson, L.U. 2001c. Exposure to flaxseed and its purified lignan reduces bone strength in young but not older male rats. *J. Toxicol. Environ. Health* **63**(Part A), 53–65.

Ward, A.T., Wittenberg, K.M., and Przybylski, R. 2002. Bovine milk fatty acid profiles produced by feeding diets containing solin, flax and canola. *J. Dairy Sci.* **85**, 1191–1196.

Warrand, J., Michaud, P., Picton, L., Muller, G., Courtois, B., Ralainirina, R., and Courtois, J. 2005. Structural investigations of the neutral polysaccharide of Linum usitatissimum L. seeds mucilage. *Int. J. Biol. Macromol.* **35**, 121–125.

Warrand, J., Michaud, P., Picton, L., Muller, G., Courtois, B., Ralainirina, R., and Courtois, B. 2005a. Flax (*Linum usitatissimum*) seed cake: A potential source of high molecular weight arabinoxylans. *J. Agric. Food Chem.* **53**, 1449–1452.

Warrand, J., Michaud, P., Picton, L., Muller, G., Courtois, B., Ralainirina, R., and Courtois, B. 2005b. Structural investigations of the neutral polysaccahride of *Linum usitatissimum* L. seeds mucilage. *Int. J. Bio. Macro.* **35**, 121–125.

Waters, A.P. and Knowler, J.T. 1982. Effect of a lignan (HPMF) on RNA synthesis in the rat uterus. *J. Rep. Fertility* **66**, 379–381.

Westcott, N.D. and Muir, A.D. 1996. Variation in flax seed lignan concentration with variety, location and year. *In* "Proceedings of the 56th Flax Institute of the United States", pp. 77–80. Fargo, ND.

White, N.D.G. and Jayas, D.S. 1991. Factors affecting the deterioration of stored flaxseed including the potential insect infestation. *Canadian J. Plant Sci.* **71**, 327–335.

WHO/FAO Expert Consultation. 2003. Diet, nutrition and the prevention of chronic diseases. *World Health Organization Technical Report Series* **916**, 89–90.

Wiesenborn, D.P., Tostenson, K., Kangas, N., and Osowski, C. 2002. Mechanical fractionation of flaxseed for edible uses. *In* "Proceedings of the 59th Flax Institute of the United States", pp. 25–29. Fargo, ND.

Wiesenborn, D.P., Tostenson, K., and Kangas, N. 2003. Continuous abrasive method for mechanically fractionating flaxseed. *J. Am. Oil Chem. Soc.* **80**, 1–7.

Wiesenborn, D.P., Zheng, Y., Kangas, N., Tostenson, K., Hall, C., III., and Chang, K.C. 2004. Quality of screw-pressed flaxseed oil. *In* "Proceedings of the 60th Flax Institute of the United States", pp. 8–14. Fargo, ND.

Wiesenborn, D., Kangas, N., Tostenson, K., Hall, C., III., and Chang, K. 2005. Sensory and oxidative quality of screw-pressed flaxseed oil. *J. Am. Oil Chem. Soc.* **82**, 887–892.

Wiesenfeld, P.W., Babu, U.S., Collins, T.F.X., Sprando, R., O'Donnell, M.W., Flynn, T.J., Black, T., and Olejnik, N. 2003. Flaxseed increased α-linolenic and eicosapen-taenoic acid and decreased arachidonic acid in serum and tissues of rat dams and offsprings. *Food Chem. Toxicol.* **41**, 841–855.

Wilkenson, P., Leach, C., Ah-Sing, E.E., Hussain, N., Miller, G.J., Millward, D.J., and Griffin, B.A. 2005. Influence of α-linolenic acid and fish-oil on markers of cardiovascular risk in subjects with an atherogenic lipoprotein phenotype. *Atheroslerosis* **181**, 115–124.

Xu, Y., Hall III, C., Wolf-Hall, C., and Manthey, F. 2006. Antifungal activity of flaxseed flours. *In* "Proceedings of the 60th Flax Institute of the United States" (J. Carter, ed.), (in press). Published by North Dakota State University, Fargo, ND.

Yang, H., Mao, Z., and Tan, H. 2004. Determination and removal methods for cyanogenic glucoside in flaxseed. ASAE/CSAE Annual International Meeting, Ottawa, Ontario, Canada, August (2004).

Yoshida, H. and Takagi, S. 1997. Effects of seed roasting temperature and time on the quality characteristics of sesame (Sesamum indicum) oil. *J. Sci. Food Agric.* **75**, 19–26.

Yoshida, H., Hirooka, N., and Kajimoto, G. 1990. Microwave energy effects on quality of some seed oils. *J. Food Sci.* **55**, 1412–1416.

Youle, R. and Huang, A. 1981. Occurrence of low-molecular-weight and high cysteine containing albumin storage proteins in oilseeds of diverse species. *Amer. J. Botany* **68**, 44–48.

Yuan, Y.V., Rickard, S.E., and Thompson, L.U. 1999. Short-term feeding of flaxseed or its lignan has minor influence on *in vivo* hepatic antioxidants status in young rats. *Nutr. Res.* **19**, 1233–1243.

Zeleniuch-Jacquotte, A., Adlercreutz, H., Shore, R.E., Koenig, K.L., Kato, I., Arslan, A.A., and Toniolo, P. 2004. Circulating enterolactone and risk of breast cancer: A prospective study in New York. *Br. J. Cancer* **91**, 99–105.

Zhang, J. 1994. Extraction and functionalities of flaxseed protein. M.Sc. Thesis. Wuxi University of Light Industry. Wuxi, China.

Zhao, G., Etherton, T.D., and Martin, K.R. 2004. Dietary alpha-linolenic acid reduces inflammatory and lipid cardiovascular risk factors in hypercholesterolemic men and women. *J. Nutr.* **134**, 2991–2997.

Zheng, Y., Wiesenborn, D.P., Tostenson, K., and Kangas, N. 2002. Bench scale screw pressing of flaxseed and flaxseed embryos. *In* "Proceedings of the 59th Flax Institute of the United States", pp. 30–37. Fargo, ND.

Zheng, Y., Wiesenborn, D.P., Tostenson, K., and Kangas, N. 2003. Screw pressing of whole and dehulled flaxseed for organic oil. *J. Am. Oil Chem. Soc.* **80**, 1039–1045.

Zheng, Y., Wiesenborn, D.P., Tostenson, K., and Kangas, N. 2005. Energy analysis in the screw pressing of whole and dehulled flaxseed. *J. Food Eng.* **66**, 193–202.

Zhuang, H., Hamilton-Kemp, T.R., Andersen, R.A., and Hildebrand, D.F. 1992. Developmental change in C6–aldehyde formation by soybean leaves. *Plant Physiol.* **100**, 80–87.

Zimmerman, D.C. and Vick, B.A. 1970. Specificity of flaxseed lipoxidase. *Lipids* **5**, 392–397.

Zimmermann, R., Bauermann, U., and Morales, F. 2006. Effects of growing site and nitrogen fertilization on biomass production and lignan content of linseed (*Linum usitatissimum* L.). *J. Sci. Food Agric.* **86**, 415–419.

Zohary, D. and Hopf, M. 2000. Oil and fibre crops. *In* "Domestication of Plants in the Old World" (D. Zohary and M. Hopf, eds), 3rd Ed., pp. 125–132. Oxford University Press, Oxford.

LYCOPENE

A. V. RAO,* M. R. RAY,[†] AND L. G. RAO[‡]

*Department of Nutritional Sciences, Faculty of Medicine
University of Toronto, Toronto, Ontario, Canada
[†]London Regional Cancer Program, London, Ontario, Canada
[‡]Department of Medicine, Calcium Research Laboratory, St. Michael's Hospital
University of Toronto, Toronto, Ontario, Canada

Oxidative stress is now recognized as an important etiological factor in the causation of several chronic diseases including cancer, cardiovascular diseases, osteoporosis, and diabetes. Antioxidants play an important role in mitigating the damaging effects of oxidative stress on cells. Lycopene, a carotenoid antioxidant, has received considerable scientific interest in recent years. Epidemiological, tissue culture, and animal studies provide

ADVANCES IN FOOD AND NUTRITION RESEARCH VOL 51

ISSN: 1043-4526
DOI: 10.1016/S1043-4526(06)51002-2

convincing evidence supporting the role of lycopene in the prevention of chronic diseases. Human intervention studies are now being conducted to validate epidemiological observations and to understand the mechanisms of action of lycopene in disease prevention. To obtain a better understanding of the role of lycopene in human health, this chapter reviews the most recent information pertaining to its chemistry, bioavailability, metabolism, role in the prevention of prostate cancer and cancer of other target organs, its role in cardiovascular diseases, osteoporosis, hypertension, and male infertility. A discussion of the most relevant molecular markers of cancer is also included as a guide to future researchers in this area. The chapter concludes by reviewing global intake levels of lycopene, suggested levels of intake, and future research directions.

I. INTRODUCTION

Lycopene, as a dietary source of a carotenoid antioxidant, has attracted considerable interest in recent years as an important phytochemical with a beneficial role in human health. Chronic diseases including cancer, cardiovascular disease, diabetes, and osteoporosis are the major causes of morbidity and mortality in the Western World. Along with genetic factors and age, lifestyle factors and diet are also considered important risk factors for these diseases (Agarwal and Rao, 2000a). The role of oxidative stress induced by reactive oxygen species (ROS) and the oxidative damage of important biomolecules is one of the main foci of research related to human diseases. Oxidative stress is thought to be involved in the cause and progression of several chronic diseases (Rao and Rao, 2004). Antioxidants are agents that inactivate ROS and provide protection from oxidative damage. Dietary guidelines for the prevention of chronic diseases, recognizing the importance of antioxidants, have recommended an increase in the consumption of fruits and vegetables that are good sources of dietary antioxidants. In addition to traditional antioxidant vitamins, such as vitamins A, E, and C, fruits and vegetables also contain several phytonutrient antioxidants. The two important classes of phytonutrient antioxidants include the carotenoids and polyphenols. *In vitro* cell culture studies, laboratory animal studies, case control and cohort studies, and dietary intervention studies have all provided evidence in support of the role of antioxidants in the prevention of cancer and other chronic diseases. The focus of this chapter will be on lycopene, a potent carotenoid antioxidant. Since animal and human experimental studies on the role of lycopene in cancer prevention are beginning to be undertaken, a brief overview of some important molecular markers of cancer, which could be used in these studies, is also reviewed.

The authors consider this to be important information in planning future animal and human clinical and intervention studies.

II. OXIDATIVE STRESS AND CHRONIC DISEASES

ROS are highly reactive oxidant molecules that are generated endogenously through regular metabolic activity, lifestyle activities, such as smoking and exercise, environmental factors, such as pollution and ultra violet radiation, and diet. Being highly reactive, they can cause oxidative damage to cellular components such as lipids, proteins, and DNA (Agarwal and Rao, 2000a). Under normal circumstances, presence of endogenous repair mechanisms, such as the antioxidant enzymes including superoxide dismutase (SOD), glutathione peroxidase (GP_x) and catalase and antioxidant vitamins, minerals, and phytonutrients, help repair the damaged biomolecules. However, when these defense mechanisms are overwhelmed by ROS, it leads to permanent damage of the biomolecules resulting in increased risk of chronic diseases. Lipid peroxidation products are increased in a variety of oxidative stress conditions (Witztum, 1994). Premenopausal women with high mammographic tissue densities at high risk for breast cancer excrete higher levels of the lipid peroxidation product malondialdehyde (Boyd and McGuire, 1990; Boyd *et al.*, 1995). A 20% increase in the levels of lipid peroxides were found in prostate cancer patients compared to their matched controls with retinoic acid therapy (Rigas *et al.*, 1994). Oxidation of the low-density lipoprotein (LDL) has been associated with increased formation of atherosclerotic plaque, leading to coronary heart disease (CHD) (Arab and Steck, 2000; Parthasarathy *et al.*, 1992; Rao, 2002b). Similarly, DNA oxidation increases the risk of cancers and other pathological disorders. Cancer tissues contain increased levels of oxidized DNA adducts (Loft and Poulsen, 1996; Musarrat *et al.*, 1996; Wang *et al.*, 1996). Oxidation of intercellular proteins results in functional changes of enzymes that modulate cellular metabolism (Rao and Agarwal, 2000). Both DNA and protein oxidation are associated with the aging process (Ames *et al.*, 1993). ROS also induce the expression of a wide variety of transcription factors, such as NFkB, AP1, and oncogenes, such as *c-fos* and *c-jun*. Conformational changes are found in p53 protein, which mimic the mutant phenotype induced by ROS (Hainaut and Miller, 1993; Toledano and Leonard, 1991; Wasylyk and Wasylyk, 1993; Wei, 1992). By inducing these alterations, ROS can influence cell cycle mechanisms and ultimately lead to human health disorders. Increases in the tissue and body fluid levels of several biomarkers of oxidized lipids, proteins, and DNA have been shown during aging and also in patients with cancer and cardiovascular diseases (Ames *et al.*, 1996; Halliwell *et al.*, 1995; Pincemail, 1995; Stadtman, 1992;

Witztum, 1994). Dietary antioxidants, such as lycopene, can protect lipids, proteins, and DNA from oxidation and play a role in the prevention of such diseases.

III. CHEMISTRY AND DIETARY SOURCES OF LYCOPENE

Lycopene belongs to the family of carotenoid compounds found in fruits, vegetables and green plants. In plants these compounds are part of the photosynthetic machinery and are responsible for the yellow, orange, and red colors of fruits and vegetables. They are synthesized by plants and microorganisms but not by animals and humans. They are important dietary sources of vitamin A and are also excellent antioxidants (Paiva and Ressell, 1999). More than 600 carotenoids are found in nature, about 40 of which are present in a typical human diet, and about 20 have been identified in blood and tissues (Agarwal and Rao, 2000b). The principal carotenoids present in the diet and human body are β-carotene, α-carotene, lycopene, α-cryptoxanthine, lutein, and zeaxanthin, accounting for over 90% of all carotenoids (Gerster, 1997). All carotenoids posses certain common chemical features consisting of a polyisoprenoid structure, a long conjugated chain of double bonds in the central position of the molecules, and a near bilateral symmetry around the central double bond (Britton, 1995). However, modifications in the base structure by cyclization of the end groups and by introduction of oxygen functions yield different carotenoids giving them characteristic colors and antioxidant properties (Agarwal and Rao, 2000b). The antioxidant properties of carotenoids are due to their ability to quench singlet oxygen species. Since carotenoids contain several double bonds, they can undergo *cis–trans* isomerization. The *trans* form is considered to be more stable and is the most common form present in foods. The biological significance of the isomeric forms of carotenoids is not fully understood at the present time. Chemical structures of some common carotenoids are shown in Figure 1.

Lycopene, like other carotenoids, is a natural pigment synthesized by plants and microorganisms to absorb light during photosynthesis and to protect them against photosensitization. It is a noncyclic carotenoid having a molecular formula of $C_{40}H_{56}$ and a molecular weight of 536.85 daltons. It is a lipophylic compound that is insoluble in water. It is a red pigment absorbing light in the visible range and a petroleum ether solution of lycopene has λ_{max} of 472 nm and $\varepsilon^{\%}$ 3450 (Rao and Agarwal, 1999). It is an open chain hydrocarbon containing 11 conjugated and 2 nonconjugated double bonds arranged in a linear array. As with other carotenoids, the double bonds in lycopene can undergo isomerization from *trans* to mono or ply-*cis* isomers by light, thermal energy, and chemical reactions. All-*trans*,

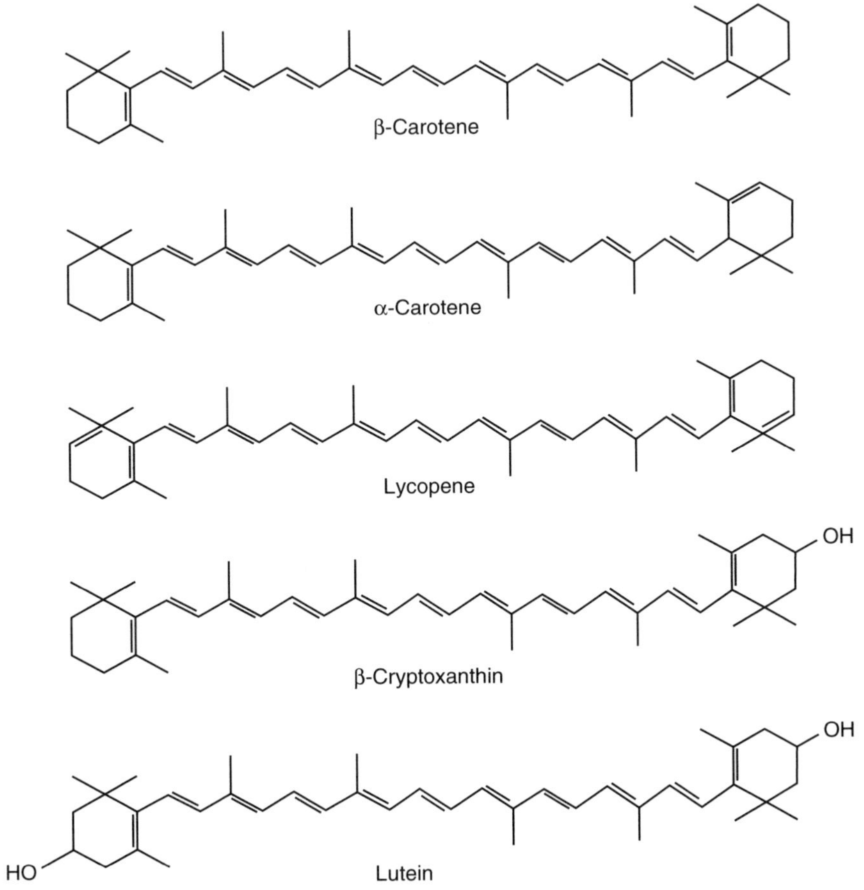

β-Carotene

α-Carotene

Lycopene

OH

β-Cryptoxanthin

OH

HO

Lutein

FIG. 1 Structures of major dietary carotenoids.

5-*cis*, 9-*cis*, 13-*cis*, and 15-*cis* are the most commonly identified isomeric forms of lycopene. Different isomeric forms of lycopene are shown in Figure 2.

Since lycopene lacks the β-ionic ring structure, unlike β-carotene, it lacks provitamin A activity. The biological activity of lycopene is thought to be primarily due to its antioxidant properties. However, other mechanisms, such as facilitating gap junction communication (GJC) (Aust *et al.*, 2003; Heber, 2002; Wertz *et al.*, 2004; Zhang *et al.*, 1991, 1992), stimulation of the immune system (Chew and Park, 2004; Heber, 2002; Heber and Lu, 2002; Kim *et al.*, 2004; Wertz *et al.*, 2004), endocrine-mediated pathways

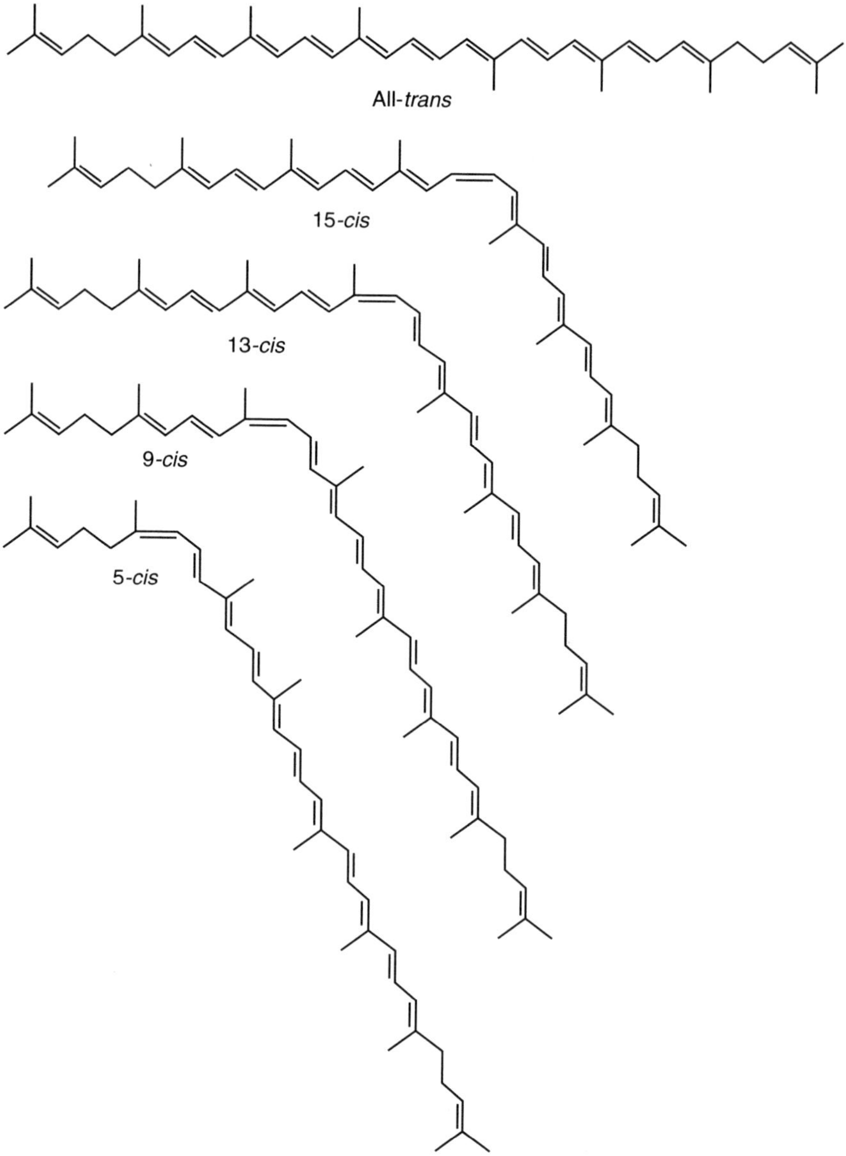

All-*trans*

15-*cis*

13-*cis*

9-*cis*

5-*cis*

FIG. 2 Structures of *trans* and *cis* isomeric forms of lycopene.

TABLE I

LYCOPENE CONTENT OF COMMON FRUITS AND VEGETABLES

Fruits and vegetables	Lycopene (μg/g wet weight)
Tomatoes	8.8–42.0
Watermelon	23.0–72.0
Pink guava	54.0
Pink grapefruit	33.6
Papaya	20.0–53.0
Apricot	<0.1

Source: Lycopene content of tomato products and their contribution to dietary lycopene. Reprinted from Food Research International. 1999; **31**, pp. 737–741 by permission of Elsevier.

TABLE II

LYCOPENE CONTENT OF COMMON TOMATO BASED FOODS

Tomato products	Lycopene (μg/g weight)
Fresh tomatoes	8.8–42.0
Cooked tomatoes	37.0
Tomato sauce	62.0
Tomato paste	54.0–1500.0
Tomato soup (condensed)	79.9
Tomato powder	1126.3–1264.9
Tomato juice	50.0–116.0
Pizza sauce	127.1
Ketchup	99.0–134.4

Source: Lycopene content of tomato products and their contribution to dietary lycopene. Reprinted from Food Research International. 1999; **31**, pp. 737–741 by permission of Elsevier.

(Heber, 2002; Heber and Lu, 2002; Wertz *et al.*, 2004), and cell cycle regulations, have also been demonstrated.

Although red-colored fruits and vegetables are the most common sources of dietary lycopene, not all red-colored plants contain lycopene. Common food sources of lycopene are the tomatoes, processed tomato products, watermelons, pink guava, pink grapefruits, papaya, and apricots. The lycopene content of these foods are shown in Table I.

Tomatoes and tomato-based products account for more than 85% of the dietary lycopene in North America. Lycopene content of some common tomato-based foods is shown in Table II.

IV. ANALYTICAL METHODS OF MEASURING LYCOPENE IN FOOD AND OTHER BIOLOGICAL MATERIALS

Spectrophotometric methods and high-pressure liquid chromatography (HPLC) are used most commonly in the quantitative estimations of total lycopene in food and biological samples. It is first extracted from the samples using various organic solvents. Typically, lycopene from tomato products is extracted with hexane:methanol:acetone (2:1:1) mixture containing 2.5% butylated hydroxytoluene (BHT). The optical density of the hexane extract is then measure spectrophotometrically at 502 nm against a hexane blank. Concentrations of lycopene are then calculated using the extinction coefficient ($E^{\%}$) of 3150 (Rao and Agarwal, 1999). Results are reported as parts per million (ppm) of lycopene or as μg per unit weight of the food product. Alternatively, the hexane extract is analyzed by HPLC using reverse-phase C18 column and an absorbance detector (Agarwal *et al.*, 2001; Rao *et al.*, 1999). Lycopene is quantified from the HPLC profile by using purified lycopene standard available from several commercial sources. Rao *et al.* (1999) compared the spectrophotometric and HPLC methods and found the results to be in good agreement. The spectrophotometric method offers a convenient, fast, and less expensive procedure for the detection of total lycopene compared to the HPLC procedure. A large number of samples can be processed by this method in a relatively short period of time without compromising the accuracy. For the detection of the *cis* isomeric forms of lycopene, the HPLC system with an absorbance or electrochemical detectors is used (Agarwal *et al.*, 2001; Clinton *et al.*, 1996; Ferruzzi *et al.*, 2001). Typically, food samples are homogenized and then extracted with the hexane: methanol:acetone (2:1:1) mixture containing 2.5% BHT. The extracts are then analyzed by reverse-phase HPLC using a C30 polymeric HPLC column. Peaks are eluted with methanol:methyltert-butyl ether (62:38) and monitored at 460 nm using an absorbance detector (Agarwal *et al.*, 2001). Lycopene content in the serum and plasma samples is estimated by extracting the lycopene with hexane:methyl chloride (5:1) containing 0.015% BHT and analyzed using a Vydac 201HS54 reverse-phase analytical HPLC column and an absorbance detector set at 460 nm. The mobile phase used is a mixture of acetonitrile, methanol, methyl chloride, and water (7:7:2:0.16). Lycopene peaks are identified and quantified with the use of external standards. Analysis of lycopene in tissue samples requires that the samples first undergo saponification by incubation in sodium hydroxide. Samples are then extracted and analyzed as before for the serum and plasma samples (Rao *et al.*, 1999).

TABLE III

EFFECT OF PROCESSING ON LYCOPENE CONTENT OF TOMATO JUICE

Stage of processing	Processing temperature (°C)	Lycopene content (ppm) Mean ± SEM
Raw tomatoes	–	124.5 ± 1.6[a]
Scalded pulp	76	105.6 ± 0.8[a]
Salting tank	–	107.9 ± 0.3[a]
Sterilization tank	120	102.6 ± 0.4[a]
Processed juice	–	102.3 ± 0.8[a]

[a]Numbers with different letters are statistically significant ($p < 0.05$).
Source: Lycopene content of tomato products: Its stability, bioavailability and *in vivo* antioxidant properties. Reprinted from Journal of Medicinal Food. 2001; **4**, pp. 9–15 by permission of Mary Ann Liebert, Inc., Publishers.

TABLE IV

EFFECT OF STORAGE ON LYCOPENE CONTENT OF TOMATO JUICE

Storage condition	Lycopene content (ppm) Mean ± SEM
Fresh juice	91.1 ± 0.8[a]
4°C, 12 months	89.5 ± 0.6[a]
25°C, 12 months	92.0 ± 0.7[a]
37°C, 12 months	92.6 ± 0.6[a]

[a]Numbers with different letters are statistically significant ($p < 0.05$).
Source: Lycopene content of tomato products: Its stability, bioavailability and *in vivo* antioxidant properties. Reprinted from Journal of Medicinal Food. 2001; **4**, pp. 9–15 by permission of Mary Ann Liebert, Inc., Publishers.

V. STABILITY AND ANTIOXIDANT PROPERTIES OF LYCOPENE AND ITS ISOMERS

Although lycopene is a fairly stable molecule, it can undergo oxidative, thermal, and photodegradation. Studies evaluating the thermal stability of lycopene have shown it to be stable under the conditions of industrial processing. Agarwal *et al.* (2001) studied the effect of processing temperatures and storage on lycopene stability and its isomerization. They showed that the lycopene content of tomatoes remained unchanged during the multistep processing operations for the production of juice or paste and remained stable for up to 12 months of storage at ambient temperatures (Tables III and IV).

TABLE V

EFFECT OF HEATING TOMATO JUICE IN THE PRESENCE OF LIPIDS ON LYCOPENE ISOMERS

	Corn oil	Olive oil	Butter
Lycopene (n = 3)	[mean ± SEM]	[mean ± SEM]	[mean ± SEM]
% *Trans*	83.82 ± 1.81^a	76.74 ± 0.3^b	84.53 ± 1.67^a
% *Cis*	16.18 ± 1.81^a	23.26 ± 0.3^b	15.47 ± 1.67^a
	Corn oil	Olive oil	Butter
Lycopene (n = 3)	[mM/L]	[mM/L]	[mM/L]
All-*trans*	11666.98^a	10681.51^b	11765.81^a
Cis	2266.65^a	3258.48^b	2167.19^a

^{a,b}Values are means ± SEM, n = 3. Values with different superscripts in a given row are significantly different, $P < 0.05$ (One-way ANOVA and Turkeys Test).

In the same study (Agarwal *et al.*, 2001), processed tomato products, such as tomato paste, ketchup, and juice, were shown to have a similar distribution of the *cis* and *trans* isomeric forms of lycopene. However, when the tomato juice was subjected to cooking temperatures in the presence of different oils, a noticeable increase in the formation of the *cis* isomers was observed in the presence of olive oil (Table V). Similar observations were also reported by Nguyen *et al.* (Nguyen and Schwartz, 1998). Conversion of the all-*trans* lycopene present in raw tomatoes to its *cis* isomeric form is of interest since the *cis* forms are generally considered to be more bioavailable. In another study (Chasse *et al.*, 2001), the stability of lycopene isomers were studied using an ab initio computational modeling. 5-*Cis* lycopene was found to be the most stable followed by all-*trans*. Lycopene is one of the most potent antioxidants, with a singlet-oxygen-quenching ability twice as high as that of β-carotene and ten times higher than that of α-tocopherol (Di Mascio *et al.*, 1989). In the computational model study (Chasse *et al.*, 2001), 5-*cis* lycopene had the highest antioxidant property as indicated by the ionization potential, followed by 9-*cis*. All-*trans* lycopene had the least antioxidant potential (Figure 3).

VI. BIOAVAILABILITY, TISSUE DISTRIBUTION, METABOLISM, AND SAFETY OF LYCOPENE

Lycopene levels in plasma and human tissues reflect dietary intake. In a study, when subjects consumed a tomato-free diet for 2 weeks, their lycopene levels dropped significantly (Rao and Agarwal, 1998a). Ingested lycopene is

Configurational stability of lycopene isomers
established at two levels of ab initio computation

5-*cis* > all-*trans* > 9-*cis* > 13-*cis* > 15-*cis* > 7-*cis* > 11-*cis*

Antioxidant properties of lycopene isomers
as indicated by their ionization potential

5-*cis* > 9-*cis* > 7-*cis* > 13-*cis* > 15-*cis* >11-*cis* > all-*trans*

FIG. 3 Stability and antioxidant properties of lycopene isomers (Chasse *et al.*, 2001).

incorporated into dietary lipid micelles and absorbed into the intestinal mucosal lining via passive diffusion. They are then incorporated into chylomicrons and released into the lymphatic system for transport to the liver. Lycopene is transported by the lipoproteins into the plasma for distribution to the different organs (Parker, 1996). Owing to the lipophilic nature of lycopene, it was found to concentrate in the LDL and very low density lipoprotein (VLDL) fractions and not the high density lipoprotein (HDL) fraction of the serum (Rao and Agarwal, 1998a; Stahl and Sies, 1996). A schematic diagram of lycopene absorption and transportation is shown in Figure 4.

In humans, lycopene absorption is in the range of 10–30% with the remainder being excreted. Many biological and lifestyle factors influence the absorption of dietary lycopene including age, gender, hormonal status, body mass and composition, blood lipid levels, smoking, alcohol, and the presence of other carotenoids in the food (Rao and Agarwal, 1999; Stahl and Sies, 1992). Fasting serum lycopene levels were found to be higher and more reproducible than postprandial levels indicating that diet induced metabolic stress (Rao and Agarwal, 1998b). Although the lycopene levels in blood do not differ significantly between men and women (Brady *et al.*, 1997; Olmedilla *et al.*, 1994), in women, they were found to be influenced by the phases of menstrual cycle with a peak during midluteal phase (Forman *et al.*, 1996). Inconsistent results are reported in the literature regarding the effect of smoking on serum lycopene levels (Brady *et al.*, 1997; Peng *et al.*, 1995; Ross *et al.*, 1996). A study showed no significant differences in serum lycopene levels between smokers and non-smokers (Rao and Agarwal, 1998a). However, the serum lycopene levels in the smokers fell by 40% immediately after smoking three cigarettes. Exposing fresh plasma under *in vitro* conditions to cigarette smoke resulted in the depletion of the lycopene and other lipid-soluble antioxidants

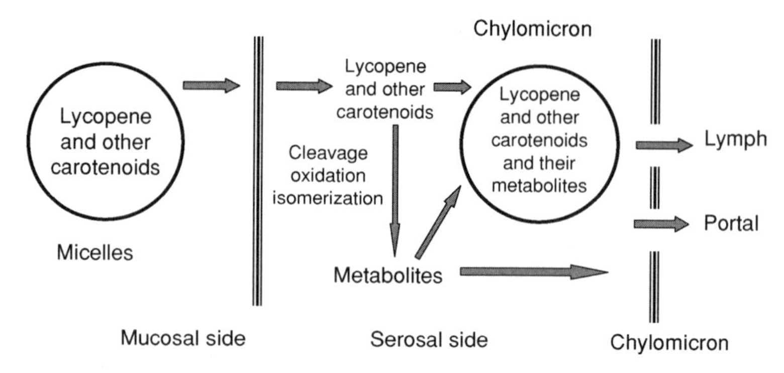

FIG. 4 Absorption and translocation of lycopene.

(Handelman *et al.*, 1996). Alcohol consumption was also shown to alter serum lycopene levels (Brady *et al.*, 1997). Other factors that influence the bioavailability of lycopene are its release from the food matrix due to processing, presence of dietary lipids, and heat-induced isomerization from the all-*trans* to *cis* conformation. They all enhance lycopene absorption into the body. Ingestion of cooked tomato juice in oil medium increased serum lycopene levels threefold whereas consumption of an equal amount of unprocessed juice did not have any effect (Stahl and Sies, 1992).

In another study, lycopene from tomato paste was shown to be 3.8 times more bioavailable than that from fresh tomato (Gärtner *et al.*, 1997). It is generally believed that conversion of all-*trans* lycopene to its *cis* isomeric form enhances its absorption (Stahl and Sies, 1992). Presence of β-carotene was shown to increase the absorption of lycopene in some studies, while the presence of canthaxanthin appears to decrease lycopene absorption (Blakely *et al.*, 1994; Gaziano *et al.*, 1995; Wahlqvist *et al.*, 1994). In a recent study (Rao and Shen, 2002), low levels of lycopene (5, 10, and 20 mg) were provided in the form of either ketchup or oleoresin capsules for 2 weeks to healthy human subjects. Serum lycopene levels at the beginning and end of the 2 weeks of treatment were compared. A significant increase in the serum lycopene levels was observed for both ketchup and capsules at all three levels of intake. Although the serum lycopene levels increased in a dose-dependent fashion with dietary intake, biomarkers of lipid and protein oxidation did not differ significantly between the treatments. The absorption efficiency of lycopene was observed to be greater at the lower levels of dietary intake.

Lycopene is the most predominant carotenoid in human plasma with a half-life of about 2–3 days (Stahl and Sies, 1996). Although the most prominent geometric isomers of lycopene in plants are the all-*trans*, in human plasma, lycopene is present as an isomeric mixture containing 50% of the

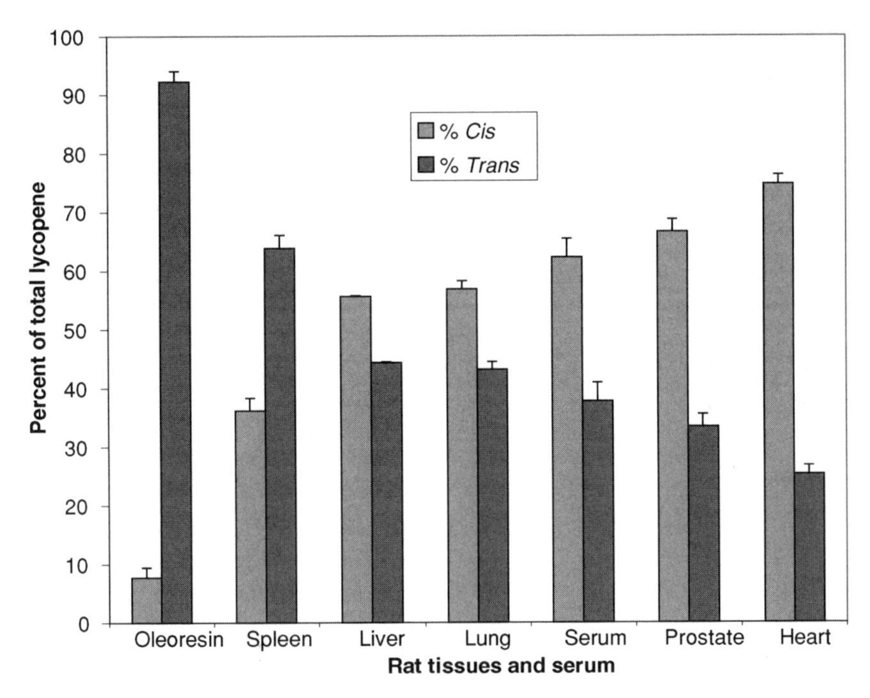

FIG. 5 *Cis* and *trans* lycopene isomers in rat serum and tissues. Values expressed are mean ± SEM, n = 3.

total lycopene as *cis* isomer (Rao and Agarwal, 2000). When animals were fed lycopene containing predominantly the all-*trans* isomeric form, serum and tissue lycopene showed the presence of *cis* lycopene (Jain *et al.*, 1999). Similar results were also observed in human serum (Figure 5).

There are also some indications of *in vivo trans* to *cis* isomerization reactions. Very little is known about the *in vivo* metabolism of lycopene. In a recent study, dos Anjos Ferreira *et al.* (2004) used the postmitochondrial fraction of rat intestinal mucosa to study lycopene metabolism. They identified two types of metabolic products of lycopene, cleavage products and oxidation products. Identified among the cleavage products were: 3-keto-apo-13-lycopenone and 3,4-dehydro-5,6-dihydro-15-apo-lycopenal. The oxidation products included: 2-ene-5,8-lycopenal-furanoxide, lycopene-5,6.5′,6′-diepoxide, lycopene-5,8-furanoxide isomer (I), lycopene-5,8-epoxide isomer (II), and 3-keto-lycopene-5′,8′-furanoxide. It is possible that similar metabolites of lycopene are also present *in vivo* in the presence of lipoxygenase enzymes. Nagao (2004) on the other hand showed *in vitro* cleavage of lycopene to acycloretinal, acycloretinoic acid, and apolycopenals in a nonenzymatic manner. They also showed that these cleaved products of lycopene induced apoptosis of HL-60 human promyelocytic leukemia cells. Only a few metabolites, such as 5,6-dihydroxy-5,6-dihydro lycopene, have been detected in human plasma (Khachik *et al.*, 1995, 1997, 2002). It is suggested that lycopene may undergo *in vivo* oxidation to form epoxides, which then may be converted to the polar 5,6-dihydoxy-5,6 dihydro-lycopene through metabolic reduction. A controversy exists as to the role of primary lycopene or its polar metabolites that are the biologically active forms. Further research is needed to better understand this aspect of lycopene.

Following the absorption of lycopene, it is transported to various organs and accumulates in tissues. Tissue distribution of lycopene varies considerably suggesting unique biological effects on some tissues and not on others. The highest concentrations of lycopene are in the testes, adrenal glands, liver, and prostate. Table VI shows the lycopene levels in human tissues (Rao and Agarwal, 1999).

Lycopene and its oxidation products are present in human milk and other body fluids (Khachik *et al.*, 1997). Human seminal plasma also contains lycopene and its levels were lower in immunoinfertile men compared to normal individuals (Palan and Naz, 1996). Although the plasma or serum levels of lycopene are used commonly to assess its bioavailability, adipose tissue has been suggested as a better tissue for the assessment of body lycopene status (Kohlmeier *et al.*, 1997).

Historically, lycopene-containing fruits and vegetables have been consumed by humans without any safety problems. Several studies were undertaken to evaluate the safety of both natural and synthetic lycopene (Jonker *et al.*, 2003; Matulka *et al.*, 2004; Mellert *et al.*, 2002). In our studies we evaluated intake levels of lycopene, ingested in the form of tomato juice, tomato sauce, and nutritional supplement from 5 to 75 mg/day (Rao and Agarwal, 1998a) in healthy human subjects. No adverse effects of consuming lycopene were observed in these studies. In another study (Mellert *et al.*, 2002) two synthetic crystalline lycopene sources, BASF lycopene 10 CWD and Lyco Vit 10%, each containing approximately 10% lycopene were tested in rats. After ingesting the test products for 13 weeks at intake levels of up to 3000 mg/kg body weight/day, no adverse effects were observed.

Lycopene derived from a fungal biomass of *Blakeslea trispos*, suspended in sunflower oil at a concentration of 20% w/w, was tested for subchronic toxicity at concentrations of 0%, 0.25%, 0.50%, and 1.0% in rats for 90 days (Jonker *et al.*, 2003). No evidence of toxicity of lycopene at dietary intake levels up to 1.0% was observed in this study. The authors suggest the no-observed-effect level (NOEL) for this lycopene to be 1.0% in the diet, the highest dietary concentration tested. McClain and Bausch (2003) published

TABLE VI
LYCOPENE LEVELS IN HUMAN TISSUES

Tissue	Lycopene (nmol/g wet weight)
Adrenal	1.90–21.60
Breast	0.78
Colon	0.31
Kidney	0.15–0.62
Liver	1.28–5.72
Lung	0.22–0.57
Ovary	0.3
Pancreas	0.7
Prostate	0.8
Skin	0.42
Stomach	0.2
Testes	4.34–21.36

a summary of additional safety studies using synthetic lycopene. No teratogenic effects were observed in the two-generation rat study. Deposition of lycopene in plasma, liver, and other tissues also had no adverse effects. The red coloration of skin associated with the intake of high levels of lycopene disappeared after 13 weeks showing the reversibility of this effect. A condition identified as lycopenemia was reported when tomato juice was ingested in excess for prolonged periods of time resulting in increased serum lycopene levels and coloration of skin (Reich *et al.*, 1960). No other adverse effects were associated with the skin coloration. In consideration of the studies establishing the safety of lycopene for human consumption, the Food and Drug Administration in the United States of America granted generally recognized as safe (GRAS) status to lycopene as a nutritional supplement.

VII. MECHANISMS OF ACTION OF LYCOPENE

The biological activities of carotenoids, such as β-carotene, are related to their provitamin A activity within the body (Clinton, 1998). Since lycopene lacks the β-ionic ring structure, it does not have any provitamin A activity (Stahl and Sies, 1996). The biological effects of lycopene in humans have therefore been attributed to mechanisms other than vitamin A. Two major hypotheses have been proposed to explain the anticarcinogenic and anti-atherogenic activities of lycopene: oxidative and nonoxidative mechanisms. The proposed mechanisms for the role of lycopene in the prevention of chronic diseases are summarized in Figure 6.

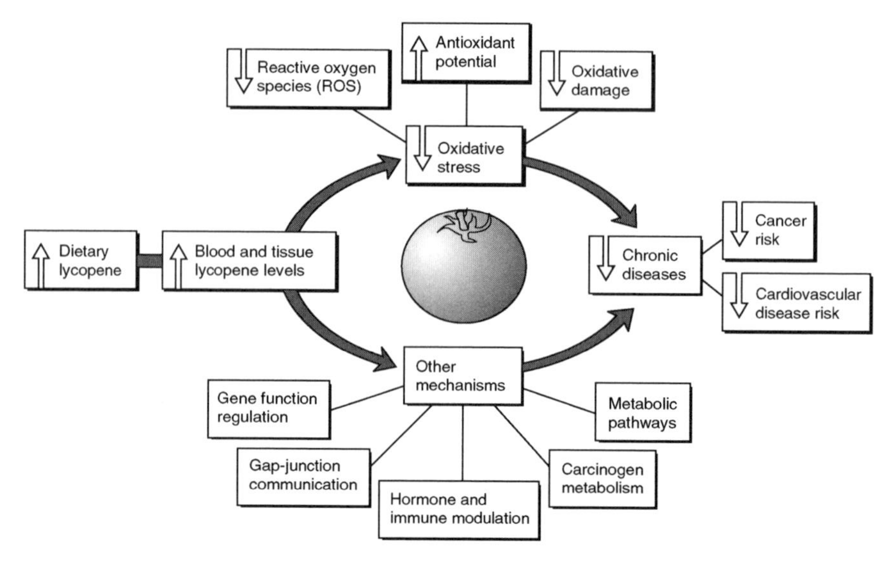

FIG. 6 Proposed mechanisms for the role of lycopene in chronic diseases. (Tomato lycopene and its role in human health and chronic diseases. Reprinted from CMAJ. 2000; **163**(6), pp. 739–744 by permission of 2000 CMA Media Inc.)

Heber (Heber and Lu, 2002) and Wertz *et al.* (2004) provided an overview of the mechanisms of action of lycopene. The antioxidant properties of lycopene constitute the major focus of research with regards to its biological effects. Dietary intake of lycopene has been shown to increase circulatory and tissue levels of lycopene. Acting as an antioxidant, it can trap ROS and reduce oxidative stress and damage to cellular components including lipids, proteins, and DNA (Agarwal and Rao, 2000a). Since oxidative damage of lipids, proteins, and DNA have all been implicated in the development of chronic diseases, such as cardiovascular diseases, cancer, and osteoporosis, lycopene acting as a potent antioxidant can reduce the risk of these diseases.

Included among the nonoxidative mechanisms are: inhibition of insulin-like growth factor-I (IGF-I) signaling, interleukin-6 (IL-6) expression, androgen signaling, improving GJC, induction of phase II drug-metabolizing enzymes and oxidative defense genes, and improving the immune response (Agarwal and Rao, 2000a; Wertz *et al.*, 2004). GJC between cells is thought to be one of the protective mechanisms related to cancer prevention. Many human tumors are deficient in GJC and its restoration or upregulation is associated with decreased proliferation of tumor cells. The anticarcinogenic effects of lycopene may be due to regulation of GJC as shown in mouse embryo fibroblast cells (Zhang *et al.*, 1991, 1992). Aust *et al.* (2003) reported that the *in vitro* oxidation product of lycopene, 2,7,11-trimethyl-tetradeca-hexaene-1,14-dial, stimulated GJC in rat liver epithelial WB-F344 cells. Studies using human and animal cells have identified the expression of the connexin 43 geneas being upregulated by lycopene allowing for a direct intercellular GJC (Heber and Lu, 2002). Suppression of the carcinogen-induced phosphorylation of the regulatory proteins, such as p53 and R_b antioncogenes by lycopene, may also play an important role stopping cell division at the G-G_1 cell cycle phase (Matsushima-Nishiwaki *et al.*, 1995). Astrog *et al.* (1997) emphasize the lycopene-induced modulation of the liver-metabolizing enzymes as the underlying mechanism of protection against carcinogen-induced preneoplastic lesions in the rat liver. There is evidence to suggest that lycopene reduces cellular proliferation induced by insulin-like growth factors, which are potent mitogens, in various cancer cell lines (Pincemail, 1995). T cell differentiation (immunomodulation) was suggested to be the mechanism for the suppression of mammary tumor growth by lycopene treatments in SHN-retired mice (Kobayashi *et al.*, 1996; Nagasawa *et al.*, 1995).

In a study, Chew and Park (2004) showed that lycopene and other nonprovitamin A carotenoids induce immunoenhancement in animals. Bone marrow-derived dendritic cells are the most potent of the antigen-presenting cells. They initiate the immune response by presenting antigens to naïve T lymphocytes. Kim *et al.* (2004) using these dendritic cells showed that lycopene can significantly increase the phenotypic and functional maturation of these cells, especially in lipopolysaccharide-induced dendritic cell maturation. The authors suggest therapeutic application for lycopene via the manipulation of the dendritic cells. Lycopene also interacts synergistically with 1,25-dihydroxyvitamin D3 [1,25$(OH)_2D_3$] and leutin to regulate cell cycle progression, suggesting some interactions at a nuclear or subcellular level and specific positioning of different carotenoids in cell membranes (Amir *et al.*, 1999). Lycopene also acts as a hypocholesterolemic agent by inhibiting HMG-CoA (3-hydroxy-3-methylglutaryl-coenzyme A) reductase that may be related to reducing the risk of cardiovascular diseases (Fuhramn *et al.*, 1997).

VIII. LYCOPENE AND HUMAN DISEASES

A. LYCOPENE AND THE PREVENTION OF CHRONIC DISEASES: THE HYPOTHESIS

The underlying hypothesis of oxidative stress, antioxidants, and chronic diseases is illustrated in Figure 7.

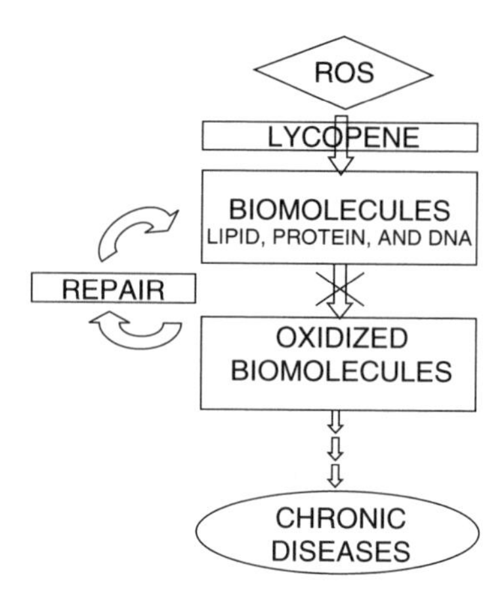

FIG. 7 Lycopene and the prevention of chronic diseases: The hypothesis. (Lycopene and the prevention of chronic diseases. Reprinted from Major findings from five international conferences. 2002. A.V. Rao, D. Heber, eds., p. 4. By permission of Caledonian Science Press.)

B. CANCER

1. Epidemiological evidence

Various epidemiological studies have suggested that a diet rich in a variety of fruits and vegetables results in a lower risk of cancer and other chronic diseases (Giovannucci, 1999). An early epidemiological study on elderly Americans indicated that high intake of tomatoes was associated with a significant reduction in mortality from cancers of all sites (Colditz *et al.*, 1985; Franceschi *et al.*, 1994). Since then several other articles have appeared in recent years reviewing the epidemiological data pertaining to the relationship between the intake of tomatoes, tomato products, and lycopene and the prevention of cancer. The Mediterranean diet, rich in tomatoes and tomato products, maybe responsible for the lower cancer incidence in that region (LaVecchia, 1997). In other epidemiological studies, serum and tissue levels of lycopene were inversely associated with the risk of prostate cancer, breast, cervical, and ovarian cancers, gastrointestinal tract cancers including stomach, colon, and the rectum, and lung cancer (Giovannucci, 1999; Giovannucci *et al.*, 1995, 2002). A review (Giovannucci, 1999) reported on the epidemiological studies including estimation of dietary intake of tomatoes and lycopene and the circulatory levels of lycopene in relation to the risk of cancers of various sites. The results were found to be consistent for a variety of cancers across numerous diverse populations and with the use of several different study designs. None of the studies analyzed reported any adverse effects of consuming high levels of tomatoes or lycopene. Although a majority of the epidemiological studies reported inverse relationship between the consumption of tomatoes, lycopene and circulating levels of lycopene and the risk of cancers, other studies (Giovannucci, 1999) found no protective effect of serum and dietary lycopene on cancer risk.

2. *Molecular markers of cancer*

With increasing interest in undertaking animal experimental and human clinical and intervention studies to evaluate the role of lycopene in cancer prevention, it is important that well-established molecular and clinical markers of cancer be used in these studies. In general, the main clinical end points used in animal and human experiments are the tumor burden and volume and survival rates. Now that our understanding of cancer pathology has advanced, several molecular events are beginning to be recognized and used in research to evaluate the outcomes from intervention studies. A brief overview of some of the more important molecular markers of cancer that can and should be used in future studies is presented in this section.

Clinical evaluation of tumors by the conventional tumor lymph node metastases (TNM) staging system is used to determine prognosis and choice of treatment. This system takes into account size of the primary tumor (T), lymph node involvement (N), and occurrence of distant metastases (M). Tumor grading is a measure of cellular differentiation and is also used as a prognostic indicator. Significant effort to identify novel markers for clinical assessment or as a research tool has given rise to several putative molecular markers for diagnosis, prognosis, and response to therapy. In fact, over 85 markers for prostate cancer alone have been reported in the literature, although the clinical value of the majority of these markers has not yet been determined (Tricoli *et al.*, 2004).

Broad-spectrum molecular markers include those involved in proliferation, cell cycle, apoptosis, and vascularization. Several of these are well-established markers that have varying degrees of clinical value. One of the most reliable markers for cellular proliferation is Ki-67, and increased expression is associated with poor prognosis for prostate, breast, lung, and bladder cancer (Claudio *et al.*, 2002; Esteva *et al.*, 2004; Martin *et al.*, 2004; Santos *et al.*, 2003). In breast cancer, Ki-67 expression correlates well with the proliferating cell nuclear antigen (PCNA), which is also a marker for proliferation (Keshgegian and Cnaan, 1995). Both proteins are nuclear antigens that are expressed at high levels in rapidly dividing cells. However, the clinical value of Ki-67 and PCNA is inconsistent, and their correlation with prognosis varies with cancer type.

Proteins related to cell cycle and apoptosis, such as p53, survivin, cyclin D1, and cyclin E, have also been associated with malignancy and may serve as molecular markers of disease. The p53 tumor suppressor protein is involved in cell cycle checkpoint control and prevents cells from progressing through the cell cycle in the event of DNA damage. Although immunohistochemical analysis shows increased p53 expression in prostate cancer, inactivating mutations are associated with disease progression (Moul, 1999; Theodorescu *et al.*, 1997). Similarly, p53-inactivating mutations are associated with poor prognosis in breast and lung cancer (Powell *et al.*, 2000; Steels *et al.*, 2001). On the other hand, increased expression of survivin, a member of the inhibitor of apoptosis (IAP) family of proteins, has been observed in multiple tumor types (reviewed in Altieri, 2003). Clinical studies show that survivin expression is associated with poor prognosis for breast cancer (Span *et al.*, 2004) and is associated with more aggressive forms of prostate and liver cancer (Morinaga *et al.*, 2004; Shariat *et al.*, 2004).

More recently, within this group of cell cycle regulatory proteins cyclin D1 and cyclin E, which promote the G1/S transition, have emerged as useful molecular markers for clinical cancer research. Overexpression of cyclin D1 is correlated with malignancy, metastasis, and progression for many types of cancer (reviewed in Fu *et al.*, 2004), while high levels of cyclin E are linked to poor prognosis of ovarian, breast, and lung cancer (Farley *et al.*, 2003; Lindahl *et al.*, 2004; Muller-Tidow *et al.*, 2001). Although the cyclins serve as molecular markers, therapeutic strategies that target cell cycle control have focused on inhibition of cyclin-dependent kinases (CDKs), which are the effector molecules associated with cyclins (Shapiro, 2004). However, strategies that decrease cyclin D1 or cyclin E expression or increase stability of CDK inhibitors, such as $p27^{kip1}$ and $p21^{cip1}$, are also of value (Swanton, 2004).

Tumor vascularization is an unfavorable predictor of metastasis and survival. Vascular endothelial growth factor (VEGF), which promotes neovascularization, is indicative of angiogenesis for several types of cancer (Callagy *et al.*, 2000; Gaffney *et al.*, 2003; Paley *et al.*, 1997; Strohmeyer *et al.*, 2000). A study demonstrated that increased VEGF expression correlated with increased microvesicle density in prostate tumor biopsies and that VEGF expression also correlated with disease progression (Strohmeyer *et al.*, 2004). Vascularization of tumors increases chances of metastasis by connecting tumors with the circulatory system. Consequently, brain metastases of breast cancer show increased VEGF expression in xenograft tumor models (Kim *et al.*, 2004).

Prostate-specific antigen (PSA) is one of the most reliable cancer-specific molecular markers. PSA expression is, for the most part, restricted to the prostate and elevated serum PSA indicates the possibility of malignant prostate cancer. Use of serum PSA as a diagnostic tool is still controversial since PSA levels are also elevated in benign prostatic hyperplasia (BPH), which is a nonmalignant growth of the prostate gland (Gittes, 1991). Nevertheless, serum PSA levels are used effectively to monitor prostate cancer recurrence and progression following androgen withdrawal therapy (Sadar *et al.*, 1999). The prostate-specific membrane antigen (PSMA) has emerged as a reliable marker for prostate cancer diagnosis, while prostate stem cell antigen (PSCA) has provided a prostate-specific therapeutic target (reviewed in Tricoli *et al.*, 2004).

Molecular markers for breast cancer include osteopontin (OPN), estrogen and progesterone receptors (ER and PR, respectively), and Her2/neu. Although the role of OPN in breast cancer is not fully understood, convincing clinical evidence demonstrates that increased OPN expression in breast carcinoma cells is associated with malignancy and that plasma OPN is elevated in women with breast cancer metastasis (Singhal *et al.*, 1997; Tuck and Chambers, 2001; Tuck *et al.*, 1998). OPN is a secreted glycoprotein that is particularly abundant in bone and its binding to cell-surface integrins results in integrin-mediated signaling, which may be important for adhesion and migration of cancer cells to secondary sites (Allan *et al.*, 2005).

The ER, PR, and Her2/neu status of breast cancer are taken into consideration when choosing treatment strategies. Tumors that express ER and PR are more likely to respond to hormone therapy (Early Breast Cancer Trialists' Collaborative Group, 1992; Bardou *et al.*, 2003; McGuire, 1978), while overexpression of Her2/neu may indicate increased sensitivity to the therapeutic anti-Her2/neu antibody trastuzumab (Herceptin) (Baselga *et al.*, 1996; Pegram *et al.*, 1998). Isoforms of ER (ER-α and ER-β) and PR (PR-A and PR-B) have been identified, and expression of receptor isoforms must be taken into account when choosing therapeutic strategies (Fuqua and Cui, 2004). Her2/neu overexpression in breast cancer correlates with higher histological tumor grade, more aggressive disease, and poor prognosis (Menard *et al.*, 2001; Slamon *et al.*, 1987), possibly by increasing cell proliferation, invasiveness, and/or tumor angiogenesis (Ignatoski *et al.*, 2000; Petit *et al.*, 1997).

The growing number of putative molecular markers of cancer suggests that a single marker is not likely to be sufficient for predicting either disease outcome or response to treatment. Consequently, high-throughput genomic and proteomic approaches that have the capacity to assess expression of multiple biomarkers at the same time are becoming increasingly important in determining prognostic signatures of disease.

In a key study by van't Veer *et al.* (2002), DNA micro array analysis of 117 primary breast cancer tumors was used to identify a "poor prognosis" gene expression signature, consisting of 70 genes, that was highly prognostic for development of metastases and for overall survival. Minn *et al.* (2005) demonstrated that this poor prognosis genetic signature was unable to predict organ-specific metastatic potential. However, an experimental metastasis xenograft model was used to identify a bone metastasis gene signature that had clinical relevance when applied retrospectively to genetic profiles of metastatic primary human tumors (Kang *et al.*, 2003; Minn *et al.*, 2005).

Although still in its infancy, clinical cancer proteomics also aims to identify molecular markers for detection, prognosis, and response to treatment, as well as novel therapeutic targets. Both matrix-assisted laser desorption ionization time-of-flight mass spectrometry (MALDI-TOF MS) (Celis *et al.*, 2004; Le Naour *et al.*, 2001; Yanagisawa *et al.*, 2003) and surface-enhanced laser desorption/ionization-mass spectrometry (SELDI-MS)–based protein chip arrays (Carter *et al.*, 2002; Liu *et al.*, 2005; Shiwa *et al.*, 2003; Wong *et al.*, 2004) have been used to identify cancer biomarkers and to generate baseline protein expression profiles of malignant tumors.

The complexity of the signaling events in cancer progression and the multidimensional nature of the disease suggest that molecular portraits of tumors are likely to be more relevant than individual biomarkers as diagnostic and prognostic indicators. Ultimately, these molecular profiles will allow not only a more precise evaluation of dietary intervention strategies, such as lycopene, but also diagnostic and prognostic assessment. It will also allow clinicians to monitor patients' responses to therapy and to optimize treatment strategies.

3. *Prostate cancer*

Of all the cancers, the role of lycopene in the prevention of prostate cancer has been studied the most. The supporting evidence for lycopene comes from tissue culture, animal, epidemiological, and human experimental studies and was reviewed by Rao and Agarwal (1999). Lycopene, β-carotene, canthaxanthin, and retinoic acid were all shown to inhibit the growth of the DU145 prostate cancer cells (Hall, 1996). Kotake-Nasra *et al.* (2001) evaluated the effects of several carotenoids including lycopene on the viability of three human prostate cancer cell lines, PC-3, DU145, and LNCaP. The viability of all three cell lines was significantly reduced by lycopene and other carotenoids present in foods. Kim *et al.* (2002) measured the effect of lycopene on the proliferation of LNCaP human prostate cancer cells in culture. A new, water-dispersible lycopene was used in this study at concentrations of 10^{-6}, 10^{-5}, and 10^{-4} M. Lycopene at concentrations of 10^{-6} and 10^{-5} M significantly reduced the growth of LNCaP cells after 48, 72, and 96 hours of incubation by 24.5–42.8%. In a follow-up study, the authors expanded the concentration range of lycopene tested to 10^{-9} and 10^{-7} M. A dose-dependent decrease in cell growth was observed. The authors suggested that lycopene as an antioxidant may play an important role in treating prostate cancer. In a review article (Heber and Lu, 2002) it was pointed out that lycopene, at physiological concentrations, can inhibit cancer cell growth by interfering with growth factor receptor signaling and cell cycle progression, specifically in prostate cancer cells, without any evidence of toxic effects or apoptosis of cells. Overall, *in vitro* tissue culture studies suggest that lycopene at physiological concentrations can reduce the growth of both androgen-dependent and androgen-independent prostate cancer cells. In an *in vivo* study, when men with localized prostate adenocarcinoma consumed tomato sauce-based pasta dishes providing 30 mg of lycopene per day for 3 weeks, the cells from prostate biopsies at the baseline and post-intervention resected tissues showed significant reduction in DNA damage (Bowen *et al.*, 2002).

Animal models have provided excellent systems to investigate *in vivo* biochemical consequences of administering lycopene in a well-defined, controlled environment, where the confounding variables could be kept to a minimum. Using laboratory mice, the radioprotective as well as antibacterial activities of lycopene were established almost 40 years ago (Forssberg *et al.*, 1959; Lingen *et al.*, 1959). Guttenplan *et al.* (2001) studied the effect of ingesting lycopene-rich tomato oleoresin at two doses on the *in vivo* mutagenesis in prostate cells of lacZ mice. Both short-term benzopyrene (BaP)-induced and long-term spontaneous mutagenesis was monitored. A nonsignificant inhibition of spontaneous mutagenesis in the prostate was observed only at the higher dose. However, lycopene was shown to inhibit mutagenesis slightly by the oleoresin. In another study (Boileau *et al.*, 2003), the effect of whole tomato powder (13 mg lycopene per kg diet), lycopene beadlets (161 mg lycopene per kg diet), and control beadlets (0 mg lycopene per kg diet) were evaluated for their effect on prostate cancer in a rat model. The results showed that the consumption of tomato powder, but not lycopene, inhibited prostate carcinogenesis. Based on these observations, the authors suggested that tomato products may contain compounds in addition to lycopene that modify prostate carcinogenesis. However, Limpens *et al.* (2004) pointed out that the levels of lycopene used in the form of beadlets may have been too high. In support of their argument, they tested the effect of two doses of synthetic lycopene (5 and 50 mg per kg body weight) in an orthotopic model of human prostate cancer cell line PC-346C in nude mice. Lycopene inhibited tumor growth and decreased PSA levels. This effect reached statistical significance only in the low-dose lycopene group suggesting that lycopene dosages might be relevant to their effects. Two other studies also reported on the effect of lycopene in prostate cancer. One study was conducted in mice using the 12T-10 Lady transgenic model, which spontaneously developed localized prostatic adenocarcinoma and neuroendocrine cancer followed by metastases that resembled the pathogenesis in humans (Venkateswaran *et al.*, 2004). Administration of lycopene, vitamin E, and selenium in combination was shown to dramatically inhibit prostate cancer development and increase the disease-free survival time. A strong correlation between the disease-free state and increased levels of the prognostic marker $p27^{Kip1}$ and a marked decrease in PCNA expression was observed giving some clues as to the mechanisms of action of the chemopreventive agents. In another study (Siler *et al.*, 2004), lycopene (200 ppm) alone, vitamin E (540 ppm) alone, or both of them in combination were tested in the MatLyLu Dunning prostate cancer model. They were used to supplement the diets for 4 weeks. Both lycopene and vitamin E accumulated in the tumor tissue and increased the necrotic area in the tumor. Lycopene interfered with local testosterone activation by downregulating 5-α-reductase and reducing the expression of steroid target genes. In addition, lycopene downregulated prostatic IGF-I and IL-6 expression. Based on these observations, the authors suggest that lycopene and vitamin E contribute to the reduction of prostate cancer by interfering with internal autocrine or paracrine loops of sex steroid hormones and growth factor activation and/or synthesis and signaling in the prostate. In contrast to these studies, lycopene had no effect on carcinogeneses with male F344 rats treated with 3,2′-dimethyl-4-aminobiphenol (DMAB) and 2-amino-1-methylimidazo[4,5-b]pyridine (PhIP) to induce prostate cancer (Imaida *et al.*, 2001).

Several epidemiological studies reported the inverse association between dietary lycopene intake and prostate cancer. In one of the earliest prospective trials (Mills *et al.*, 1989), a cohort of Seventh-Day Adventist men consuming high levels of tomato products, more than five times per week, had significantly decreased prostate cancer risk compared to the men consuming lower intakes of tomato products less than one time per week. The largest published Health Professionals Follow-up Study (Giovannucci *et al.*, 1995, 2002) reported an inverse relationship between the consumption of various tomato products and prostate cancer incidence. A 35% reduction in the risk of prostate cancer was observed for a consumption frequency of 10 or more servings of tomato products per week. The protective effect was stronger with more advanced or aggressive stages of prostate cancer. Of all the foods evaluated, tomato sauce was the strongest dietary predictor of reduced prostate cancer risk and serum lycopene levels. In another large nested case-control study, the Physicians Health Study (Gann *et al.*, 1999), prediagnostic plasma lycopene levels between confirmed cancer cases and controls were compared. A 41% reduction in the overall prostate cancer risk and a 60% reduction in patients with aggressive prostate cancer was observed in men with the highest plasma lycopene. All lycopene-supplemented groups together had an overall 37% reduction in risk. In this study, lycopene was the only carotenoid that significantly and consistently reduced the risk of prostate cancer. Another nested case-control study (Hsing *et al.*, 1990) compared serum lycopene levels from men who developed prostate cancer with those of a matched group of control subjects. Results showed a modest 6.2% lower mean serum lycopene levels in subjects developing prostate cancer compared to the control group. Overall, of the several epidemiological studies, close to 60% of the studies observed a significant association between lycopene intake or plasma levels and the risk for prostate cancer. Comparing different studies is made difficult by the use of different methods for estimating the intake of lycopene, use of different databases, differentiating between the consumption of lycopene itself from the intake of lycopene-containing fruits and vegetables, and differences in the bioavailability of lycopene from different dietary sources.

There are only a few randomized, controlled clinical trials reported in the literature investigating the role of lycopene in prostate cancer. A recent case-control study assessed the status of oxidative stress and antioxidants in prostate cancer patients and compared them against age-matched control subjects (Rao *et al.*, 1999). Their dietary history was recorded in order to estimate the intake of lycopene. Fasting blood samples and biopsy tissue samples were obtained for analyses. Serum and tissue lycopene levels in prostate cancer patients were significantly lower by 44% than in the control subjects. None of the other carotenoids and vitamins A and E showed any differences (Figure 8). Analysis of the serum biomarkers of oxidation showed a significantly higher level of protein oxidation in the cancer patients compared to the control subjects. Significantly higher levels of serum PSA and lower levels of protein thiols were observed in the prostate cancer patients compared to their age-matched controls (Figure 9). Based on the observation that only lycopene levels were significantly lower in the prostate cancer patients, authors of the study suggested that lycopene is used preferentially as the dietary antioxidant. In a follow-up pharmacokinetic study (Rao, 2002a), use of lycopene by prostate cancer patients was investigated and compared with their age-matched controls. Following the collection of a fasting blood sample, 50 mg of lycopene in the form of tomato juice as the source of lycopene was given as a single dose. Hourly blood samples were then collected and analyzed for serum lycopene levels. Results are shown in Figure 10. It was concluded that the utilization rates of lycopene by both groups were similar. However, absorption of lycopene by prostate cancer patients seemed lower than the controls. The reasons for this observation are not fully understood. Two new studies addressed the role of lycopene in treating prostate cancer. In one trial (Kucuk *et al.*, 2001) newly diagnosed prostate cancer patients were randomly assigned to receive 15 mg lycopene or no supplement for a period of 3 weeks before radical prostatectomy. Seventy-three percent of the men in the intervention group receiving

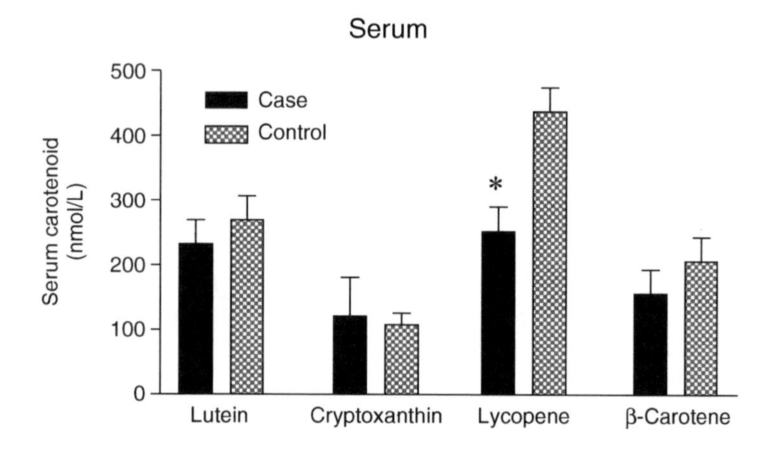

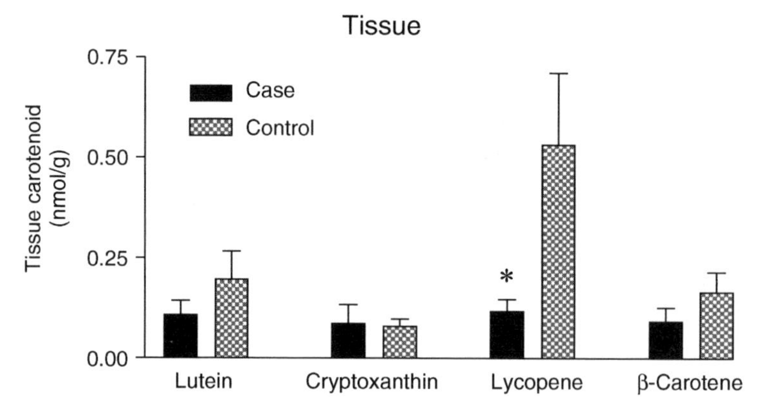

FIG. 8 Serum and tissue carotenoid levels in prostate cancer cases and controls (*$p < 0.05$). (Lycopene, tomatoes and health: New Perspectives 2000. Reprinted from Lycopene and the prevention of chronic diseases. Major findings from five international conferences. 2002. A.V. Rao, D. Heber, eds., p. 22. By permission of Caledonian Science Press.)

lycopene compared to 18% in the control group showed no involvement of surgical margins and/or extraprostate tissue with cancer. A statistically significant lower percentage of diffuse involvement of the prostate by high-grade prostatic intraepithelial neoplasms was also observed in the lycopene-treated group compared to the controls. Prostatic lycopene levels were significantly higher in the intervention group with a modest decrease in the serum PSA levels. A second study by the same authors (Kucuk *et al.*, 2002) also confirmed the previous observation that lycopene supplements reduced the growth of prostate cancer. These observations raise the possibility that lycopene may be used to treat patients with established prostate cancer.

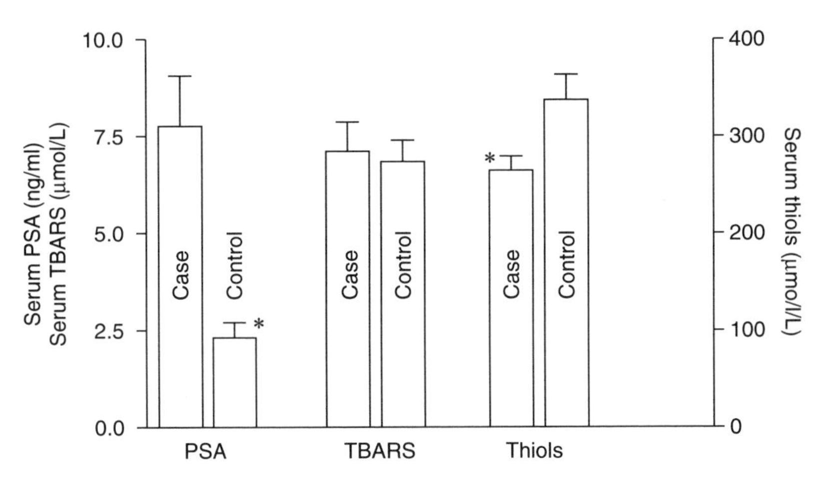

FIG. 9 Serum PSA, thiobarbituric acid substances (TBARS), and thiols in prostate cancer cases and controls (*$p < 0.05$). (Lycopene, tomatoes and health: New Perspectives 2000. Reprinted from Lycopene and the prevention of chronic diseases. Major findings from five international conferences. 2002. A.V. Rao, D. Heber, eds., p. 23. By permission of Caledonian Science Press.)

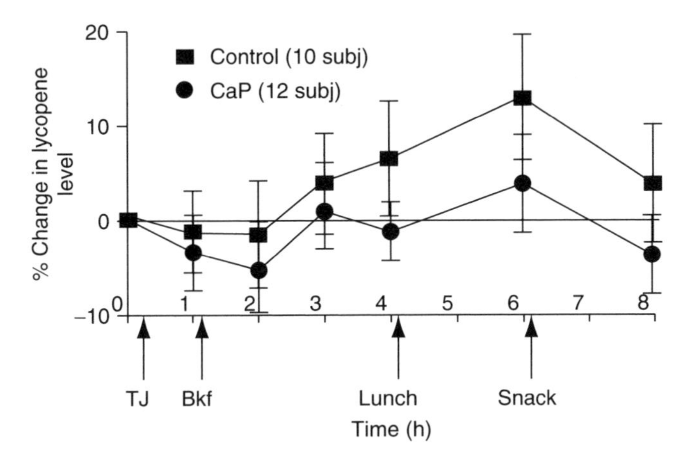

FIG. 10 Lycopene absorption kinetics in prostate cancer patients and controls. CaP, cancer patients; TJ, tomato juice; Bkf, breakfast. (Lycopene, tomatoes and health: New Perspectives 2000. Reprinted from Lycopene and the prevention of chronic diseases. Major findings from five international conferences. 2002. A.V. Rao, D. Heber, eds., p. 24. By permission of Caledonian Science Press.)

Other studies also provide supporting data for the role of lycopene in treating prostate cancer. When tomato sauce was used as a source of lycopene, providing 30 mg lycopene per day for 3 weeks preceding prostatectomy in men diagnosed with prostate cancer, serum and prostate lycopene levels were elevated significantly (Bowen *et al.*, 2002). Oxidative damage to DNA was reduced and serum PSA levels declined significantly by 20% with lycopene treatment. In a study from India (Ansari and Gupta, 2003), where patients with metastasized prostate cancer were given either orchidectomy alone or in combination with 2 mg lycopene twice daily and followed for 2 years, serum PSA levels in the lycopene group were reduced more markedly than in the group with orchidectomy alone. Improvement was also observed in urine flow rates, bone scans, and survival. However, some questions regarding the relatively low doses of lycopene and the lack of dietary controls have been raised with regards to this study, requiring further studies to be undertaken in the future. Two new clinical trials sponsored by the National Cancer Institute are currently under way. The first of these two studies is a phase I investigation of lycopene for the chemoprevention of prostate cancer. In this study, healthy subjects with a baseline serum lycopene level of less than 600 nM will be given different doses of lycopene to evaluate dose-limiting toxicity and the maximum tolerated dose of lycopene given orally. The second study, a randomized control trial, will compare the effectiveness of lycopene and isoflavone administered in different doses with multivitamin supplements prior to surgery for the treatment of patients with stage I and stage II prostate cancer. A multivitamin supplementation group will be used as the control. Results from these studies will contribute significantly to our understanding of the role of lycopene in the treatment of established prostate cancer.

4. *Other cancers*

In addition to prostate cancer, there is growing evidence to indicate that lycopene may also play a role in the prevention of cancer of other sites including breast, lung, gastrointestinal, cervical, ovarian, and pancreatic cancers (Giovannucci, 1999). As with prostate cancer, the main evidence in support of the role of lycopene in the prevention of these cancers comes from cell culture, animal, and epidemiological studies. No clinical and dietary intervention studies have so far been reported. Lycopene was shown to interfere with cell cycle progression and IGF-I signaling in MCF-7 mammary cancer cells (Karas *et al.*, 2000). Inhibition of the proliferation of estrogen-dependent and estrogen-independent human breast cancer cells, MCF-7 and MDA-MB-231, treated with lycopene and other carotenoids was reported by Prakash *et al.* (2001). In another study, lycopene caused only a modest inhibition of the MCF-7 human mammary cancer cells compared to an open chain analogue of retinoic acid (Ben-Dor *et al.*, 2001). Inhibition of human breast and endometrial cancer cells by lycopene is generally associated with the inhibition of cell cycle progression at the G (1) phase. When human breast cancer cells synchronized in the G (1) phase were treated with lycopene, the reduction in cell cycle progression was associated with reduction in the cyclin D levels and retention of p27, leading to the inhibition of G (1) CDK activities (Ben-Dor *et al.*, 2001). Multidrug resistance (MDR) of a majority of human tumor cells is responsible for the failure of therapeutic treatments. When mouse lymphoma and human breast cancer cells transfected with the MDR-1 gene were treated with lycopene, apoptosis of the cancer cells was induced, suggesting it as a possible drug resistance modifier in cancer therapy (Molnar *et al.*, 2004). In another study (Amir *et al.*, 1999), lycopene reduced cell cycle progression and differentiation in HL-60 promyelocytic leukemia cells. In the same study, lycopene had a synergistic effect with $1{,}25(OH)_2D_3$ on cell proliferation and differentiation and an additive effect on cell cycle progression. Oxidized lycopene, in a separate study, enhanced the inhibition of the growth of leukemia cells much more than the unoxidized lycopene (Nara *et al.*, 2001). Lycopene had similar inhibitory effects on the proliferation and differentiation of oral cancer cells (Livny *et al.*, 2002, 2003) and rat ascites hepatoma cells (Kozuki *et al.*, 2000).

Several animal studies have reported on the role of lycopene in cancers other than prostate. Lycopene inhibited the growth and development of C6 glioma cells (malignant brain cells) transplanted into rats (Wang *et al.*, 1959). The growth inhibition was more pronounced when it was given before the inoculation of glioma cells. Chronic intake of lycopene was shown to markedly delay the onset and reduced spontaneous mammary tumor growth and development in SHN virgin mice (Nagasawa *et al.*, 1995). This effect was associated with reduced mammary gland thymidylate synthetase activity and lowered levels of serum-free fatty acids and prolactin, a hormone known to be involved in breast cancer development by stimulating cell division. In another study (Sharoni *et al.*, 1997), a DMBA-induced rat mammary tumor model was used to compare the effect of lycopene-enriched tomato oleoresin with β-carotene on the initiation and progression of tumors. The lycopene-treated rats developed significantly fewer tumors, and the tumor area was smaller than the unsupplemented rats. β-Carotene showed no protection against the development of mammary cancer. However, when *N*-methylnitrosourea (MNU) was used to induce mammary tumorigenesis, lycopene had no effect on tumor incidence, latency, multiplicity, volume, or total tumor per group in rats (Cohen *et al.*, 1999). Colon cancer has also been studied using animal models. Several chemical carcinogens including azoxymethane (AOM), dimethylhydrazine (DMH), and MNU were used to induce carcinogenesis. The preneoplastic marker, aberrant crypt foci (ACF), was used in most of these experiments as the preneoplastic marker of carcinogenesis. In one study, lycopene and tomato juice showed no effect on the incidence of colon cancer in B6C3F_1 mice (Kim *et al.*, 1998). Narisawa *et al.* (1998) reported that tomato juice but not lycopene significantly reduced the incidence of colon cancer in female F344/NSlc rats treated with MNU. In a study conducted in our laboratory (Jain *et al.*, 1999), the effect of lycopene on colon cancer was evaluated. Male Fischer 344 rats were treated with AOM to induce ACF in the colon. Lycopene in the form of 6% oleoresin was incorporated into an AIN93M diet at a concentration of 10 ppm lycopene. The ACF in lycopene-fed rats showed reduced number and size compared to the control group of rats. These effects were more pronounced when lycopene was fed during the promotion stage of carcinogenesis than during the initiation stage. In the same study, ingestion of lycopene by the rats reduced lipid and protein oxidation. Other animal studies showed similar protective effect of lycopene against lung (Kim *et al.*, 1997, 2000), liver (Gradelet *et al.*, 1998), urinary bladder (Okajima *et al.*, 1998), and hamster cheek pouch cancers (Bhuvaneswari *et al.*, 2002).

An extensive review of the epidemiological studies on the role of tomatoes, tomato-based products, and lycopene in cancer was published by Giovannucci (Giovannucci, 1999; Giovannucci *et al.*, 2002). Other studies showing the effect of lycopene on reducing the risk of stomach, other digestive tract, lung, breast, and cervix cancers were reviewed in the same articles (Giovannucci, 1999; Giovannucci *et al.*, 2002). Overall, consumption of lycopene-rich sources of foods, such as tomatoes and tomato-based products, show encouraging outcomes. However, the number of such studies at present is few, requiring further research in this important area of diet and cancer prevention.

C. CARDIOVASCULAR DISEASES

CHD is the major cause of death in North America and the rest of the Western World. It is also recognized as an important contributor of morbidity and mortality in the developing countries of the world. The emphasis so far has been on the relationship between serum cholesterol level as a biomarker for the risk of CHD. Oxidative stress induced by ROS is also considered to play an important part in the etiology of several chronic diseases including CHD (Ames *et al.*, 1993, 1996; Halliwell *et al.*, 1992, 1995; Pincemail, 1995; Stadtman, 1992; Witztum, 1994). Oxidation of the circulating LDL that carries cholesterol into the blood stream to oxidized LDL (LDL_{ox}) is thought to play a key role in the pathogenesis of arteriosclerosis, which is the underlying disorder leading to heart attack and ischemic strokes (Heller *et al.*, 1998; Parthasarathy *et al.*, 1992; Witztum, 1994) (Figure 11).

Antioxidant nutrients are believed to slow the progression of arteriosclerosis because of their ability to inhibit the damaging oxidative processes

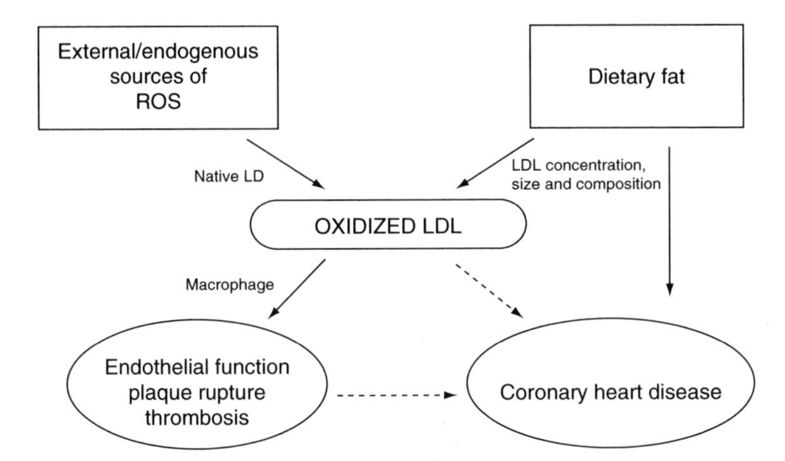

FIG. 11 LDL oxidation and coronary heart disease (Rao and Balachandran, 2004). (Role of antioxidant lycopene in heart disease. Reprinted from Antioxidants and cardiovascular disease. 2004. R. Nath, M. Khullar, Singal P.K., eds., pp. 62–83. By permission of Narosa Publishing House.)

(Heller *et al.*, 1998; Morris *et al.*, 1994; Parthasarathy *et al.*, 1992). Several retrospective and prospective epidemiological studies have shown that consumption of antioxidant vitamins, such as vitamin E, and β-carotene may reduce the risk of CHD (Agarwal and Rao, 2000a; Kohlmeier *et al.*, 1997). Randomized clinical trials also support the reduced risk for CHD with vitamin E supplementation (Stephens *et al.*, 1996; Virtamo *et al.*, 1998; Zock and Katan, 1998). The protective effect of vitamin E observed in these studies has been ascribed to its antioxidant properties. Support for the protective effect of antioxidants also comes from the observations that men and women with CHD exhibit lower levels of circulating antioxidants (Parthasarathy, 1998). However, some large-scale human trials have failed to confirm the protective effect of β-carotene and reported inconclusive results with vitamin E. In the recently completed Heart Outcomes Prevention Evaluation (HOPE) Study, supplementation with 400 IU/d of vitamin E for 4.5 years did not result in any beneficial effects on cardiovascular events in patients at high risk (Hodis *et al.*, 1995).

Lycopene as a dietary antioxidant has received much attention. Epidemiological studies have shown an inverse relationship between the incidence of CHD and the intake of tomatoes and lycopene and serum and adipose tissue lycopene levels (Arab and Steck, 2000; Kohlmeier *et al.*, 1997). These observations have generated scientific interest in lycopene as a preventative agent for CHD. Unlike the number of studies on vitamin E and β-carotene, only a few similar studies have been performed with lycopene (Agarwal and Rao, 2000a; Arab and Steck, 2000; Kohlmeier and Hastings, 1995; Rao and Agarwal, 2000) in the prevention of CHD. A number of *in vitro* studies have shown that lycopene can protect native LDL from oxidation and suppress cholesterol synthesis (Dugas *et al.*, 1998; Fuhramn *et al.*, 1997). In the J-774 A.1 macrophage-like cell line, lycopene at a concentration of 10 μM induced 73% inhibition in cholesterol synthesis from acetate. A slightly lower inhibition was also observed with β-carotene (Fuhramn *et al.*, 1997). In this study, both the carotenoids augmented the activity of the macrophage LDL receptor. However, in another study, dietary enrichment of endothelial cells with β-carotene but not lycopene inhibited the oxidation of LDL (Dugas *et al.*, 1999). The predictability of *in vitro* LDL oxidation as a marker of arteriosclerosis or CHD has been questioned (Zock and Katan, 1998). Similarly in animal model studies, the increased resistance of extracted LDL *in vitro* to oxidation does not necessarily correlate to reduced risk of arteriosclerosis (Diaz *et al.*, 1997).

The evidence in support of the role of lycopene in the prevention of CHD stems primarily from the epidemiological observations on normal and at risk populations. Present knowledge largely relies on the data obtained from dietary estimates or plasma values in relation to the risk of CHD. Epidemiological studies have suggested that a diet rich in a variety of fruits and vegetables results in lower risk of CHD. Fruits and vegetables are in general good sources of dietary carotenoids including lycopene. The antioxidant properties of lycopene may be responsible for the beneficial effects of these food products. Mediterranean diets rich in tomatoes, tomato products, lycopene, and other carotenoids are associated with the lower incidence of arteriosclerosis and CHD. One of the earlier studies that investigated the relationship between serum antioxidant status including lycopene and myocardial infarctions (Street *et al.*, 1994) reported an odds ratio of 0.75. However, in this study there were no controls for other variables such as age, health status, and diet. The strongest population-based evidence comes from a recently reported multicenter case-control study (EURAMIC) that evaluated the relationship between adipose tissue antioxidant status and acute myocardial infarction (Kohlmeier *et al.*, 1997). Subjects (662 cases and 717 controls) from 10 European countries were recruited to maximize the variability in exposure within the study. Needle aspiration biopsy samples of the adipose tissue were taken shortly after the infarction and the levels of α- and β-carotenes, lycopene, and α-tocopherol measured. Adipose lycopene levels expressed as mg/g of fatty acids varied from the lowest 0.21 to the highest 0.36. After adjusting for age, body mass index, socioeconomic status, smoking, hypertension, and maternal and paternal history of the disease only lycopene, and not β-carotene levels, was found to be protective with an odds ratio of 0.52 for the contrast of the 10th and 90th percentiles with a p value of 0.005. The results also showed a dose–response relationship between each quintile of adipose tissue lycopene and the risk of myocardial infarction.

The protective potential of lycopene was maximal among individuals with highest polyunsaturated fat stores. The odds ratios for lycopene of subjects who never smoked, ex-smokers, and smokers were 0.33, 0.41, and 0.63, respectively. These findings seem to support the antioxidant hypothesis. A component of this larger EUREMIC study representing the Malaga region of Spain was analyzed further (Gomez-Aracena *et al.*, 1997). In this case-control study consisting of 100 cases and 102 controls, adipose tissue lycopene levels showed an odds ratio of 0.39 with a 95% confidence interval of 0.13 and 1.19. The p value for the trend was 0.04. In the Arteriosclerosis Risk in Communities (ARIC) case-control study, fasting serum antioxidant levels of 231 cases and an equal number of control subjects were assessed in relationship to the intima-media thickness as an indicator of asymptotic early arteriosclerosis (Iribarren *et al.*, 1997). After controlling for other variables, an odds ratio of 0.81 was observed but the p value for the association was not significant. Statistical significance was observed only for β-cryptoxanthin, lutein, and zeaxanthin. In a cross-sectional study comparing Lithuanian and Swedish populations showing diverging mortality rates from CHD, lower blood lycopene levels were found to be associated with increased risk and mortality from CHD (Kristenson *et al.*, 1997). In the Austrian stroke prevention study, lower levels of serum lycopene and α-tocopherol were reported in individuals from an elderly population at high risk for microangiopathy-related cerebral damage, which is considered as a risk factor for cerebrovascular disease (Schmidt *et al.*, 1997).

Although the epidemiological studies done so far provide evidence for the role of lycopene in CHD prevention, it is at best only suggestive and not proof of a causal relationship between lycopene intake and the risk of CHD. Such a proof can be obtained only by performing controlled clinical dietary intervention studies in which both the biomarkers of the status of oxidative stress and the disease are measured. To date, very few such intervention studies have been reported in the literature. In one study when healthy human subjects consumed a lycopene-free diet for a period of 2 weeks, their serum lycopene levels decreased by 50% by the end of the 2 weeks and at the same time an increase of 25% in the *in vivo* lipid oxidation was observed (Rao and Agarwal, 1998a). In a small dietary supplementation study, six healthy male subjects consumed 60 mg/day lycopene for 3 months. At the end of the treatment period, a significant 14% reduction in their plasma LDL cholesterol levels was observed (Fuhramn *et al.*, 1997). As part of the same study, the authors investigated the effect of lycopene, *in vitro*, on the activity of macrophage HMG-CoA reductase, a rate-limiting enzyme in cholesterol synthesis *in vitro*. Based on these observations, they concluded that dietary supplements of carotenoids may act as moderate hypocholesterolemic agents (Fuhramn *et al.*, 1997). In a randomized, crossover, dietary intervention study 19 healthy human subjects (10 males and 9 females), nonsmokers and not on any medication and vitamin supplements, consumed lycopene from traditional tomato products and nutritional supplements for 1 week. The levels of lycopene consumption ranged from 20 to 150 mg/day. Lycopene was absorbed readily from all dietary sources resulting in a significant increase in serum lycopene levels and lower levels of lipid, protein, and DNA oxidation (Rao and Agarwal, 1998b). In the same study, serum lipoproteins and LDL oxidation were also evaluated. LDL oxidation was estimated by measuring the levels of thiobarbituric acid-reactive substances (TBARS) and conjugated dienes (CD). Although there were no changes in serum total cholesterol and LDL and HDL cholesterols, serum lipid peroxidation and LDL cholesterol oxidation were significantly decreased as the serum lycopene levels increased (Agarwal and Rao, 1988) (Figure 12).

The rationale for the protective role of dietary antioxidants, such as lycopene, is scientifically valid. However, only a few studies have so far been performed with lycopene. Cardiovascular and cerebrovascular diseases are progressively degenerative diseases consisting of several stages that eventually lead to death (Rao, 2002b). Lycopene has been shown to protect LDL oxidation that characterizes the early events of the disease. Population-based epidemiological studies that use death due to cardiovascular disease (CVD)

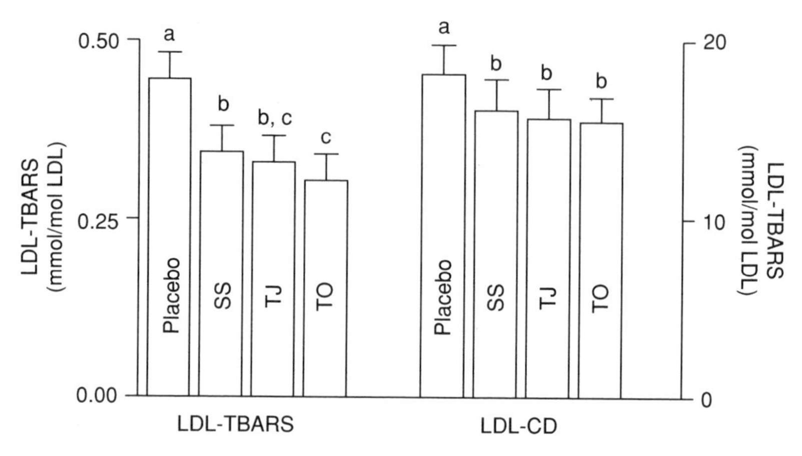

FIG. 12 Effect of dietary lycopene supplements on serum LDL oxidation (SS, spaghetti sauce; TJ, tomato juice; TO, tomato oleoresin). Bars with different letters are statistically different ($p < 0.05$). (Tomato lycopene and low density lipoprotein oxidation: A human dietary intervention study. Reprinted from Lipids. 1988; **33**, pp. 981–984 by permission of AOCS Press.)

as the end point have also provided evidence in support of the role of lycopene in the prevention of CVD. However, the link between early events in the disease, such as LDL oxidation, and the terminal outcome of death needs further studies. In the future, well-controlled clinical and dietary intervention studies evaluating the effectiveness of lycopene will provide useful information in the management of CVD. Important aspects of such studies will be to use well-defined subject populations, standardized outcome measures of oxidative stress and the disease, and lycopene ingestion that is representative of normal healthy dietary intakes.

D. OSTEOPOROSIS AND OTHER BONE DISORDERS

The role of lycopene in bone health to date is based on its potent antioxidant properties, the well-known role of oxidative stress in bone health, and the limited reported studies on the effects of lycopene in bone cells in culture. Therefore, in order to understand the role that lycopene can play in bone health, we have included a review of the reported studies on the role of oxidative stress in bone health and bone cells.

Bone is a dynamic tissue that is continuously being renewed throughout life by the process of bone remodeling involving the coupled events of removal of old bone by osteoclasts and formation of new bone by osteoblasts (for review see Chan and Duque, 2002; Kenny and Raisz, 2002; Mundy, 1999). The remodeling process is the result of the interactions of these cells with multiple molecular agents including hormones, growth factors, and cytokines. Disturbances in the remodeling process can lead to metabolic bone diseases (Lindsay and Cosman, 1990; Raisz, 1993). As will be reviewed later, oxidative stress, shown to control the functions of both osteoclasts and osteoblasts, may contribute to the pathogenesis of skeletal system including osteoporosis, the most prevalent metabolic bone disease.

1. Evidence associating oxidative stress and antioxidants with osteoporosis

ROS-induced oxidative stress is associated with the pathogenesis of osteoporosis. Epidemiological evidence suggests that certain antioxidants including vitamins C and E and β-carotene may reduce the risk of osteoporosis (Leveille *et al.*, 1997; Melhus *et al.*, 1999; Morton *et al.*, 2001; Singh, 1992) and counteract the adverse effects of oxidative stress on bone that is produced during strenuous exercise (Singh, 1992) and among smokers (Melhus *et al.*, 1999). Vitamins C, E, and A, uric acid, the antioxidant enzymes SOD in plasma and erythrocytes and GP_x in plasma were consistently lower in osteoporotic than in control subjects, while plasma levels of malondialdehyde did not differ between groups. These results showed that antioxidant defenses are markedly decreased in osteoporotic women (Maggio *et al.*, 2003). Increased oxidative stress biomarker 8-*iso*-prostaglandin F alpha (8-iso-PGFα) is biochemically linked with reduced bone density (Basu *et al.*, 2001; Sontakke and Tare, 2002). The severity of osteoporosis was positively correlated with the level of the oxidative stress marker, lactic acid, in two men with mitochondrial DNA (mtDNA) deletions (Varanasi *et al.*, 1999), and a study of severe osteoporotic syndrome in relatively young males linked osteoporosis to an increase in oxidative stress (Polidori *et al.*, 2001). A correlation between serum glutathione reductases and bone densitometry values has been reported (Avitabile *et al.*, 1991). In spite of these reports, the cellular and molecular mechanisms involved in the role of oxidative stress in osteoporosis remain poorly defined.

Low bone density is also associated with oxidative stress in lower species. Thus, in ovariectomized rats melatonin has a bone-protective effect, which depends in part on its free radical-scavenging properties (Cardinali *et al.*, 2003). A mouse model that has been used to study the role of ROS in age-related disorders including osteoporosis is the accelerated mouse-senescence-prone P/2 (SAM-P/2) that generates increased oxygen radicals (Hosokawa, 2002; Udagawa, 2002). This model could be very useful in studying the role of lycopene in osteoporosis.

2. *Evidence associating oxidative stress and antioxidants with osteoblasts*

Very little work has been reported on the role of oxidative stress in osteoblasts. However, osteoblasts can be induced to produce intracellular ROS (Cortizo *et al.*, 2000; Liu *et al.*, 1999), which can cause a decrease in alkalinephosphatase (ALP) activity that is partially inhibited by vitamin E and cause cell death (Cortizo *et al.*, 2000; Liu *et al.*, 1999). Treatment of rat osteosarcoma ROS 17/2.8 cells with tumor necrosis factor-alpha (TNF-α) suppressed bone sialoprotein (BSP) gene transcription through a tyrosine kinase-dependent pathway that generates ROS (Samoto *et al.*, 2002). H_2O_2 modulated intracellular calcium (Ca^{2+}) activity in osteoblasts by increasing Ca^{2+} release from the intracellular Ca^{2+} stores (Nam *et al.*, 2002).

3. *Evidence associating oxidative stress and antioxidants with osteoclasts*

The mechanisms involved in the differentiation of osteoclasts and their ability to resorb bone are poorly understood. ROS may be involved in this process (Silverton, 1994). Both the H_2O_2 produced by endothelial cells (Zaidi *et al.*, 1993) intimately associated with osteoclasts and the H_2O_2 that is produced by osteoclasts (Bax *et al.*, 1992) increase osteoclastic activity and bone resorption. H_2O_2 may also be involved in the regulation of osteoclast formation (Suda *et al.*, 1993), differentiation of osteoclast precursors (Steinbeck *et al.*, 1998), and osteoclast motility (Bax *et al.*, 1992). The tartrate-resistant acid phosphatase (TRAP), found on the surface of osteoclasts, reacts with H_2O_2 to produce highly destructive ROS that target the degradation of collagen and other proteins (Halleen *et al.*, 1999). Superoxide was localized both intracellularly and at the osteoclast-bone interface using nitroblue tetrazolium (NBT), which is reduced to purple-colored formazan by ROS, suggesting the participation of superoxide in bone resorption (Key *et al.*, 1990) and in the formation and activation of osteoclasts (Garrett *et al.*, 1990). Osteoclastic superoxide is produced by NADPH oxidase (Darden *et al.*, 1996; Steinbeck *et al.*, 1994). However, Fraser *et al.* (1996) suggested that H_2O_2, not superoxide, stimulates bone resorption in mouse calvaria and that the earlier finding of stimulation by superoxide (Garrett *et al.*, 1990; Key *et al.*, 1994) may be due in part to conversion of this radical to H_2O_2. 1,25-Dihydroxyvitamin D_3 had a direct nongenomic effect on the generation of superoxide anion (O_2^-), which was inhibited by estrogen (Berger *et al.*, 1999). Estrogen has been reported to have an antioxidant property (Clarke *et al.*, 2001; Wagner *et al.*, 2001). Hormones known to stimulate bone resorption, such as parathyroid hormone (PTH) (Datta *et al.*, 1996) and $1,25(OH)_2D_3$, have stimulatory effects on ROS production in osteoclasts (Berger *et al.*, 1999) and hormones known to have inhibitory effects on bone resorption, such as calcitonin, inhibit ROS production (Berger *et al.*, 1999; Datta *et al.*, 1995).

Antioxidants also play a role in osteoclast activity. Osteoclasts produce the antioxidant enzyme SOD in the plasma membrane (Oursler *et al.*, 1991). ROS production in osteoclasts was inhibited after treating the cells with antioxidant enzymes such as SOD (Key *et al.*, 1990) and catalase (Suda *et al.*, 1993). ROS production in osteoclasts was also inhibited by estrogen (Berger *et al.*, 1999), the superoxide scavenger deferoxamine mesylate-manganese complex (Key *et al.*, 1994; Ries *et al.*, 1992), pyrrolidine dithiocarbamate (PDTC), and *N*-acetyl cysteine (NAC) (Hall *et al.*, 1995). The use of antioxidants from natural sources, such as fruits and vegetables, could be another way of inhibiting ROS. The use of lycopene in this regard is reviewed later.

4. In vitro *studies of lycopene in bone cells*

a. Effects of lycopene on osteoblasts. The studies on the effects of lycopene on osteoblasts are limited to two reports (Kim *et al.*, 2003; Park *et al.*, 1997). Kim *et al.* (2003) studied the effect of incubating osteoblast-like

SaOS-2 cells in 10^{-6} and 10^{-5} M lycopene or their respective vehicle controls on cell proliferation. They showed that lycopene stimulated cell proliferation as shown in Figure 13.

They also reported that lycopene had a stimulatory effect on ALP activity, a marker of osteoblastic differentiation in more mature cells, but depending on the time of addition, it had an inhibitory or no effect on younger SaOS-Dex cells (Figure 14). These findings constituted the first report on the effect of lycopene on human osteoblasts. In another study by Park *et al.* (1997), the effect of lycopene on MC3T3 cells (the osteoblastic cells of mice) was contrary to the findings of Kim *et al.* (2003). Park demonstrated that lycopene had an inhibitory effect on cell proliferation. Both studies, however, reported that ALP activity was stimulated. The discrepancy in the effect of lycopene on cell proliferation could be a result of species differences or experimental conditions. More studies are required to clarify the role of lycopene in osteoblasts.

b. Effects of lycopene on osteoclasts. To date, there are only two reported studies on the effects of lycopene in osteoclasts (Ishimi *et al.*, 1999; Rao *et al.*, 2003). Rao *et al.* (2003) cultured cells from bone marrow prepared from rat femur in 16-well calcium phosphate-coated Osteologic^{TM} multitest slides (Millenium Biologix Inc.). Varying concentrations of lycopene in the

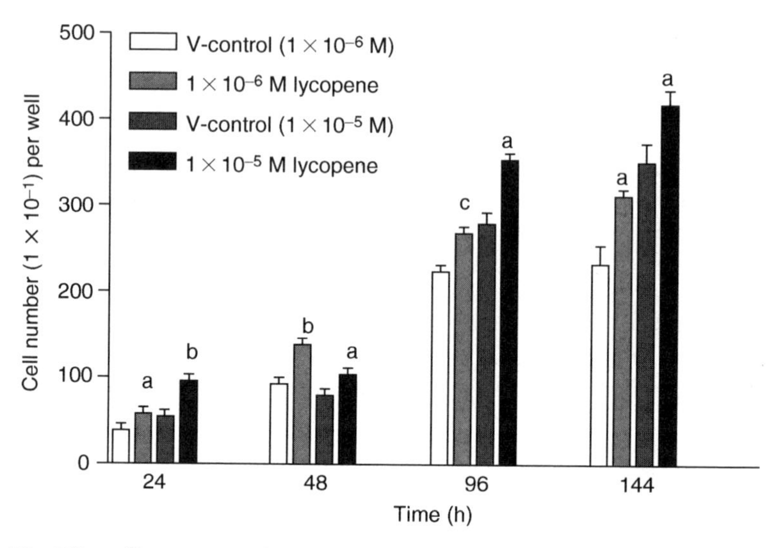

FIG. 13 Effect of lycopene on the proliferation of SaOS-2 cells. Compared with respective vehicle control of the same dilution: a $= p < 0.05$, b $= p < 0.001$, $c = p < 0.005$ (Kim *et al.*, 2003). (Lycopene II – Effect on osteoblasts: The carotenoid lycopene stimulates cell proliferations and alkaline phosphotase activity of SaOS-2 cells. Reprinted from Journal of Medicinal Food. 2003; **6**, pp. 79–86 by permission of Mary Ann Liebert, Inc., Publishers.)

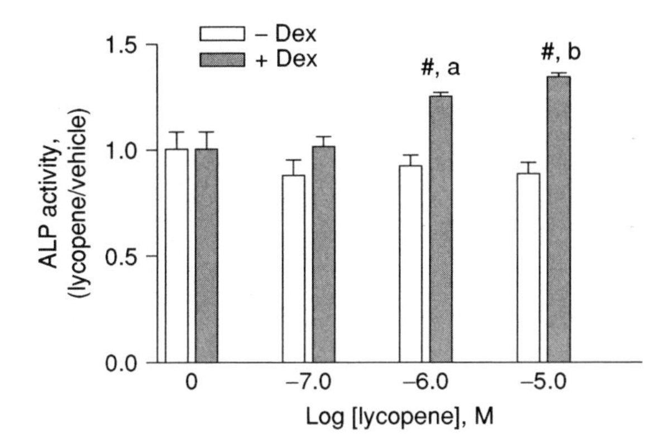

FIG. 14 Effect of lycopene on ALP activity # $p < 0.005$, comparison with zero control: a < 0.05, b < 0.01 (Kim *et al.*, 2003). (Lycopene II – Effect on osteoblasts: The carotenoid lycopene stimulates cell proliferation and alkaline phosphotase activity of SaOS-2 cells. Reprinted from Journal of Medicinal Food. 2003; **6**, pp. 79–86 by permission of Mary Ann Liebert, Inc., Publishers.)

absence or presence of the resorbing agent PTH-(1–34) were added at the start of culture and at each medium change every 48 hours. The effects of lycopene on mineral resorption are shown in Figure 15.

Lycopene inhibited the TRAP+ multinucleated cell formation in both vehicle- and PTH-treated cultures. The cells that were stained with the NBT reduction product formazan were decreased by treatment with 10^{-5} M lycopene, indicating that lycopene inhibited the formation of ROS-secreting osteoclasts (Figure 16).

Rao *et al.* (2003) concluded that lycopene inhibited basal and PTH-stimulated osteoclastic mineral resorption and formation of TRAP+ multinucleated osteoclasts, as well as the ROS produced by osteoclasts. These findings are new and may be important in the pathogenesis, treatment, and prevention of osteoporosis.

The effects of lycopene on osteoclast formation and bone resorption were also reported by Ishimi *et al.* (1999) in murine osteoclasts formed in coculture with calvarial osteoblasts (Ishimi *et al.*, 1999). Their results differed from those of Rao *et al.* (2003) in that they found that lycopene inhibited the PTH-induced, but not the basal, TRAP+ multinucleated cell formation. Furthermore, they could not demonstrate any effect of lycopene on bone resorption. They also did not study the effect of lycopene on ROS production.

5. *Clinical studies on the role of lycopene in postmenopausal women at risk of osteoporosis*

Postmenopause is associated with a global increase in bone turnover markers (Kushida *et al.*, 1995; Vernejoul, 1998). These markers predict bone loss

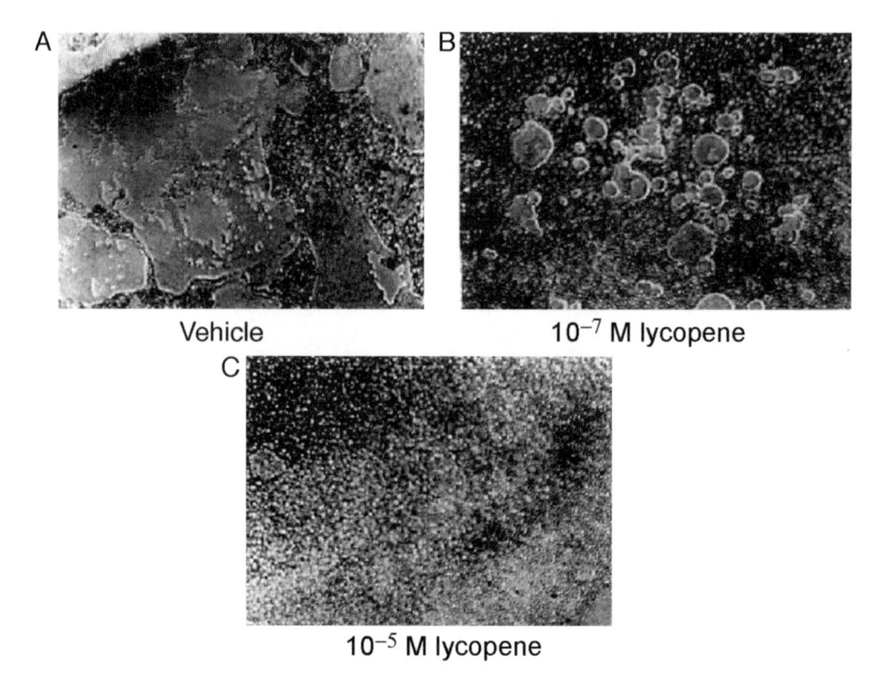

FIG. 15 Effect of lycopene on resorption of the calcium phosphate substrate coating of osteologic multitest slides in the presence of osteoclasts (Rao *et al.*, 2003). (Lycopene I – Effect on osteoclasts: Lycopene inhibits basal and parathyroid hormone-stimulated osteoclast formation and mineral resorption mediated by reactive oxygen species in ray bone marrow cultures. Reprint from Journal of Medicinal Food. 2003; **6**, pp. 69–78 by permission of Mary Ann Liebert, Inc., Publishers.)

and osteoporosis in postmenopausal women (Garnero *et al.*, 1996). One of the objectives of our current clinical study at St. Michael's Hospital is to test whether the serum lycopene correlates inversely with the oxidative stress parameters and bone turnover markers in postmenopausal women who are at risk for osteoporosis. Thirty-three women aged 50–60 were recruited and asked to complete a 7-day food intake record prior to giving fasting blood samples. Oxidative stress parameters, total antioxidant capacity, serum lycopene, and the bone turnover markers bone ALP (bone formation) and cross-linked N-telopeptides of type I collagen (NTx) (bone resorption) were measured from serum samples. The participants were grouped into quartiles according to their serum lycopene per kilogram body weight (nM/kg) and correlation analyses were carried out using the Newman-Keuls posttest. The most important and interesting findings to date are the correlation between lycopene intake and serum lycopene (Figure 17) indicating that lycopene is readily absorbed in the body, significant decreases in protein

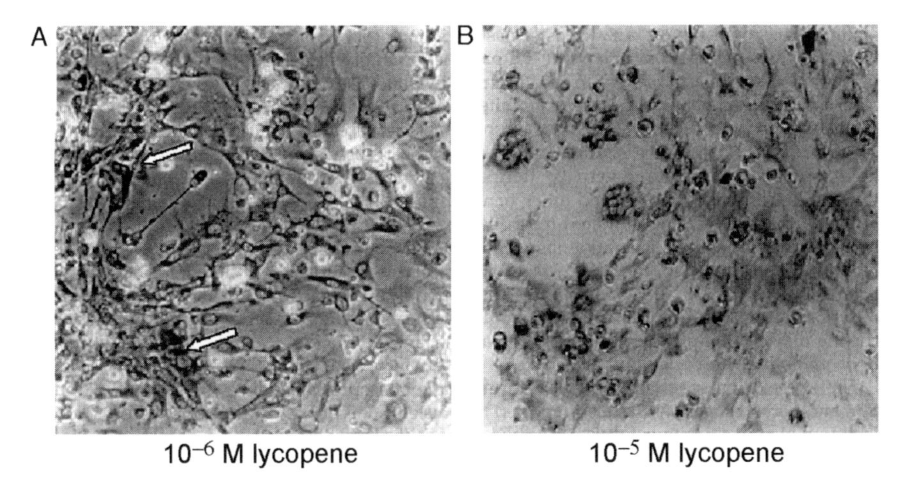

FIG. 16 Effect of lycopene on ROS production in osteoclasts (Rao *et al.*, 2003). (Lycopene I – Effect on osteoclasts: Lycopene inhibits basal and parathyroid hormone-stimulated osteoclast formation and mineral resorption mediated by reactive oxygen species in ray bone marrow cultures. Reprint from Journal of Medicinal Food. 2003; **6**, pp. 69–78 by permission of Mary Ann Liebert, Inc., Publishers.)

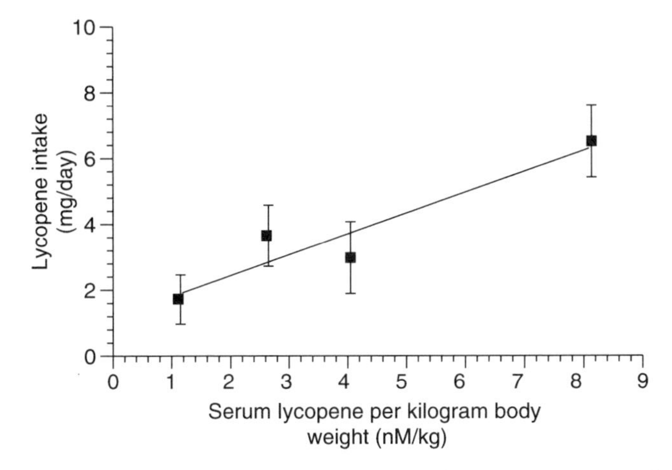

FIG. 17 Effect of lycopene intake on serum lycopene in 33 postmenopausal volunteers.

oxidation as indicated by increased thiols ($p < 0.05$) and decreased NTx values ($p < 0.005$) as levels of serum lycopene increase (Figure 18) (Rao *et al.*, 2005).

Since there was a significant positive correlation between serum lycopene levels and dietary lycopene intake as determined from the estimated food

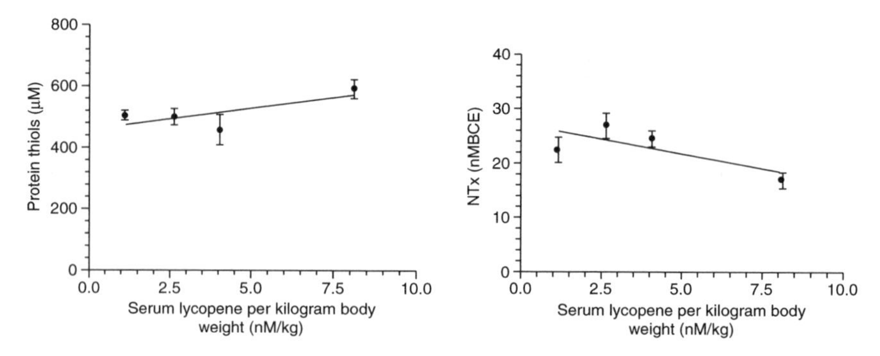

FIG. 18 Effect of serum lycopene on protein oxidation and bone turnover markers in 33 postmenopausal volunteers.

records ($p < 0.01$) (Figure 18), our results support the hypothesis that dietary lycopene acts as an effective antioxidant, reducing oxidative stress and bone turnover markers. Our observations suggest an important role for lycopene mediated via its antioxidant property in reducing the risk of osteoporosis. Dietary intervention studies with varying doses and sources of lycopene are currently being conducted to demonstrate the beneficial effects of lycopene in the prevention and management of osteoporosis.

E. HYPERTENSION

High blood pressure or hypertension is a major health problem effecting close to 25% of the adult population in North America. It is a condition commonly associated with narrowing of the arteries. It is known as the "silent killer" because there may be no symptoms until a person develops a fatal complication of the disease. Although the exact cause of hypertension is unknown, there are several factors and conditions that may contribute to its occurrence, including genetic factors, family history of hypertension, obesity, sedentary lifestyle, excess salt intake, alcohol, smoking, stress, age, hormone levels, abnormalities in the nervous and circulatory systems and kidneys, and the salt and water content in the body. Oxidative stress has also been implicated in the causation of hypertension (Friedman *et al.*, 2003; Lassegue and Griendling, 2004; Zini *et al.*, 1993). The ROS generated endogenously can effect multiple tissues, either directly or through nitric oxide depletion and includes contraction and endothelial dysfunction in the vasculature, hypertrophic remodeling in the blood vessels and myocardium, reabsorption of salt and decreased glomerular filtration in the kidney, and increased efferent sympathetic activity from the central nervous system (CNS). Although several pharmaceutical agents are used in the effective management of hypertension, there has been considerable interest in the use of naturally occurring food components. Antioxidant polyphenols derived from green tea have been studied in this context and have been shown to be effective in controlling high blood pressure. Because of the potent antioxidant property of lycopene, it has also been studied for its role in hypertension. Paran and Engelhard (2001) using a single-blind, placebo-controlled trial studied the effect of tomato lycopene on blood pressure. Thirty, grade one hypertensive patients between the ages of 40–65 years, not requiring any blood pressure and lipid-lowering medications, were recruited into the study. After a 2-week run-in period for baseline evaluation, the patients were placed on 4-week placebo and 8-week treatment periods. The treatment consisted of ingesting tomato extract dietary supplement capsules that provided 15 mg lycopene every day. Results showed no significant changes in the diastolic blood pressure after the 8 weeks of treatment but did show a considerable reduction in the systolic blood pressure from the baseline value of 144 mm Hg to 134 mm Hg at the end of the lycopene treatment (Figure 19).

In another study (Moriel *et al.*, 2002), 11 patients with mild essential hypertension were compared with 11 healthy subjects for water- and lipid-soluble antioxidants and the concentrations of nitric oxide derivatives in the plasma. A significant reduction in plasma lycopene was observed in the hypertensive patients compared to the normal subjects. Similar reductions in ascorbate, urate, and β-carotene were also observed in this study. However, there were no differences in the nitrous oxide derivatives between the two groups. Hypertension and lymphatic circulation impairment are associated with liver cirrhosis. When patients with liver cirrhosis were compared to healthy matched controls, a significant reduction in serum lycopene, other carotenoid antioxidants, retinol, and α-tocopherol were observed in the cirrhotic patients. Based on these observations, the authors recommend thorough screening for the antioxidants and improved diet in the

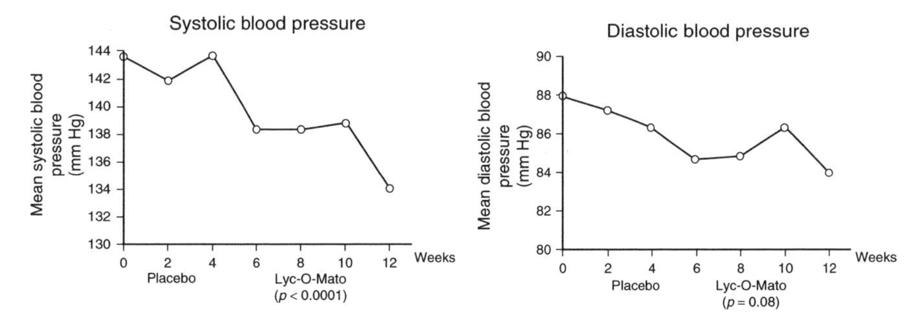

FIG. 19 Effect of tomato extract on systolic and diastolic blood pressure.

follow-up of liver cirrhosis patients and the link to hypertension (Rocchi *et al.*, 1991). A dietary approaches to stop hypertension (DASH) diet is recommended for lowering high blood pressure (Most, 2004). The DASH diet was designed to give beneficial levels of fibre, potassium, magnesium, and calcium. As such it contains more fruits, vegetables, and whole grains compared to control diets and is substantially higher in antioxidant phytochemicals. When the DASH diet was compared with the control diet, it was found to contain substantially higher levels of lycopene and other carotenoids, polyphenols, flavanols, flavanones, and flavan-3-ols. The beneficial effects of these phytochemicals in the management of blood pressure are now being recognized. Further clinical and intervention studies are required to better understand the role of lycopene in hypertension.

F. MALE INFERTILITY

Infertility affects 15% of all couples and at least 30–50% of these couples will have an abnormality detectable in the male partner, which contributes to their difficulty achieving a pregnancy. Therefore, it is estimated that about 7–10% of adult men in their reproductive years (20–50 years of age) are infertile and 25% of all men with infertility will have nonspecific or idiopathic infertility (Dubin and Amelar, 1971; Greenberg *et al.*, 1978; Johnson, 1975). For these men, medical therapy has generally been ineffective in improving sperm quality and fertility. Oxidative stress has been suggested as an important contributory factor in male infertility. Significant levels of ROS are detectable in the semen of up to 25% of infertile men, whereas fertile men do not produce detectable levels of ROS in their semen (Iwasaki and Gagnon, 1992; Zini *et al.*, 1993). The role of oxidative stress and antioxidants in male infertility is shown in Figure 20.

The identification of novel, less invasive treatments, such as vitamins and antioxidants, for male infertility could potentially have a great impact on the management of the infertile couple. To date, a small number of studies have evaluated the role of vitamins and antioxidants (mostly as single agents) in male infertility. In general, these studies suggest a beneficial role of antioxidant therapy in the treatment of male-factor infertility but additional studies are needed. In 1991, Fraga *et al.* (1991) demonstrated that dietary vitamin C has a beneficial effect on the integrity of sperm DNA (DNA oxidation) in male smokers. In a small placebo-controlled trial, Dawson *et al.* (1993) found that supplemental vitamin C improves sperm quality.

A number of investigators have evaluated the role of vitamin E and have reported improved sperm quality in controlled and uncontrolled trials (Geva *et al.*, 1996; Kessopoulou *et al.*, 1995; Suleiman *et al.*, 1996). In a small placebo-controlled trial, Lenzi *et al.* (1998) have reported that glutathione

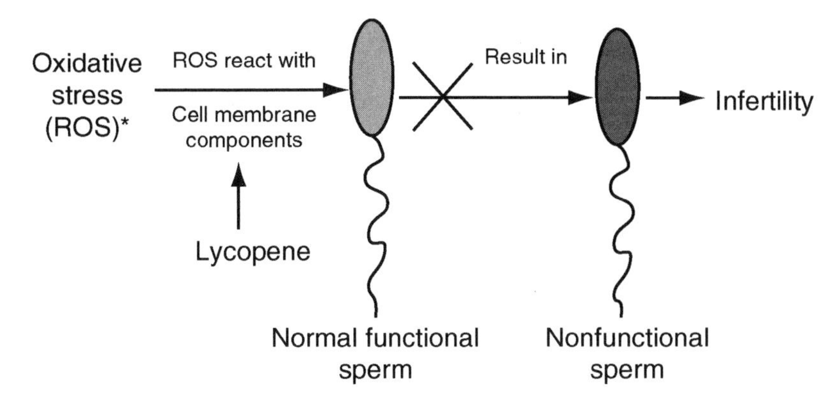

FIG. 20 Effect of oxidative stress and lycopene on sperm functionality.

(intramuscular) improves sperm quality significantly. The role of l-carnitine has been evaluated in uncontrolled trials with early promising results (Moncada *et al.*, 1992). Vitamins C and E and other antioxidants, including taurine (Alvarez and Storey, 1983), L-carnitine (Moncada *et al.*, 1992), and coenzyme Q10 (Alleva and Scaaraamucci, 1997; Lewin and Lavon, 1997), protect spermatozoa from oxidative stress *in vitro*. Since the recognition of lycopene as a potent antioxidant, and its preventive role in oxidative stress-mediated chronic diseases, researchers are beginning to investigate its role in protecting sperm from oxidative damage leading to infertility. Men with antibody-mediated infertility were found to have lower semen lycopene levels than fertile controls (Palan and Naz, 1996). In another study (Mohanty *et al.*, 2001), 50 infertile male volunteers between the ages of 21–50 years were recruited. The subjects had a normal hormonal profile of antisperm antibody titre and without any history of having taken any therapy for infertility or having obstructive azospermia. They consumed a daily dose of 8 mg lycopene in capsule form. The treatment was continued until sperm analysis showed optimal levels or until pregnancy of their partners was achieved. After a 12-month follow-up, it was reported that lycopene treatment resulted in significant increases in serum lycopene concentration. Significant improvements in sperm motility, sperm motility index, sperm morphology, and functional sperm concentration were also observed. The partners of 18 of the 50 subjects had successful pregnancies, accounting for a 36% success rate. Other studies are now in progress and their results will further advance our knowledge of the beneficial role of lycopene in male infertility (Figure 21).

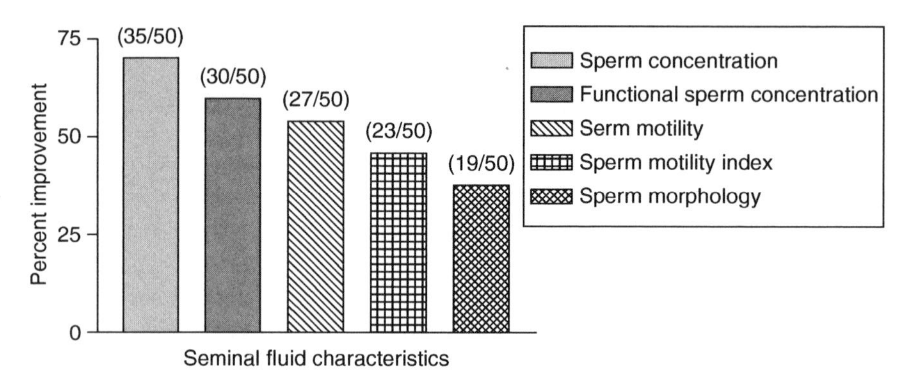

FIG. 21 Effect of lycopene on the sperm quality in infertile men (Mohanty *et al.*, 2001).

G. NEURODEGENERATIVE DISEASES

Neurodegenerative diseases (NDD) are a group of degenerative disorders of the nervous system, including the brain, spinal cord, and peripheral nerves. They include Alzheimer's disease, Parkinson's disease, Huntington's disease, amyotrophic lateral sclerosis, and epilepsy (Rao and Balachandran, 2003). They are a group of disorders with varied clinical importance and etiologies. Oxidative stress is now established as being an important causative factor as well as an ancillary factor in the pathogenesis of the NDD (Ebadi *et al.*, 1996; Singh *et al.*, 2004). The brain and nervous system are particularly vulnerable to free radical damage for a number of reasons as shown in Table VII.

The high lipid content of the nervous system, low antioxidant capacity, and the presence of iron, coupled with its high aerobic metabolic activity, make it particularly susceptible to oxidative damage (Halliwell, 1989; Rao and Balachandran, 2003; Rao and Rao, 2004). Several antioxidant systems have been shown to be effective in mitigating the neurotoxin effect of ROS. Important among them are the free radical-deactivating enzymes, SOD, glutathione peroxidase and catalase, free radical-scavenging agents that include vitamins A, C, and E, iron chelators, and selenium, and phytonutrients such as the carotenoids, flavonoids, and terpenoids (Rao and Balachandran, 2003; Singh *et al.*, 2004). Activity of the antioxidant enzymes was shown to be reduced in patients with Parkinson's disease (Ambani *et al.*, 1975; Fahn and Cohen, 1992; Kish *et al.*, 1985). As in Parkinson's, Alzheimer's, and Huntington's disease, and amyotrophic lateral sclerosis, increased levels of lipid peroxidation and oxidation of DNA were observed compared to controls, suggesting diets high in antioxidants might be effective in reducing the

TABLE VII

VULNERABILITY OF BRAIN TO FREE RADICAL DAMAGE

Brain consumes large quantities of oxygen for its relatively small weight contributing to the formation of ROS.
Membrane lipids in brain contain high levels of polyunsaturated fatty acid side chains that are prone to free radical attack.
Presence of iron and other transition metals in the brain can also contribute to the production of ROS.
Brain contains lower levels of antioxidant vitamins such as vitamins A, C, and E.
Brain contains low to moderate amounts of antioxidant enzymes, such as catalase, superoxide dismutase, and glutathione peroxidase, which play an important role in the inactivation of ROS.

risk of these diseases (Borlongan *et al.*, 1996; Ferrante *et al.*, 1997; Grant, 1997; Jenner, 1996; Retz *et al.*, 1998). Several *in vitro* studies have demonstrated the effectiveness of antioxidants in protecting nervous tissue from damage by the free radicals. Among the dietary antioxidants, most of the studies were directed at the role of vitamins A, C, and E and β-carotene. They were shown to prevent neuronal damage due to oxidative stress (Mitchell *et al.*, 1999). Similarly, in clinical trials, high doses of vitamins C and E were shown to reduce the rate of Parkinson's disease progression to the point of drug intervention by 2½ years (Fahn, 1991). Other epidemiological studies also provide supporting evidence for the role of vitamins A, C, and E and β-carotene in reducing the risk of NDD. Relatively small number of studies have been reported for the role of lycopene in NDD. In one study, significant reduction in lycopene levels was observed in Parkinson's disease and in vascular dementia (Foy *et al.*, 1999). The Austrian Stroke Prevention study showed lower serum lycopene and α-tocopherol levels to be associated with a high risk of microangiopathy (Schmidt *et al.*, 1997). In another case-control study of dietary risk factors for amyotrophic lateral sclerosis conducted in New England, a modest protective association was suggested for lycopene (Longnecker *et al.*, 2000). A correlation between high blood lycopene levels and functional capacity, such as the ability to perform self-care tasks, was reported in an elderly population in another study (Snowdon *et al.*, 1966). It is suggested that antioxidants, such as lycopene, may act directly on the neurons in an indirect manner affecting peripheral markers of oxidative stress (Sinclair *et al.*, 1998). The levels of lycopene in the CNS were present in much lower concentrations than in the other human tissues (Clinton, 1998). However, it is believed that lycopene can cross the blood–brain barrier and be effective in reducing the damage caused by ROS.

Although the present epidemiological, *in vitro*, and *in vivo* studies suggest an important role for lycopene in the prevention of NDD, further research needs to be done to gain a better understanding of its role in the managements of neuronal disorders that constitute an important human health problem globally.

H. OTHER HUMAN DISEASES

Since the recognition of the biological role of lycopene in the prevention of chronic diseases, the emphasis of the scientific community has been in the area of cancer, with special focus on prostate cancer. However, based on the hypothesis that oxidative stress may be an important etiological factor in the causation of most of the degenerative diseases and that lycopene is a potent antioxidant, the scientific community has started to study its role in diseases other than the ones reviewed in this chapter. These health disorders include skin and ocular diseases, rheumatoid arthritis, periodontal diseases, and inflammatory disorders. The scientific information pertaining to the role of lycopene in these diseases is still in its infancy. However, the rationale for undertaking these studies is scientifically valid and it is hoped that in the next 3–5 years several studies will be reported in the literature.

IX. DIETARY INTAKE LEVELS OF LYCOPENE

Since humans do not synthesize lycopene, it has to be provided through the diet. Estimating the dietary intakes of lycopene by various populations around the world is made difficult due to the variability in the reported levels of lycopene in food sources. However, several reports have appeared in the literature showing average daily intake levels of lycopene. Table VIII summarizes the reported levels of lycopene intake in different regions of the world (Block, 1994; Frorman *et al.*, 1993; Rao *et al.*, 1999; U.S. Department of Agriculture and Agricultural Research Service, 1998; U.S. Department of Agriculture and CSFII, 1994–1996). Other studies have indicated the average intake of lycopene in North America to be 5.3 mg/day. However, 50% of the population was shown to consume 1.86 mg/day or even less in some cases (Agarwal and Rao, 2000a; Rao and Agarwal, 1999).

Although there is a need to make more accurate estimations of lycopene intake by various populations, the general thinking among scientists is that the average intake levels of lycopene in North America are lower than the levels required for its beneficial biological effects. Hence, there may be a need to incorporate sources of lycopene, such as the tomato products, as part of a

TABLE VIII
LYCOPENE INTAKE LEVELS

Country	Average daily intake (mg)
United States of America	16.15
	5.93
	5.20
	3.70
Canada	25.20
Germany	1.30
United Kingdom	1.10
Finland	0.70

Source: Rao (2002a). Future Directions and intake recommendations. Reprinted from Lycopene and the prevention of chronic diseases. Major findings from five international conferences. 2002. A.V. Rao, D. Heber, eds., p. 43. By permission of Caledonian Science Press.

healthy diet and lifestyle. Lycopene supplements may also have a role in contributing to beneficial levels.

Until recently, nutritionists failed to recognize the importance of lycopene in human health due to its lack of provitamin A activity. As a result, at present, it is not considered as an "essential" nutrient and as such there are no established "recommended daily intake" (RDI) or "recommended nutrient intake" (RNI) levels for lycopene. However, with the recognition of the role of lycopene in human health, there is considerable interest now among the nutritionists and other health professionals to make "suggestions" based on scientific knowledge about daily intake levels. The main assumption regarding the role of lycopene in the prevention of cancer and other chronic diseases is that it has to be absorbed from the diet and be present at the site of its action (Rao and Agarwal, 1999). Information about the absorption and *in vivo* antioxidant properties of lycopene is therefore essential in formulating dietary guidelines. Serum and plasma levels of lycopene have been used extensively to assess the absorption of lycopene from dietary sources and to assess its biological significance. Adipose tissue levels of lycopene have also been used and are considered a more accurate reflection of lycopene status in the body. However, due to the invasive nature of collecting adipose tissue samples, circulatory levels are still considered as standard procedure to assess lycopene absorption. In addition to the levels of lycopene in serum/plasma, measuring the biomarkers of oxidative stress is also used to assess the biological activity of lycopene including its effect on the prevention of prostate cancer. The rationale being that in the presence of the antioxidant lycopene, the biomarkers of oxidative stress will be lower.

Since oxidative stress is related to the risk of chronic diseases, lower levels of oxidative stress are considered consistent with lower risk of these diseases. A few studies have also measured biochemical and pathobiological markers of cancer in patients after lycopene intervention (Kucuk *et al.*, 2001). However, the levels of lycopene used in these studies were based on preliminary studies investigating the absorption and oxidative stress status in healthy and at risk for cancer patients. Dose–response studies have not been undertaken.

In a recently reported study, when human subjects refrained from the consumption of lycopene-containing foods, their serum lycopene levels fell significantly within 2 weeks (Rao, 2002a; Rao and Agarwal, 1999). This observation was consistent with the fact that humans do not synthesize lycopene and have to be provided with dietary sources to maintain circulatory lycopene levels. Another study was undertaken to investigate the absorption of lycopene in healthy human subjects (Rao and Agarwal, 1998a). A total of 19 subjects (male and female) underwent a washout period during which they restrained from consuming foods that were known sources of lycopene. Following the washout period, subjects consumed tomato juice, tomato sauce, or lycopene supplements for a period of 1 week in a randomized crossover study design. Subjects also underwent a washout period between the treatments. Lycopene levels varied from 28 to 150 mg/day. Lycopene was found to be absorbed equally efficiently from all three sources of lycopene. Increased levels of serum lycopene paralleled significant reductions in lipid, protein, and DNA oxidation. Based on these studies, an intake level of 30–35 mg of lycopene per day was suggested. In a later study, lower levels of lycopene (5, 10, and 20 mg/day) from either tomato ketchup or lycopene supplement capsules were studied for their effect on serum lycopene levels and oxidative biomarkers (Rao and Shen, 2002). Once again, lycopene was absorbed equally well from both sources. Based on the results from this study, the previously "suggested" levels of lycopene intake of 30–35 mg/day were lowered to 7–8 mg/day. At these levels of intake, a maximum level of serum lycopene and reduction in lipid, protein, and DNA oxidation was observed.

Although the suggested level of 7–8 mg of lycopene per day is based on one recent study (Rao and Shen, 2002), a majority of other studies reported in the literature use a single level of intake of 15, 30, or 75 mg of lycopene to study its effect on prostate cancer and other diseases. Systematic dose-response studies have not yet been carried out using disease biomarkers as the end point. Under these conditions, the best estimate of lycopene intake is based on serum/plasma concentrations and biomarkers of oxidation.

X. CONCLUSIONS

Lycopene is a carotenoid antioxidant that has received considerable scientific interest in recent years. Although the chemistry and *in vitro* antioxidant properties of lycopene have been known for several years, its role in human health has just begun to be investigated. A publication in 1995 reported a significant inverse relationship between the intake and circulatory levels of lycopene and the risk of prostate cancer. We started our studies to investigate the bioavailability and *in vivo* antioxidant properties of lycopene and demonstrated that lycopene was absorbed readily from food sources and maintained its antioxidant properties *in vivo* as indicated by reduced levels of lipid, protein, and DNA oxidation. Till recently, most of the evidence in support of the role of lycopene in reducing the risk of cancers and other chronic diseases was based on epidemiological observations. Although these studies provide convincing evidence to suggest the beneficial role of lycopene in human health, they are at best indicative. More recently, several clinical and dietary intervention studies are beginning to be undertaken. It is to be expected that results from these and other studies that are yet to be undertaken will enhance our knowledge about lycopene and disease prevention and perhaps even the treatment of some chronic diseases. Equally important, these studies will help us understand the mechanisms of action of lycopene.

Oxidative stress is now being recognized as an important causative agent of many chronic diseases. Oxidation of lipids and in particular oxidation of LDL cholesterol is being associated with increasing the risk of CVD. Similarly, oxidation of DNA is associated with increased risk of cancers. Since most of the enzymes involved in metabolic and regulatory processes are proteins, its oxidation can lead to inactivation or improper activity of the enzymes, resulting in increased risk of diseases. The significance of these observations perhaps is to suggest that oxidative stress may be the main common feature of several diseases. Lycopene, being one of the most potent naturally occurring antioxidant, can be effective in reducing the risk of these diseases. Although the initial interest in lycopene was due to its role in the prevention of prostate cancer, it is now being investigated for its possible role in other diseases such as CVD, osteoporosis, hypertension, male infertility, macular degeneration, and diseases related to inflammation. Although the antioxidant property of lycopene is the major mechanism being investigated, evidence is beginning to appear which would suggest that the biological role of lycopene may also involve other mechanisms such as stimulating the GJC, immune stimulation, and metabolic regulations. These mechanisms need further research in the future.

Many questions still remain unclear regarding the biological role of lycopene. However, based on the current evidence, it would be prudent to include sources of lycopene as part of a healthy diet.

REFERENCES

Agarwal, A., Shen, H., Agarwal, S., and Rao, A. 2001. Lycopene content of tomato products: Its stability, bioavailability and *in vivo* antioxidant properties. *J. Med. Food* **4**, 9–15.

Agarwal, S. and Rao, A.V. 1988. Tomato lycopene and low density lipoprotein oxidation: A human dietary intervention study. *Lipids* **33**, 981–984.

Agarwal, S. and Rao, A.V. 2000a. Tomato lycopene and its role in human health and chronic diseases. *CMAJ* **163**, 739–744.

Agarwal, S. and Rao, A.V. 2000b. Carotenoids and chronic diseases. *Drug Metabol. Drug Interact.* **17**, 189–210.

Allan, A.L., Tuck, A.B., Bramwell, V.H.C., Vandenberg, T.A., Winquist, E.W., and Chambers, A.F. 2005. Contribution of osteopontin to the development of bone metastasis. *In* "Bone Metastasis: Experimental and Clinical Therapeutics" (G. Singh and S.A. Rabbani, eds), pp. 107–124. Humana Press Inc., Totowa, NJ.

Alleva, R., Scararmucci, A., Mantero, F., Bompadre, S., Leoni, L., and Littarru, G.P. 1997. The protective role of ubiquinol-10 against formation of lipid hydroperoxides in human seminal fluid. *Mol. Aspects Med.* **18**(Suppl.), S221–S228.

Altieri, D.C. 2003. Validating survivin as a cancer therapeutic target. *Nat. Rev. Cancer* **3**, 46–54.

Alvarez, J.G. and Storey, B.T. 1983. Taurine, hypotaurine, epinephrine and albumin inhibit lipid peroxidation in rabbit spermatozoa and protect against loss of motility. *Biol. Reprod.* **29**, 548–555.

Ambani, L.M., Van Woert, M.H., and Murphy, S. 1975. Brain peroxidase and catalase in Parkinson's disease. *Arch. Neurol.* **32**, 114–118.

Ames, B.N., Shigenaga, M.K., and Hagen, T.M. 1993. Oxidants, antioxidants and the degenerative diseases of aging. *Proc. Natl. Acad. Sci. USA* **90**, 7915–7922.

Ames, B.N., Gold, L.S., and Willet, W.C. 1996. Causes and prevention of cancer. *Proc. Natl. Acad. Sci. USA* **92**, 5258–5265.

Amir, H., Karas, M., Giat, J., Danilenko, M., Levy, R., Yermiahu, T., Levy, J., and Sharoni, Y. 1999. Lycopene and $1{,}25(OH)_2D_3$ cooperate in the inhibition of cell cycle progression and induction of HL-60 leukemic cells. *Nutr. Cancer* **33**, 105–112.

Ansari, M.S. and Gupta, N.P. 2003. A comparison of lycopene and orchidectomy vs orchidectomy alone in the management of advanced prostate cancer. *BJU Int.* **92**, 375–378.

Arab, L. and Steck, S. 2000. Lycopene and cardiovascular disease. *Am. J. Clin. Nutr.* **71**(Suppl.), S1691–S1695.

Astrog, P., Gradelet, S., Berges, R., and Suschetet, M. 1997. Dietary lycopene decreases initiation of liver preneoplastic foci by diethylnitrosamine in rat. *Nutr. Cancer* **29**, 60–68.

Aust, O., Ale-Agha, N., Zhang, L., Wollersen, H., Sies, H., and Stahl, W. 2003. Lycopene oxidation product enhances gap junction communication. *Food Chem. Toxicol.* **41**, 1399–1407.

Avitabile, M., Campagna, N.E., Magri, G.A., Vinci, M., Sciacca, G., Alia, G., and Ferro, A. 1991. Correlation between serum glutathione reductases and bone densitometry values. [Italian]. *Bollettino–Societa Italiana Biologia Sperimentale* **67**, 931–937.

Bardou, V.-J., Arpino, G., Elledge, R.M., Osborne, C.K., and Clark, G.M. 2003. Progesterone receptor status significantly improves outcome prediction over estrogen receptor status alone for adjuvant endocrine therapy in two large breast cancer databases. *J. Clin. Oncol.* **21**, 1973–1979.

Baselga, J., Tripathy, D., Mendelsohn, J., Baughman, S., Benz, C., Dantis, L., Sklarin, N., Seidman, A., Hudis, C., Moore, J., Rosen, P., Twaddell, T. *et al.* 1996. Phase II study of weekly intravenous recombinant humanized anti-p185HER2 monoclonal antibody in patients with HER2/neu-overexpressing metastatic breast cancer. *J. Clin. Oncol.* **14**, 737–744.

Basu, S., Michaelsson, K., Olofsson, H., Johansson, S., and Melhus, H. 2001. Association between oxidative stress and bone mineral density. *Biochem. Biophys. Res. Commun.* **288**, 275–279.

Bax, B.E., Alam, A.S., Banerji, B., Bax, C.M., Bevis, P.J., Stevens, C.R., Moonga, B.S., Blake, D.R., and Zaidi, M. 1992. Stimulation of osteoclastic bone resorption by hydrogen peroxide. *Biochem. Biophys. Res. Commun.* **183**, 1153–1158.

Ben-Dor, A., Nahum, A., Danilenko, M., Giat, Y., Stahl, W., Martin, H.D., Emmerich, T., Noy, N., Levy, J., and Sharoni, Y. 2001. Effects of acyclo-retinoic acid and lycopene on activation of the retinoic acid receptor and proliferation of mammary cancer cells. *Arch. Biochem. Biophys.* **391**, 295–302.

Berger, C.E., Horrocks, B.R., and Datta, H.K. 1999. Direct non-genomic effect of steroid hormones on superoxide generation in the bone resorbing osteoclasts. *Mol. Cell. Endocrinol.* **149**, 53–59.

Bhuvaneswari, V., Velmurugan, B., and Nagini, S. 2002. Induction of glutathione-dependent hepatic biotransformation enzymes by lycopene in the hamster cheek pouch carcinogenesis model. *J. Biochem. Mol. Biol. Biophys.* **6**, 257–260.

Blakely, S., Brown, E., Babu, U., Grundel, E., and Mitchell, G. 1994. Bioavailability of carotenoids in tomato paste and dried spinach and their interactions with canthaxanthin. *FASEB J.* **8**, 192.

Block, G. 1994. Health Habits and History Questionaire: Diet history and other risk factors. *Dietary Analysis System Packet*. Bsathesda, Md.: National Cancer Institute.

Boileau, T., Liao, Z., Kim, S., Lemeshow, S., Erdman, J., and Clinton, S. 2003. Prostate carcinogenesis in N-methyl-N-nitrosourea (NMU)-testosterone-treated rats fed tomato powder, lycopene, or energy-restricted diets. *J. Natl. Cancer Inst.* **95**, 1578–1586.

Borlongan, C.V., Kanning, K., Poulos, S.G., Freeman, T.B., Cahill, D.W., and Sanberg, P.R. 1996. Free radical damage and oxidative stress in Huntington's disease. *J. Fla. Med. Assoc.* **83**, 335–341.

Bowen, P., Chen, L., Stacewicz-Sapuntzakis, M., Duncan, C., Sharifi, R., Ghosh, L., Kim, H.S., Christov-Tzelkov, K., and van Breemen, R. 2002. Tomato sauce supplementation and prostate cancer: Lycopene accumulation and modulation of biomarkers of carcinogenesis. *Exp. Biol. Med. (Maywood)* **227**, 886–893.

Boyd, N.F. and McGuire, V. 1990. Evidence of lipid peroxidation in premenopausal women with mammographic dysplasia. *Cancer Lett.* **50**, 31–37.

Boyd, N.F., Connelly, P., Byng, J., Yaffe, M., Draper, H., Little, L., Jones, D., Martin, L., Lockwood, G., and Tritchler, D. 1995. Plasma lipids, lipoproteins, and mammographic densities. *Cancer Epidemiol. Biomarkers Prev.* **4**, 727–733.

Brady, W.E., Mares-Perlman, J.A., Bowen, P., and Stacewicz-Sapuntzakis, M. 1997. Human serum carotenoid concentrations are related to physiologic and lifestyle factors. *J. Nutr.* **126**, 129–137.

Britton, G. 1995. Structure and propertieds of carotenoids in relation to function. *FASEB J.* **9**, 1551–1558.

Callagy, G., Dimitriadis, E., Harmey, J., Bouchier-Hayes, D., Leader, M., and Kay, E. 2000. Immunohistochemical measurement of tumor vascular endothelial growth factor in breast cancer. A more reliable predictor of tumor stage than microvessel density or serum vascular endothelial growth factor. *Appl. Immunohistochem. Mol. Morphol.* **8**, 104–109.

Cardinali, D.P., Ladizesky, M.G., Boggio, V., Cutrera, R.A., and Mautalen, C. 2003. Melatonin effects on bone: Experimental facts and clinical perspectives. *J. Pineal Res.* **34**, 81–87.

Carter, D., Douglass, J.F., Cornellison, C.D., Retter, M.W., Johnson, J.C., Bennington, A.A., Fleming, T.P., Reed, S.G., Houghton, R.L., Diamond, D.L., and Vedvick, T.S. 2002. Purification and characterization of the mammaglobin/lipophilin B complex, a promising diagnostic marker for breast cancer. *Biochemistry* **41**, 6714–6722.

Celis, J.E., Gromov, P., Cabezon, T., Moreira, J.M., Ambartsumian, N., Sandelin, K., Rank, F., and Gromova, I. 2004. Proteomic characterization of the interstitial fluid perfusing the breast tumor microenvironment: A novel resource for biomarker and therapeutic target discovery. *Mol. Cell. Proteomics* **3**, 327–344.

Chan, G.K. and Duque, G. 2002. Age-related bone loss: Old bone, new facts. *Gerontology* **48**, 62–71.

Chasse, G.A., Mak, M.L., Deretey, E., Farkas, I., Torday, L.L., Papp, J.G., Sarma, D.S.R., Agarwal, A., Chakravarthi, S., Agarwal, S., and Rao, A.V. 2001. An *ab initio* computational study on selected lycopene isomers. *J. Mol. Struc. (Theochem.)* **571**, 27–37.

Chew, B. and Park, J. 2004. Carotenoid action on immune response. *J. Nutr.* **134**, S257–S261.

Clarke, R., Leonessa, F., Welch, J.N., and Skaar, T.C. 2001. Cellular and molecular pharmacology of antiestrogen action and resistance. *Pharmacol. Rev.* **53**, 25–71.

Claudio, P.P., Zamparelli, A., Garcia, F.U., Claudio, L., Ammirati, G., Farina, A., Bovicelli, A., Russo, G., Giordano, G.G., McGinnis, D.E., Giordano, A., and Cardi, G. 2002. Expression of cell-cycle-regulated proteins pRb2/p130, p107, p27(kip1), p53, mdm-2, and Ki-67 (MIB-1) in prostatic gland adenocarcinoma. *Clin. Cancer Res.* **8**, 1808–1815.

Clinton, S., Emenhoser, C., and Schwartz, S. 1996. *Cis-trans* lycopene isomers, carotenoids and retinol in human prostate. *Cancer Epidemiol. Biomarkers Prev.* **5**, 823–833.

Clinton, S.K. 1998. Lycopene: Chemistry, biology, and implications for human health and disease. *Nutr. Rev.* **1**, 35–51.

Cohen, L.A., Zhao, Z., Pittman, B., and Khachik, F. 1999. Effect of dietary lycopene on N-methylnitrosourea-induced mammary tumorigenesis. *Nutr. Cancer* **34**, 153–159.

Colditz, G.A., Branch, L.G., Lipnick, R.J., Willett, W.C., Rosner, B., Posner, B.M., and Hennekens, C.H. 1985. Increased green and yellow vegetable intake and lowered cancer deaths in an elderly population. *Am. J. Clin. Nutr.* **41**, 32–36.

Cortizo, A.M., Bruzzone, L., Molinuevo, S., and Etcheverry, S.B. 2000. A possible role of oxidative stress in the vanadium-induced cytotoxicity in the MC3T3E1 osteoblast and UMR106 osteosarcoma cell lines. *Toxicology* **147**, 89–99.

Darden, A.G., Ries, W.L., Wolf, W.C., Rodriguiz, R.M., and Key, L.L., Jr. 1996. Osteoclastic superoxide production and bone resorption: Stimulation and inhibition by modulators of NADPH oxidase. *J. Bone Miner. Res.* **11**, 671–675.

Datta, H.K., Manning, P., Rathod, H., and McNeil, C.J. 1995. Effect of calcitonin, elevated calcium and extracellular matrices on superoxide anion production by rat osteoclasts. *Exp. Physiol.* **80**, 713–719.

Datta, H.K., Rathod, H., Manning, P., Turnbull, Y., and McNeil, C.J. 1996. Parathyroid hormone induces superoxide anion burst in the osteoclasts: Evidence of the direct instantaneous activation of the osteoclast by the hormone. *J. Endocrinol.* **149**, 269–275.

Dawson, V.L., Dawson, T.M., Bartley, D.A., Uhl, G.R., and Snyder, S.H. 1993. Mechanisms of nitric oxide-mediated neurotoxicity in primary brain cultures. *J. Neurosci.* **13**, 2651–2661.

Di Mascio, P., Kaiser, S., and Sies, H. 1989. Lycopene as the most efficient biological carotenoid singlet oxygen quencher. *Arch. Biochem. Biophys.* **274**, 532–538.

Diaz, M., Frei, B., Vita, J.A., and Keaney, J.F. 1997. Antioxidants and atherosclerotic heart disease. *N. Engl. J. Med.* **337**, 408–416.

dos Anjos Ferreira, A.L., Yeum, K.J., Russell, R.M., Krinsky, N.I., and Tang, G. 2004. Enzymatic and oxidative metabolites of lycopene. *J. Nutr. Biochem.* **15**, 493–502.

Dubin, L. and Amelar, R.D. 1971. Etiologic factors in 1294 consecutive cases of male infertility. *Fertil. Steril.* **22**, 469–474.

Dugas, T.R., Morel, D.W., and Harrison, E.H. 1998. Impact of LDL carotenoid and alpha-tocopherol content on LDL oxidation by endothelial cells in culture. *J. Lipid Res.* **39**, 999–1007.

Dugas, T.R., Morel, D.W., and Harrison, E.H. 1999. Dietary supplementation with b-carotene, but not with lycopene inhibits endothelial cell-mediated oxidation of low-density lipoprotein. *Free Radic. Biol. Med.* **26**, 1238–1244.

Early Breast Cancer Trialists' Collaborative Group 1992. Systemic treatment of early breast cancer by hormonal, cytotoxic, or immune therapy. 133 randomised trials involving 31,000 recurrences and 24,000 deaths among 75,000 women. *Lancet* **339**, 1–15.

Ebadi, M., Srinivasan, S.K., and Baxi, M.D. 1996. Oxidative stress and antioxidant therapy in Parkinson's disease. *Prog. Neurobiol.* **48**, 1–19.

Esteva, F.J., Sahin, A.A., Smith, T.L., Yang, Y., Pusztai, L., Nahta, R., Buchholz, T.A., Buzdar, A. U., Hortobagyi, G.N., and Bacus, S.S. 2004. Prognostic significance of phosphorylated P38 - mitogen-activated protein kinase and HER-2 expression in lymph node-positive breast carcinoma. *Cancer* **100**, 499–506.

Fahn, S. 1991. An open trial of high-dosage antioxidants in early Parkinson's disease. *Am. J. Clin. Nutr.* **53**, S380–S382.

Fahn, S. and Cohen, G. 1992. The oxidant stress hypothesis in Parkinson's disease: Evidence supporting it. *Ann. Neurol.* **32**, 804–812.

Farley, J., Smith, L.M., Darcy, K.M., Sobel, E., O'Connor, D., Henderson, B., Morrison, L.E., and Birrer, M.J. 2003. Cyclin E expression is a significant predictor of survival in advanced, suboptimally debulked ovarian epithelial cancers: A gynecologic oncology group study. *Cancer Res.* **63**, 1235–1241.

Ferrante, R.J., Browne, S.E., Shinobu, L.A., Bowling, A.C., Baik, M.J., MacGarvey, U., Kowall, N. W., Brown, R.H., Jr., and Beal, M.F. 1997. Evidence of increased oxidative damage in both sporadic and familial amyotrophic lateral sclerosis. *J. Neurochem.* **69**, 2064–2074.

Ferruzzi, M., Nguyen, M., Sander, L., Rock, C., and Schwartz, S. 2001. Analysis of lycopene geometrical isomers in biological microsamples by liquid chromatography with coulometric array detection. *J. Chromatogr.* **760**, 289–299.

Forman, M.R., Beecher, G.R., Muesing, R., Lanza, E., Olson, B., Campbell, W.S., McAdam, P., Raymond, E., Schulman, J.D., and Graubard, B.I. 1996. The fluctuation of plasma carotenoid concentrations by phase of menstrual cycle: A controlled diet study. *Am. J. Clin. Nutr.* **64**, 559–565.

Forssberg, A., Lingen, C., Ernster, L., and Lindenberg, O. 1959. Modification of x-irradiated syndrom by lycopene. *Exp. Cell Res.* **16**, 7–14.

Foy, C.J., Passmore, A.P., Vahidassr, M.D., Young, I.S., and Lawson, J.T. 1999. Plasma chain-breaking antioxidants in Alzheimer's disease, vascular dementia and Parkinson's disease. *QJM* **92**, 39–45.

Fraga, C.G., Motchnik, P.A., Shigenaga, M.K., Helbock, H.J., Jacob, R.A., and Ames, B.N. 1991. Ascorbic acid protects against endogenous oxidative DNA damage in human sperm. *Proc. Natl. Acad. Sci. USA* **88**, 11003–11006.

Franceschi, S., Bidoli, E., La Vecchia, C., Talamini, R., D'Avanzo, B., and Negri, E. 1994. Tomatoes and risk of digestive-tract cancers. *Int. J. Cancer* **59**, 181–184.

Fraser, J.H., Helfrich, M.H., Wallace, H.M., and Ralston, S.H. 1996. Hydrogen peroxide, but not superoxide, stimulates bone resorption in mouse calvariae. *Bone* **19**, 223–226.

Friedman, J., Peleg, E., Kagan, T., Shnizer, S., and Rosenthal, T. 2003. Oxidative stress in hypertensive, diabetic, and diabetic hypertensive rats. *Am. J. Hypertens.* **16**, 1049–1052.

Frorman, M.R., Lanza, E., Yong, L.C., Holden, J.M., Graubard, B.I., Beecher, G.R., Meltiz, M., Brown, E.D., and Smith, J.C. 1993. The correlation between two dietary assessments of carotenoid intake and plasma carotenoid concentrations: Application of a carotenoid food composition database. *Am. J. Clin. Nutr.* **58**, 519–524.

Fu, M., Wang, C., Li, Z., Sakamaki, T., and Pestell, R.G. 2004. Minireview: Cyclin D1: Normal and abnormal functions. *Endocrinology* **145**, 5439–5447.

Fuhramn, B., Elis, A., and Aviram, M. 1997. Hypocholesterolemic effect of lycopene and b-carotene is related to suppression of cholesterol synthesis and augmentation of LDL receptor activity in macrophage. *Biochem. Biophys. Res. Commun.* **233**, 658–662.

Fuqua, S.A. and Cui, Y. 2004. Estrogen and progesterone receptor isoforms: Clinical significance in breast cancer. *Breast Cancer Res. Treat.* **87**(Suppl. 1), S3–S10.

Gaffney, D.K., Haslam, D., Tsodikov, A., Hammond, E., Seaman, J., Holden, J., Lee, R.J., Zempolich, K., and Dodson, M. 2003. Epidermal growth factor receptor (EGFR) and vascular endothelial growth factor (VEGF) negatively affect overall survival in carcinoma of the cervix treated with radiotherapy. *Int. J. Radiat. Oncol. Biol. Phys.* **56**, 922–928.

Gann, P., Ma, J., Giovannucci, E., Willett, W., Sacks, F.M., and Hennekens, C.H. 1999. Lower prostate cancer risk in men with elevated plasma lycopene levels: Results of a prospective analysis. *Cancer Res.* **59**, 1225–1230.

Garnero, P., Sornay-Rendu, E., Chapuy, M.-C, and Delmas, P.D. 1996. Increased bone turnover in late postmenopausal women is a major determinant of osteoporosis. *J. Bone Miner. Res.* **11**, 337–349.

Garrett, I.R., Boyce, B.F., Oreffo, R.O.C., Bonewald, L., Pser, J., and Mundy, G.R. 1990. Oxygen-derived free radicals stimulate osteoclastic bone resorption in rodent bone *in vitro* and *in vivo*. *J. Clin. Invest.* **85**, 632–639.

Gärtner, C., Stahl, W., and Sies, H. 1997. Lycopene is more bioavailable from tomato paste than from fresh tomatoes. *Am. J. Clin. Nutr.* **66**, 116–122.

Gaziano, J., Johnson, E., Russell, R., Manson, J., Stampfer, M., Ridker, P., Frei, B., Hennekens, C., and Krinsky, N. 1995. Discrimination in absorption or transport of beta-carotene isomers after oral supplementation with either all-trans or 9-cis-beta-carotene. *Am. J. Clin. Nutr.* **61**, 1248–1252.

Gerster, H. 1997. The potential role of lycopene for human health. *J. Am. Coll. Nutr.* **16**, 109–126.

Geva, E., Bartoov, B., Zabludovsky, N., Lessing, J.B., Lerner-Geva, L., and Amit, A. 1996. The effect of antioxidant treatment on human spermatozoa and fertilization rate in an *in vitro* fertilization program. *Fertil. Steril.* **66**, 4320–4434.

Giovannucci, E. 1999. Tomatoes, tomato-based products, lycopene, and cancer: Review of the epidemiologic literature. *J. Natl. Cancer Inst.* **91**, 317–331.

Giovannucci, E., Ascherio, A., Rimm, E.B., Stampfer, M.J., Colditz, G.A., and Willett, W.C. 1995. Intake of carotenoids and retinol in relation to risk of prostate cancer. *J. Natl. Cancer Inst.* **87**, 1767–1776.

Giovannucci, E., Rimm, E.B., Liu, Y., Stampfer, M.J., and Willett, W.C. 2002. A prospective study of tomato products, lycopene, and prostate cancer risk. *J. Natl. Cancer Inst.* **94**, 391–398.

Gittes, R.F. 1991. Carcinoma of the prostate. *N. Engl. J. Med.* **324**, 236–245.

Gomez-Aracena, J., Sloots, J., and Garcia-Rodriguez, A. 1997. Antioxidants in adipose tissue and myocardial infarction in Mediterranean area. The EURAMIC study in Malaga. *Nutr. Metab. Cardiovasc. Dis.* **7**, 376–382.

Gradelet, S., LeBon, A.M., Berges, R., Suschetet, M., and Astorg, P. 1998. Dietary carotenoids inhibit aflatoxin B1-induced liver preneoplastic foci and DNA damage in rats: Role of modulation of aflatoxin B1 metabolism. *Carcinogenesis* **19**, 403–411.

Grant, W.B. 1997. Dietary links to Alzheimer's disease. *Alzheimer's Dis. Rev.* **2**, 42–55.

Greenberg, S.H., Lipshuitz, L.L., and Wein, A.J. 1978. Experience with 425 subfertile male patients. *J. Urol.* **119**, 507–510.

Guttenplan, J., Chen, M., Kosinska, W., Thompson, S., Zhao, Z., and Cohen, L. 2001. Effects of a lycopene-rich diet on spontaneous and benzo[a]pyrene-induced mutagenesis in prostate, colon and lungs of the lacZ mouse. *Cancer Lett.* **164**, 1–6.

Hainaut, P. and Miller, J. 1993. Redox modulation of p53 conformation and sequence-specific DNA binding. *Cancer Res.* **53**, 4469–4473.

Hall, A. 1996. Liarozole amplifies retinoid-induced apoptosis in human prostate cancer cells. *Anticancer Drugs* **7**, 12–20.

Hall, T.J., Schaeublin, M., Fuller, K., and Chambers, T.J. 1995. The role of oxygen intermediates in osteoclastic bone resorption. *Biochem. Biophys. Res. Commun.* **207**, 280–287.

Halleen, J.M., Raisanen, S., Salo, J.J., Reddy, S.V., Roodman, G.D., Hentunen, T.A., Lehenkari, P.P., Kaija, H., Vihko, P., and Vaananen, H.K. 1999. Intracellular fragmentation of bone resorption products by reactive oxygen species generated by osteoclastic tartrate-resistant acid phosphatase. *J. Biol. Chem.* **274**, 22907–22910.

Halliwell, B. 1989. Oxidants and the central nervous system: Some fundamental questions. *Acta Neurol. Scand.* **126**, 23–33.

Halliwell, B., Cross, C.E., and Gutteridge, J.M.C. 1992. Free radicals, antioxidants and human diseases: Where are we now? *J. Lab. Clin. Med.* **119**, 598–620.

Halliwell, B., Murcia, M.A., Chirico, S., and Aruoma, O.I. 1995. Free radicals and antioxidants in food and *in vivo*: What they do and how they work. *Crit. Rev. Food Sci. Nutr.* **35**, 7–20.

Handelman, G.J., Packer, L., and Cross, C.E. 1996. Destruction of tocopherols, carotenoids and retinol in human plasma by cigarette smoke. *Am. J. Clin. Nutr.* **63**, 559–565.

Heber, D. 2002. Mechanisms of action of lycopene: Overview. *In* "Lycopene and the Prevention of Chronic Diseases" (A.V. Rao and D. Heber., eds), Vol. 1, pp. 41–42. Caledonian Science Press, Scotland.

Heber, D. and Lu, Q.-L. 2002. Overview of mechanisms of action of lycopene. *Exp. Biol. Med. (Maywood)* **227**, 920–923.

Heller, F.R., Descamps, O., and Hondekijn, J.C. 1998. LDL oxidation: Therapeutic perspectives. *Atherosclerosis* **137**, S25–S31.

Hodis, H.N., Mack, W.J., LaBree, L., Cashin-Hemphill, L., Sevanian, A., Johnson, R., and Azen, S.P. 1995. Serial coronary angiographic evidence that antioxidant vitamin intake reduces progression of coronary artery atheroscloresis. *JAMA* **273**, 1849–1854.

Hosokawa, M. 2002. A higher oxidative status accelerates senescence and aggravates age-dependent disorders in SAMP strains of mice. *Mech. Ageing Dev.* **123**, 1553–1561.

Hsing, A.W., Comstock, G.W., Abbey, H., and Polk, B.F. 1990. Seriologic precursors of cancer. Retinol, carotenoids, and tocopherol and risk of prostate cancer. *J. Natl. Cancer Inst.* **82**, 941–946.

Ignatoski, K.M., Maehama, T., Markwart, S.M., Dixon, J.E., Livant, D.L., and Ethier, S.P. 2000. ERBB-2 overexpression confers PI 3′ kinase-dependent invasion capacity on human mammary epithelial cells. *Br. J. Cancer* **82**, 666–674.

Imaida, K., Tamano, S., Kato, K., Ikeda, Y., Asamoto, M., Takahashi, S., Nir, Z., Murakoshi, M., Nishino, H., and Shirai, T. 2001. Lack of chemopreventive effects of lycopene and curumin on experimental rat prostate carcinogenesis. *Carcinogenesis* **22**, 467–472.

Iribarren, C., Folsom, A.R., Jacobs, D.R., Gross, M.D., Belcher, J.D., and Eckfeldt, J.H. 1997. Association of serum vitamin levels, LDL susceptibility to oxidation, and autoantibodies against MDA-LDL with carotid atherosclerosis. *Arterioscler. Thromb. Vasc. Biol.* **17**, 1171–1177.

Ishimi, Y., Ohmura, M., Wang, X., Yamaguchi, M., and Ikegami, S. 1999. Inhibition by carotenoids and retinoic acid of osteoclast-like cell formation induced by bone-resorbing agents *in vitro*. *J. Clin. Biochem. Nutr.* **27**, 113–122.

Iwasaki, A. and Gagnon, C. 1992. Formation of reactive oxygen species in spermatozoa of infertile patients. *Fertil. Steril.* **57**, 409–416.

Jain, C.K., Agarwal, S., and Rao, A.V. 1999. The effect of dietary lycopene on bioavailability, tissue distribution, *in-vivo* antioxidant properties and colonic preneoplasia in rats. *Nutr. Res.* **19**, 1383–1391.

Jenner, P. 1996. Oxidative stress in Parkinson's disease and other neurodegenerative disorders. *Pathol. Biol.* **44**, 57–64.

Johnson, W. 1975. Proceedings. 120 Infertile men. *Br. J. Urol.* **47**, 230.

Jonker, D., Kuper, C., Fraile, N., Estrella, A., and Rodriguez, O. 2003. Ninety-day oral toxicity study of lycopene from *Blakeslea trispora* in rats. *Regul. Toxicol. Pharmacol.* **37**, 396–406.

Kang, Y., Siegel, P.M., Shu, W., Drobnjak, M., Kakonen, S.M., Cordon-Cardo, C., Guise, T.A., and Massague, J. 2003. A multigenic program mediating breast cancer metastasis to bone. *Cancer Cell* **3**, 537–549.

Karas, M., Amir, H., Fishman, D., Danilenko, M., Segal, S., Nahum, A., Koifmann, A., Giat, Y., Levy, J., and Sharoni, Y. 2000. Lycopene interferes with cell cycle progression and insulin-like growth factor I signaling in mammary cancer cells. *Nutr. Cancer* **36**, 101–111.

Kenny, A.M. and Raisz, L.G. 2002. Mechanism of bone remodelling: Implications for clinical practice. *J. Rep. Med.* **47**, 63–70.

Keshgegian, A.A. and Cnaan, A. 1995. Proliferation markers in breast carcinoma. Mitotic figure count, S-phase fraction, proliferating cell nuclear antigen, Ki-67 and MIB-1. *Am. J. Clin. Pathol.* **104**, 42–49.

Kessopoulou, E., Powers, H.J., Sharma, K.K., Pearson, M.J., Russell, J.M., Cooke, I.D., and Barratt, C.L. 1995. A double-blind randomized placebo cross-over controlled trial using the antioxidant vitamin E to treat reactive oxygen species associated male infertility. *Fertil. Steril.* **64**, 825–831.

Key, L.L., Ries, W.L., Taylor, R.G., Hays, B.D., and Pitzer, B.L. 1990. Oxygen derived free radicals in osteoclasts: The specificity and location of the nitroblue tetrazolium reaction. *Bone* **11**, 115–119.

Key, L.L., Wolf, W.C., Gundberg, C.M., and Ries, W.L. 1994. Superoxide and bone resorption. *Bone* **15**, 431–436.

Khachik, F., Beecher, G.R., and Smith, J.C., Jr. 1995. Lutein, lycopene and their oxidative metabolite in chemoprevention of cancer. *J. Cell. Biochem.* **22**, 236–246.

Khachik, F., Spangler, C.J., Smith, J.C., Jr., Canfield, L.M., Steck, A., and Pfander, H. 1997. Identification, quantification, and relative concentrations of carotenoids and their metabolites in human milk and serum. *Anal. Chem.* **69**, 1873–1881.

Khachik, F., Carvallo, L., Bernstein, P.S., Muir, G.J., Zhao, D.Y., and Katz, N.B. 2002. Chemistry, distribution and metabolism of tomato carotenoids and their impact on human health. *Exp. Biol. Med. (Maywood)* **227**, 845–851.

Kim, D.J., Takasuka, N., Nishino, H., and Tsuda, H. 2000. Chemoprevention of lung cancer by lycopene. *Biofactors* **13**, 95–102.

Kim, G.Y., Kim, J.H., Ahn, S.C., Lee, H.J., Moom, D.O., Lee, C.M., and Park, Y.M. 2004. Lycopene suppresses the lipopolysaccharide-induced phenotypic and functional maturation of murine dendritic cells through inhibition of mutogen-activated protein kinases and nuclear factor-kappaB. *Immunology* **113**, 203–211.

Kim, J.M., Takasuka, N., Kim, J.M., Sekine, K., Ota, T., Asamoto, M., Murakoshi, M., Nishino, H., Nir, Z., and Tsuda, H. 1997. Chemoprevention by lycopene of mouse lung neoplasia after combined initiation treatment with DEN, MNU and DMH. *Cancer Lett.* **120**, 15–22.

Kim, J.M., Araki, S., Kim, D.J., Park, C.B., Takasuka, N., Baba-Toriyama, H., Ota, T., Nir, Z., Khachik, F., Shimidzu, N., Tanaka, Y., Osawa, T. *et al.* 1998. Chemopreventive effects of carotenoids and curcumins on mouse colon carcinogenesis after 1,2-dimethylhydrazine initiation. *Carcinogenesis* **19**, 81–85.

Kim, L., Rao, A.V., and Rao, L.G. 2002. Effect of lycopene on prostate LNCaP cancer cells in culture. *J. Med. Food* **5**, 181–187.

Kim, L., Rao, A.V., and Rao, L.G. 2003. Lycopene II—Effect on osteoblasts: The caroteroid lycopene stimulates cell proliferation and alkaline phosphatase activity of SaOS-2 cells. *J. Med. Food* **6**, 79–86.

Kim, L.S., Huang, S., Lu, W., Lev, D.C., and Price, J.E. 2004. Vascular endothelial growth factor expression promotes the growth of breast cancer brain metastases in nude mice. *Clin. Exp. Metastasis* **21**, 107–118.

Kish, S.J., Morito, C., and Hornykiewicz, O. 1985. Glutathione peroxidase activity in Parkinson's disease. *Neurosci. Lett.* **58**, 343–346.

Kobayashi, T., Lijima, K., Mitamura, T., Toriizuka, K., Cyong, J., and Nagasawa, H. 1996. Effects of lycopene, a carotenoid, on intrathymic T cell differentiation and peripheral CD4/CD8 ratio in high mammary tumor strain of SHN retired mice. *Anticancer Drugs* **7**, 195–198.

Kohlmeier, L. and Hastings, S.B. 1995. Epidemiologic evidence of a role of carotenoids in cardiovascular disease prevention. *Am. J. Clin. Nutr.* **62**(Suppl. 6), S1370–S1376.

Kohlmeier, L., Kark, J.D., Gomez-Garcia, E., Martin, B.C., Steck, S.E., Kardinaal, A.F.M., Ringstad, J., Thamm, M., Masaev, V., Riemersma, R., Martin-Moreno, J.M., Huttunen, J.K. *et al.* 1997. Lycopene and myocardial infarction risk in the EURAMIC study. *Am. J. Epidemiol.* **146**, 618–626.

Kotake-Nasra, E., Kushiro, M., Zhang, H., Sugawara, T., Miyashita, K., and Nagao, A. 2001. Carotenoids affect proliferation of human prostate cancer cells. *J. Nutr.* **131**, 3303–3306.

Kozuki, Y., Miura, Y., and Yagasaki, K. 2000. Inhibitory effects of carotenoids on the invasion of rat ascites hepatoma cells in culture. *Cancer Lett.* **151**, 111–115.

Kristenson, M., Ziedén, B., Kucinskienë, Z., Abaravicius, A., Razinkovienë, L., Elinder, L.S., Bergdahl, B., Elwing, B., Calkauskas, H., and Olsson, A.G. 1997. Antioxidant state and mortality from coronary heart disease in Lithuanian and Swedish men: Concomitant cross sectional study of men aged 50. *BMJ* **314**, 629–633.

Kucuk, O., Sarkar, F.H., Sakr, W., Djuric, Z., Pollak, M.N., Khachik, F., Li, Y.-W, Banerjee, M., Grignon, D., Bertram, J.S., Crissman, J.D., Pontes, E.J. *et al.* 2001. Phase II randomized clinical trial of lycopene supplementation before radical prostatectomy. *Cancer Epidemiol. Biomarkers Prev.* **10**, 861–868.

Kucuk, O., Sarkar, F., Djuric, Z., Sakr, W., Pollak, M., Khachik, F., Banerjee, M., Bertram, J., and Wood, D.P., Jr. 2002. Effects of lycopene supplementation in patients with localized prostate cancer. *Exp. Biol. Med.* **227**, 881–885.

Kushida, K., Takahashi, M., Kawana, K., and Inoue, T. 1995. Comparison of markers for bone formation and resorption in premenopausal and postmenopausal subjects, and osteoporosis patients. *J. Clin. Endocrinol. Metab.* **80**, 2447–2450.

Lassegue, B. and Griendling, K.K. 2004. Reactive oxygen species in hypertension. *Am. J. Hypertens.* **17**, 852–860.

LaVecchia, C. 1997. Mediterranean epidemiological evidence on tomatoes and the prevention of digestive tract cancers. *Proc. Soc. Exp. Biol. Med.* **218**, 125–128.

Le Naour, F., Misek, D.E., Krause, M.C., Deneux, L., Giordano, T.J., Scholl, S., and Hanash, S.M. 2001. Proteomics-based identification of RS/DJ-1 as a novel circulating tumor antigen in breast cancer. *Clin. Cancer Res.* **7**, 3328–3335.

Lenzi, A., Gandini, L., and Picardo, M. 1998. A rationale for glutathione therapy. Debate on: Is antioxidant therapy a promising strategy to improve human reproduction? *Hum. Reprod.* **13**, 1419–1424.

Leveille, S.G., LaCroix, A.Z., Koepsell, T.D., Beresford, S.A., VanBelle, G., and Buchner, D.M. 1997. Dietary vitamin C and bone mineral density in postmenopausal women in Washington State, USA. *J. Epidemiol. Community Health* **51**, 479–485.

Lewin, A. and Lavon, H. 1997. The effect of coenzyme Q10 on sperm motility and function. *Mol. Aspects Med.* **18**(Suppl.), S213–S219.

Limpens, J., Weerden, W., Kramer, K., Pallapies, D., Obermuller-Jevic, U., and Schroder, F. 2004. Re: Prostate carcinogenesis in N-methyl-N-nitrosourea (NMU)-testosterone-treated rats fed tomato powder, lycopene, or energy restricted diets. *J. Natl. Cancer Inst.* **96**, 554–557.

Lindahl, T., Landberg, G., Ahlgren, J., Nordgren, H., Norberg, T., Klaar, S., Holmberg, L., and Bergh, J. 2004. Overexpression of cyclin E protein is associated with specific mutation types in the *p53* gene and poor survival in human breast cancer. *Carcinogenesis* **25**, 375–380.

Lindsay, R. and Cosman, F. 1990. Prevention of osteoporosis. *In* "Primer on the Metabolic Bone Diseases and Disorders of Mineral Metabolism" (M.J. Favus, ed.), pp. 264–270. Lippincott Williams & Wilkins, New York.

Lingen, C., Ernster, L., and Lindenberg, O. 1959. The promoting effects of lycopene on the non-specific resistance of animals. *Exp. Cell. Res.* **16**, 384–393.

Liu, A.Y., Zhang, H., Sorensen, C.M., and Diamond, D.L. 2005. Analysis of prostate cancer by proteomics using tissue specimens. *J. Urol.* **173**, 73–78.

Liu, H.-C., Cheng, R.-M., Lin, F.-H., and Fang, H.-W. 1999. Sintered beta-dicalcium phosphate particles induce intracellular reactive oxygen species in rat osteoblasts. *Biomed. Eng. Appl. Basis Commun.* **11**, 259–264.

Livny, O., Kaplan, I., Reifen, R., Polak-Charcon, S., Madar, Z., and Schwartz, B. 2002. Lycopene inhibits proliferation and enhances gap-junction communication of KB-1 human oral tumor cells. *J. Nutr.* **132**, 3754–3759.

Livny, O., Kaplan, I., Reifen, R., Polak-Charcon, S., Madar, Z., and Schwartz, B. 2003. Oral cancer cells differ from normal oral epithelial cells in tissue like organization and in response to lycopene treatment: An organotypic cell culture study. *Nutr. Cancer* **47**, 195–209.

Loft, S. and Poulsen, H.E. 1996. Cancer risk and oxidative DNA damage in man. *J. Mol. Med.* **74**, 297–312.

Longnecker, M.P., Kamel, F., Umbach, D.M., Munsal, T.L., Shefuer, J.M., Lansdell, L.W., and Sandler, D.P. 2000. Dietary intake of calcium, magnesium and antioxidants in relation to risk of amyotrophic lateral sclerosis. *Neuroepidemiology* **19**, 210–216.

Maggio, D., Barabani, M., Pierandrei, M., Polidori, M.C., Catani, M., Mecocci, P., Senin, U., Pacifici, R., and Cherubini, A. 2003. Marked decrease in plasma antioxidants in aged osteoporotic women: Results of a cross-sectional study. *J. Clin. Endocrinol. Metabol.* **88**, 1523–1527.

Martin, B., Paesmans, M., Mascaux, C., Berghmans, T., Lothaire, P., Meert, A.P., Lafitte, J.J., and Sculier, J.P. 2004. Ki-67 expression and patients survival in lung cancer: Systematic review of the literature with meta-analysis. *Br. J. Cancer* **91**, 2018–2025.

Matsushima-Nishiwaki, R., Shidoji, Y., Nishiwaki, S., Yamada, T., Moriwaki, H., and Muto, Y. 1995. Suppression by carotenoids of microcystin-induced morphological changes in mouse hepatocytes. *Lipids* **30**, 1029–1034.

Matulka, R., Hood, A., and Griffiths, J. 2004. Safety evaluation of a natural tomato oleoresin extract derived from food-processing tomatoes. *Regul. Toxicol. Pharmacol.* **39**, 390–402.

McClain, R. and Bausch, J. 2003. Summary of safety studies conducted with synthetic lycopene. *Regul. Pharmacol. Toxicol.* **37**, 274–285.

McGuire, W.L. 1978. Hormone receptors: Their role in predicting prognosis and response to endocrine therapy. *Semin. Oncol.* **5**, 428–433.

Melhus, H., Michaelsson, K., Holmberg, L., Wolk, A., and Ljunghall, S. 1999. Smoking, antioxidant vitamins, and the risk of hip fracture. *J. Bone Miner. Res.* **14**, 129–135.

Mellert, W., Deckardt, K., Gembardt, C., Schulte, S., and Van Ravenzwaay, B. 2002. Thirteen-week oral toxicity study of synthetic lycopene products in rats. *Food Chem. Toxicol.* **40**, 1581–1588.

Menard, S., Fortis, S., Castiglioni, F., Agresti, R., and Balsari, A. 2001. HER2 as a prognostic factor in breast cancer. *Oncology* **61**(Suppl. 2), 67–72.

Mills, P.K., Beeson, W.L., Phillips, R.L., and Fraser, G.E. 1989. Cohort study of diet, lifestyle, and prostate cancer in Adventist men. *Cancer Epidemiol. Biomarkers Prev.* **64**, 598–604.

Minn, A.J., Kang, Y., Serganova, I., Gupta, G.P., Giri, D.D., Doubrovin, M., Ponomarev, V., Gerald, W.L., Blasberg, R., and Massague, J. 2005. Distinct organ-specific metastatic potential of individual breast cancer cells and primary tumors. *J. Clin. Invest.* **115**, 44–55.

Mitchell, J.J., Paiva, M., and Heaton, M.B. 1999. Vitamin E and beta-carotene protect against ethanol combined with ischemia in an embryonic rat hippacapal culture model of fetal alcohol syndrome. *Neurosci. Lett.* **263**, 189–192.

Mohanty, N.K., Kumar, R., and Gupta, N.P. 2001. Lycopene therapy in the management of idiopathic oligoasthenospermia. *Ind. J. Urol.* **56**, 102–103.

Molnar, J., Gyemant, N., Mucsi, I., Molnar, A., Szabo, M., Kortvelyesi, T., Varga, A., Molnar, P., and Toth, G. 2004. Modulation of multidrug resistance and apoptosis of cancer cells by selected carotenoids. *In Vivo* **18**, 237–244.

Moncada, M.L., Vicari, E., Cimino, C., Calogero, A.E., Mongioi, A., and D'Agata, R. 1992. Effect of acetylcarnitine in oligoasthenospermic patients. *Acta Eur. Fertil.* **23**, 221–224.

Moriel, P., Sevanian, A., Ajzen, S., Zanella, M.T., Plavnik, F.L., Rubbo, H., and Abdalla, D.S. 2002. Nitric oxide, cholesterol oxides and endothelium-dependent vasodialation in plasma of patients with essential hypertension. *Braz. J. Med. Biol. Res.* **35**, 1301–1309.

Morinaga, S., Nakamura, Y., Ishiwa, N., Yoshikawa, T., Noguchi, Y., Yamamoto, Y., Rino, Y., Imada, T., Takanashi, Y., Akaike, M., Sugimasa, Y., and Takemiya, S. 2004. Expression of survivin mRNA associates with apoptosis, proliferation and histologically aggressive features in hepatocellular carcinoma. *Oncol. Rep.* **12**, 1189–1194.

Morris, D.L., Kritchevsky, S.B., and Davis, C.E. 1994. Serum carotenoids and coronary heart disease: The lipid research clinics coronary primary prevention trial and follow-up study. *JAMA* **272**, 1439–1441.

Morton, D.J., Barrett-Connor, E.L., and Schneider, D.L. 2001. Vitamin C supplement and bone mineral density in postmenopausal women. *J. Bone Miner. Res.* **16**, 135–140.

Most, M.M. 2004. Estimated phytochemical content of the dietary approaches to stop hypertension (DASH) diet is higher than in the control study diet. *J. Am. Diet. Assoc.* **104**, 1725–1727.

Moul, J.W. 1999. Angiogenesis, p53, bcl-2 and Ki-67 in the progression of prostate cancer after radical prostatectomy. *Eur. Urol.* **35**, 399–407.

Muller-Tidow, C., Metzger, R., Kugler, K., Diederichs, S., Idos, G., Thomas, M., Dockhorn-Dworniczak, B., Schneider, P.M., Koeffler, H.P., Berdel, W.E., and Serve, H. 2001. Cyclin E is the only cyclin-dependent kinase 2-associated cyclin that predicts metastasis and survival in early stage non-small cell lung cancer. *Cancer Res.* **61**, 647–653.

Mundy, G.R. 1999. Bone remodeling. *In* "Primer on the Metabolic Bone Diseases and Disorders of Mineral Metabolism" (M.J. Favus, ed.), pp. 30–38. Lippincott Williams & Wilkins, New York.

Musarrat, J., Arezinawilson, J., and Wani, A.A. 1996. Prognostic and etiologic relevance of 8-hydroxyguanosine in human breast carcinogenesis. *Eur. J. Cancer* **32**, 1209–1214.

Nagao, A. 2004. Oxidative conversion of carotenoids to retinoids and other products. *J. Nutr.* **134**, S237–S240.

Nagasawa, H., Mitamura, T., Sakamoto, S., and Yamamoto, K. 1995. Effects of lycopene on spontaneous mammary tumor development in SHN virgin mice. *Anticancer Res.* **15**, 1173–1178.

Nam, S.H., Jung, S.Y., Yoo, C.M., Ahn, E.H., and Suh, C.K. 2002. H_2O_2 enhances Ca^{2+} release from osteoblast internal stores. *Yonsei Med. J.* **43**, 229–235.

Nara, E., Hayashi, H., Kotake, M., Miyashita, K., and Nagao, A. 2001. Acyclin carotenoids and their oxidation mixtures inhibit the growth of HL-60 human promyelocytic leukemia cells. *Nutr. Cancer* **39**, 273–283.

Narisawa, T., Fukaura, Y., Hasebe, M., Nomura, S., Oshima, S., Sakamoto, H., Inakuma, T., Ishiguro, Y., Takayasu, J., and Nishino, H. 1998. Prevention of N-methylnitrosourea-induced colon carcinogenesis in F344 rats by lycopene and tomato juice rich in lycopene. *Jpn. J. Cancer Res.* **89**, 1003–1008.

Nguyen, M. and Schwartz, S. 1998. Lycopene stability during food processing. *Proc. Soc. Exp. Biol. Med.* **218**, 101–105.

Okajima, E., Tsutsumi, M., Ozono, S., Akai, H., Denda, A., Nishino, H., Oshima, S., Sakamoto, H., and Konishi, Y. 1998. Inhibitory effect of tomato juice on rat urinary bladder carcinogenesis after N-butyl-N-(4hydroybutyl) nitrosamine initiation. *Jpn. J. Cancer Res.* **89**, 22–26.

Olmedilla, B., Granado, F., Blanco, I., and Rojas-Hidalgo, E. 1994. Seasonal and sex related variations in six serum carotenoids, retinol and a-tocopherol. *Am. J. Clin. Nutr.* **60**, 106–110.

Oursler, M.J., Collin-Osdoby, P., Li, L., Schmitt, E., and Osdoby, P. 1991. Evidence for an immunological and functional relationship between superoxide dismutase and a high molecular weight osteoclast plasma membrane glycoprotein. *J. Cell. Biochem.* **46**, 331–344.

Paiva, S. and Ressell, R. 1999. Beta carotene and other carotenoids as antioxidants. *J. Am. Coll. Nutr.* **18**, 426–433.

Palan, P. and Naz, R. 1996. Changes in various antioxidant levels in human seminal plasma related to immunofertility. *Arch. Androl.* **36**, 139–143.

Paley, P.J., Staskus, K.A., Gebhard, K., Mohanraj, D., Twiggs, L.B., Carson, L.F., and Ramakrishnan, S. 1997. Vascular endothelial growth factor expression in early stage ovarian carcinoma. *Cancer* **80**, 98–106.

Paran, E. and Engelhard, Y. 2001. Effect of Lyc-O-Mato, standardized tomato extract on blood pressure, serum lipoproteins, plasma homocysteine and oxidative stress markers in grade 1 hypertensive patients. *In* "Proceedings of the 16th Annual Scientific Meeting of the Society of Hypertension", San Francisco, USA.

Park, C.K., Ishimi, Y., Ohmura, M., Yamaguchi, M., and Ikegami, S. 1997. Vitamin A and carotenoids stimulate differentiation of mouse osteoblastic cells. *J. Nutr. Sci. Vitaminol.* **43**, 281–296.

Parker, R.S. 1996. Absorption, metabolism and transport of carotenoids. *FASEB J.* **10**, 542–551.

Parthasarathy, S. 1998. Mechanisms by which dietary antioxidants may prevent cadiovascular diseases. *J. Med. Food* **1**, 45–51.

Parthasarathy, S., Steinberg, D., and Witztum, J.L. 1992. The role of oxidized low-density lipoproteins in pathogenesis of atherosclerosis. *Ann. Rev. Med.* **43**, 219–225.

Pegram, M.D., Pauletti, G., and Slamon, D.J. 1998. HER-2/neu as a predictive marker of response to breast cancer therapy. *Breast Cancer Res. Treat.* **52**, 65–77.

Peng, Y.M., Peng, Y.S., Lin, Y., Moon, T., Roe, D.J., and Ritenbaugh, C.H. 1995. Concentrations and plasma-tissue-diet relationships of carotenoids, retinoids, and tocopherols in humans. *Nutr. Cancer* **23**, 233–246.

Petit, A., Rak, J., Hung, M., Rockwell, P., Goldstein, N., Fendly, B., and Kerbel, R. 1997. Neutralizing antibodies against epidermal growth factor and ErbB-2/neu receptor tyrosine kinases downregulate vascular endothelial growth factor production by tumor cells *in vitro* and *in vivo*: Angiogenic implications for signal transduction therapy of solid tumors. *Am. J. Pathol.* **151**, 1523–1530.

Pincemail, J. 1995. "Free Radicals and Antioxidants in Human Disease". Birkhäuser Verlag, Basel.

Polidori, M.C., Stahl, W., Eichler, O., Niestroj, I., and Sies, H. 2001. Profiles of antioxidants in human plasma. *Free Radic. Biol. Med.* **30**, 456–462.

Powell, B., Soong, R., Iacopetta, B., Seshadri, R., and Smith, D.R. 2000. Prognostic significance of mutations to different structural and functional regions of the *p53* gene in breast cancer. *Clin. Cancer Res.* **6**, 443–451.

Prakash, P., Russell, R.M., and Krinsky, N.I. 2001. *In vitro* inhibition of proliferation of estrogen-dependent and estrogen-independent human breast cancer cells treated with carotenoids or retinoids. *J. Nutr.* **131**, 1574–1680.

Raisz, L.G. 1993. Bone cell biology: New approaches and unanswered questions. *J. Bone Miner. Res.* **8**, S457–S465.

Rao, A. and Rao, L. 2004. Lycopene and human health. *Curr. Top. Nutr. Res.* **2**, 127–136.

Rao, A., Waseem, Z., and Agarwal, S. 1999. Lycopene content of tomatoes and tomato products and their contribution to dietary lycopene. *Food Res. Int.* **31**, 737–741.

Rao, A.V. 2002a. Lycopene, tomatoes and health: New perspectives (2000). *In* "Lycopene and the Prevention of Chronic Diseases: Major Findings from Five International Conferences" (A.V. Rao and D. Heber, eds), pp. 19–28. Caledonian Science Press, Scotland.

Rao, A.V. 2002b. Lycopene, tomatoes and the prevention of coronary heart disease. *Exp. Biol. Med.* **227**, 908–913.

Rao, A.V. and Agarwal, S. 1998a. Bioavailability and *in vivo* antioxidant properties of lycopene from tomato products and their possible role in the prevention of cancer. *Nutr. Cancer* **31**, 199–203.

Rao, A.V. and Agarwal, S. 1998b. Effect of diet and smoking on serum lycopene and lipid peroxidation. *Nutr. Res.* **18**, 713–721.

Rao, A.V. and Agarwal, S. 1999. Role of lycopene as antioxidant carotenoid in the prevention of chronic diseases: A review. *Nutr. Res.* **19**, 305–323.

Rao, A.V. and Agarwal, S. 2000. Role of antioxidant lycopene in cancer and heart disease. *J. Am. Coll. Nutr.* **19**, 563–569.

Rao, A.V. and Balachandran, B. 2003. Role of oxidative stress and antioxidants in neurodegenerative diseases. *Nutr. Neurosci.* **5**, 291–309.

Rao, A.V. and Shen, H.L. 2002. Effect of low dose lycopene intake on lycopene bioavailability and oxidative stress. *Nutr. Res.* **22**, 1125–1131.

Rao, A.V., Fleshner, N., and Agarwal, S. 1999. Serum and tissue lycopene and biomarkers of oxidation in prostate cancer patients: A case-control study. *Nutr. Cancer* **33**, 159–162.

Rao, L.G., Krishnadev, N., Banasikowska, K., and Rao, A.V. 2003. Lycopene I—Effect on osteoclasts: Lycopene inhibits basal and parathyroid hormone-stimulated osteoclast formation and mineral resorption mediated by reactive oxygen species in rat bone marrow cultures. *J. Med. Food* **6**, 69–78.

Rao, A.V. and Balachandran, A.V. 2004. Role of antioxidant lycopene in heart disease. *In* "Antioxidants and Cardiovascular Disease" (R. Nath, M. Khullar, and P.K. Singal, eds), pp. 62–83. Narosa Publishing House, New Delhi.

Rao, L.G., Collins, E.S., Josse, R.G., Strauss, A., and Rao, A.V. 2005. Lycopene consumption significantly decreases oxidative stress and bone resorption marker in postmenopausal women at risk of osteoporosis. *Joint Meeting of the ECTS and IBMS* June 25–29. Geneva, Switzerland.

Reich, P., Shwachman, H., and Craig, J.M. 1960. Lycopenemia: A variant of carotenemia. *N. Engl. J. Med.* **262**, 263–269.

Retz, W., Gsell, W., Munch, G., Rosler, M., and Riederer, P. 1998. Free radicals in Alzheimer's disease. *J. Neural. Transm. Suppl.* **54**, 221–236.

Ries, W.L., Key, L.L., and Rodriguiz, R.M. 1992. Nitroblue tetrazolium reduction and bone resorption by osteoclasts *in vitro* inhibited by a manganese-based superoxide dismutase mimic. *J. Bone Miner. Res.* **7**, 931–938.

Rigas, J.R., Warrell, R.P., Jr., and Young, C.W. 1994. Elevated plasma lipid peroxide content correlates with rapid plasma clearance of all-trans-retinoic acid in patients with advanced cancer. *Cancer Res.* **54**, 2125–2128.

Rocchi, E., Borghi, A., Paolillo, F., Pradelli, M., and Casalgrandi, G. 1991. Carotenoids and liposoluble vitamins in liver cirrhosis. *J. Lab. Clin. Med.* **118**, 176–185.

Ross, M.A., Crosely, L.K., Brown, M.K., Duthie, S.J., Collins, A.C., Arthur, J.R., and Duthie, G.G. 1996. Plasma concentrations of carotenoids and antioxidant vitamins in Scottish males: Influence of smoking. *Eur. J. Clin. Nutr.* **49**, 861–865.

Sadar, M.D., Hussain, M., and Bruchovsky, N. 1999. Prostate cancer: Molecular biology of early progression to androgen independence. *Endocr. Relat. Cancer* **6**, 487–502.

Samoto, H., Shimizu, E., Matsuda-Honjo, Y., Saito, R., Yamazaki, M., Kasai, K., Furuyama, S., Sugiya, H., Sodek, J., and Ogata, Y. 2002. TNF-alpha suppresses bone sialoprotein (BSP) expression in ROS17/2.8 cells. *J. Cell. Biochem.* **87**, 313–323.

Santos, L., Amaro, T., Costa, C., Pereira, S., Bento, M.J., Lopes, P., Oliveira, J., Criado, B., and Lopes, C. 2003. Ki-67 index enhances the prognostic accuracy of the urothelial superficial bladder carcinoma risk group classification. *Int. J. Cancer* **105**, 267–272.

Schmidt, R., Fazekas, F., Hayn, M., Schmidt, H., Kapeller, P., Toob, G., Offenbacher, H., Schumacher, M., Eber, B., Weinrauch, V., Kostner, G.M., and Esterbauer, H. 1997. Risk factors for microangiopathy-related cerebral damage in Aistrian stroke prevention study. *J. Neurol. Sci.* **152**, 15–21.

Shapiro, G.I. 2004. Preclinical and clinical development of the cyclin-dependent kinase inhibitor flavopiridol. *Clin. Cancer Res.* **10**, S4270–S4275.

Shariat, S.F., Lotan, Y., Saboorian, H., Khoddami, S.M., Roehrborn, C.G., Slawin, K.M., and Ashfaq, R. 2004. Survivin expression is associated with features of biologically aggressive prostate carcinoma. *Cancer* **100**, 751–757.

Sharoni, Y., Giron, E., Rise, M., and Levy, J. 1997. Effects of lycopene-enriched tomato oleoresin on 7,12-dimethyl-benz[a]anthracene-induced rat mammary tumors. *Cancer Detect. Prev.* **21**, 118–123.

'Shiwa, M., Nishimura, Y., Wakatabe, R., Fukawa, A., Arikuni, H., Ota, H., Kato, Y., and Yamori, T. 2003. Rapid discovery and identification of a tissue-specific tumor biomarker from 39 human cancer cell lines using the SELDI ProteinChip platform. *Biochem. Biophys. Res. Commun.* **309**, 18–25.

Siler, U., Barella, L., Spitzer, V., Schnorr, J., Lein, M., Goralczyk, R., and Wertz, K. 2004. Lycopene and vitamin E interfere with autocrine/paracrine loops in the Dunning prostate cancer model. *FASEB J.* **18**, 1019–1021.

Silverton, S. 1994. Osteoclast radicals. *J. Cell. Biochem.* **56**, 367–373.

Sinclair, A.J., Bayer, A.J., Johnston, J., Warner, C., and Maxwell, S.R. 1998. Altered plasma antioxidant status in subjects with Alzheimer's disease and vascular dementia. *Int. J. Geriatr. Psychiatry* **13**, 840–845.

Singh, R.P., Sharad, S., and Singh, S.K. 2004. Free radicals and oxidative stress in neurodegenerative diseases: Relevance of dietary antioxidants. *J. Indian Acad. Clin. Med.* **5**, 218–225.

Singh, V.N. 1992. A current perspective on nutrition and exercise. *J. Nutr.* **122**, 760–765.

Singhal, H., Bautista, D.S., Tonkin, K.S., O'Malley, F.P., Tuck, A.B., Chambers, A.F., and Harris, J. F. 1997. Elevated plasma osteopontin in metastatic breast cancer associated with increased tumor burden and decreased survival. *Clin. Cancer Res.* **3**, 605–611.

Slamon, D.J., Clark, G.M., Wong, S.G., Levin, W.J., Ullrich, A., and McGuire, W.L. 1987. Human breast cancer: Correlation of relapse and survival with amplification of the HER-2/neu oncogene. *Science* **235**, 177–182.

Snowdon, D.A., Gross, M.D., and Butler, S.M. 1966. Antioxidants and reduced functional capacity in the elderly: Finding from the Nun study. *J. Gerontol. A Biol. Sci. Med. Sci.* **51**, M10–M16.

Sontakke, A.N. and Tare, R.S. 2002. A duality in the roles of reactive oxygen species with respect to bone metabolism. *Clin. Chim. Acta* **318**, 145–148.

Span, P.N., Sweep, F.C., Wiegerinck, E.T., Tjan-Heijnen, V.C., Manders, P., Beex, L.V., and de Kok, J.B. 2004. Survivin is an independent prognostic marker for risk stratification of breast cancer patients. *Clin. Chem.* **50**, 1986–1993.

Stadtman, E.R. 1992. Protein oxidation and aging. *Science* **257**, 1220–1224.

Stahl, W. and Sies, H. 1992. Uptake of lycopene and its geometrical isomers is greater from heat-processed than from unprocessed tomato juice in humans. *J. Nutr.* **122**, 2161–2166.

Stahl, W. and Sies, H. 1996. Lycopene: A biologically important carotenoid for humans? *Arch. Biochem. Biophys.* **336**, 1–9.

Steels, E., Paesmans, M., Berghmans, T., Branle, F., Lemaitre, F., Mascaux, C., Meert, A.P., Vallot, F., Lafitte, J.J., and Sculier, J.P. 2001. Role of p53 as a prognostic factor for survival in lung cancer: A systematic review of the literature with a meta-analysis. *Eur. Respir. J.* **18**, 705–719.

Steinbeck, M.J., Appel, W.H., Jr., Verhoeven, A.J., and Karnovsky, M.J. 1994. NADPH-oxidase expression and *in situ* production of superoxide by osteoclasts actively resorbing bone. *J. Cell Biol.* **126**, 765–772.

Steinbeck, M.J., Kim, J.K., Trudeau, M.J., Hauschka, P.V., and Karnovsky, M.J. 1998. Involvement of hydrogen peroxide in the differentiation of clonal HD-11EM cells into osteoclast-like cells. *J. Cell. Physiol.* **176**, 574–587.

Stephens, N.G., Parsons, A., Schodiel, P.M., Kelly, F., Cheeseman, K., and Mitchinson, M.J. 1996. Randomised controlled trial of vitamin E in patients with coronary disease: Cambridge heart antioxidant study (CHAOS). *Lancet* **347**, 781–786.

Street, D.A., Comstock, G.W., Salkeld, R.M., Schuep, W., and Klag, M.J. 1994. Serum antioxidant and myocardial infarction: Are low levels of carotenoids and alpha-tocopherol risk factors for myocardial infarction? *Circulation* **90**, 1154–1161.

Strohmeyer, D., Rossing, C., Bauerfeind, A., Kaufmann, O., Schlechte, H., Bartsch, G., and Loening, S. 2000. Vascular endothelial growth factor and its correlation with angiogenesis and p53 expression in prostate cancer. *Prostate* **45**, 216–224.

Strohmeyer, D., Strauss, F., Rossing, C., Roberts, C., Kaufmann, O., Bartsch, G., and Effert, P. 2004. Expression of bFGF, VEGF and c-met and their correlation with microvessel density and progression in prostate carcinoma. *Anticancer Res.* **24**, 1797–1804.

Suda, N., Morita, I., Kuroda, T., and Murota, S. 1993. Participation of oxidative stress in the process of osteoclast differentiation. *Biochim. Biophys. Acta* **1157**, 318–323.

Suleiman, S.A., Ali, M.E., Zaki, Z.M., el-Malik, E.M., and Nasr, M.A. 1996. Lipid peroxidation and human sperm motility: Protective role of vitamin E. *J. Androl.* **17**, 530–537.

Swanton, C. 2004. Cell-cycle targeted therapies. *Lancet Oncol.* **5**, 27–36.

Theodorescu, D., Broder, S.R., Boyd, J.C., Mills, S.E., and Frierson, H.F., Jr. 1997. p53, bcl-2 and retinoblastoma proteins as long-term prognostic markers in localized carcinoma of the prostate. *J. Urol.* **158**, 131–137.

Toledano, M.B. and Leonard, W.J. 1991. Modulation of transcription factor NF6B binding activity by oxidation-reduction *in vitro*. *Proc. Natl. Acad. Sci. USA* **88**, 4328–4332.

Tricoli, J., Schoenfeldt, M., and Conley, B. 2004. Detection of prostate cancer and predicting progression: Current and future diagnostic markers. *Clin. Cancer Res.* **10**, 3943–3953.

Tuck, A.B. and Chambers, A.F. 2001. The role of osteopontin in breast cancer: Clinical and experimental studies. *J. Mammary Gland Biol. Neoplasia* **6**, 419–429.

Tuck, A.B., O'Malley, F.P., Singhal, H., Harris, J.F., Tonkin, K.S., Kerkvliet, N., Saad, Z., Doig, G. S., and Chambers, A.F. 1998. Osteopontin expression in a group of lymph node negative breast cancer patients. *Int. J. Cancer* **79**, 502–508.

U.S. Department of Agriculture and Agricultural Research Service 1998. USDA-NCC carotenoid database for U.S. Foods. 1998 (1998) Nutrient Data Laboratory Home Page. www.nal.usda.gov/fnic/foodcomp.

U.S. Department of Agriculture and CSFII, A. R. S. (1994–1996). Food Surveys Research Group Home Page. www.sun.ars-rin.gov/ars/Beltsville/barc/bhnrc/foodsurvey/home.

Udagawa, N. 2002. Mechanisms involved in bone resorption. *Biogerontology* **3**, 79–83.

van 't Veer, L.J., Dai, H., van de Vijver, M.J., He, Y.D., Hart, A.A.M., Mao, M., Peterse, H.L., van der Kooy, K., Marton, M.J., Witteveen, A.T., Schreiber, G.J., Kerkhoven, R.M. *et al.* 2002. Gene expression profiling predicts clinical outcome of breast cancer. *Nature* **415**, 530–536.

Varanasi, S.S., Francis, R.M., Berger, C.E., Papiha, S.S., and Datta, H.K. 1999. Mitochondrial DNA deletion associated oxidative stress and severe male osteoporosis. *Osteoporos. Int.* **10**, 143–149.

Venkateswaran, V., Fleshner, N., Sugar, L., and Klotz, L. 2004. Antioxidants block prostate cancer in lady transgenic mice. *Cancer Res.* **64**, 5891–5896.

Vernejoul, M-C de. 1998. Markers of bone remodelling in metabolic bone disease. *Drugs Aging* **1**(Suppl. 1), 9–14.

Virtamo, J., Rapola, J.M., Ripatti, S., Heinonen, O.P., Taylor, P.R., Albanes, D., and Huttunen, J.K. 1998. Effect of vitamin E and beta carotene on the incidence of primary nonfatal myocardial infarction and fatal coronary heart disease. *Arch. Intern. Med.* **158**, 668–675.

Wagner, A.H., Schroeter, M.R., and Hecker, M. 2001. 17b-Estradiol inhibition of NADPH oxidase expression in human endothelial cells. *FASEB J.* **15**, 2121–2130.

Wahlqvist, M., Wattanapenpaiboon, N., Macrae, F., Lambert, J., MacLennan, R., and Hsu-Hage, B. 1994. Changes in serum carotenoids in subjects with colorectal adenomas after 24 mo of beta-carotene supplementation. *Am. J. Clin. Nutr.* **60**, 936–943.

Wang, C.J., Chou, M.Y., and Lin, J.K. 1959. Inhibition of growth and development of the transplantable C-6 glioma cells incoculated in rats by retinoids and carotenoids. *Cancer Lett.* **48**, 135–142.

Wang, M., Dhingra, K., Hittelman, W.N., Liehr, J.G., de Andrade, M., and Donghui, L. 1996. Lipid peroxidation-induced putative malondialdehyde-DNA adducts in human breast tissues. *Cancer Epidemiol. Biomarkers Prev.* **5**, 705–710.

Wasylyk, C. and Wasylyk, B. 1993. Oncogene conversion of Ets affects redox regulation *in vivo* and *in vitro*. *Nucleic Acid Res.* **21**, 523–529.

Wei, H. 1992. Activation of oncogenes and/or inactivation of anti-oncogene by reactive oxygen species. *Med. Hypotheses* **39**, 267–270.

Wertz, K., Siler, U., and Goralczyk, R. 2004. Lycopene: Modes of action to promote prostate health. *Arch. Biochem. Biophys.* **430**, 127–134.

Witztum, J.L. 1994. The oxidation hypothesis of artherosclerosis. *Lancet* **344**, 793–796.

Wong, Y.F., Cheung, T.H., Lo, K.W., Wang, V.W., Chan, C.S., Ng, T.B., Chung, T.K., and Mok, S. C. 2004. Protein profiling of cervical cancer by protein-biochips: Proteomic scoring to discriminate cervical cancer from normal cervix. *Cancer Lett.* **211**, 227–234.

Yanagisawa, K., Shyr, Y., Xu, B.J., Massion, P.P., Larsen, P.H., White, B.C., Roberts, J.R., Edgerton, M., Gonzalez, A., Nadaf, S., Moore, J.H., Caprioli, R.M. *et al.* 2003. Proteomic patterns of tumor subsets in non-small-cell lung cancer. *Lancet* **362**, 433–439.

Zaidi, M., Alam, A.S., Bax, B.E., Shankar, V.S., Bax, C.M., Gill, J.S., Pazianas, M., Huang, C.L., Sahinoglu, T., and Moonga, B.S. 1993. Role of the endothelial cell in osteoclast control: New perspectives. [Review] [62 refs]. *Bone* **14**, 97–102.

Zhang, L.-X., Cooney, R.V., and Bertram, J.S. 1991. Carotenoids enhance gap junctional communication and inhibit lipid peroxidation in C3H/10T1/2 cells: Relationship to their cancer chemopreventive action. *Carcinogenesis* **12**, 2109–2114.

Zhang, L.-X, Cooney, R.V., and Bertram, J.S. 1992. Carotenoids up-regulate connexin43 gene expression independent of their provitamin A or antioxidant properties. *Cancer Res.* **52**, 5707–5712.

Zini, A., de Lamirande, E., and Gagnon, C. 1993. Reactive oxygen species in semen of infertile patients: Levels of superoxide dismutase and catalase-like activities in seminal plasma and spermatozoa. *Int. J. Androl.* **16**, 183–188.

Zock, P. and Katan, M.B. 1998. Diet, LDL oxidation, and coronary artery disease. *Am. J. Clin. Nutr.* **68**, 759–760.

FOOD COMPONENTS THAT REDUCE CHOLESTEROL ABSORPTION

TIMOTHY P. CARR AND ELLIOT D. JESCH

Department of Nutrition and Health Sciences
University of Nebraska-Lincoln, Lincoln, Nebraska 68583

I. INTRODUCTION

Cholesterol is a vital component of the human body. It stabilizes cell membranes and is the precursor of bile acids, vitamin D, and steroid hormones. The body's cells can synthesize cholesterol when needed, but excess cholesterol cannot be broken down and must be excreted from the body through the bile into the small intestine. When imbalances occur, cholesterol can accumulate in the gallbladder promoting gallstone formation. Cholesterol accumulation in the bloodstream (hypercholesterolemia) can cause atherosclerotic plaques to form within artery walls.

Absorption of cholesterol in the small intestine contributes to maintaining whole-body cholesterol homeostasis, yet the mechanisms of absorption have not been completely defined. For many years it was believed that cholesterol, a normal component of cell membranes, simply diffused through the brush border membrane of enterocytes (Grundy, 1983; Westergaard and Dietschy, 1974). However, the discovery of specific transporters, receptors,

ADVANCES IN FOOD AND NUTRITION RESEARCH VOL 51

ISSN: 1043-4526
DOI: 10.1016/S1043-4526(06)51003-4

and enzymes is quickly changing our understanding of how cholesterol and other sterols are absorbed into the body. It now appears that the transport of cholesterol into and out of the enterocyte and other cells is highly regulated and subject to modification by dietary factors including lipids, carbohydrates, and proteins.

From a health standpoint, the efficiency of cholesterol absorption is of great interest because human and animal studies have linked cholesterol absorption with plasma total and low-density lipoprotein (LDL) cholesterol concentration (Carr *et al.*, 1996, 2002; Gylling and Miettinen, 1995; Kesaniemi and Miettinen, 1987; Rudel *et al.*, 1994). So important is this link that a new family of drugs is being developed that blocks the intestinal absorption of cholesterol and, consequently, reduces plasma LDL cholesterol concentration. Experimentation with these drugs has also helped researchers elucidate some of the mechanisms of cholesterol transport at the cellular level (Altmann *et al.*, 2004; Burnett, 2004). One of the drugs, ezetimibe (sold as Zetia® and Vytorin®), received Food and Drug Administration (FDA) approval in October of 2002 and has become an important therapy in managing LDL cholesterol levels. But because drugs can produce severe side effects, it is desirable to learn more about natural food components that inhibit cholesterol absorption so that food ingredients and dietary supplements can be developed for consumers who wish to manage their plasma cholesterol levels by nonpharmacological means.

This article focuses on specific dietary components—whether naturally occurring or added as food ingredients—known to interfere with the mechanisms of cholesterol absorption. An overview of cholesterol absorption is provided and emphasizes the critical role of bile acids and micelle formation in solubilizing cholesterol for transport to the brush border membrane of enterocytes. Where applicable, information is also included about commercial food ingredients that are specifically used as cholesterol-lowering agents.

II. MECHANISMS OF CHOLESTEROL ABSORPTION

Cholesterol enters the small intestine from two sources: the diet and bile (Figure 1). Dietary intake of cholesterol is about 300 mg/day (Briefel and Johnson, 2004; Ishinaga *et al.*, 2005; Valsta *et al.*, 2004), whereas the bile contributes 800–1400 mg/day (Duane, 1993; Grundy and Metzger, 1972). The liver—not the diet—is therefore the primary source of cholesterol available for absorption, a point that is often underappreciated. Consequently, therapies that block cholesterol absorption are effective at lowering LDL cholesterol mainly because they prevent the reabsorption of endogenous

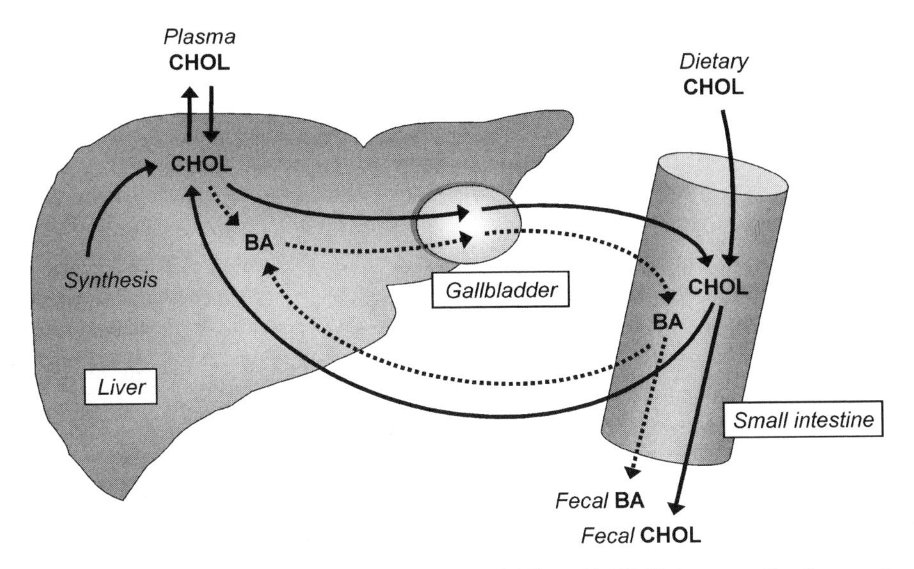

FIG. 1 Movement of cholesterol (CHOL) and bile acids (BA) between the liver and small intestine. CHOL and BA in the liver are secreted into the gallbladder where they are stored temporarily until a fat-containing meal causes their secretion into the intestinal lumen. BA are absorbed with high efficiency (95%) and are recycled back to the liver via the hepatic portal vein. CHOL is absorbed less efficiently (50–60%) and must be incorporated into lipoproteins (chylomicrons) for transport back to the liver via the systemic circulation. Accumulation of CHOL in the liver can promote secretion of CHOL into plasma, thus increasing LDL-CHOL concentration. Loss of CHOL and BA in feces represents the primary route of CHOL elimination from the body.

cholesterol back to the liver. This explains why individuals who consume no animal products (i.e., no cholesterol) will also experience reductions in LDL cholesterol when given absorption-blocking therapies.

Some cholesterol entering from the diet may be esterified to various fatty acids, although the extent of esterification is variable. For example, egg yolk cholesterol is about 10% esterified (Bitman and Wood, 1980; Tattrie, 1972); cholesterol in meat and poultry is at least 50% esterified (Kritchevsky and Tepper, 1961). Esterified cholesterol entering the intestinal tract is mostly hydrolyzed by pancreatic enzymes, yielding free cholesterol and fatty acids (Howles *et al.*, 1996). Only unesterified cholesterol is available for absorption.

Biliary cholesterol is entirely unesterified and flows into the small intestine as a component of bile. The other major components of bile are phosphatidylcholine (lecithin) and bile acids. Absorption of cholesterol and other lipids depends on their ability to form micelles within the intestinal lumen.

Micelles are lipid aggregates that form when a critical concentration of lipid from bile (i.e., bile acids, phospholipid, and cholesterol) mixes with lipid from the diet (i.e., triglyceride, phospholipid, and cholesterol). The bile acids act as detergents allowing the lipids to "dissolve" in an aqueous environment, facilitating their delivery to the brush border. Furthermore, the proportion of bile acids, phospholipid, and cholesterol in gallbladder bile must be maintained within a specific range to prevent gallstone formation and to optimize micelle formation in the intestinal lumen (Apstein and Carey, 1996; Yao *et al.*, 2002). Many dietary components that reduce cholesterol absorption do so by binding bile acids or otherwise disrupting their ability to form micelles. Under normal conditions, all of the micelle components—except cholesterol—are transported to the brush border and absorbed with high efficiency (90–100%); cholesterol absorption is typically 50–60% (Matthan and Lichtenstein, 2004). Some evidences suggest that biliary cholesterol, because of its inherent association with bile acids, is absorbed slightly more efficiently than dietary cholesterol (Wilson and Rudel, 1994), although this difference probably has little impact on overall cholesterol balance given the minor contribution of dietary cholesterol.

Bile acids play an important role in cholesterol homeostasis. In addition to being required for micellar solubilization of cholesterol, bile acids are synthesized from cholesterol in the liver (Figure 1). Following micelle formation and delivery of lipid to the proximal intestine, bile acids are reabsorbed in the ileum in a process mediated by the apical sodium-dependent bile acid transporter (ASBT) (Dawson and Oelkers, 1995). Bile acid sequestrants, such as cholestyramine (Questran®) and colestipol (Colestid®), are effective at lowering plasma LDL cholesterol because they promote intestinal bile acid excretion, which limits their reabsorption, causing the liver to use more cholesterol for bile acid synthesis. Increased demand for cholesterol causes the liver to recruit LDL cholesterol from plasma, thus reducing plasma LDL cholesterol concentration. However, micelle formation and cholesterol absorption are not affected by treatment with bile acid sequestrants or pharmacological inhibition of ASBT because the liver compensates by increasing bile acid synthesis (Hui and Howles, 2005; Packard and Shepherd, 1982). In this way, the total bile acid pool size is not diminished even though turnover of bile acids in the enterohepatic circulation may be increased by drugs or dietary factors.

Some dietary factors can also change the bile acid species and, by doing so, alter cholesterol absorption. The liver synthesizes the primary bile acids, cholic and chenodeoxycholic acid. Bacteria in the intestine can convert some of the primary bile acids into secondary bile acids, producing deoxycholic from cholic acid and lithocholic from chenodeoxycholic acid. When certain dietary components alter the intestinal microflora, the rate of secondary bile acid production can change. Secondary bile acids are more hydrophobic than primary bile acids, and their diminished presence in the enterohepatic circulation can decrease the efficiency by which micelles solubilize cholesterol (Armstrong and Carey, 1987). The "hydrophobicity index" of bile is a numeric value that quantitatively defines the hydrophilic–hydrophobic balance of bile acids in bile samples (Heuman, 1989). Hydrophobicity index is therefore a useful indicator of the capacity of bile to solubilize cholesterol in the intestinal lumen, a necessary step for cholesterol absorption.

Studies have indicated that cholesterol transport into enterocytes is mediated by a specific transporter, Neiman-Pick C1 Like 1 (NPC1L1) protein (Figure 2). NPC1L1 is highly expressed in the proximal portion of the small intestine and is specifically inhibited by the drug ezetimibe (Altmann *et al.*, 2004). NPC1L1 also transports dietary plant sterols into the enterocyte. Using NPC1L1 knockout mice, Davis *et al.* (2004) demonstrated that intestinal uptake of both cholesterol and plant sterols was significantly reduced compared to wild-type mice. Despite the ability of NPC1L1 to transport cholesterol and plant sterols, less than 1% of dietary plant sterols eventually enter the circulation, whereas 50–60% of intestinal cholesterol

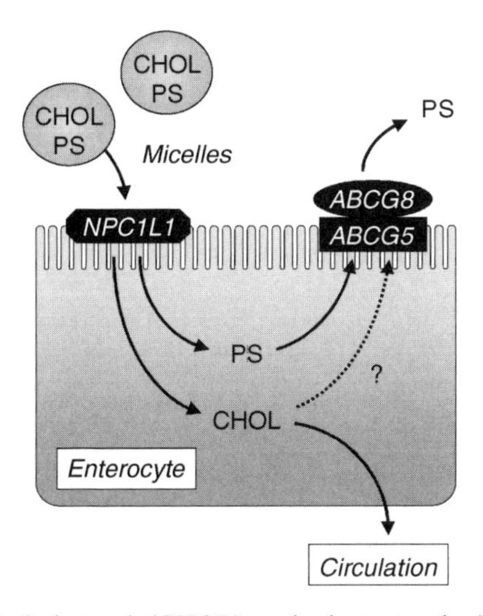

FIG. 2 Transport of cholesterol (CHOL) and plant sterols (PS) in the enterocyte. CHOL, PS, and other lipids are solubilized in micelles that deliver the lipids to the brush border membrane. CHOL and PS are transported into the enterocyte by NPC1L1. Nearly all of the PS are redirected back to the intestinal lumen by the transporters ABCG5 and ABCG8. The extent to which CHOL is transported by ABCG5 and ABCG8 is not known. CHOL within the enterocyte is packaged into lipoproteins (chylomicrons) and secreted into lymph and eventually the bloodstream for transport to the liver.

enters the circulation. It appears that cholesterol and plant sterols are handled differently once inside the enterocyte. Another transport system, involving the proteins ABCG5 and ABCG8 (Berge *et al.*, 2000; Lee *et al.*, 2001a), apparently redirects plant sterols back into the intestinal lumen, but the role of ABCG5 and ABCG8 in cholesterol transport is uncertain (Lee *et al.*, 2001b). The inability (or diminished ability) of cholesterol to interact with ABCG5 and ABCG8 is one possible explanation for the differential absorption rates between cholesterol and plant sterols.

While the discovery of NPC1L1, ABCG5, and ABCG8 represents important milestones in understanding cholesterol absorption, some questions remain. First, cholesterol absorption in NPC1L1 knockout mice was significantly reduced, but not completely eliminated (Davis *et al.*, 2004), suggesting a small proportion of cholesterol is absorbed independent of NPC1L1. What other mechanisms beside NPC1L1 account for cholesterol transport into enterocytes? Second, to what extent does cholesterol interact with ABCG5 and ABCG8, and does this transport system represent an important regulatory pathway in overall cholesterol absorption? Finally, how is cholesterol transported inside the enterocyte? The intracellular trafficking of sterols is an intense area of investigation. While there is still much to be learned, a general understanding of cholesterol absorption is useful when assessing how dietary factors impact cholesterol absorption.

III. FOOD COMPONENTS THAT REDUCE CHOLESTEROL ABSORPTION

A. PLANT STEROLS AND STANOLS

Plant sterols are essential components of cell membranes and are present in all plants. They are structurally similar to cholesterol, with the differences occurring in the side chain attached to the steroid ring (Figure 3). Dozens of plant sterols have been identified, although the most abundant are sitosterol, campesterol, and stigmasterol. Stanols are saturated sterols (i.e., no double bond in the steroid ring) and are much less abundant in nature than the corresponding sterols. Plant stanols comprise about 5–10% of the total sterol/stanol mixture naturally present in the human diet (Andersson *et al.*, 2004; Normén *et al.*, 2001; Phillips *et al.*, 1999; Valsta *et al.*, 2004). Because of their low abundance, stanols are often reported as part of the total "plant sterol" content of food or dietary intake. Individuals living in Western societies consume about 200–300 mg plant sterols per day (Andersson *et al.*, 2004; de Vries *et al.*, 1997; Morton *et al.*, 1995; Normén *et al.*, 2001; Phillips *et al.*, 1999; Schothorst and Jekel, 1999; Valsta *et al.*, 2004). Asian

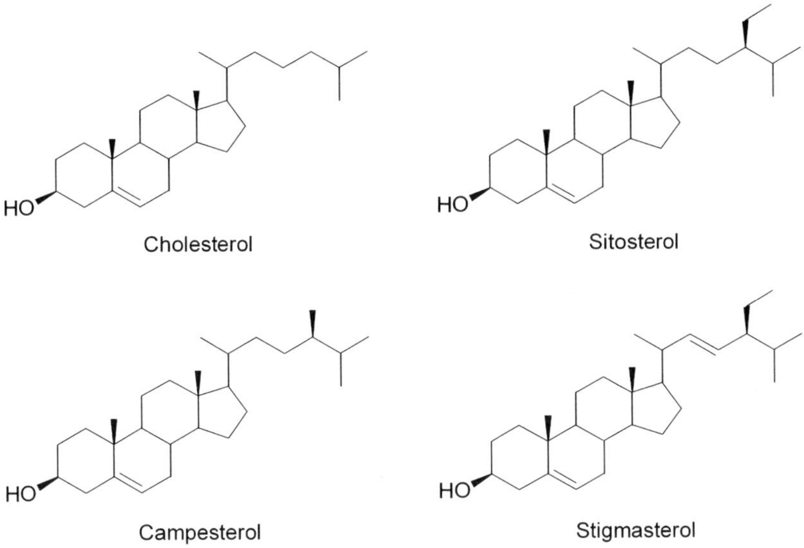

FIG. 3 Structure of cholesterol and common plant sterols.

and vegetarian diets provide higher amounts of plant sterols (Vuoristo and Miettinen, 1994; Zhou *et al.*, 2003). The total plant sterol content of several common foods is provided in Table I. Note that vegetable oils are particularly good sources of plant sterols.

Moderate levels of plant sterols consumed in usual diets probably exert some minimal effect on cholesterol absorption, although higher amounts (1–3 g/day) are needed to produce significant reductions in plasma LDL cholesterol (Law, 2000; Nguyen, 1999). Augmenting the diet with plant sterols is therefore necessary for therapeutic maintenance of LDL cholesterol at desirable levels. Contrary to popular belief, using plant sterols for this purpose is not new. Eli Lilly and Company introduced a plant sterol preparation called Cytellin™ in 1957. Cytellin™ contained free (unesterified) plant sterols suspended in fruit-flavored syrup and was widely prescribed through the 1980s before statin drugs became available. However, the poor solubility of free plant sterols in the intestinal lumen led to inconsistent and confusing results in clinical studies (Ahrens *et al.*, 1957; Denke, 1995; Lees *et al.*, 1977). Moreover, doses of Cytellin™ exceeding 25 g/day were often required to achieve significant LDL cholesterol reduction. Apparently, free plant sterols form highly stable crystals that resist solubilization, even in the presence of bile and dietary lipids. The crystalline nature and poor solubility of free sterols also limited their application in food products.

TABLE I
STEROL CONTENT OF COMMON PLANT FOODS

	Total plant sterols (mg/100 g edible portion)
Fruits	
Apple	13
Banana	14
Fig	22
Lemon	18
Orange	24
Peach	15
Pear	12
Pineapple	17
Watermelon	1
Vegetables	
Broccoli	39
Carrot	16
Cauliflower	40
Celery	17
Mushroom	18
Olive, black	50
Onion	8
Potato, white	4
Tomato	5
Refined oils	
Canola	250
Chestnut	5350
Coconut	133
Corn	952
Cottonseed	327
Olive	176
Rice bran	1055
Safflower	444
Soybean	221
Sunflower	725
Cereals	
Corn flour	52
Couscous	58
Rice flour	23
Rolled oats	39
Rye flour	86
Whole wheat flour	70
Wheat flour	28
Wheat bran	200
Wheat germ	344

Source: Weihrauch and Gardner (1978); Normén *et al.* (1999, 2002).

In the 1970s, Fred Mattson and colleagues working at Procter & Gamble discovered that esterifying plant sterols with long-chain fatty acids increased their solubility in oil from about 2% to more than 20% and that esterification did not impair their ability to inhibit cholesterol absorption (Jandacek *et al.*, 1977; Mattson *et al.*, 1977, 1982). Armed with this information, researchers at Raisio Group in Finland developed a commercial process of esterifying plant stanols with vegetable oil fatty acids (Miettinen *et al.*, 1996). Their process utilizes stanols, rather than sterols, and is based on studies indicating that free stanols are more effective than free sterols at reducing cholesterol absorption (Becker *et al.*, 1993; Heinemann *et al.*, 1988; Ikeda *et al.*, 1979; Sugano *et al.*, 1977). However, more recent studies using esterified sterols and stanols have established their equal effectiveness at reducing cholesterol absorption and plasma LDL cholesterol concentration (Hallikainen *et al.*, 2000; Jones *et al.*, 2000; Normén *et al.*, 2000; Weststrate and Meijer, 1998). Nevertheless, the plant stanol esters produced by Raisio Group are the main ingredient in Benecol® spreads, yogurts, and other food products. Benecol® spreads were launched in Finland in 1995 and in the United States in 1999 after plant stanol esters were declared "generally recognized as safe" (GRAS) for use in vegetable oil-based spreads in amounts not exceeding 20%. The stanol esters used in Benecol® products are made from plant sterols derived from tall oil, a by-product of the wood pulp industry. The sterols are isolated and purified, chemically hydrogenated to stanols, then esterified with vegetable oil fatty acids (Hicks and Moreau, 2001). A similar spread, Take Control®, is made by Unilever and contains plant sterol esters as the active ingredient. The plant sterols in Take Control® are derived from soybean oil and are used directly for esterification, eliminating the need and expense of converting sterols to stanols. The retail price of Take Control® spread is consequently lower than Benecol® spread.

The food and nutraceutical industries have responded to increasing consumer demand for products containing plant sterols. Ingredient companies, including Cargill (makers of CoroWise™) and Archer-Daniels-Midland (makers of CardioAid™), now offer a range of free and esterified plant sterol compounds that are used in a variety of foods available to consumers worldwide. Both companies have also developed technologies of emulsifying plant sterols for use in beverages and water-based foods. CoroWise™ plant sterols are found in Minute Maid® Premium Heart Wise™ orange juice and GNC's Heart Advance™ dietary supplement. CardioAid™ ingredients are available in a variety of spreads, dairy products, and dietary supplements. Another company, Forbes Medi-Tech, developed an ingredient called Reducol™ that contains a mixture of plant sterols and stanols derived from tall oil. Reducol™ is the active ingredient in Naturemade's Cholest-Off™

dietary supplement. Forbes Medi-Tech received approval in November of 2004 to market Reducol™ as a food ingredient in Europe.

Several clinical studies have been conducted in recent years, which indicate a dose of 1–3 g of plant sterol (or stanol) esters reduces plasma LDL cholesterol concentration up to 15% compared to placebo (Law, 2000; Nguyen, 1999; O'Neill *et al.*, 2005). It appears that a dose–response relationship is continuous up to about 2 g of plant sterol esters per day, with no further reductions in LDL cholesterol above that dose (Law, 2000). Accordingly, the National Cholesterol Education Program now recommends 2 g/day of plant sterol (or stanol) esters as a therapeutic option for reducing plasma LDL cholesterol concentration (National Institutes of Health, 2002). Furthermore, in response to a request by Cargill, the FDA agreed in February of 2003 to allow a heart health claim for a broad range of foods and beverages. The FDA-approved claim states that foods containing at least 0.4 g plant sterol/stanol (or 0.65 g sterol/stanol esters) per serving, consumed twice a day with meals for a daily total of at least 0.8 g plant sterol/stanol (or 1.3 g sterol/stanol ester), as part of a diet low in saturated fat and cholesterol, may reduce the risk of heart disease.

The cholesterol-lowering properties of dietary plant sterols have been known for decades (Best *et al.*, 1954; Peterson, 1951; Pollak, 1953), due specifically to reductions in cholesterol absorption. Inverse correlations between plant sterol intake and cholesterol absorption have been reported in animals (Carr *et al.*, 2002; Ntanios and Jones, 1999) and humans (Ellegård *et al.*, 2000). The exact mechanism by which plant sterols inhibit cholesterol absorption is unclear, and several mechanisms of action have been proposed, including: (1) competition with cholesterol for solubilization in micelles within the intestinal lumen, (2) cocrystallization with cholesterol to form insoluble crystals, (3) interaction with digestive enzymes, and (4) regulation of intestinal transporters of cholesterol.

First, dietary mixed micelles play a key role in dietary lipid absorption, acting as vehicles that transport both lipophilic and amphiphilic compounds toward the intestinal wall. Solubilization of cholesterol in mixed micelles is necessary for cholesterol transit to the brush border membranes of enterocytes. Cholesterol not dissolved in micelles will form a separate oil phase within the intestinal lumen, making it generally unavailable for absorption (Hofmann and Small, 1967). Using model bile solutions *in vitro*, Ikeda and coworkers reported that cholesterol solubility was significantly decreased in the presence of sitosterol (Ikeda and Sugano, 1983; Ikeda *et al.*, 1988, 1989a). They further demonstrated that sitosterol, infused with cholesterol into rat intestinal tracts as an artificial "bile" mixture, significantly reduced cholesterol absorption *in vivo* (Ikeda *et al.*, 1988). Armstrong and Carey (1987) conducted a thermodynamic analysis of micellar solubilities and found that sitosterol, compared to cholesterol, had a higher binding affinity for micelles. *In vitro* studies in our laboratory suggest that the higher affinity of plant sterols causes cholesterol to be displaced from the micelle (Jesch and Carr, 2005). Heinemann *et al.* (1991) published an intestinal perfusion study in healthy volunteers and found that both sitosterol and sitostanol reduced cholesterol absorption by disrupting cholesterol solubility in micelles. In another infusion study, Nissinen *et al.* (2002) observed that in subjects receiving either high or low amounts of plant stanol esters, cholesterol solubility in micelles was decreased due to displacement by plant stanols. Using a variety of *in vitro* techniques, Mel'nikov *et al.* (2004a) found that both sitosterol and sitostanol reduced the concentration of cholesterol in dietary mixed micelles via a dynamic competition mechanism. These investigators further concluded that cholesterol, sitosterol, and sitostanol compete equally for solubilization in micelles (Mel'nikov *et al.*, 2004a). While there is general agreement that plant sterols compete with cholesterol during micelle formation, the degree of solubilization likely depends on the composition of other lipids—dietary and biliary—present in the intestinal lumen (Yao *et al.*, 2002).

The second proposed mechanism whereby plant sterols reduce cholesterol absorption is cocrystallization of cholesterol with plant sterols. The concept that cocrystallization would render cholesterol unavailable for absorption has been considered for some time, but the data are quite limited. Christiansen *et al.* (2003) investigated the solubility and phase behavior of sitosterol and cholesterol (and mixtures thereof) in the presence and absence of water. As expected, the solubility of both sitosterol and cholesterol was significantly reduced in water–acetone solutions compared to acetone alone, but the decrease in solubility was much greater with sitosterol. When mixtures of cholesterol/sitosterol in ratios of 3:1, 1:1, and 1:3 were coprecipitated from the water–acetone solution, the total sterol solubility decreased with increasing proportions of sitosterol, suggesting that the more hydrophobic sitosterol promotes cocrystallization with cholesterol. Under more realistic conditions, Mel'nikov *et al.* (2004b) examined the cocrystallization properties of cholesterol/sitosterol and cholesterol/sitostanol mixtures from triglyceride oil that was hydrolyzed to mimic the intestinal environment during digestion. However, during lipolysis of the model dietary emulsions, no crystal formation was detected. The researchers concluded that the solubility of sterols significantly increased in the products of lipid hydrolysis and that their solubility increased in parallel with solvent polarity (free fatty acids > diglyceride oil > triglyceride oil). These results suggest that cocrystallization of plant sterols and cholesterol may not occur to a great extent *in vivo* and would not be a major contributor in reducing cholesterol absorption (Mel'nikov *et al.*, 2004b).

A third possible mechanism involves the interaction of plant sterol esters with digestive enzymes, although the extent of interaction is still uncertain. On one hand, Nissinen *et al.* (2002) demonstrated that high levels of sitostanol esters infused into healthy subjects were rapidly hydrolyzed and incorporated into micelles, causing cholesterol and its esters to accumulate in the oil phase. Preferential interaction of plant sterol esters with digestive enzymes would also preclude dietary cholesterol esters from hydrolysis, further limiting cholesterol absorption. In contrast, it has been suggested that plant sterol esters are poorly hydrolyzed by digestive enzymes (Trautwein *et al.*, 2003). If the plant sterol esters remain intact within the intestinal lumen, they could attract other lipophilic compounds, including cholesterol and cholesterol esters, and carry them to distal parts of the intestinal lumen where cholesterol absorption is much less efficient. Further research is clearly needed to resolve this issue.

A fourth proposed mechanism involves the regulation of the intestinal transporters, NPC1L1, ABCG5, and ABCG8. As described earlier (Figure 2), NPC1L1 resides in the brush border membrane and transports both cholesterol and plant sterols into the enterocyte (Davis *et al.*, 2004; Salen *et al.*, 2004), whereas ABCG5 and ABCG8 transport plant sterols and possible cholesterol back to the intestinal lumen (Lee *et al.*, 2001b). In this way, plant sterols could compete with cholesterol for binding to NPC1L1, although direct evidence for this is lacking. It is also possible that plant sterols could inhibit gene expression of NPC1L1 or, conversely, enhance expression of ABCG5 and ABCG8, which could promote cholesterol efflux if cholesterol is transported by ABCG5 and ABCG8. However, Field *et al.* (2004) reported that NPC1L1 mRNA was not changed in hamsters fed plant stanols. They also found that ABCG5 and ABCG8 mRNA was decreased by plant stanols rather than increased, suggesting that the cholesterol-lowering effect of plant stanols (and sterols) is unrelated to changes in gene expression of NPC1L1, ABCG5, or ABCG8. The study by Field *et al.* (2004) does not exclude the possibility that unknown transporters of cholesterol are regulated by plant sterols.

B. SOLUBLE FIBER

Consumption of foods rich in fiber is associated with reduced plasma LDL cholesterol concentration, diminished glycemic response, and improved bowel function. These physiological responses lead to reductions in risk of coronary heart disease, diabetes, and intestinal cancers. In 2002, dietary reference intakes (DRI) were established for fiber, ranging from 30–36 g/day for adult males and 21–29 g/day for adult females (Institute of Medicine, 2002). The actual fiber intake for American adults appears to be much lower, with estimated intakes of 17 g/day for males and 13 g/day for females (Institute of Medicine, 2002).

Fiber has traditionally been defined as nondigestible carbohydrates of plant origin, although nondigestible carbohydrates from nonplant sources may be considered "fiber." The fiber content of several common foods is provided in Table II. Because fiber can be consumed as an inherent component in native foods or as an added ingredient in manufactured foods, the Institute of Medicine's Food and Nutrition Board has defined fiber accordingly: "Dietary fiber" consists of nondigestible carbohydrates that are intrinsic and intact in plants, whereas "functional fiber" consists of isolated nondigestible carbohydrates that have beneficial physiological effects in humans (Institute of Medicine, 2002). These definitions recognize the diversity of fibers in the human food supply and allow for flexibility in

TABLE II
FIBER CONTENT OF COMMON PLANT FOODS

	Total fiber (g/100 g edible portion)	Soluble fiber (% of total fiber)
Fruits		
Apple	2.4	26.2
Banana	2.6	22.1
Orange	2.2	36.8
Peach	1.5	28.8
Pear	3.1	17.7
Pineapple	1.4	11.2
Vegetables		
Broccoli	2.6	30.9
Carrot	2.8	32.2
Corn	2.7	11.3
Potato, white	2.4	34.0
Tomato	1.2	13.8
Legumes		
Kidney beans	6.4	20.2
Navy beans	10.5	25.3
Lentils	7.9	9.3
Pinto beans	9.0	23.7
Cereals		
Bread, white	2.4	33.3
Rolled oats	9.4	34.0
Rice, brown	3.4	23.7
Wheat flour	2.7	29.8
Whole wheat flour	12.2	14.5

Source: Anderson and Bridges (1988); US Department of Agriculture (2005).

incorporating new fiber sources and ingredients into consumer products with specific functionality in mind.

Fibers may also be categorized according to their solubility in water. Insoluble fibers are found mainly in cell walls of plants and include cellulose, some hemicelluloses, and lignin. Generally speaking, good sources of insoluble fiber are vegetables, legumes, whole wheat (particularly bran), nuts, and seeds. Cellulose-based ingredients are frequently used as thickening agents, as fat replacers, or simply to reduce calories in food products. The moderate water-holding capacity of insoluble fibers and their ability to add dietary "bulk" results in increased fecal volume and faster transit time through the colon, thus promoting laxation. However, insoluble fibers generally have little or no impact on cholesterol absorption (Gallaher and Schneeman, 2001). A possible exception is chitosan, the deacetylated form of chitin, found in the exoskeleton of crustaceans and in certain fungi. Although chitosan is not derived from plants, it is a nondigestible carbohydrate and may be considered a functional fiber because of its ability to lower plasma LDL cholesterol concentration (Ylitalo *et al.*, 2002). Studies in rats have indicated that dietary chitosan inhibits cholesterol absorption (Gallaher *et al.*, 2000; Vahouny *et al.*, 1983). A study in humans indicated that a supplement containing equal amounts of chitosan and a soluble fiber (glucomannan) promoted cholesterol excretion (Gallaher *et al.*, 2002), but cholesterol absorption efficiency was not directly measured nor was it possible to isolate the independent effects of chitosan.

Most water-soluble fibers, in contrast to insoluble fibers, have the ability to inhibit cholesterol absorption and subsequently lower plasma LDL cholesterol concentration, as confirmed in meta-analyses of studies involving several types of soluble fiber (Brown *et al.*, 1999; Castro *et al.*, 2005). Many soluble fibers are considered both dietary and functional fibers because they are abundant in native foods and they are used frequently as additives in food products. These include pectin, β-glucans, fructans, gums, and resistant starch (i.e., resistant to digestion by mammalian enzymes). Pectin is actually a family of related compounds that have high-binding and gel-forming properties (Thakur *et al.*, 1997). Good sources of pectin include apples, citrus fruits, and strawberries. The majority of pectin used commercially is extracted from citrus peel or apples, and is used in food products that require gelling such as jellies, icings, frozen foods, and fat-reduced foods. β-Glucans are found in cereal brans, especially oats and barley, and in yeast, the latter being an important commercial source (Bell *et al.*, 1999). β-Glucans provide thickening properties when used as a food ingredient. Fructans are soluble in water but do not have gel-forming or ion-binding properties (Schneeman, 1999). The major fructans—oligofructose and inulin, and consumed mainly in wheat and onions—are known for their ability to support the growth of beneficial intestinal microflora, rather than contributing to the physical characteristics of food products (Boeckner *et al.*, 2001). Gums are used extensively as functional fibers in the manufacture of food products, although they are also consumed as dietary fibers in legumes, oats, and barley. Gums are derived from a variety of sources including seeds, seaweed, plant exudates, and microbial fermentation and are used for their ability to provide thickening, stability, emulsification, and glossy appearance to food products (Pszczola, 2003). An example of a specialized gum is the dietary supplement, Benefiber® (Novartis Consumer Health, Inc.), containing partially hydrolyzed guar gum, which prevents it from thickening when mixed with liquids. Resistant starches, found in a wide range of plant-based foods, can impart viscosity to foods when added as an ingredient. Several types of resistant starch are commercially available and may be used as functional fibers to improve intestinal health (Kendall *et al.*, 2004). Another soluble fiber, psyllium, is purely a functional fiber obtained from the husk of plantago seeds. It has a very high water-holding capacity and forms viscous solutions when mixed with water. Psyllium is the main component of Metamucil® (Procter & Gamble) and is used for its laxative properties.

While pectin and certain gums have received much attention regarding their impact on cholesterol absorption, information regarding β-glucans, resistant starches, and other gums is still emerging. The wide range of commercial soluble fibers available for research has also contributed to inconsistent results in the scientific literature. For example, guar gum was reported to decrease cholesterol absorption in some studies (Ebihara and Schneeman, 1989; Vahouny *et al.*, 1988) but not in others (Evans *et al.*, 1992; Miettinen and Tarpila, 1989). Similarly, pectin was shown to inhibit cholesterol absorption in some studies (Kelley and Tsai, 1978; Vahouny *et al.*, 1988), while others found no effect of pectin on absorption (Fernandez *et al.*, 1994; Mathé *et al.*, 1977). The amount of soluble fiber consumed, chemical and physical modification of the fibers, and the composition of the background diet are variables that can affect the research outcomes.

Despite these variables, it appears that the primary attribute of soluble fibers that inhibit cholesterol absorption is the ability to form a viscous matrix when hydrated. Many water-soluble fibers become viscous in the small intestine (Eastwood and Morris, 1992). It is believed that increased viscosity impedes the movement of cholesterol, bile acids, and other lipids and hinders micelle formation, thus reducing cholesterol absorption and promoting cholesterol excretion from the body. Consumption of viscous fibers was shown to increase the thickness of the unstirred water layer in humans (Flourie *et al.*, 1984; Johnson and Gee, 1981) and reduce the amount of cholesterol appearing in the lymph of cannulated rats (Ikeda *et al.*, 1989b; Vahouny *et al.*, 1988). Turley *et al.* (1991, 1994) reported that bile acid output was increased in hamsters fed psyllium, suggesting that the increased viscosity may have disrupted micelle formation by promoting bile acid excretion. In contrast, Favier *et al.* (1998) reported that fecal cholesterol excretion was increased in rats fed guar gum, whereas excretion of bile acids was not affected, indicating that viscosity can specifically effect cholesterol independent of bile acids. Isolating the effects of viscosity is somewhat hampered due to other properties of soluble fiber that influence cholesterol metabolism (i.e., fermentation, inhibition of digestive enzymes, and direct binding of cholesterol and bile acids). One tool that has proven useful in this regard is hydroxypropyl methylcellulose (HPMC), a cellulose derivative that imparts viscosity in the small intestine but, unlike other soluble fibers, is resistant to fermentation (Gallaher *et al.*, 1993). We have demonstrated in hamsters and rats that increased viscosity due to HPMC feeding was inversely correlated with cholesterol absorption efficiency and promoted fecal cholesterol excretion (Carr *et al.*, 1996, 2003).

Another possible mechanism involves the direct binding of cholesterol or bile acids, which could alter micelle formation and decrease the ability of cholesterol to incorporate into micelles. To varying degrees, soluble fibers appear to bind cholesterol (Eastwood and Mowbray, 1976; Lund, 1984) and other lipids that comprise micelles, including phospholipid and triglyceride (Vahouny *et al.*, 1980, 1981). Story and Kritchevsky (1976) reported that soluble fibers have the ability to bind primary and secondary bile acids (and their taurine and glycine conjugates). However, Gallaher and Schneeman (1986) demonstrated that while soluble fibers bound significant quantities of bile acids *in vitro*, lipid solubilization and the ability to form micelles was not impaired. The notion of lipid binding as a means of reducing cholesterol absorption may seem attractive, but its contribution is likely to be small relative to the effects of viscosity. Both psyllium and HPMC increased the intestinal contents viscosity and reduced cholesterol absorption, but neither appeared to bind bile acids (Carr *et al.*, 2003; Turley *et al.*, 1991). From a different point of view, Ellegård *et al.* (1997) indicated that oligofructose and inulin—soluble fibers that apparently do not bind bile acids nor do they become viscous—predictably had no effect on cholesterol absorption. It is possible that the binding ability of soluble fibers *in vitro* is greater than that within the physiological environment of the small intestine where viscosity is likely the major factor in preventing micelle formation.

Oligofructose and inulin are known to lower plasma LDL cholesterol, but their mode of action does not involve inhibiting cholesterol absorption (Beylot, 2005; Kaur and Gupta, 2002). As mentioned earlier, oligofructose and inulin are not viscous fibers but rather serve as excellent fuel sources for beneficial intestinal bacteria, particularly lactobacilli and bifidobacteria (Boeckner *et al.*, 2001). In this way, changes in intestinal microflora induced by oligofructose and inulin have been shown to alter the bile acid profile and promote fecal bile acid excretion (Levrat *et al.,* 1994; Trautwein *et al.*, 1998). Bile acids normally circulate within the enterohepatic circulation, so any loss of bile acids requires the liver to increase synthesis. The result is an increase in the utilization of hepatic cholesterol, the substrate for bile acid synthesis, and a parallel increase in recruitment of cholesterol from the plasma via LDL receptors (Fernandez, 2001). The relative impact of soluble fibers on this mechanism versus cholesterol absorption is uncertain.

C. SAPONINS

Saponins are a group of naturally occurring compounds found mainly in plants but are also present in lower marine animals and some bacteria (Francis *et al.*, 2002; Price *et al.*, 1987). They appear to act in the defense system of plants by repelling insects and inhibiting the growth of bacteria and mold. Saponins consist of a hydrophobic triterpenoid or steroid moiety linked to one or more hydrophilic carbohydrate molecules. Figure 4 illustrates the basic structure of representative triterpenoid and steroid saponins, where the R group represents carbohydrate linkages that often include rhamnose, xylose, arabinose, galactose, glucose, or glucuronic acid. The amphiphilic nature of saponins—having both polar and nonpolar regions within its chemical structure—provides the molecules with excellent emulsifying properties and the ability to form soap-like foams, thus giving rise to their name.

Their unique structure and surface activity not only appears largely responsible for their beneficial biological effects but also for the long-held belief that saponins are toxic. Some saponins are strongly hemolytic and are

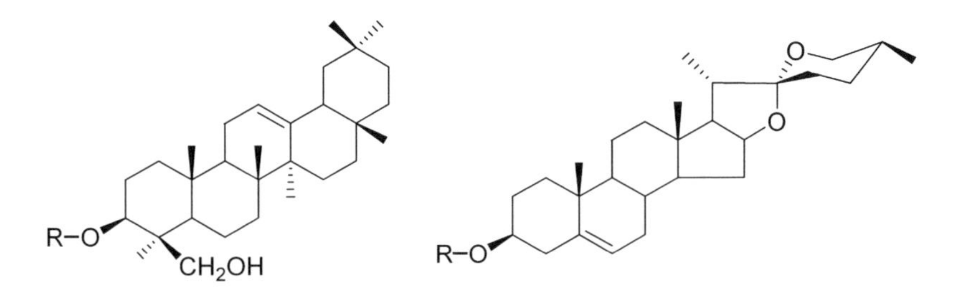

FIG. 4 Structure of representative saponins found in soybeans (triterpenoid) and eggplant (steroid). The R group represents one or more carbohydrate linkages that may contain rhamnose, xylose, arabinose, galactose, glucose, or glucuronic acid.

known toxicants in certain fish and cold-blooded animals (Francis *et al.*, 2002; Price *et al.*, 1987). Their bitter taste in high concentrations has undoubtedly fueled the suspicion of their antinutritional qualities (Liener, 1994). However, the saponins traditionally consumed by humans and other mammals appear to be safe (Oakenfull and Sidhu, 1990), although it seems likely that saponins from unusual sources, or common saponins administered in purified form or in high doses, could produce undesirable effects.

Saponins are present in most plants and are particularly abundant in legumes and alfalfa (Table III). They are also consumed, albeit in smaller amounts, with many herbs, spices, and tea. Triterpenoid saponins are generally predominant in food crops (e.g., legumes, sugar beets, spinach), while steroid saponins are more common in herbs and spices. Botanicals have long been used by ancient cultures for food and medicinal purposes, and it is now believed that many healthful properties of plants are due to saponins. Among the many health benefits attributed to saponins are reduced plasma cholesterol levels, inhibition of cancer cell growth, lower blood glucose response, reduced incidence of kidney stones, immune stimulation, anti-inflammation, and antioxidant properties (Francis *et al.*, 2002; Lacaille-Dubois and Wagner, 1996; Shi *et al.*, 2004).

The cholesterol-lowering ability of saponins was first observed in the 1950s and has since been confirmed in a number of species including humans (reviewed in Oakenfull and Sidhu, 1990). Saponins have also been shown to decrease the development of arterial atherosclerosis (Koo, 1983; Malinow

TABLE III

SAPONIN CONTENT OF COMMON PLANT FOODS

	Total saponins (g/100 g edible portion)
Alfalfa sprouts	8.0
Asparagus	1.5
Black-eye peas	0.03
Chickpeas	3.6
Kidney beans	4.1
Lentils	0.5
Navy beans	4.0
Oats	1.3
Peanuts	1.6
Pinto beans	4.1
Soybeans	5.6
Spinach	4.7
Sugar beets	5.8

Source: Price *et al.* (1987); Oakenfull and Sidhu (1989).

et al., 1978; Sautier *et al.*, 1979). A number of pharmaceutical and nutraceutical companies have developed saponin-based compounds for the purpose of lowering plasma LDL cholesterol concentration. To date, the FDA has not granted approval for a saponin-containing prescription drug intended for this purpose. However, a saponin extract from alfalfa called Esterin® (Innovative Product Solutions Group) is available over-the-counter and is the main ingredient in the dietary supplements Cholestaid™, Cholesterin™, and CholestSorb™.

Saponins appear to lower plasma LDL cholesterol concentration by interfering with cholesterol absorption. Studies in rats and monkeys fed naturally occurring saponins exhibited significant reductions in cholesterol absorption efficiency and an increase in fecal cholesterol excretion (Malinow *et al.*, 1981; Nakamura *et al.*, 1999; Sidhu *et al.*, 1987). Decreased bile acid absorption and increased excretion has also been reported in animals fed saponins (Malinow *et al.*, 1981; Nakamura *et al.*, 1999; Stark and Madar, 1993). One possible mechanism of action for decreased cholesterol absorption is the ability of saponins to form insoluble complexes with cholesterol (Gestetner *et al.*, 1972; Malinow *et al.*, 1977). In an effort to isolate the specific properties of saponins, Malinow (1985) prepared a variety of synthetic saponins in which the complex carbohydrate moieties of native plant saponins were replaced with simplified carbohydrates such as glucose or cellobiose. One of these synthetic saponins, tiqueside (Pfizer, Inc.), can effectively precipitate cholesterol from micelle solutions *in vitro* and inhibit cholesterol absorption in a variety of animals (Harwood *et al.*, 1993) and in humans (Harris *et al.*, 1997). But despite ample data showing the formation of a saponin/cholesterol complex *in vitro*, there is essentially no definitive evidence that complexation occurs in the intestinal lumen (Morehouse *et al.*, 1999).

Another possible mechanism involves the effect of saponins on micelle formation. Saponins are known to alter the size or shape of micelles (Oakenfull, 1986; Oakenfull and Sidhu, 1983), an observation that is consistent with decreased bile acid absorption (Stark and Madar, 1993) and increased fecal bile acid excretion (Malinow *et al.*, 1981; Nakamura *et al.*, 1999). Saponins may also directly bind bile acids (Oakenfull and Sidhu, 1989), which would presumably interfere with micelle formation and decrease cholesterol absorption. Other studies have found that saponins decrease the absorption of fat-soluble vitamins (Jenkins and Atwal, 1994) and triglycerides (Han *et al.*, 2002; Okuda and Han, 2001), indicating decreased micelle formation. However, direct evidence showing impaired micelle formation *in vivo* is lacking. Moreover, Harwood *et al.* (1993) reported no change in bile acid absorption or interruption of the enterohepatic circulation of bile acids in hamsters fed tiqueside, despite significant reductions in cholesterol absorption.

Because of their amphiphilic nature and high surface activity, saponins could affect cholesterol absorption by interacting with cholesterol and other lipids in the brush border membrane of enterocytes. *In vitro* studies have indicated that the permeability of the intestine can be affected by saponin treatment (Alvarez and Torres-Pinedo, 1982; Johnson *et al.*, 1986). Onning *et al.* (1996) reported that saponins altered the permeability of albumin in rat intestine *in vitro*, but the transport of glucose was not affected. Furthermore, when given *in vivo* in the same study, the saponins did not affect albumin transport across the intestinal membrane. Most common food-derived saponins are unlikely to affect intestinal cell membranes to a significant extent. Even though the detergent action of specific hemolytic saponins is responsible for their toxicity, these hazardous saponins are generally not part of the human food supply or they are consumed in negligible amounts (Francis *et al.*, 2002; Price *et al.*, 1987). The synthetic saponin, tequeside, did not alter cholesterol or bile acid absorption in everted sacs after pretreatment with saponins, indicating no significant change in enterocyte transport function (Sidhu *et al.*, 1987). Tequeside is synthesized from food-based saponins and is probably more representative of saponins commonly consumed by humans. Tequeside is also well tolerated by humans and no toxicological side effects have been observed (Harris *et al.*, 1997).

The full extent to which saponins reduce cholesterol absorption requires further study. Because of the large number of saponins present in the food supply, it is possible that all of the mechanisms discussed earlier contribute to reduced cholesterol absorption. Unlike plant sterols in which their mode of action is relatively well defined, there are probably multiple effects of saponins within the intestinal tract, including their ability to interact with other dietary constituents and the ability of some saponins to be absorbed systemically. The regulatory effects of saponins on cellular cholesterol transport have not been examined.

D. SOY PROTEIN

Replacing animal proteins in the diet with plant proteins—primarily soy protein—is known to lower plasma cholesterol concentration in both animals and humans (Forsythe *et al.*, 1986). In a large cross-study of 4838 Japanese men and women, Nagata *et al.* (1998) observed a significant trend for decreasing total cholesterol concentration with an increasing intake of soy products. A meta-analysis of 38 clinical studies indicated that an average intake of 47 g soy protein per day was associated with a 13% decrease in plasma LDL concentration (Anderson *et al.*, 1995). Based largely on the meta-analysis, the FDA approved a health claim for the association between dietary soy protein and reduced risk of coronary heart disease in October of

1999 in response to a petition submitted by Protein Technologies International, Inc. The FDA determined that a daily intake of 25 g of soy protein was required to elicit a clinically significant reduction in plasma LDL cholesterol and thus allows the following health claim: "25 grams of soy protein a day, as part of a diet low in saturated fat and cholesterol, may reduce the risk of heart disease."

Studies in monkeys (Greaves *et al.*, 2000), rabbits (Huff and Carroll, 1980), and rats (Nagata *et al.*, 1982; Vahouny *et al.*, 1984) have demonstrated significant reductions in cholesterol absorption due to dietary soy protein. In each of these studies, soy protein was compared to equivalent amounts of casein. Greaves *et al.* (2000) also compared soy protein to casein enriched with either isoflavone extract or conjugated equine estrogen, but only the soy protein caused a decrease in cholesterol absorption. Increased fecal cholesterol excretion has also been reported in experimental animals fed soy protein (Beynen, 1990; Potter, 1995). One human study is in agreement with the animal data (Duane, 1999), although another reported no change in fecal cholesterol output in hypercholesterolemic patients treated with soy protein (Fumagalli *et al.*, 1982).

Investigators originally focused on the amino acid content of soy protein, with its low lysine/arginine ratio compared to casein, as being responsible for increased fecal cholesterol excretion. Sugano *et al.* (1984) suggested that the low lysine/arginine ratio may alter pancreatic secretions (hormones and digestive enzymes), which in turn could influence cholesterol absorption. However, Gibney (1983) was unable to show any effect of the lysine/arginine ratio on cholesterol kinetics in rabbits, and the precise mechanisms involving lysine and arginine have not been subsequently defined. Furthermore, rats fed amino acid mixtures that mimic soy protein exhibited no change in cholesterol absorption or fecal cholesterol excretion relative to dietary casein, but that intact soy protein caused an inhibition in cholesterol absorption and greater fecal cholesterol output (Nagata *et al.*, 1982; Tanaka *et al.*, 1984). The latter studies have led to the belief that the specific amino acids of soy protein probably have little impact on cholesterol absorption, raising the possibility that nonprotein constituents within commercial soy protein preparations are responsible for the cholesterol-lowering action previously observed in animal and human studies.

Soy protein preparations contain a variety of biologically active compounds including saponins, fibers, trypsin inhibitors, and isoflavones (Potter, 2000). Hamsters and rats fed ethanol-extracted soy protein isolates had no ability to lower plasma cholesterol compared to intact soy protein isolates (Lucas *et al.*, 2001; Ni *et al.*, 1999). Extraction with ethanol is a treatment that would remove saponins, isoflavones, and other phytochemicals from the protein. Although one study showed that ethanol washing did not diminish the cholesterol-lowering ability of soy protein isolate (Fukui *et al.*, 2004), it seems likely that the relatively high saponin content of soybeans (Table III) could contribute to the inhibition of cholesterol absorption. Soy protein "isolates" are the most highly refined soy protein products available to consumers and contain about 90% protein. The saponin content of commercial isolates ranges from 0.3 to 2.3 g/100 g (Fenwick and Oakenfull, 1981; Ireland *et al.*, 1986). Soy protein "concentrates" contain about 70% protein and retain most of the soybean fiber, which may also contribute to inhibition of cholesterol absorption, but the saponin concentration is undetectable (Ireland *et al.*, 1986).

Another consideration is the effect of partially digested soy protein on cholesterol metabolism. Sugano *et al.* (1990) demonstrated that a soy protein hydrolysate (partially digested by porcine pepsin or microbial proteases) fed to rats decreased plasma cholesterol levels and promoted fecal cholesterol excretion to a greater extent than intact soy protein. Using the Caco-2 intestinal cell model, Nagaoka *et al.* (1997) reported that soy protein hydrolysate directly inhibited the absorption of micellar cholesterol. The same investigators later found that rats infused with soy protein hydrolysate had significantly lower rates of cholesterol absorption and increased fecal excretion of cholesterol compared to rats fed intact soy protein (Nagaoka *et al.*, 1999). In additional experiments, Nagaoka *et al.* (1999) showed that soy protein hydrolysate decreased the micellar solubility of cholesterol *in vitro*. These studies suggest that digestion products of soy protein may directly interfere with micelle formation.

Soy protein intake has also been reported to increase bile acid excretion in animals (Beynen *et al.*, 1990; Huff and Carroll, 1980; Kuyvenhoven *et al.*, 1989; Nagata *et al.*, 1982; Wright and Salter, 1998), although the effect was only marginal in a human study (Duane, 1999). As with soluble fibers and saponins, increased bile acid excretion may indicate a disruption in micelle formation, leading to reduced cholesterol solubilization and absorption, although this has not been definitively established. Lin *et al.* (2004) reported an additive effect of soy protein and plant sterol esters in promoting the excretion of both cholesterol and bile acids when fed simultaneously to hamsters, indicating the benefit of consuming a variety of cholesterol-lowering plant-based foods.

E. PHOSPHOLIPIDS

Phospholipids are a group of amphiphilic molecules in which long acyl chains form a hydrophobic region, while the presence of a hydrophilic region is due to a phosphate-containing "head" group. Figure 5 illustrates two types of phospholipids, phosphatidylcholine (PC) and sphingomyelin (SM),

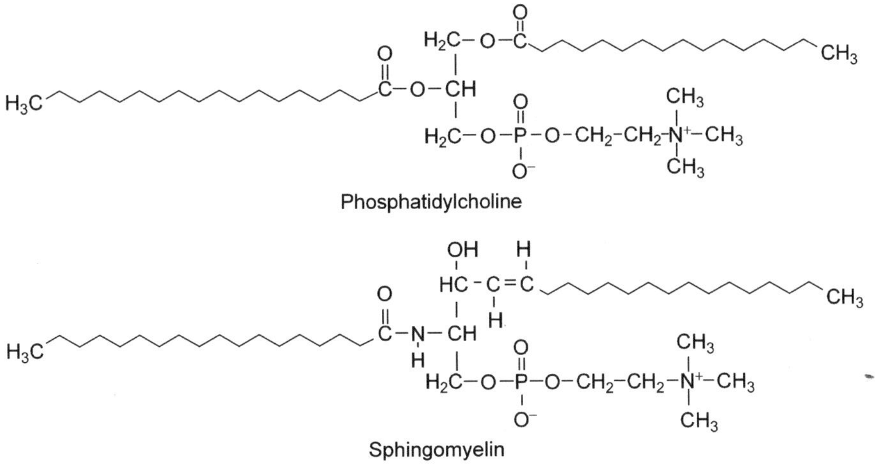

Phosphatidylcholine

Sphingomyelin

FIG. 5 Phospholipids present in the food supply.

found throughout nature and thus in the food supply. Various molecules can comprise the polar head group, although choline is frequently found in both plant and animal phospholipids. The long chain acyl moieties can be highly variable, although phospholipids from animals tend to be more saturated than those of plant origin. Because of their unique chemical structure, phospholipids spontaneously form lipid bilayers and are therefore the primary structural components of cell membranes. Most phospholipids are synthesized in the body but may also be consumed in diet. In addition to their role in cell membranes, phospholipids (specifically PC) are an important component of bile and are required for micelle formation and cholesterol solubilization. Because of biliary secretion, the principle phospholipid in the intestinal lumen is PC, which can exceed dietary amounts by as much as 5 to 1 (Eckhardt *et al.*, 2002). Food sources rich in PC include egg yolks, muscle foods, peanuts, and soybeans. SM is most abundant in egg yolks, muscle foods, soybean, and milk and dairy products. Food manufacturers frequently use purified PC (also called lecithin) as a food additive because of its excellent emulsifying properties. The PC found in commercial products is most often derived from soybeans and, to a lesser extent, egg yolk.

Dietary SM and PC are known to inhibit cholesterol absorption in experimental animals, resulting in lower plasma cholesterol levels (Imaizumi *et al.*, 1992; Jimenez *et al.*, 1990; O'Brien and Corrigan, 1988; Wilson *et al.*, 1998). Inhibition of cholesterol absorption was observed in humans infused intraduodenally with purified soy PC (Beil and Grundy, 1980). Kesaniemi and Grundy (1986) also found a small but significant reduction in cholesterol absorption in hyperlipidemic patients fed soy PC. Koo and colleagues have shown that egg PC inhibits cholesterol absorption in lymph duct cannulated rats and that egg PC is more effective at reducing absorption than soy PC (Jiang *et al.*, 2001; Koo and Noh, 2001). In fact, soy PC did not interfere with cholesterol absorption but rather produced a slight increase in absorption relative to no-PC controls (Jiang *et al.*, 2001). These researchers suggested that the higher degree of fatty acid saturation in egg PC compared to soy PC may have caused greater disruption in the micellar solubilization of cholesterol. This hypothesis was further supported in rats that had lower rates of cholesterol absorption when fed hydrogenated (i.e., fully saturated) PC compared to native PC, irrespective of whether the PC was from soy (Nyberg *et al.*, 2000) or egg (Jiang *et al.*, 2001). Naturally occurring soy PC contains 10–20% saturated fatty acids (SFA), whereas native egg PC contains 40–50% SFA.

In vitro studies using intestinal cells have suggested that intact PC—that is, PC that has not been hydrolyzed by digestive enzymes—exerts an inhibitory effect on micelle formation and, hence, cholesterol solubility. Young and Hui (1999) reported an inhibitory effect of PC on cholesterol uptake in IEC-6 intestinal cells when PC was incorporated into lipid emulsions. In another study, the addition of phospholipase A_2 eliminated the inhibitory action of PC, resulting in increased absorption of cholesterol into Caco-2 cells (Homan and Hamelehle, 1998). Studies have indicated that the presence of PC on the surface of lipid emulsions hinders the hydrolysis of triglycerides (Borgström, 1980; Patton and Carey, 1981), which also inhibits cholesterol uptake into intestinal cells (Young and Hui, 1999). These data indicate that proper digestion of dietary lipids is necessary to maximize micelle formation and cholesterol solubilization and that PC can inhibit digestive enzymes. The data further suggest that saturated egg PC is a more potent inhibitor of lipases than unsaturated soy PC mainly because saturated PC is a poor substrate for phospholipase A_2 and is not readily hydrolyzed (Jiang *et al.*, 2001; Kinkaid and Wilton, 1991). Still unknown is the effect of PC fatty acid chain length on micellar solubility of cholesterol.

SM also inhibits cholesterol absorption in experimental animals, and this effect has been shown to be dose dependent (Eckhardt *et al.*, 2002; Noh and Koo, 2003; Nyberg *et al.*, 2000). However, there is little information regarding how SM from different food sources affects cholesterol absorption. Noh and Koo (2004) reported that milk SM was a more potent inhibitor than egg SM, suggesting that the higher degree of saturation and longer chain length of milk SM may be important factors, although further investigation is needed.

SM is similar to PC to the extent that they share the same phosphatecholine head group and have long chain hydrophobic moieties (Figure 5), but SM differs in several ways from naturally occurring PC. The backbone of SM is a sphingoid base, whereas the backbone of PC is a glycerol base, which increases the polarity of SM and allows for stronger intra- and intermolecular hydrogen bonding (Yeagle *et al.*, 1976). The main acyl chain of SM is usually longer than the fatty acyl chains of PC and is frequently saturated. These characteristics contribute to a stronger interaction between cholesterol and SM in cell membranes compared to other phospholipids (Demel *et al.*, 1977; McIntosh *et al.*, 1992). Eckhardt *et al.* (2002) indicated that milk SM compared to egg PC was more effective in reducing cholesterol solubility in micelles and in limiting cholesterol uptake in Caco-2 cells. They also observed that milk SM significantly reduced cholesterol absorption in mice, whereas egg PC had no effect. On an equal molar basis, it appears that SM is a more effective inhibitor of cholesterol absorption than PC. Dietary intake of SM in humans is estimated to be 0.3–0.4 g/day (Vesper *et al.*, 1999), whereas PC intake is about 1–2 g/day (Åkesson, 1982). When taking into account biliary secretion of PC, there is considerably more PC than SM in the intestinal lumen under normal conditions. Addition of SM to the diets of experimental animals was still able to reduce cholesterol absorption in the presence of normal biliary PC, indicating a higher potency of SM relative to PC (Eckhardt *et al.*, 2002; Nyberg *et al.*, 2000). Further evidence of the effectiveness of SM is indicated by the case of the 88-year-old "egg man," who habitually ate 25 eggs/day, yet had normal plasma cholesterol levels (Kern, 1991). Eggs contain large amounts of SM (about 0.17% by weight), which could interact with cholesterol and decrease its availability for micellar solubilization. Foods that have little or no cholesterol but high SM, such as soybeans, have a greater potential to bind endogenous cholesterol, thus preventing its absorption.

F. STEARIC ACID

Stearic acid is a long chain SFA present, to varying degrees, in virtually all edible fats and oils. Table IV provides the fatty acid composition of fats and oils commonly consumed by humans. The most abundant food sources of stearic acid in the American diet are beef fat and cocoa butter (chocolate). Cocoa butter is valued by chocolate manufacturers because it remains solid at room temperature but dissolves quickly at body temperature, a unique characteristic of chocolate that is due largely to stearic acid. During the last few decades as cocoa butter prices and supplies have fluctuated, food companies began looking for alternative oils that could provide equivalent amounts of stearic acid in order to retain the desirable physical characteristics. Several

TABLE IV

FATTY ACID COMPOSITION (%) OF EDIBLE FATS AND OILS

	Medium chain fatty acids (8:0–12:0)[a]	Myristic acid (14:0)	Palmitic acid (16:0)	Palmit-oleic acid (16:1)	Stearic acid (18:0)	Oleic acid (18:1)	Linoleic acid (18:2)	Linolenic acid (18:3)	Others
Beef fat (tallow)	0.9	3.7	24.9	4.2	18.9	36.0	3.1	0.6	7.7
Butter oil	8.3	10.0	26.2	2.2	12.1	25.0	2.2	1.4	12.6
Canola oil	–	0.1	4.0	0.2	1.8	56.1	20.3	9.3	8.2
Chicken fat	0.1	0.9	21.6	5.7	6.0	37.3	19.5	1.0	7.9
Cocoa butter	–	0.1	25.4	0.2	33.2	32.6	2.8	0.1	5.6
Coconut oil	58.7	16.8	8.2	–	2.8	5.8	1.8	0.1	5.8
Corn oil	–	0.1	10.9	0.2	1.8	24.2	58.0	0.7	4.1
Olive oil	–	–	11.0	0.8	2.2	72.5	7.9	0.6	5.0
Palm oil	0.1	1.0	43.5	0.3	4.3	36.6	9.1	0.2	4.9
Palm kernel oil	54.2	16.4	8.1	–	2.8	11.4	1.6	–	5.5
Peanut oil	–	0.1	9.5	0.1	2.2	46.8	32.0	–	9.3
Pork fat (lard)	0.3	1.3	23.8	2.7	13.5	41.2	10.2	1.0	6.0
Safflower oil	–	0.1	4.3	0.1	1.9	14.4	74.6	0.4	4.2
Soybean oil	–	0.1	10.3	0.2	3.8	22.8	51.0	6.8	5.0

[a]Abbreviations represent the number of carbon atoms (left side of colon) followed by the number of double bonds.

Source: US Department of Agriculture (2005).

tree nuts and seeds indigenous to West Africa, India, and Southeast Asia were found to be rich in stearic acid, including dhupa, illipe, kokum, mango kernel, sal, and shea (Bhattacharyya, 2002). Shea nut oil (shea butter) and sal oil have proven to be the most feasible and are mainly used in Europe and Japan as cocoa butter substitutes. These stearic acid-rich oils are also used in cosmetics, candles, and other industrial products worldwide, in addition to providing a local source of cooking oil. Shea nut oil contains about 38% stearic acid and sal oil contains about 34% stearic acid. Hydrogenating vegetable oils can increase the stearic acid content by converting unsaturated 18-carbon fatty acids to stearic acid, but the process also yields undesirable *trans* fatty acids.

The impact of dietary fatty acids on plasma cholesterol levels has been studied for decades, summarized by the classic prediction equations of Keys *et al.* (1965) and Hegsted *et al.* (1965) and later refined to focus on LDL cholesterol (Hegsted *et al.*, 1993; Howell *et al.*, 1997). The purpose of the equations is to predict changes in plasma LDL cholesterol concentration in response to changes in dietary intake of SFA, polyunsaturated fatty acids (PUFA), and cholesterol. Dietary monounsaturated fatty acids are believed to have "neutral" effects on plasma cholesterol and are excluded from the equations. One common feature of all the prediction equations is that the strongest determinant of plasma cholesterol concentration is SFA (increases plasma cholesterol) followed by PUFA (decreases cholesterol), while dietary cholesterol has relatively little effect on plasma cholesterol concentrations (in contrast to dietary recommendations to limit cholesterol intake). The strong influence by SFA has led to the well-known recommendation from the National Cholesterol Education Program, the American Heart Association, and other health organizations to limit SFA intake (Krauss *et al.*, 2000; National Institutes of Health, 2002). But despite the blanket condemnation of SFA, the original data of Keys *et al.* (1965) and Hegsted *et al.* (1965) clearly showed that stearic acid was unique among dietary SFA because it did not raise plasma cholesterol levels. The neutral or cholesterol-lowering effect of dietary stearic acid has since been confirmed repeatedly in animal and human studies (Grundy and Denke, 1990; Kris-Etherton and Yu, 1997; Sanders, 2003).

Feldman *et al.* (1979a,b) were the first to demonstrate a reduction in cholesterol absorption due to dietary stearic acid in rats. The investigators used three different methods to quantify cholesterol absorption (i.e., plasma isotope ratio method, fecal dual isotope method, and lymph duct cannulation), and in each case absorption was significantly decreased by stearic acid. Other studies using lymph duct cannulated rats fed stearic acid-enriched diets showed significant reductions in cholesterol absorption (Chen *et al.*, 1989; Ikeda *et al.*, 1994). Using a more realistic dietary approach, we conducted a study in hamsters fed NIH-07 cereal-based diet specifically enriched in single fatty acids (stearic, palmitic, oleic, linoleic, or *trans* fatty acids). Cholesterol absorption was similar among hamsters fed palmitic, oleic, linoleic, or *trans* fatty acids, however, cholesterol absorption was significantly reduced in hamsters fed stearic acid compared to the other fatty acids (Schneider *et al.*, 2000). Dietary stearic acid has also been shown to increase fecal cholesterol excretion in rats and hamsters (Imaizumi *et al.*, 1993; Kamei *et al.*, 1995; Schneider *et al.*, 2000).

Schmidt and Gallaher (1997) reported that cholesterol solubilization was decreased within the intestinal contents of rats fed stearic acid-enriched diets. One possible mechanism of action of stearic acid is its ability to interfere with micelle formation through its incorporation into phospholipids. Unlike most of the other food components that inhibit cholesterol absorption, stearic acid is relatively well absorbed into the bloodstream and body tissues. Wang and Koo (1993a,b) reported that after absorption, stearic acid was preferentially incorporated into hepatic and biliary phospholipids compared to other dietary fatty acids. Cohen and Carey (1991) demonstrated that micelle stability and cholesterol solubility were impaired when micellar phospholipids contained stearic acid compared to unsaturated fatty acids. Another possible mode of action may involve alterations in the bile acid species present in the enterohepatic circulation. We have shown that dietary stearic acid decreased the proportion of secondary bile acids in the gallbladder compared to primary bile acids, thus decreasing the overall hydrophobicity index (Cowles *et al.*, 2002). Similarly, Hassel *et al.* (1997) reported significantly lower proportions of secondary bile acids in feces of hamsters fed stearic acid. Secondary bile acids are more hydrophobic than primary bile acids and their diminished presence in the enterohepatic circulation can decrease the efficiency by which micelles solubilize cholesterol (Armstrong and Carey, 1987). It is also possible the stearic acid exerts some regulatory effect on cholesterol transport into (or within) the enterocyte, although this has not been reported. Nevertheless, dietary stearic acid most likely inhibits cholesterol absorption through systemic mechanisms rather than disrupting micelle formation through physical interactions within the intestinal lumen.

IV. CONCLUSIONS

Many cholesterol-lowering compounds are naturally present in the human food supply. Each of these compounds can be isolated, purified, and subsequently used as additives in food products and dietary supplements designed specifically for reducing plasma LDL cholesterol concentration. This chapter focuses only on those compounds known to lower plasma cholesterol by inhibiting cholesterol absorption in the small intestine, including plant sterols and stanols, soluble fibers, saponins, soy protein, phospholipids (SM and PC), and stearic acid. All of these compounds—except, perhaps, stearic acid—appear to exert their effects mainly by interfering with micellar solubilization of cholesterol within the intestinal lumen. This can be the result of displacing cholesterol from the micelle, binding or precipitating cholesterol, impeding the movement of cholesterol by forming a viscous matrix, inhibiting digestive enzymes, binding bile acids and decreasing their participation in micelle formation, or downregulating cholesterol transporters within the enterocyte. Stearic acid appears to work systemically by incorporating into hepatic and biliary phospholipids, which destabilizes micelles and reduces cholesterol solubility.

These compounds are attractive to food and nutraceutical companies because, in most cases, they are regulated as foods and not drugs. The FDA currently allows a heart health claim for certain products containing plant sterols or soy protein, while the American Heart Association endorses certain products rich in soluble fiber. Many of these compounds also contribute important functional properties when added to foods, such as emulsification, improved texture, and calorie reduction. Some soluble fibers may function as beneficial prebiotics that promote intestinal health. Furthermore, most of the compounds work entirely within the intestine and are poorly absorbed, if at all, thus significantly reducing (or eliminating) the risk of toxicity. In view of these desirable characteristics, manufacturers are likely to develop a greater diversity of food and nutraceutical products in the coming years giving consumers more choices to manage their plasma cholesterol levels through nonpharmacological means.

REFERENCES

Ahrens, E.H., Jr., Hirsch, J., Insull, W., Tsaltas, T.T., Blomstrand, R., and Peterson, M.L. 1957. The influence of dietary fats on serum-lipid levels in man. *Lancet* **1**, 943–953.

Åkesson, B. 1982. Content of phospholipids in human diets studied by the duplicate-portion technique. *Br. J. Nutr.* **47**, 223–229.

Altmann, S.W., Davis, H.R., Jr., Zhu, L.J., Yao, X., Hoos, L.M., Tetzloff, G., Iyer, S. P. N., Maguire, M., Golovko, A., Zeng, M., Wang, L., Murgolo, N., *et al.* 2004. Niemann-Pick C1 Like 1 protein is critical for intestinal cholesterol absorption. *Science* **303**, 1201–1204.

Alvarez, J.R. and Torres-Pinedo, R. 1982. Interactions of soybean lectin, soyasaponins, and glycinin with rabbit jejunal mucosa *in vitro*. *Pediatr. Res.* **16**, 728–731.

Anderson, J.W. and Bridges, S.R. 1988. Dietary fiber content of selected foods. *Am. J. Clin. Nutr.* **47**, 440–447.

Anderson, J.W., Johnstone, B.M., and Cook-Newell, M.E. 1995. Meta-analysis of the effects of soy protein intake on serum lipids. *N. Engl. J. Med.* **333**, 276–282.

Andersson, S.W., Skinner, J., Ellegård, L., Welch, A.A., Bingham, S., Mulligan, A., Andersson, H., and Khaw, K.T. 2004. Intake of dietary plant sterols is inversely related to serum cholesterol concentration in men and women in the EPIC Norfolk population: A cross-sectional study. *Eur. J. Clin. Nutr.* **58**, 1378–1385.

Apstein, M.D. and Carey, M.C. 1996. Pathogenesis of cholesterol gallstones: A parsimonious hypothesis. *Eur. J. Clin. Invest.* **26**, 343–352.

Armstrong, M.J. and Carey, M.C. 1987. Thermodynamic and molecular determinants of sterol solubilities in bile salt micelles. *J. Lipid Res.* **28**, 1144–1155.

Becker, M., Staab, D., and von Bergmann, K. 1993. Treatment of severe familial hypercholesterolemia in childhood with sitosterol and sitostanol. *J. Pediatr.* **122**, 292–296.

Beil, F.U. and Grundy, S.M. 1980. Studies on plasma lipoproteins during absorption of exogenous lecithin in man. *J. Lipid Res.* **21**, 525–536.

Bell, S., Goldman, V.M., Bistrian, B.R., Arnold, A.H., Ostroff, G., and Forse, R.A. 1999. Effect of beta-glucan from oats and yeast on serum lipids. *Crit. Rev. Food Sci. Nutr.* **39**, 189–202.

Berge, K.E., Tian, H., Graf, G.A., Yu, L., Grishin, N.V., Schultz, J., Kwiterovich, P., Shan, B., Barnes, R., and Hobbs, H.H. 2000. Accumulation of dietary cholesterol in sitosterolemia caused by mutations in adjacent ABC transporters. *Science* **290**, 1771–1775.

Best, M.M., Duncan, C.H., Van Loon, E.J., and Wathen, J.D. 1954. Lowering of serum cholesterol by the administration of a plant sterol. *Circulation* **10**, 201–206.

Beylot, M. 2005. Effects of inulin-type fructans on lipid metabolism in man and in animal models. *Br. J. Nutr.* **93**(Suppl. 1), S163–S168.

Beynen, A.C. 1990. Comparison of the mechanisms proposed to explain the hypocholesterolemic effect of soybean protein versus casein in experimental animals. *J. Nutr. Sci. Vitaminol. (Tokyo)* **36**(Suppl. 2), S87–S93.

Beynen, A.C., West, C.E., Spaaij, C.J., Huisman, J., Van Leeuwen, P., Schutte, J.B., and Hackeng, W. H. 1990. Cholesterol metabolism, digestion rates and postprandial changes in serum of swine fed purified diets containing either casein or soybean protein. *J. Nutr.* **120**, 422–430.

Bhattacharyya, D.K. 2002. Lesser-known Indian plant sources for fats and oils. *Inform* **13**, 151–157.

Bitman, J. and Wood, D.L. 1980. Cholesterol and cholesteryl esters of eggs from various avian species. *Poult. Sci.* **59**, 2014–2023.

Boeckner, L.S., Schnepf, M.I., and Tungland, B.C. 2001. Inulin: A review of nutritional and health implications. *Adv. Food Nutr. Res.* **43**, 1–63.

Borgström, B. 1980. Importance of phospholipids, pancreatic phospholipase A2, and fatty acid for the digestion of dietary fat: *In vitro* experiments with the porcine enzymes. *Gastroenterology* **78**, 954–962.

Briefel, R.R. and Johnson, C.L. 2004. Secular trends in dietary intake in the United States. *Annu. Rev. Nutr.* **24**, 401–431.

Brown, L., Rosner, B., Willett, W.W., and Sacks, F.M. 1999. Cholesterol-lowering effects of dietary fiber: A meta analysis. *Am. J. Clin. Nutr.* **69**, 30–42.

Burnett, D.A. 2004. β-Lactam cholesterol absorption inhibitors. *Curr. Med. Chem.* **11**, 1873–1887.

Carr, T.P., Gallagher, D.D., Yang, C.-H., and Hassel, C.A. 1996. Increased intestinal contents viscosity reduces cholesterol absorption efficiency in hamsters fed hydroxypropyl methylcellulose. *J. Nutr.* **126**, 1463–1469.

Carr, T.P., Cornelison, R.M., Illston, B.J., Stuefer-Powell, C.L., and Gallaher, D.D. 2002. Plant sterols alter bile acid metabolism and reduce cholesterol absorption in hamsters fed a beef-based diet. *Nutr. Res.* **22**, 745–754.

Carr, T.P., Wood, K.J., Hassel, C.A., Bahl, R., and Gallaher, D.D. 2003. Raising intestinal contents viscosity leads to greater excretion of neutral steroids but not bile acids in hamsters and rats. *Nutr. Res.* **23**, 91–102.

Castro, I.A., Barroso, L.P., and Sinnecker, P. 2005. Functional foods for coronary heart disease risk reduction: A meta-analysis using a multivariate approach. *Am. J. Clin. Nutr.* **82**, 32–40.

Chen, I.S., Subramaniam, S., Vahouny, G.V., Cassidy, M.M., Ikeda, I., and Kritchevsky, D. 1989. A comparison of the digestion and absorption of cocoa butter and palm kernel oils and their effects on cholesterol absorption in rats. *J. Nutr.* **119**, 1569–1573.

Christiansen, L., Karjalainen, M., Seppanen-Laakso, T., Hiltunen, R., and Yliruusi, J. 2003. Effect of beta-sitosterol on precipitation of cholesterol from non-aqueous and aqueous solutions. *Int. J. Pharm.* **254**, 155–166.

Cohen, D.E. and Carey, M.C. 1991. Acyl chain unsaturation modulates distribution of lecithin molecular species between mixed micelles and vesicles in model bile. Implications for particle structure and metastable cholesterol solubilities. *J. Lipid Res.* **32**, 1291–1302.

Cowles, R.L., Lee, J.-Y., Gallaher, D.D., Stuefer-Powell, C.L., and Carr, T.P. 2002. Dietary stearic acid alters gallbladder bile acid composition in hamsters fed cereal-based diets. *J. Nutr.* **132**, 3119–3122.

Davis, H.R., Jr., Zhu, L.J., Hoos, L.M., Tetzloff, G., Maguire, M., Liu, J., Yao, X., Iyer, S. P. N., Lam, M.H., Lund, E.G., Detmers, P.A., Graziano, M.P., *et al.* 2004. Niemann-Pick C1 like 1 (NPC1L1) is the intestinal phytosterol and cholesterol transporter and a key modulator of whole-body cholesterol homeostasis. *J. Biol. Chem.* **279**, 33586–33592.

Dawson, P.A. and Oelkers, P. 1995. Bile acid transporters. *Curr. Opin. Lipidol.* **6**, 109–114.

Demel, R.A., Jansen, J.W., van Dijck, P.W., and van Deenen, L.L. 1977. The preferential interaction of cholesterol with different classes of phospholipids. *Biochim. Biophys. Acta* **465**, 1–10.

Denke, M.A. 1995. Lack of efficacy of low-dose sitostanol therapy as an adjunct to a cholesterol-lowering diet in men with moderate hypercholesterolemia. *Am. J. Clin. Nutr.* **61**, 392–396.

de Vries, J. H. M., Jansen, A., Kromhout, D., van de Bovenkamp, P., van Staveren, W.A., Mensink, R.P., and Katan, M.B. 1997. The fatty acid and sterol content of food composites of middle-aged men in seven countries. *J. Food Compost. Anal.* **10**, 115–141.

Duane, W.C. 1993. Effects of lovastatin and dietary cholesterol on sterol homeostasis in healthy human subjects. *J. Clin. Invest.* **92**, 911–918.

Duane, W.C. 1999. Effects of soybean protein and very low dietary cholesterol on serum lipids, biliary lipids, and fecal sterols in humans. *Metabolism* **48**, 489–494.

Eastwood, M. and Mowbray, L. 1976. The binding of the components of mixed micelle to dietary fiber. *Am. J. Clin. Nutr.* **29**, 1461–1467.

Eastwood, M.A. and Morris, E.R. 1992. Physical properties of dietary fiber that influence physiological function: A model for polymers along the gastrointestinal tract. *Am. J. Clin. Nutr.* **55**, 436–442.

Ebihara, K. and Schneeman, B.O. 1989. Interaction of bile acids, phospholipids, cholesterol and triglyceride with dietary fibers in the small intestine of rats. *J. Nutr.* **119**, 1100–1106.

Eckhardt, E. R. M., Wang, D. Q. H., Donovan, J.M., and Carey, M.C. 2002. Dietary sphingomyelin suppresses intestinal cholesterol absorption by decreasing thermodynamic activity of cholesterol monomers. *Gastroenterology* **122**, 948–956.

Ellegård, L., Andersson, H., and Bosaeus, I. 1997. Inulin and oligofructose do not influence the absorption of cholesterol, or the excretion of cholesterol, Ca, Mg, Zn, Fe, or bile acids but increases energy excretion in ileostomy subjects. *Eur. J. Clin. Nutr.* **51**, 1–5.

Ellegård, L., Bosaeus, I., and Andersson, H. 2000. Will recommended changes in fat and fibre intake affect cholesterol absorption and sterol excretion? An ileostomy study. *Eur. J. Clin. Nutr.* **54**, 306–313.

Evans, A.J., Hood, R.L., Oakenfull, D.G., and Sidhu, G.S. 1992. Relationship between structure and function of dietary fibre: A comparative study of the effects of three galactomannans on cholesterol metabolism in the rat. *Br. J. Nutr.* **68**, 217–229.

Favier, M.L., Bost, P.E., Demigne, C., and Remesy, C. 1998. The cholesterol-lowering effect of guar gum in rats is not accompanied by an interruption of bile acid cycling. *Lipids* **33**, 765–771.

Feldman, E.B., Russell, B.S., Schnare, F.H., Miles, B.C., Doyle, E.A., and Moretti-Rojas, I. 1979a. Effect of tristearin, triolein and safflower oil diets on cholesterol balance in rats. *J. Nutr.* **109**, 2226–2236.

Feldman, E.B., Russell, B.S., Schnare, F.H., Moretti-Rojas, I., Miles, B.C., and Doyle, E.A. 1979b. Effect of diets of homogenous saturated triglycerides on cholesterol balance in rats. *J. Nutr.* **109**, 2237–2246.

Fenwick, D.E. and Oakenfull, D.G. 1981. Saponin content of soya beans and some commercial soya bean products. *J. Sci. Food Agric.* **32**, 273–278.

Fernandez, M.L. 2001. Soluble fiber and nondigestible carbohydrate effects on plasma lipids and cardiovascular risk. *Curr. Opin. Lipidol.* **12**, 35–40.

Fernandez, M.L., Lin, E.C., Trejo, A., and McNamara, D.J. 1994. Prickly pear (Opuntia sp.) pectin alters hepatic cholesterol metabolism without affecting cholesterol absorption in guinea pigs fed a hypercholesterolemic diet. *J. Nutr.* **124**, 817–824.

Field, F.J., Born, E., and Mathur, S.N. 2004. Stanol esters decrease plasma cholesterol independently of intestinal ABC sterol transporters and Niemann-Pick C1–like 1 protein gene expression. *J. Lipid Res.* **45**, 2252–2259.

Flourie, B., Vidon, N., Florent, C.H., and Bernier, J.J. 1984. Effect of pectin on jejunal glucose absorption and unstirred layer thickness in normal man. *Gut* **25**, 936–941.

Forsythe, W.A., Green, M.S., and Anderson, J. J. B. 1986. Dietary protein effects on cholesterol and lipoprotein concentrations: A review. *J. Am. Coll. Nutr.* **5**, 533–549.

Francis, G., Kerem, Z., Makkar, H. P. S., and Becker, K. 2002. The biological action of saponins in animal systems: A review. *Br. J. Nutr.* **88**, 587–605.

Fukui, K., Tachibana, N., Fukuda, Y., Takamatsu, K., and Sugano, M. 2004. Ethanol washing does not attenuate the hypocholesterolemic potential of soy protein. *Nutrition* **20**, 984–990.

Fumagalli, R., Soleri, L., Farina, R., Musanti, R., Mantero, O., Noseda, G., Gatti, E., and Sirtori, C.R. 1982. Fecal cholesterol excretion studies in type II hypercholesterolemic patients treated with the soybean protein diet. *Atherosclerosis* **43**, 341–353.

Gallaher, C.M., Munion, J., Hesslink, R., Jr., Wise, J., and Gallaher, D.D. 2000. Cholesterol reduction by glucomannan and chitosan is mediated by changes in cholesterol absorption and bile acid and fat excretion in rats. *J. Nutr.* **130**, 2753–2759.

Gallaher, D.D. and Schneeman, B.O. 1986. Intestinal interaction of bile acids, phospholipids, dietary fibers, and cholestyramine. *Am. J. Physiol.* **250**, G420–G426.

Gallaher, D.D. and Schneeman, B.O. 2001. Dietary fiber. *In* "Present Knowledge in Nutrition," (B.A. Bowman and R.M. Russell, eds), 2nd Ed., pp. 83–91. ILSI Press, Washington, D.C.

Gallaher, D.D., Hassel, C.A., and Lee, K.J. 1993. Relationships between viscosity of hydroxypropyl methylcellulose and plasma cholesterol in hamsters. *J. Nutr.* **123**, 1732–1738.

Gallaher, D.D., Gallaher, C.M., Mahrt, G.J., Carr, T.P., Hollingshead, C.H., Hesslink, R., Jr., and Wise, J. 2002. A glucomannan and chitosan fiber supplement decreases plasma cholesterol and increases cholesterol excretion in overweight normocholesterolemic humans. *J. Am. Coll. Nutr.* **21**, 428–433.

Gestetner, B., Assa, Y., Henis, Y., Tencer, Y., Rotman, M., Birk, Y., and Bondi, A. 1972. Interaction of leucerne saponins with steroids. *Biochim. Biophys. Acta* **270**, 181–187.

Gibney, M.J. 1983. The effect of dietary lysine to arginine ratio on cholesterol kinetics in rabbits. *Atherosclerosis* **47**, 263–270.

Greaves, K.A., Wilson, M.D., Rudel, L.L., Williams, J.K., and Wagner, J.D. 2000. Consumption of soy protein reduces cholesterol absorption compared to casein protein alone or supplemented with an isoflavone extract or conjugated equine estrogen in ovariectomized cynomolgus monkeys. *J. Nutr.* **130**, 820–826.

Grundy, S.M. 1983. Absorption and metabolism of dietary cholesterol. *Annu. Rev. Nutr.* **3**, 71–96.

Grundy, S.M. and Denke, M.A. 1990. Dietary influences on serum lipids and lipoproteins. *J. Lipid Res.* **31**, 1149–1172.

Grundy, S.M. and Metzger, A.L. 1972. A physiological method for estimation of hepatic secretion of biliary lipids in man. *Gastroenterology* **62**, 1200–1217.

Gylling, H. and Miettinen, T.A. 1995. The effect of cholesterol absorption inhibition on low density lipoprotein cholesterol level. *Atherosclerosis* **117**, 305–308.

Hallikainen, M.A., Sarkkinen, E.S., Gylling, H., Erkkila, A.T., and Uusitupa, M.I. 2000. Comparison of the effects of plant sterol ester and plant stanol ester-enriched margarines in lowering serum cholesterol concentrations in hypercholesterolaemic subjects on a low-fat diet. *Eur. J. Clin. Nutr.* **54**, 715–725.

Han, L.K., Zheng, Y.N., Xu, B.J., Okuda, H., and Kimura, Y. 2002. Saponins from platycodi radix ameliorate high fat diet-induced obesity in mice. *J. Nutr.* **132**, 2241–2245.

Harris, W.S., Dujovne, C.A., Windsor, S.L., Gerrond, L.L., Newton, F.A., and Gelfand, R.A. 1997. Inhibiting cholesterol absorption with CP-88,818 β-tigogenin cellobioside; tiqueside: Studies in normal and hyperlipidemic subjects. *J. Cardiovasc. Pharmacol.* **30**, 55–60.

Harwood, H.J., Jr., Chandler, C.E., Pellarin, L.D., Bangerter, F.W., Wilkins, R.W., Long, C.A., Cosgrove, P.G., Malinow, M.R., Marzetta, C.A., Pettini, J.L., Savoy, Y.E., and Mayne, J.T. 1993. Pharmacologic consequences of cholesterol absorption inhibition: Alteration in cholesterol metabolism and reduction in plasma cholesterol concentration induced by the synthetic saponin β-tigogenin cellobioside (CP-88818; tiqueside). *J. Lipid Res.* **34**, 377–395.

Hassel, C.A., Mensing, E.A., and Gallaher, D.D. 1997. Dietary stearic acid reduces plasma and hepatic cholesterol concentrations without increasing bile acid excretion in cholesterol-fed hamsters. *J. Nutr.* **127**, 1148–1155.

Hegsted, D.M., McGandy, R.B., Myers, M.L., and Stare, F.J. 1965. Quantitative effects of dietary fat on serum cholesterol in man. *Am. J. Clin. Nutr.* **17**, 281–295.

Hegsted, D.M., Ausman, L.M., Johnson, J.A., and Dallal, G.E. 1993. Dietary fat and serum lipids: An evaluation of the experimental data. *Am. J. Clin. Nutr.* **57**, 875–883. Erratum in: *Am. J. Clin. Nutr.* **58,** 245.

Heinemann, T., Pietruck, B., Kullak-Ublick, G., and von Bergmann, K. 1988. Comparison of sitosterol and sitostanol on inhibition of intestinal cholesterol absorption. *Agents Actions Suppl.* **26**, 117–122.

Heinemann, T., Kullak-Ublick, G.-K., Pietruck, B., and von Bergmann, K. 1991. Mechanisms of action of plant sterols on inhibition of cholesterol absorption. Comparison of sitosterol and sitostanol. *Eur. J. Clin. Pharmacol.* **40**, S50–S63.

Heuman, D.M. 1989. Quantitative estimation of the hydrophilic-hydrophobic balance of mixed bile salt solutions. *J. Lipid Res.* **30**, 719–730.

Hicks, K.B. and Moreau, R.A. 2001. Phytosterols and phytostanols: Functional food cholesterol busters. *Food Technol.* **55**, 63–67.

Hofmann, A.F. and Small, D.M. 1967. Detergent properties of bile salts: Correlation with physiological function. *Annu. Rev. Med.* **18**, 333–376.

Homan, R. and Hamelehle, K.L. 1998. Phospholipase A2 relieves phosphatidylcholine inhibition of micellar cholesterol absorption and transport by human intestinal cell line Caco-2. *J. Lipid Res.* **39**, 1197–1209.

Howell, W.H., McNamara, D.J., Tosca, M.A., Smith, B.T., and Gaines, J.A. 1997. Plasma lipid and lipoprotein responses to dietary fat and cholesterol: A meta-analysis. *Am. J. Clin. Nutr.* **65**, 1747–1764.

Howles, P.N., Carter, C.P., and Hui, D.Y. 1996. Dietary free and esterified cholesterol absorption in cholesterol esterase (bile salt-stimulated lipase) gene-targeted mice. *J. Biol. Chem.* **271**, 7196–7202.

Huff, M.W. and Carroll, K.K. 1980. Effects of dietary protein on turnover, oxidation, and absorption of cholesterol, and on steroid excretion in rabbits. *J. Lipid Res.* **21**, 546–558.

Hui, D.Y. and Howles, P.N. 2005. Molecular mechanisms of cholesterol absorption and transport in the intestine. *Semin. Cell Dev. Biol.* **16**, 183–192.

Ikeda, I. and Sugano, M. 1983. Some aspects of mechanism of inhibition of cholesterol absorption by β-sitosterol. *Biochim. Biophys. Acta* **732**, 651–658.

Ikeda, I., Morioka, H., and Sugano, M. 1979. The effect of dietary β-sitosterol and β-sitostanol on the metabolism of cholesterol in rats. *Agric. Biol. Chem.* **43**, 1927–1933.

Ikeda, I., Tanaka, K., Sugano, M., Vahouny, G.V., and Gallo, L.L. 1988. Inhibition of cholesterol absorption in rats by plant sterols. *J. Lipid Res.* **29**, 1573–1582.

Ikeda, I., Tanabe, Y., and Sugano, M. 1989a. Effects of sitosterol and sitostanol on micellar solubility of cholesterol. *J. Nutr. Sci. Vitaminol.* **35**, 361–369.

Ikeda, I., Tomari, Y., and Sugano, M. 1989b. Interrelated effects of dietary fiber and fat on lymphatic cholesterol and triglyceride absorption in rats. *J. Nutr.* **119**, 1383–1387.

Ikeda, I., Imasato, Y., Nakayama, M., Imaizumi, K., and Sugano, M. 1994. Lymphatic transport of stearic acid and its effect on cholesterol transport in rats. *J. Nutr. Sci. Vitaminol. (Tokyo)* **40**, 275–282.

Imaizumi, K., Tominaga, A., Sato, M., and Sugano, M. 1992. Effects of dietary sphingolipids on levels of serum and liver lipids in rats. *Nutr. Res.* **12**, 543–548.

Imaizumi, K., Abe, K., Kuroiwa, C., and Sugano, M. 1993. Fat containing stearic acid increases fecal neutral steroid excretion and catabolism of low density lipoproteins without affecting plasma cholesterol concentration in hamsters fed a cholesterol-containing diet. *J. Nutr.* **123**, 1693–1702.

Institute of Medicine 2002. Dietary reference intakes for energy, carbohydrate, fiber, fat, fatty acids, cholesterol, protein, and amino acids. Food and Nutrition Board. National Academy Press, Washington, D.C.

Ireland, P.A., Dziedzic, S.Z., and Kearsley, M.W. 1986. Saponin content of soya and some commercial soya products by means of high-performance liquid chromatography of the sapogenins. *J. Sci. Food Agric.* **37**, 694–698.

Ishinaga, M., Ueda, A., Mochizuki, T., Sugiyama, S., and Kobayashi, T. 2005. Cholesterol intake is associated with lecithin intake in Japanese people. *J. Nutr.* **135**, 1451–1455.

Jandacek, R.J., Webb, M.R., and Mattson, F.H. 1977. Effect of an aqueous phase on the solubility of cholesterol in an oil phase. *J. Lipid Res.* **18**, 203–210.

Jenkins, K.J. and Atwal, A.S. 1994. Effects of dietary saponins on fecal bile acids and neutral sterols, and availability of vitamins A and E in the chick. *J. Nutr. Biochem.* **5**, 134–137.

Jesch, E.D. and Carr, T.P. 2005. Plant sterols inhibit cholesterol solubilization in micelles. *FASEB J.* **19**, A1011.

Jiang, Y., Noh, S.K., and Koo, S.I. 2001. Egg phosphatidylcholine decreases the lymphatic absorption of cholesterol in rats. *J. Nutr.* **131**, 2358–2363.

Jimenez, M.A., Scarino, M.L., Vignolini, F., and Mengheri, E. 1990. Evidence that polyunsaturated lecithin induces a reduction in plasma cholesterol level and favorable changes in lipoprotein composition in hypercholesterolemic rats. *J. Nutr.* **120**, 659–667.

Johnson, I.T. and Gee, J.M. 1981. Effect of gel-forming gums on the intestinal unstirred layer and sugar transport *in vitro*. *Gut* **22**, 398–403.

Johnson, I.T., Gee, J.M., Price, K.R., Curl, C., and Fenwick, G.R. 1986. Influence of saponins on gut permeability and active nutrient transport *in vitro*. *J. Nutr.* **116**, 2270–2277.

Jones, P. J. H., Raeini-Sarjaz, M., Ntanios, F.Y., Vanstone, C.A., Feng, J.Y., and Parsons, W.E. 2000. Modulation of plasma lipid levels and cholesterol kinetics by phytosterol versus phytostanol esters. *J. Lipid Res.* **41**, 697–705.

Kamei, M., Ohgaki, S., Kanbe, T., Niiya, I., Mizutani, H., Matsui-Yuasa, I., Otani, S., and Morita, S. 1995. Effects of highly hydrogenated soybean oil and cholesterol on plasma, liver cholesterol, and fecal steroids in rats. *Lipids* **30**, 533–539.

Kaur, N. and Gupta, A.K. 2002. Applications of inulin and oligofructose in health and nutrition. *J. Biosci.* **27**, 703–714.

Kelley, J.J. and Tsai, A.C. 1978. Effect of pectin, gum arabic and agar on cholesterol absorption, synthesis, and turnover in rats. *J. Nutr.* **108**, 630–639.

Kendall, C.W., Emam, A., Augustin, L.S., and Jenkins, D.J. 2004. Resistant starches and health. *J. AOAC Int.* **87**, 769–774.

Kern, F., Jr. 1991. Normal plasma cholesterol in an 88-year-old man who eats 25 eggs a day. Mechanisms of adaptation. *N. Engl. J. Med.* **324**, 896–899.

Kesaniemi, Y.A. and Grundy, S.M. 1986. Effects of dietary polyenylphosphatidylcholine on metabolism of cholesterol and triglycerides in hypertriglyceridemic patients. *Am. J. Clin. Nutr.* **43**, 98–107.

Kesaniemi, Y.A. and Miettinen, T.A. 1987. Cholesterol absorption efficiency regulates plasma cholesterol level in the Finnish population. *Eur. J. Clin. Invest.* **17**, 391–395.

Keys, A., Anderson, J.T., and Grande, F. 1965. Serum cholesterol response to changes in the diet. IV. Particular saturated fatty acids in the diet. *Metabolism* **14**, 776–787.

Kinkaid, A. and Wilton, D.C. 1991. Comparison of the catalytic properties of phospholipase A2 from pancreas and venom using a continuous fluorescence displacement assay. *Biochem. J.* **278**, 843–848.

Koo, J.H. 1983. The effect of ginseng saponin on the development of experimental atherosclerosis. *Hanyang Uidae Haksulchi* **3**, 273–286.

Koo, S.I. and Noh, S.K. 2001. Phosphatidylcholine inhibits and lysophosphatidylcholine enhances the lymphatic absorption of α-tocopherol in adult rats. *J. Nutr.* **131**, 717–722.

Krauss, R.M., Eckel, R.H., Howard, B., Appel, L.J., Daniels, S.R., Deckelbaum, R.J., Erdman, J.W., Jr., Kris-Etherton, P., Goldberg, I.J., Kotchen, T.A., Lichtenstein, A.H., Mitch, W.E., *et al.* 2000. AHA dietary guidelines. Revision 2000: A statement for healthcare professionals from the Nutrition Committee of the American Heart Association. *Circulation* **102**, 2296–2311.

Kris-Etherton, P.M. and Yu, S. 1997. Individual fatty acid effects on plasma lipids and lipoproteins: Human studies. *Am. J. Clin. Nutr.* **65**(Suppl. 5), 1628S–1644S.

Kritchevsky, D. and Tepper, S.A. 1961. The free and ester sterol content of various foodstuffs. *J. Nutr.* **74**, 441–444.

Kuyvenhoven, M.W., Roszkowski, W.F., West, C.E., Hoogenboom, R.L., Vos, R.M., Beynen, A.C., and van der Meer, R. 1989. Digestibility of casein, formaldehyde-treated casein and soya-bean protein in relation to their effects on serum cholesterol in rabbits. *Br. J. Nutr.* **62**, 331–342.

Lacaille-Dubois, M.A. and Wagner, H. 1996. A review of the biological and pharmacological activities of saponins. *Phytomedicine* **2**, 363–386.

Law, M. 2000. Plant sterol and stanol margarines and health. *Br. Med. J.* **320**, 861–864.

Lee, M.H., Lu, K., Hazard, S., Yu, H., Shulenin, S., Hidaka, H., Kojima, H., Allikmets, R., Sakuma, N., Pegoraro, R., Srivastava, A.K., Salen, G., *et al.* 2001a. Identification of a gene, ABCG5, important in the regulation of dietary cholesterol absorption. *Nat. Genet.* **27**, 79–83.

Lee, M.H., Lu, K., and Patel, S.B. 2001b. Genetic basis of sitosterolemia. *Curr. Opin. Lipidol.* **12**, 141–149.

Lees, A.M., Mok, H. Y. I., Lees, R.S., McCluskey, M.A., and Grundy, S.M. 1977. Plant sterols as cholesterol-lowering agents: Clinical trials in patients with hypercholesterolemia and studies of sterol balance. *Atherosclerosis* **28**, 325–338.

Levrat, M.A., Favier, M.L., Moundras, C., Remesy, C., Demigne, C., and Morand, C. 1994. Role of dietary propionic acid and bile acid excretion in the hypocholesterolemic effects of oligosaccharides in rats. *J. Nutr.* **124**, 531–538.

Liener, I.E. 1994. Implications of antinutritional components in soybean foods. *Crit. Rev. Food Sci. Nutr.* **34**, 31–67.

Lin, Y., Meijer, G.W., Vermeer, M.A., and Trautwein, E.A. 2004. Soy protein enhances the cholesterol-lowering effect of plant sterol esters in cholesterol-fed hamsters. *J. Nutr.* **134**, 143–148.

Lucas, E.A., Khalil, D.A., Daggy, B.P., and Arjmandi, B.H. 2001. Ethanol-extracted soy protein isolate does not modulate serum cholesterol in golden Syrian hamsters: A model of postmenopausal hypercholesterolemia. *J. Nutr.* **131**, 211–214.

Lund, E.D. 1984. Cholesterol binding capacity of fiber from tropical fruits and vegetables. *Lipids* **19**, 85–90.

Malinow, M.R. 1985. Effects of synthetic glycosides on cholesterol absorption. *Ann. NY Acad. Sci.* **454**, 23–27.

Malinow, M.R., McLaughlin, P., Papworth, L., Stafford, C., Kohler, G.O., Livingston, A.L., and Cheeke, P.R. 1977. Effect of alfalfa saponins on intestinal cholesterol absorption in rats. *Am. J. Clin. Nutr.* **30**, 2061–2067.

Malinow, M.R., McLaughlin, P., Naito, H.K., Lewis, L.A., and McNulty, W.P. 1978. Effect of alfalfa meal on shrinkage (regression) of atherosclerotic plaques during cholesterol feeding in monkeys. *Atherosclerosis* **30**, 27–43.

Malinow, M.R., Connor, W.E., McLaughlin, P., Stafford, C., Lin, D.S., Livingston, A.L., Kohler, G. O., and McNulty, W.P. 1981. Cholesterol and bile acid balance in Macaca fascicularis. Effects of alfalfa saponins. *J. Clin. Invest.* **67**, 156–162.

Mathé, D., Lutton, C., Rautureau, J., Coste, T., Gouffier, E., Sulpice, J.C., and Chevallier, F. 1977. Effects of dietary fiber and salt mixtures on the cholesterol metabolism of rats. *J. Nutr.* **107**, 466–474.

Matthan, N.R. and Lichtenstein, A.H. 2004. Approaches to measuring cholesterol absorption in humans. *Atherosclerosis* **174**, 197–205.

Mattson, F.H., Volpenhein, R.A., and Erickson, B.A. 1977. Effect of plant sterol esters on the absorption of dietary cholesterol. *J. Nutr.* **107**, 1139–1146.

Mattson, F.H., Grundy, S.M., and Crouse, J.R. 1982. Optimizing the effect of plant sterols on cholesterol absorption in man. *Am. J. Clin. Nutr.* **35**, 697–700.

McIntosh, T.J., Simon, S.A., Needham, D., and Huang, C.H. 1992. Structure and cohesive properties of sphingomyelin/cholesterol bilayers. *Biochemistry (Mosc.)* **31**, 2012–2020.

Mel'nikov, S.M., Seijen ten Hoorn, J.W., and Eijkelenboom, A.P. 2004a. Effect of phytosterols and phytostanols on the solubilization of cholesterol by dietary mixed micelles: An *in vitro* study. *Chem. Phys. Lipids* **127**, 121–141.

Mel'nikov, S.M., Seijen ten Hoorn, J.W., and Bertrand, B. 2004b. Can cholesterol absorption be reduced by phytosterols and phytostanols via a cocrystallization mechanism? *Chem. Phys. Lipids* **127**, 15–33.

Miettinen, T.A. and Tarpila, S. 1989. Serum lipids and cholesterol metabolism during guar gum, plantago ovata and high fibre treatments. *Clin. Chim. Acta* **183**, 253–262.

Miettinen, T., Vanhanen, H., and Wester, I. 1996. Use of a stanol fatty acid ester for reducing serum cholesterol level. U.S. patent no. 5,502,045.

Morehouse, L.A., Bangerter, F.W., DeNinno, M.P., Inskeep, P.B., McCarthy, P.A., Pettini, J.L., Savoy, Y.E., Sugarman, E.D., Wilkins, R.W., Wilson, T.C., Woody, H.A., Zaccaro, L.M., *et al.* 1999. Comparison of synthetic saponin cholesterol absorption inhibitors in rabbits: Evidence for a non-stoichiometric, intestinal mechanism of action. *J. Lipid Res.* **40**, 464–474.

Morton, G.M., Lee, S.M., Buss, D.H., and Lawrence, P. 1995. Intakes and major dietary sources of cholesterol and phytosterols in the British diet. *J. Hum. Nutr. Diet.* **8**, 429–440.

Nagaoka, S., Awano, T., Nagata, N., Masaoka, M., Hori, G., and Hashimoto, K. 1997. Serum cholesterol reduction and cholesterol absorption inhibition in CaCo-2 cells by a soyprotein peptic hydrolyzate. *Biosci. Biotechnol. Biochem.* **61**, 354–356.

Nagaoka, S., Miwa, K., Eto, M., Kuzuya, Y., Hori, G., and Yamamoto, K. 1999. Soy protein peptic hydrolysate with bound phospholipids decreases micellar solubility and cholesterol absorption in rats and Caco-2 cells. *J. Nutr.* **129**, 1725–1730.

Nagata, Y., Ishiwaki, N., and Sugano, M. 1982. Studies on the mechanism of antihypercholesterolemic action of soy protein and soy protein-type amino acid mixtures in relation to the casein counterparts in rats. *J. Nutr.* **112**, 1614–1625.

Nagata, C., Takatsuka, N., Kurisu, Y., and Shimizu, H. 1998. Decreased serum total cholesterol concentration is associated with high intake of soy products in Japanese men and women. *J. Nutr.* **128**, 209–213.

Nakamura, Y., Tsumura, Y., Tonogai, Y., and Shibata, T. 1999. Fecal steroid excretion is increased in rats by oral administration of gymnemic acids contained in *Gymnema sylvestre* leaves. *J. Nutr.* **129**, 1214–1222.

National Institutes of Health 2002. Third report of the National Cholesterol Education Program (NCEP) expert panel on detection, evaluation, and treatment of high blood cholesterol in adults (Adult Treatment Panel III): Final report. *Circulation* **106**, 3143–3421.

Nguyen, T.T. 1999. The cholesterol-lowering action of plant stanol esters. *J. Nutr.* **129**, 2109–2112.

Ni, W., Yoshida, S., Tsuda, Y., Nagao, K., Sato, M., and Imaizumi, K. 1999. Ethanol-extracted soy protein isolate results in elevation of serum cholesterol in exogenously hypercholesterolemic rats. *Lipids* **34**, 713–716.

Nissinen, M., Gylling, H., Vuoristo, M., and Miettinen, T.A. 2002. Micellar distribution of cholesterol and phytosterols after duodenal plant stanol ester infusion. *Am. J. Physiol. Gastrointest. Liver Physiol.* **282**, G1009–G1015.

Noh, S.K. and Koo, S.I. 2003. Egg sphingomyelin lowers the lymphatic absorption of cholesterol and α-tocopherol in rats. *J. Nutr.* **133**, 3571–3576.

Noh, S.K. and Koo, S.I. 2004. Milk sphingomyelin is more effective than egg sphingomyelin in inhibiting intestinal absorption of cholesterol and fat in rats. *J. Nutr.* **134**, 2611–2616.

Normén, L., Johnsson, M., Andersson, H., van Gameren, Y., and Dutta, P. 1999. Plant sterols in vegetables and fruits commonly consumed in Sweden. *Eur. J. Nutr.* **38**, 84–89.

Normén, L., Dutta, P., Lia, A., and Andersson, H. 2000. Soy sterol esters and beta-sitostanol ester as inhibitors of cholesterol absorption in human small bowel. *Am. J. Clin. Nutr.* **71**, 908–913.

Normén, L., Brants, H. A. M., Voorrips, L.E., Andersson, H.A., van den Brandt, P.A., and Goldbohm, R.A. 2001. Plant sterol intakes and colorectal cancer risk in the Netherlands cohort study on diet and cancer. *Am. J. Clin. Nutr.* **74**, 141–148.

Normén, L., Bryngelsson, S., Johnsson, M., Evheden, P., Ellegård, L., Brants, H., Andersson, H., and Dutta, P. 2002. The phytosterol content of some cereal foods commonly consumed in Sweden and in the Netherlands. *J. Food Compos. Anal.* **15**, 693–704.

Ntanios, F.Y. and Jones, P. J. H. 1999. Dietary sitostanol reciprocally influences cholesterol absorption and biosynthesis in hamsters and rabbits. *Atherosclerosis* **143**, 341–351.

Nyberg, L., Duan, R.D., and Nilsson, A. 2000. A mutual inhibitory effect on absorption of sphingomyelin and cholesterol. *J. Nutr. Biochem.* **11**, 244–249.

O'Brien, B.C. and Corrigan, S.M. 1988. Influence of dietary soybean and egg lecithins on lipid responses in cholesterol-fed guinea pigs. *Lipids* **23**, 647–650.

O'Neill, F.H., Sanders, T. A. B., and Thompson, G.R. 2005. Comparison of efficacy of plant stanol ester and sterol ester: Short-term and longer-term studies. *Am. J. Cardiol.* **96**(Suppl.), 29D–36D.

Oakenfull, D.G. 1986. Aggregation of saponins and bile acids in aqueous solution. *Aust. J. Chem.* **39**, 1671–1683.

Oakenfull, D.G. and Sidhu, G.S. 1983. A physico-chemical explanation for the effects of dietary saponins on cholesterol and bile salt metabolism. *Nutr. Rep. Int.* **27**, 1253–1259.

Oakenfull, D.G. and Sidhu, G.S. 1989. Saponins. *In* "Toxicants of Plant Origin. Vol. II. Glycosides" (P. Cheeke, ed.), pp. 97–141. CRC Press, Baco Raton, Florida.

Oakenfull, D.G. and Sidhu, G.S. 1990. Could saponins be a useful treatment for hypercholesterolemia? *Eur. J. Clin. Nutr.* **44**, 79–88.

Okuda, H. and Han, L.K. 2001. Medicinal plant and its related metabolic modulators. *Nippon Yakurigaku Zasshi* **118**, 347–351.

Onning, G., Wang, Q., Westrom, B.R., Asp, N.G., and Karlsson, B.W. 1996. Influnce of oat saponins on intestinal permeability *in vitro* and *in vivo* in the rat. *Br. J. Nutr.* **76**, 141–151.

Packard, C.J. and Shepherd, J. 1982. The hepatobiliary axis and lipoprotein metabolism: Effects of bile acid sequestrants and ileal bypass surgery. *J. Lipid Res.* **23**, 1081–1098.

Patton, J.S. and Carey, M.C. 1981. Inhibition of human pancreatic lipase-colipase activity by mixed bile salt-phospholipid micelles. *Am. J. Physiol.* **241**, G328–G336.

Peterson, D.W. 1951. Effect of soybean sterols in the diet on plasma and liver cholesterol in chicks. *Proc. Soc. Exp. Biol. Med.* **78**, 143–147.

Phillips, K.M., Tarragó-Trani, M.T., and Stewart, K.K. 1999. Phytosterol content of experimental diets differing in fatty acid composition. *Food Chem.* **64**, 415–422.

Pollak, O.J. 1953. Reduction of blood cholesterol in man. *Circulation* **7**, 702–706.

Potter, S.M. 1995. Overview of proposed mechanisms for the hypocholesterolemic effect of soy. *J. Nutr.* **125**, 606S–611S.

Potter, S.M. 2000. Soy: New health benefits associated with an ancient food. *Nutr. Today* **35**, 53–60.

Price, K.R., Johnson, I.T., and Fenwick, G.R. 1987. The chemistry and biological significance of saponins in foods and feeding stuffs. *Crit. Rev. Food Sci. Nutr.* **26**, 27–135.

Pszczola, D.E. 2003. Plot thickens, as gums add special effects. *Food Technol.* **57**, 34–47.

Rudel, L.L., Deckelman, C., Wilson, M.D., Scobey, M., and Anderson, R. 1994. Dietary cholesterol and downregulation of cholesterol 7α-hydroxylase and cholesterol absorption in African green monkeys. *J. Clin. Invest.* **93**, 2463–2472.

Salen, G., von Bergmann, K., Lutjohann, D., Kwiterovich, P., Kane, J., Patel, S.B., Musliner, T., Stein, P., and Musser, B. 2004. Ezetimibe effectively reduces plasma plant sterols in patients with sitosterolemia. *Circulation* **109**, 966–971.

Sanders, T. A. B. 2003. Dietary fat and postprandial lipids. *Curr. Atheroscler. Rep.* **5**, 445–451.

Sautier, C., Doucet, C., Flament, C., and Lemonnier, D. 1979. Effects of soy protein and saponins on serum, tissue and feces steroids in rat. *Atherosclerosis* **34**, 233–241.

Schmidt, K. and Gallaher, D. 1997. Reduced cholesterol absorption and intestinal solubilization by stearic acid-rich fats in rats. *FASEB J.* **11**, A378.

Schneeman, B.O. 1999. Fiber, inulin and oligofructose: Similarities and differences. *J. Nutr.* **129**, 1424S–1427S.

Schneider, C.L., Cowles, R.L., Stuefer-Powell, C.L., and Carr, T.P. 2000. Dietary stearic acid reduces cholesterol absorption and increases endogenous cholesterol excretion in hamsters fed cereal-based diets. *J. Nutr.* **130**, 1232–1238.

Schothorst, R.C. and Jekel, A.A. 1999. Oral sterol intake in the Netherlands: Evaluation of the results obtained by GC analyses of duplicate 24–h diet samples collected in (1994). *Food Chem.* **64**, 561–566.

Shi, J., Arunasalam, K., Yeung, D., Kakuda, Y., Mittal, G., and Jiang, Y. 2004. Saponins from edible legumes: Chemistry, processing, and health benefits. *J. Med. Food* **7**, 67–78.

Sidhu, G.S., Upson, B., and Malinow, M.R. 1987. Effect of soy saponins and tigogenin cellobioside on intestinal uptake of cholesterol, cholate, and glucose. *Nutr. Rep. Int.* **35**, 615–623.

Stark, A. and Madar, Z. 1993. The effect of an ethanol extract derived from fenugreek (*Trigonella foenum-graecum*) on bile acid absorption and cholesterol levels in rats. *Br. J. Nutr.* **69**, 277–287.

Story, J.A. and Kritchevsky, D. 1976. Comparison of the binding of various bile acids and bile salts *in vitro* by several types of fiber. *J. Nutr.* **106**, 1292–1294.

Sugano, M., Morioka, H., and Ikeda, I. 1977. A comparison of hypocholesterolemic activity of β-sitosterol and β-sitostanol in rats. *J. Nutr.* **107**, 2011–2019.

Sugano, M., Ishiwaki, N., and Nakashima, K. 1984. Dietary protein-dependent modification of serum cholesterol level in rats. Significance of the arginine/lysine ratio. *Ann. Nutr. Metab.* **28**, 192–199.

Sugano, M., Goto, S., Yamada, Y., Yoshida, K., Hashimoto, Y., Matsuo, T., and Kimoto, M. 1990. Cholesterol-lowering activity of various undigested fractions of soybean protein in rats. *J. Nutr.* **120**, 977–985.

Tanaka, K., Aso, B., and Sugano, M. 1984. Biliary steroid excretion in rats fed soybean protein and casein or their amino acid mixtures. *J. Nutr.* **114**, 26–32.

Tattrie, N.H. 1972. Isolation and identification of egg yolk cholesteryl esters. *Can. J. Biochem.* **50**, 1414–1416.

Thakur, B.R., Singh, R.K., and Handa, A.K. 1997. Chemistry and uses of pectin: A review. *Crit. Rev. Food Sci. Nutr.* **37**, 47–73.

Trautwein, E.A., Rieckhoff, D., and Erbersdobler, H.F. 1998. Dietary inulin lowers plasma cholesterol and triacylglycerol and alters biliary bile acid profile in hamsters. *J. Nutr.* **128**, 1937–1943.

Trautwein, E.A., Duchateau, G. S. M. J.E., Lin, Y., Molhuizen, S.M., Mel'nikov, H. O. F., and Ntanios, F.Y. 2003. Proposed mechanisms of cholesterol lowering action of plant sterols. *Eur. J. Lipid Sci. Technol.* **105**, 171–185.

Turley, S.D., Daggy, B.P., and Dietschy, J.M. 1991. Cholesterol-lowering action of psyllium mucilloid in the hamster: Sites and possible mechanisms of action. *Metabolism* **40**, 1063–1073.

Turley, S.D., Daggy, B.P., and Dietschy, J.M. 1994. Psyllium augments the cholesterol-lowering action of cholestyramine in hamsters by enhancing sterol loss from the liver. *Gastroenterology* **107**, 444–452.

U.S. Department of Agriculture, Agricultural, Research Service 2005. USDA Nutrient Database for Standard Reference, Release 17. Nutrient Data Laboratory Home Pagehttp://www.nal.usda.gov/fnic/foodcomp.

Vahouny, G.V., Tombes, R., Cassidy, M.M., Kritchevsky, D., and Gallo, L.L. 1980. Dietary fibers. V. Binding of bile salts, phospholipids and cholesterol from mixed micelles by bile acid sequestrants and dietary fibers. *Lipids* **15**, 1012–1018.

Vahouny, G.V., Tombes, R., Cassidy, M.M., Kritchevsky, D., and Gallo, L.L. 1981. Dietary fibres. VI. Binding of fatty acids and monolein from mixed micelles containing bile salts and lecithin. *Proc. Soc. Exp. Biol. Med.* **166**, 12–16.

Vahouny, G.V., Satchithanandam, S., Cassidy, M.M., Lightfoot, F.B., and Furda, I. 1983. Comparative effects of chitosan and cholestyramine on lymphatic absorption of lipids in the rat. *Am. J. Clin. Nutr.* **38**, 278–284.

Vahouny, G.V., Chalcarz, W., Satchithanandam, S., Adamson, I., Klurfeld, D.M., and Kritchevsky, D. 1984. Effect of soy protein and casein intake on intestinal absorption and lymphatic transport of cholesterol and oleic acid. *Am. J. Clin. Nutr.* **40**, 1156–1164.

Vahouny, G.V., Satchithanandam, S., Chen, I., Tepper, S.A., Kritchevsky, D., Lightfoot, F.G., and Cassidy, M.M. 1988. Dietary fiber and intestinal adaptation: Effects on lipid absorption and lymphatic transport in the rat. *Am. J. Clin. Nutr.* **47**, 201–206.

Valsta, L.M., Lemström, A., Ovaskainen, M.L., Lampi, A.M., Toivo, J., Korhonen, T., and Piironen, V. 2004. Estimation of plant sterol and cholesterol intake in Finland: Quality of new values and their effect on intake. *Br. J. Nutr.* **92**, 671–678.

Vesper, H., Schmelz, E.M., Nikolova-Karakashian, M.N., Dillehay, D.L., Lynch, D.V., and Merrill, A.H., Jr. 1999. Sphingolipids in food and the emerging importance of sphingolipids to nutrition. *J. Nutr.* **129**, 1239–1250.

Vuoristo, M. and Miettinen, T.A. 1994. Absorption, metabolism, and serum concentrations of cholesterol in vegetarians: Effects of cholesterol feeding. *Am. J. Clin. Nutr.* **59**, 1325–1331.

Wang, S. and Koo, S.I. 1993a. Evidence for distinct metabolic utilization of stearic acid in comparison with palmitic and oleic acids in rats. *J. Nutr. Biochem.* **4**, 594–601.

Wang, S. and Koo, S.I. 1993b. Plasma clearance and hepatic utilization of stearic, myristic and linoleic acids introduced via chylomicrons in rats. *Lipids* **28**, 697–703.

Weihrauch, J.L. and Gardner, J.M. 1978. Sterol content of foods of plant origin. *J. Am. Diet. Assoc.* **73**, 39–47.

Westergaard, H. and Dietschy, J.M. 1974. Delineation of the dimensions and permeability characteristics of the two major diffusion barriers to passive mucosal uptake in the rabbit intestine. *J. Clin. Invest.* **54**, 718–732.

Weststrate, J.A. and Meijer, G.W. 1998. Plant sterol-enriched margarines and reduction of plasma total- and LDL-cholesterol concentrations in normocholesterolaemic and mildly hypercholesterolaemic subjects. *Eur. J. Clin. Nutr.* **52**, 334–343.

Wilson, M.D. and Rudel, L.L. 1994. Review of cholesterol absorption with emphasis on dietary and biliary cholesterol. *J. Lipid Res.* **35**, 943–955.

Wilson, T.A., Meservey, C.M., and Nicolosi, R.J. 1998. Soy lecithin reduces plasma lipoprotein cholesterol and early atherogenesis in hypercholesterolemic monkeys and hamsters: Beyond linoleate. *Atherosclerosis* **140**, 147–153.

Wright, S.M. and Salter, A.M. 1998. Effects of soy protein on plasma cholesterol and bile acid excretion in hamsters. *Comp. Biochem. Physiol. B Biochem. Mol. Biol.* **119**, 247–254.

Yao, L., Heubi, J.E., Buckley, D.D., Fierra, H., Setchell, K.D., Granholm, N.A., Tso, P., Hui, D.Y., and Woollett, L.A. 2002. Separation of micelles and vesicles within lumenal aspirates from healthy humans: Solubilization of cholesterol after a meal. *J. Lipid Res.* **43**, 654–660.

Ylitalo, R., Lehtinen, S., Wuolijoki, E., Ylitalo, P., and Lehtimaki, T. 2002. Cholesterol-lowering properties and safety of chitosan. *Arzneimittelforschung* **52**, 1–7.

Yeagle, P.L., Hutton, W.C., Huang, C.H., and Martin, R.B. 1976. Structure in the polar head region of phospholipid bilayers: A 31P [1H] nuclear Overhauser effect study. *Biochemistry (Mosc.)* **15**, 2121–2124.

Young, S.C. and Hui, D.Y. 1999. Pancreatic lipase/colipase-mediated triacylglycerol hydrolysis is required for cholesterol transport from lipid emulsions to intestinal cells. *Biochem. J.* **339**, 615–620.

Zhou, B.F., Stamler, J., Dennis, B., Moag-Stahlberg, A., Okuda, N., Robertson, C., Zhao, L., Chan, Q., and Elliott, P. 2003. Nutrient intakes of middle-aged men and women in China, Japan, United Kingdom, and United States in the late 1990s: The INTERMAP study. *J. Hum. Hypertens.* **17**, 623–630.

IMAGING TECHNIQUES FOR THE STUDY OF FOOD MICROSTRUCTURE: A REVIEW

PASQUALE M. FALCONE,* ANTONIETTA BAIANO,* AMALIA CONTE,* LUCIA MANCINI,† GIULIANA TROMBA,† FRANCO ZANINI,† AND MATTEO A. DEL NOBILE*

**Department of Food Science, University of Foggia*
Foggia, FG 71100, Italy
†Sincrotrone Trieste S.C.p.A. in Area Science Park I
Basovizza, TS 34012, Italy

ADVANCES IN FOOD AND NUTRITION RESEARCH VOL 51

ISSN: 1043-4526
DOI: 10.1016/S1043-4526(06)51004-6

ABBREVIATIONS

S_v	Number of solid voxels. It corresponds to the solid phase (crumb) within the 3D testing sphere used in the stereological calculus.
V_v	Number of void voxels. It corresponds to the void phase (air) within the 3D testing sphere.
T_v	Total number of voxels. It is calculated by summing S_v and V_v.
δ	Crumb density (solid volume fraction). It was calculated as a ratio between S_v and T_v.
P	Crumb porosity (void volume fraction).
W_N	Cell wall number.
W_Th	Cell wall thickness.
W_Sp	Cell wall spacing.
SS	Specific surface. It is the solid phase surface per unit of crumb volume.
$\mathrm{PrinMIL}_{1,2,3}$	Are the magnitude of the principal mean intercept length vectors (MILv) along the three main eigenvectors ($\vec{e}_1, \vec{e}_2, \vec{e}_3$). This parameter is related to the amount of solid phase along the main orientations of the cell structure cell walls.
Is_Ix	Is the isotropic index according to Falcone *et al.* (2004b), Benn (1994) and Ryan and Ketcham (2002). This parameter indicates the degree of difference of the MILv ellipsoid from the spherical shape and it is related to the degree of randomization of the solid phase around the primary orientation of the cell walls ($\vec{e}_1$).
El_Ix	Is the elongation index according to Falcone *et al.* (2004b), Benn (1994) and Ryan and Ketcham (2002). This parameter indicates the degree of elongation of the MILv ellipsoid and it is related to the solid phase spread around the tertiary orientations of the cell walls, ($\vec{e}_3$).
DA	Is the degree of anisotropy according to Falcone *et al.*, (2004b), Lim and Barigou (2004), Benn (1994), and Ryan and Ketcham (2002). This parameter corresponds to the inverse of the isotropic index and indicates the flatting degree of the MIL ellipsoid along the secondary favorite orientation of cell walls ($\vec{e}_2$).
$\mathrm{DA}_{1,2,3}$	Are the relative degrees of anisotropy (Falcone *et al.*, 2004b). This parameter indicates the mean spread degree of the solid phase along the primary (1), secondary (2),

and tertiary (3) favorite orientation of the cell walls ($\vec{e}_1, \vec{e}_2, \vec{e}_3$).

FA Is the fractional anisotropy according to Matusani (2003). It is a complex index of anisotropy and takes into account both the isotropic index and the elongation index of the structure.

I. INTRODUCTION

A. IMPORTANCE OF MICROSTRUCTURE STUDIES

Quality of a food product is related to its sensorial (shape, size, color) and mechanical (texture) characteristics. These features are strongly affected by the food structural organization (Stanley, 1987) that, according to Fardet *et al.* (1998), can be studied at molecular, microscopic, and macroscopic levels. In particular, microstructure and interactions of components, such as protein, starch, and fat, determine the texture of a food that could be defined as the "external manifestation of this structure" (Allan-Wojtas *et al.*, 2001).

Concerning bread, characteristics such as cell wall thickness, cell size, and uniformity of cell size affect the texture of bread crumb (Kamman, 1970) and also the appearance, taste perception, and stability of the final product (Autio and Laurikainen, 1997). Crumb elasticity can be predicted from its specific volume and is strongly affected by the amylose-rich regions joining partially gelatinized starch granules in the crumb cell walls (Scanlon and Liu, 2003).

The structural organization of the components of a cheese, especially the protein network, affect the cheese texture; in particular the stress at fracture, the modulus, and work at fracture could be predicted very well from the size of the protein aggregates (Wium *et al.*, 2003). Cheeses having a regular and close protein matrix with small and uniform (in size and shape) fat globules show a more elastic behavior than cheeses with open structure and numerous and irregular cavities (Buffa *et al.*, 2001).

The mechanical properties of cocoa butter are strongly dependent on its morphology at microscopic level and, in particular, on the polymorphic transformation of the fat crystals and the coexistence of different polymorphic forms (Brunello *et al.*, 2003). Thorvaldsson *et al.* (1999) studied the influence of heating rate on rheology and structure of heat-treated pasta dough. They found that the fast-heated samples had pores smaller than the slow-heated ones and that the pore dimensions affect the energy required to cause a fracture. In particular, the energy required to determine a fracture in the samples having the smallest pores was more than that for the samples having the highest pores.

A study carried out on the effects of grind size on peanut butter texture demonstrated that an increase of that variable decreased sensory smoothness, spreadability, and adhesiveness (Crippen *et al.*, 1989).

Langton *et al.* (1996) studied correlations existing between microstructure and texture of a particulate proteic gel (spherical particles joined together to form strands). They found that the texture, as measured with destructive methods, was sensitive to pore size and particle size, whereas it was sensitive to the strand characteristics if measured with nondestructive methods.

Martens and Thybo (2000) investigated the relationships among microstructure and quality attributes of potatoes. They found that volume fractions of raw starch, gelatinized starch, and dry matter were positively correlated to reflection, graininess, mealiness, adhesiveness, and chewiness and negatively correlated to moistness.

From the evidence that microstructure affects food sensorial properties, an important consideration derives: foods having a similar microstructure also have a similar behavior (Kaláb *et al.*, 1995). Since microstructure is determined both by nature and by processing, food processing can be considered as the way for obtaining the desired microstructure (and consequently the desired properties) from the available food components (Aguilera, 2000). As a consequence, knowledge of microstructure must precede the regulation of texture (Ding and Gunasekaran, 1998) and other food attributes.

B. SUITABLE TECHNIQUES FOR MICROSTRUCTURE STUDIES

Studies on food structure at a microscopic level can be performed by using a large variety of microscopic techniques including light microscopy (LM), scanning electron microscopy (SEM), transmission electron microscopy (TEM), and confocal laser scanning microscopy (CLSM). Although magnification of LM is modest if compared to electron microscopy techniques, LM allows the specific staining of the different chemical components of a food (proteins, fat droplets, and so on). For this reason, it represents a technique suitable for the investigation of multicomponent or multiphase foods such as cereal-based foods (Autio and Salmenkallio-Marttila, 2001). LM, SEM, and TEM can be used to put in evidence different aspects of particulate structures. For example, in a study on microporous, particulate gels (Langton *et al.*, 1996), LM was used to visualize pores, TEM was applied to evaluate particle size, and SEM allowed to detect how the particles were linked together, that is, the three-dimensional (3D) structure. SEM and TEM allow a higher resolution if compared to LM but, as the latter, require sample preparation procedures (freezing, dehydration, and so on) that may lead to artifacts (Kaláb, 1984). CLSM represents a suitable alternative method in food microstructure evaluation because it require a minimum sample preparation. In fact, the CLSM technique includes the optical slicing of the sample (Autio and Laurikainen, 1997; Rao *et al.*, 1992). CLSM can be used for examining the 3D structure of the protein network of pasta samples (Fardet *et al.*, 1998), doughs (Thorvaldsson *et al.*, 1999), or high-fat foods (Wendin *et al.*, 2000) that cannot be prepared for conventional microscopy without the loss of fat (Autio and Laurikainen, 1997).

The microstructure of bread and other microporous foods can be conveniently studied by applying synchrotron radiation X-ray microtomography (X-MT) (Falcone *et al.*, 2004a; Maire *et al.*, 2003) to centimeter- or millimeter-sized samples (Lim and Barigou, 2004). X-MT application only requires the presence of areas of morphological or mass density heterogeneity in the sample materials. The use of this technique for food microstructure detection is of recent date. It was traditionally used for the analysis of bone quality (Peyrin *et al.*, 1998, 2000; Ritman *et al.*, 2002).

Among the imaging techniques, atomic force microscopy (AFM) and magnetic resonance imaging (MRI) have been introduced into food science as nondestructive techniques. The first one is particularly suitable for studying surface roughness, especially in fresh foods (Kaláb *et al.*, 1995). The second one can be successfully applied for studying processing such as frying, foam drainage, fat crystallization, and other operations in which a dynamic study of food structure is needed (Kaláb *et al.*, 1995). Examples of MRI application are the researches of Takano *et al.* (2002) and Grenier *et al.* (2003) that qualitatively and quantitatively studied the local porosity in dough during proving, a stage in which invasive analytical methods may cause the dough collapse. MRI is based on the nuclear magnetic resonance (NMR) technique that has recently been introduced in investigations of dough and bread. It allows the study of the changes in distribution and mobility of different types of protons (fat, structural and bulk water) in a sample and, thanks to the high correlations between texture parameters and NMR relaxation data, NMR is suitable to predict firmness and elasticity of these foods (Engelsen *et al.*, 2001).

Ultrasound represents another promising technique able to investigate some structural properties of foods. Since the agreement between rheological (storage and loss moduli) and ultrasound measurements (velocity of ultrasound propagation and attenuation), ultrasound could be used as on-line quality control technique for dough (Ross *et al.*, 2004). Ultrasound can also be used to distinguish crystalline fats from liquid fats (McClements and Povey, 1988) or to determine a food composition (Chanamai and McClements, 1999). In fact, the speed of sound decreases with temperature in fat whereas increases with temperature in aqueous solutions (Coupland, 2004).

Information about the crystallinity level and crystal size in a food can be obtained by submitting the food matrix to X-ray diffraction. For example, this technique supplies useful information about crystalline and gelatinized starch (Severini *et al.*, 2004).

Fourier transform infrared microspectroscopy (FTIR) and Raman microspectroscopy provide quantitative information about the chemical microstructure of heterogeneous solid foods (Cremer and Kaletunç, 2003; Piot *et al.*, 2000; Thygesen *et al.*, 2003) without sample destruction.

Another interesting technique able to investigate the microstructure of dense microemulsions is represented by the small-angle neutron scattering (SANS) (de Campo *et al.*, 2004).

Since many food characteristics are strongly dependent on microstructure, it is possible to obtain microstructural information by studying mechanical and viscoelastic properties of foods. A food sample submitted to mechanical tests gives rise to a force–time curve from which several parameters related to microstructure can be extrapolated: hardness, cohesiveness, springiness, chewiness, gumminess, stickiness (Martinez *et al.*, 2004). When submitted to a stress (under compression, tension, or shear conditions), food samples suffer a strain. The elastic modulus or Young's modulus of the analyzed sample can be obtained from the stress–strain curve (Del Nobile *et al.*, 2003; Liu *et al.*, 2003). The viscoelastic properties of a food can be expressed in terms of G', G'', and tan δ parameters. G' takes into account the elastic (solid-like) behavior of a material, G'' is a measure of the viscous (fluid-like) behavior of a material, and tan δ represents the ratio between G'' and G'. These parameters can be evaluated by performing dynamic mechanical and rheological tests (Brunello *et al.*, 2003; Kokelaar *et al.*, 1996; Ross *et al.*, 2004; Wildmoser *et al.*, 2004).

C. WHY IMAGING TECHNIQUES FOR MICROSTRUCTURE STUDIES?

Since sensory and mechanical properties of a food depend on its microstructure, the knowledge of microstructure must precede any operation aimed to the attainment of a specific texture (Ding and Gunasekaran, 1998). The instrumental measurements of mechanical and rheological properties represent the food responses to the forces acting on the food structure and, for this reason, are affected by the way in which these analyses are performed. Furthermore, mechanical and rheological tests are always destructive and make impossible the execution of other analyses.

As reported earlier, several types of microscopy techniques can be used for the observation of food microstructure. They allow the generation of data in the form of images (Kaláb *et al.*, 1995). Because of the artifacts due to the preparation of samples before microscopy analysis, it is advisable to apply a variety of techniques to the same samples in order to compare the results obtained. Another limit of microscopic techniques is represented by the so-called "optical illusions," that is, the tendency of human eye to see what one is looking for (Aguilera and Stanley, 1999). This means that human eye is not suitable for objective evaluation.

The application of the traditional microscopic techniques is also affected by the difficulty of quantifying the structural features. Computer image analysis allows to process images in order to extract numerical data referred to the microstructure (Ding and Gunasekaran, 1998; Inoue, 1986; Kaláb *et al.*, 1995).

In the last years, the use of image analysis techniques has increased. Novaro *et al.* (2001) applied the image analysis to whole durum wheat grains in order to predict semolina yield. Quevedo *et al.* (2002) used the fractal image texture analysis to quantitatively analyze the surface of several foods (potato chips, chocolate, and so on) and to evaluate the potato starch gelatinization during frying and the chocolate blooming. Basset *et al.* (2000) and Li *et al.* (2001) applied the image texture analysis for the classification of meat.

Generally, microstructural features (<100 μm) obtained from digital images without any intrusion opened new opportunities in the field of the food evaluation. Several possibilities and instrumental facilities are available in order to acquire internal details at a microscopic level by means of optical, electronic, and more sophisticated systems. Furthermore, engineers try to develop physical models and numerical algorithms to determine accurate and quantitative information from digital images. At a first level of analysis there are several commercial softwares able to perform basic tasks such as image editing, segmentation, object selection, and measurement of global geometrical features such as volume fraction and specific surface area. A second level of analysis is performed by algorithms for the shape recognition and statistical classification of objects into specific classes.

With the rapid growth of computing and imaging tools, such as the X-ray–computed tomographic scanners, both 2D and 3D digital images of the internal structure can be readily acquired with high resolution and contrast and without any sample preparation. These images can be processed by means of the fractal and stereological analysis in order to quantify a number of *structural elements*. Fractal analysis allows to investigate the fractal geometry in both 2D and 3D digital images. Stereology, instead, allows to obtain 3D features from 2D images. These data-processing techniques represent new promising approaches to a full characterization of complex internal structures. Such advances in food evaluation open new horizons for the development of mathematical and computational models able to individuate the interactions between product microstructure and their mechanical properties.

D. STEPS OF THE IMAGE-PROCESSING ANALYSIS

Independently on the way of image acquisition, the image-processing analysis consists of the following five steps: image acquisition, preprocessing, segmentation, object measurement, and classification.

1. Image acquisition

Images are the spatial representations of objects (Gunasekaran, 1996). They are stored as matrixes of x columns by y rows containing thousands of cells called "pixels" (picture elements). Each pixel contains a numerical value (digital number) that represents the sensor reading (www.geog.ubc.ca/courses/geog470/notes/image_processing/). The image acquisition just consists in the capture of an image in this digital form. The prerequisite for the image quality is represented by lighting conditions during the image acquisition. Although there are a variety of image acquisition techniques (described in section II), a few types of devices are generally used for capturing images. They are represented by video cameras, scanners, magnetic resonance imagers, ultrasound scanners, and tomographs (Aguilera and Stanley, 1999).

2. Preprocessing

The aim of this operation is the improvement of the quality of an image, in order to remove distortions or enhances some image characteristics. Preprocessing includes pixel preprocessing and local preprocessing (Du and Sun, 2004).

a. Pixel preprocessing. Pixel preprocessing consists of a color space transformation. Color images are, in fact, normally acquired as 24 bit RGB (red, green, blue) images. But most programs are able to operate on gray scale (8 bit, monochrome), so the first step after acquisition consists in the transformation of the RGB digital color image in a gray scale image or in three monochrome images (monochrome red, monochrome green, and monochrome blue). Since 256 gray levels are available, each of the image pixels can have an integer value ranging from 0 (black) to 255 (white). In food image analysis, to better distinguish among the different parts of an object, it is preferable to transform the RGB color space in the HSI (hue, saturation, and intensity) channels (Li *et al.*, 2001; Sun and Brosnam, 2003; Tao *et al.*, 1995). Another color space is represented by the L*a*b* space that separates lightness (gray scale intensity) from two orthogonal color axes, a* that takes into account the content of red or green and b* that considers the content of yellow and blue. HSL and HSV are two similar sets of coordinates that separate the gray scale brightness (L for luminance or V for value) from the hue (that is the distinction between red, orange, yellow, green, and so on) and from the saturation (the amount of color, e.g., the difference between pink and red).

b. Local preprocessing. Sometimes it is necessary to improve the image because of a nonuniform illumination or brightness across the image or the insufficient signal. These problems are generally solved using the so-called "point processes" that replace pixel values based on the individual values with new values based on the averaging brightness values of points having similar properties to the processed points. Different types of filters are used as a function of the noise magnitude (Goodrum and Elster, 1992; Leemans *et al.*, 1998; So and Wheaton, 1996; Utku and Koksel, 1998).

An alternative to filtering is represented by binarization that allows the transformation of the color or gray level image into a black-and-white image. In this way a value of black or white is assigned to each pixel (Aguilera and Stanley, 1999). After binarization, images can be manually edited so as to remove artifacts and noise and to apply other functions. The binary image is then ready for quantitative analysis performed by using stereological or morphometric methods.

3. Segmentation

It is the step that allows the separation of the image into features and background. Obviously, pixels contained in an object or features have values similar to those of the pixels belonging to the same category. Segmentation may be done manually or automatically.

Segmentation can be performed according to four different approaches: thresholding based, region based, gradient based, and classification based. In current applications, the first two methods are generally preferred (Du and Sun, 2004).

The thresholding-based segmentation is the simplest segmentation method but it works well if the objects have uniform gray level clearly distinguishable from the background that must have a different but uniform gray level with respect to the objects.

The region-based segmentation is a more powerful segmentation approach and includes two methods. The first, called "region growing and merging," acts by grouping pixels into larger regions as a function of homogeneity criteria. Successively, the second method, called "region splitting and merging," divides the image into smaller regions according to other criteria (Du and Sun, 2004).

The gradient-based segmentation, instead, allows to directly find the edges by their high-gradient magnitudes.

The classification-based method uses statistic or other techniques to assign each pixel to the different regions of the image.

4. Object measurement

After segmentation, it is possible to measure the features of interest for each of the individuated objects. The measurable features regard size (area, perimeter, length, width), shape (circularity, eccentricity, compactness, extent, and so on), color, and texture (smoothness, coarseness, graininess) (Du and Sun, 2004). Texture, in particular, is evaluated on the base of the gray level variation within the object (Aguilera and Stanley, 1999).

5. Classification

It allows to attribute a new object to one of the individuated categories by comparing the measured characteristics of the new object to those of a known one. Also in this case, several approaches are available: statistical, fuzzy, neural network.

The statistical classification uses probability models to classify objects.

The fuzzy classification method groups objects into categories without defined boundaries so as to take into account the degree of similarity of the considered object with respect to the others (Du and Sun, 2004).

Finally, the artificial neural network methods try to imitate human intelligence with the power of statistic.

II. IMAGE ACQUISITION TECHNIQUES

A. LIGHT MICROSCOPY

Microscopes used in LM are composed of a beam of visible light (photons) that represents the illumination source (probe), a system for focusing the source onto the sample (the condenser or condense glass lens), a location to place the sample or specimen, and the objective.

There are different types of LM: bright field (dark field viewing, phase contrast, oil immersion microscopy, differential interference contrast), polarizing, and fluorescence microscopy.

In the conventional bright field microscopy, light from an incandescent source is sequentially transmitted through the condenser, the specimen, an objective lens, and a second magnifying lens, the ocular or eyepiece, prior to reaching the eye. Some microscopes have an internal illuminator, while others use a mirror to reflect light from an external source. If the specimen is not very colored, several mechanisms for the formation of contrast can be performed. For example, it is possible to use dyes or stains specific for the different components of the specimen (fast green and acid fuchsin for proteins, toluidine blue O for pectin and lignin in vegetables and for muscle and fibroblasts in meat, oil red O for fats) (Kaláb *et al.*, 1995).

Bright field microscopy is suitable to observe stained bacteria, thick tissue sections, thin sections with condensed chromosomes, large protists or metazoans, living protists or metazoans, algae, and other microscopic plant material.

Limitations of the bright field microscopy are little related to magnification, whereas they are highly dependent on resolution, illumination, and contrast. Resolution can be improved using oil immersion lenses, whereas lighting and contrast can be greatly improved using modifications of the technique such as dark field, phase contrast, oil immersion microscopy, differential interference contrast (DIC). Dark field viewing is obtained by placing an opaque disk in the light path between source and condenser so that only light that is scattered by particles on the slide can reach the eye. In this way, light does not pass through the specimen but is reflected by it. Sometimes, neither bright field nor dark field can be used: the first due to the little contrast between the structures belonging to the same object, the second due the too thin section of the samples. In these cases, the phase contrast microscopy is applied by exploiting the differences in the refractive index of the various parts of the object. Oil immersion microscopy is an interesting alternative to bright field. In fact, when light passes from a material having a refractive index to a material having another refractive index, light bends and causes a loss in resolution, in particular at high magnifications. It is possible to improve resolution by placing a drop of oil with the same refractive index as glass between the cover slip and objective lens so as to eliminate two refractive surfaces (www.ruf.rice.edu/~bioslabs/methods/microscopy/microscopy.html). In DIC microscopy, a minute difference in refractive indexes of light passing through an unstained specimen is transformed into a monochromatic shadow-cast image so as to allow the observation of living and thick specimens (www.nikon-instruments.jp/eng/tech/1-0-4.aspx).

An alternative to the bright field microscopy is represented by the polarizing microscopy obtained by inserting two polarizers in the light path, the first between light source and specimen and the second between the objective and the eye. The light produced by the first polarizer vibrates in one of the planes perpendicular to the direction of travel. By opportunely rotating the second polarizer (called analyzer), it is possible to distinguish within a specimen the amorphous region (that appears dark) from the crystalline domains (that appears bright because of their birefringence).

In fluorescence microscopy, specimen itself represents the light source. This method is based on the phenomenon that certain material can absorb light of a specific wavelength and emit energy detectable as light of a longer wavelength and lower intensity. The sample can either be fluorescing in its natural form (autoflorescence) or treated with fluorescing chemicals. In vegetable tissues, autofluorescent molecules are, for example, chlorophyll, carotenoids, lignin, and ferulic acid. In animal tissues, the main sources of fluorescence are some fats.

B. CONFOCAL LASER SCANNING MICROSCOPY

CLSM allows the observation of thin optical sections in thick, intact specimens. CLSM represents an advanced technology with respect to fluorescence microscopy. In fact, in conventional fluorescence microscopy, out-of-focus fluorescence can cover the image details. Instead, to induce fluorescence, CLSM uses a laser spot that is scanned in lines across the field of view resembling image formation by an electron beam in a computer screen. In this way, sample is illuminated and imaged one point at a time, whereas in the LM the object is uniformly illuminated. Fluorescence is detected by a highly sensitive photomultiplier. Out-of-focus fluorescence is excluded by the presence of a pinhole at the focal plane of the image able to produce a sufficiently thin laser beam (Kaláb *et al.*, 1995). Resolution increases as the open degree of the confocal pinhole decreases. Furthermore, the intensity of the laser beam decreases with the third potency above and below the focal plane (www.plbio.kvl.dk/~als/confocal.htm). Also, in CLSM it is possible to use fluorescent labels to put in evidence specific food components. CLSM allows the extraction of topographic information from a set of confocal images acquired over a number of focal planes. Multiple confocal slices of the image can be obtained. In this way, a 3D topographic map of the object is available (Ding and Gunasekaran, 1998).

Light microscopes are limited by the physics of light to 500x or 1000x magnification and a resolution of 0.2 μm. The image resolution is related to the beam wavelength that in the case of LM and CLSM ranges from 400 to 600 nm, whereas in the case of electron microscopy is of 0.0037 nm (Hermansson *et al.*, 2000). The very short wavelength of electrons allows the resolution improvement and modern technology makes it easy to obtain a resolution of 3 nm or lower (www.hei.org/research/depts/aemi/emt.htm).

C. ELECTRON MICROSCOPY

It allows the obtainment of information about topography (surface features of an object, i.e., its texture and relation among these features and the material properties), morphology (shape and size of the particles and relation among these structures and the materials properties), and composition (elements, compounds, and their relative amounts, relationship between composition and materials properties).

Since in electron microscopy the illumination source is represented not by light but by a focused beam of electrons, the resulting micrographs cannot be in color but in various shades of gray. Colors can be successively added to the obtained micrographs. Individual components can be distinguished thanks to the differences in their affinity for various heavy metals such as osmium, ruthenium, lead, and uranium. Gold granules of dimensions of few nanometers attached to antibodies may be used to identify macromolecules (enzymes, protein-based hormones, and so on) through immunological reactions.

In an electron microscope, a stream of electrons is formed by the electron source and is accelerated toward the specimen using a positive electric potential. The stream is focused into a thin, monochromatic beam by using metal apertures and magnetic lenses. The electron beam interacts with the specimen and the effects of these interactions are detected and transformed into an image.

Electron microscopy works under vacuum conditions because air absorbs electrons. For these reasons, wet samples cannot be analyzed by electron microscopy without previous dehydration, freezing, or freeze-drying due to the sublimation phenomena (Bache and Donald, 1998).

Electron microscopy can be divided into SEM and TEM. These two types of electron microscopy differ from each other in the way in which the image is formed. The transmission electron microscope was the first type of electron microscope to be developed (1931) and it works exactly as a light transmission microscope except for the focused beam of electrons used in place of light. The first scanning electron microscope was built in 1942 but was commercially available in 1965 due to the complicated electronics involved in "scanning" the beam of electrons across the sample.

1. Transmission electron microscopy

It allows to determine the internal structure of materials. The structure of a transmission electron microscope constitutes of an evacuated metal column with the source of illumination, a tungsten filament (the cathode), at the top. When the filament is heated at a high voltage, the filament emits electrons. These electrons are accelerated to an anode (positive charge) and pass through a tiny hole in it to form an electron beam that passes down the column. Electromagnets placed in the column work as magnetic lens. The electron beam is focused onto the specimen. Some electrons pass through the sample and the image, magnified by the intermediate lens, is observed, thanks to the projector lens, on a fluorescent screen at the base of the microscope column or photographed (www.hei.org/research/depts/aemi/emt.htm). The image formation is due to the differences in electron density of the analyzed samples or due to the thickness of the metal replica (Kaláb *et al.*, 1995).

To allow electrons to transmit through the materials, samples have to be very thin (50–100 nm). The energy of the electrons in the TEM determines the relative degree of penetration of electrons in a specific sample. So energy of 400 kV provides high resolution and high penetration in samples of medium thickness. TEM resolution often exceeds 0.3 nm (Yada *et al.*, 1995). TEM allows to obtain magnifications of 350,000 times and over (www.uq.edu.au/nanoworld/tem_gen.html).

In TEM, thin sections of samples are embedded in epoxy resins or, alternatively, platinum-carbon replicas of the samples are produced in order to the avoid release of vapor or gases.

Contrast in the TEM increases as the atomic number of the atoms in the specimen increases. Since biological molecules are composed of atoms of very low atomic number (carbon, hydrogen, nitrogen, and so on), contrast is increased with a selective staining, obtained by exposure of the specimen to salts of heavy metals, such as uranium, lead, and osmium, which are electron opaque (www.hei.org/research/depts/aemi/emt.htm).

2. *Scanning electron microscopy*

It allows the detection of the sample surface that emits or reflects the electron beam (Hermansson *et al.*, 2000). To provide electron conductivity, a 5- to 20-nm coating is applied on the sample surface (Kaláb *et al.*, 1995).

In conventional SEM to avoid the vapor release samples are previously dried, whereas in cryo-SEM samples are frozen and analyzed at low temperature.

Wet samples can be analyzed without a previous preparation by the so-called environmental scanning electron microscopy (ESEM). In this technique, instead of the vacuum conditions, the sample chamber is kept in a modest gas pressure (Bache and Donald, 1998). The upper part of the column (illumination source) is kept in high vacuum conditions. A system of differential pumps allows to create a pressure gradient through the column (Bache and Donald, 1998; Stokes and Donald, 2000). The choice of the gas depends on the kind of food: hydrated food is kept under water vapor.

D. MAGNETIC RESONANCE IMAGING

MRI is an analytical imaging technique primarily used in medical settings in order to produce high-quality images of the insides of the human body. Nevertheless, food applications of the MRI technique have been developed in the last years. MRI is based on the principles of the NMR, a spectroscopic technique used by scientists to obtain chemical and physical information about molecules. It was named magnetic resonance imaging rather than nuclear magnetic resonance imaging (NMRI) because of the negative connotations associated with the term "nuclear" in the late 1970s. MRI started out as a tomography imaging technique, which allowed the conversion of the NMR signal deriving from a thin slice through the material in an image. This technique is based on the absorption and emission of energy in the radio frequency range of the electromagnetic spectrum (Bows *et al.*, 2001; Van Duynhoven *et al.*, 2003). Magnetic resonance images are based on proton density and proton relaxation dynamics, differing from those produced by using X-rays that are associated to the absorption of X-ray energy instead. The generation of magnetic resonance images can be controlled by the radio frequency pulse sequence used for exciting the nuclear spins. In the NMR, hydrogen nuclei are subjected to a strong magnetic field that determines their alignment in a spin-up or spin-down orientation. If a radio frequency pulse is applied to this system, the nucleus alignment is inverted and their individual processions will be brought into phase. When the radio frequency is switched off, the phase relationship decays with a characteristic time constant referred to as T2, and the nucleus alignment relaxes with a different time constant named T1. Both these parameters are temperature dependent. The chemical shift is another parameter that results from the electron clouds that shield each nucleus from the applied magnetic field, thus altering their frequency of NMR precession. A magnetic resonance image is composed of several picture elements called pixels. The intensity of a pixel is proportional to the NMR signal intensity of the contents of the corresponding volume element. The main advantage of the MRI technique is that it allows the obtainment of 2D or 3D images of the inner part of a material in a noninvasive and nondestructive way and without any preliminary sample preparation (Martinez *et al.*, 2003). MRI technique permits the spatial distribution of water, fat, and salt content in foods. In particular, its ability to study the spatial distribution and mobility of water and its dependence on temperature has led to a new approach to the validation of thermal processing in food manufacture. For example, Bows *et al.* (2001) used MRI to map the temperature distribution induced in water-based foods by microwave and conductive heating in order to evaluate their suitability as potential tools for microbiological assurance. NMR parameters, such as the relaxation times and diffusion coefficients, allow the definition of the interactions among water and other molecules. The attributes that can be quantified by applying MRI range from the composition of a material (moisture and fat content) to its physical (color, size, shape, volume) and chemical (density, viscosity, pH, water activity) properties. MRI can be used to control food processes and to understand the changes occurring in food during processing. MRI not only permits the detection of internal defects of products, such as fruits and vegetables (hollow heart in potatoes, brown center and bruises in apples, freeze damage in oranges), but also more complex analysis regarding the grading of the quality of some foodstuffs. Internal structures, differentiated by water or fat content, can be highlighted by MRI because water, lipid, and proteins contain nuclei distinguishable by their NMR chemical shifts. The signals can be localized by the imposition of magnetic field gradients. The use of NMR microimaging to characterize water properties in cooked and high-pressure-treated beef as a function of the length of the ageing period was proposed by Bertram *et al.* (2004). Ishida *et al.* (2001) introduced the use of MRI to study the architecture of baked bread made of fresh or frozen dough. The NMR imaging represented an alternative to SEM technique to provide a quantitative estimation of the network structure of bread as one of the main elements determining the quality of the product. The method, providing information about the internal structure with a spatial resolution of 100 μm, was suitable for depicting crust surface and gluten network. The quality of frozen dough is generally lower than that of fresh dough because of the degradation of the gluten structure, the partial disruption of gluten fibrils, the separation of starch granules, and the deterioration of starch consequently to the ice crystal formation. The study of local porosity by means of the MRI gray level has been also presented and validated at whole dough scale by other authors (Grenier *et al.*, 2003). Bonny *et al.* (2004) used MRI to examine the dough fermentation process in terms of bubble size distribution and cell wall thickness. The classification of objects with different internal structure with respect to MR-image gray tone distribution has been used for the simple analysis of potato structure and texture (Thybo *et al.*, 2004). Another field of application of MRI technique is the visualization of lipid migration or oil distribution in food products. In a paper by Miquel *et al.* (2001), MRI was used to study the migration of hazelnut oil into chocolate in a composite confectionery. Yan *et al.* (1996) presented a work on the oil distribution in two types of crackers, laminated and non-laminated. NMR images represented the proton density map of the oil distribution. MRI has been successfully used to visualize the phase transition within food products during freezing. In this area, the mathematical models fail to give information on the variations of heat transfer that, instead, are detected by MRI from the differences in the product sugar concentration (McCarthy and McCarthy, 1996). Kerr *et al.* (1998) applied MRI technique to follow the ice formation in several foods during freezing. They used an image resolution of 350 μm, suitable for viewing macroscopic movement at the ice interface. Kuo *et al.* (2003) observed the ice formation during freezing of pasta filata and nonpasta filata mozzarella cheeses by mapping the distribution of water through MRI. Renou *et al.* (2003) investigated the NMR parameters during freezing process of meat. A decrease of signal strength means a reduction of proton mobility during phase transitions. So the transition from water to ice can be inferred from a decrease in signal strength, whereas the time required for the disappearance of the NMR signal corresponds to that required for reaching a steady state enthalpy value. The major limitation of MRI is that it can be applied to material investigations only if they have a sufficient water content. Furthermore, the equipment required for MRI measurement is expensive.

E. ULTRASONIC IMAGING

The basis of the ultrasonic analysis of foods is the relationship between their ultrasonic properties (velocity, attenuation coefficient, and impedance) and their physical and microstructural properties (Coupland, 2004; Povey and McClements, 1988). Ultrasonic waves propagate more or less easily depending on material density and elastic modulus. Ultrasonic properties are also frequency dependent, particularly in the case of highly structured materials. The attenuation coefficient is a measure of the decrease in amplitude of an ultrasonic wave and it is expressed as the logarithm of the relative change in energy after traveling unit distance. It is a consequence of absorption and scattering. In the first case, the energy stored as ultrasound is converted into heat. In the second case, the energy is still stored as ultrasound but it is not detected because its propagation direction and phase have been altered. Like the ultrasonic velocity and attenuation coefficient, the acoustic impedance is a fundamental physical characteristic that depends on the composition and microstructure of a material so its measure can be used to provide valuable information about the properties of foods. The relationship between ultrasonic parameters and microstructural properties of a material can be empirically established by a calibration curve that relates the property of interest to the measured ultrasonic property or, theoretically, by using equations that describe the propagation of ultrasound through the material. Ultrasonic waves are similar to sound waves, but they have frequencies that are too high to be detected by the human ear. An ultrasonic wave is transmitted as a series of small deformations in the medium. When an oscillatory force is applied to the surface of the material, it is transmitted through it. If the force is perpendicularly applied, a compression wave is generated. Finally, if it is applied parallel to it, a shear wave is generated (Povey and McClements, 1988).

The key elements of an ultrasonic measurement system are: a transducer, which converts an electrical impulse into mechanical vibration, a signal generator to produce the original electrical excitation signal, and a display system to record and measure the echo patterns produced. The pulser-receiver generates an electrical pulse that is sent to an ultrasonic transducer, where it is converted into an ultrasonic pulse that travels into the sample being analyzed. The signal received from the sample is converted back into an electrical pulse by the transducer and sent to the analog-to-digital converter where it is digitized. The two ways for characterizing the encompassment of these elements are the pulse-echo and the resonance techniques (Coupland, 2004). The first one is a useful way to measure the surface ultrasonic properties of a sample. Pulsed methods are commonly used as the basis for ultrasonic imaging devices.

By measuring distance, velocity, and attenuation as a function of the transducer position, it is possible to generate a 2D image of the sample properties. By rotating the sample, a 3D image can be reconstructed (Coupland, 2004).

Unlike light-scattering studies, for which dilution is often a prerequisite, ultrasound can measure food properties at concentrations that are technologically relevant. This aspect has obvious benefits for the analysis of inhomogeneous foods such as solidifying fats, dynamically changing dairy food systems, dough, and emulsions.

In the food industry, the applications of ultrasound can be divided into two distinct categories, depending on whether they use low-intensity or high-intensity ultrasound. The low-intensity ultrasound is a nondestructive tool because it uses power levels so small that no physical or chemical alterations in the material occur (Javanaud, 1998). In contrast, the power level used in the high-intensity ultrasound is so large that it causes physical disruption or promotes chemical reactions (McClements, 1995). The low-intensity ultrasound is commonly applied to provide information about properties of foods such as composition, structure, and physical state. Low-intensity ultrasound offers the possibility of acquiring images of the internal structure of foods for their quality evaluation. A small size ultrasonic probe (2 MHz) equipped with a LCD display was used for evaluating meat quality (Ozutsumi *et al.*, 1996). The picture signals were fed into a computer for the estimation of the fat content and other chemical characteristics of the meat. The results obtained were in agreement with the actual carcass measurements. An automatic classification equipment, based on the use of ultrasound imaging, was developed in order to measure the texture of cheese (Benedito *et al.*, 2000), correlate some meat textural features with the intramuscular fat content (Kim *et al.*, 1998), and measure fat and meat depth in carcasses (Busk *et al.*, 1999). Unlike meat classification systems, ultrasonic imaging is a noninvasive method for the on-line determination of lean meat percentage in carcasses and has low costs of maintenance. A novel approach to grading pork carcasses was proposed by Fortin *et al.* (2003). In this study, ultrasound imaging was use to scan a cross-section of the loin muscle and capture 2D and 3D images of the carcasses. By coupling muscle area measures and fat thickness obtained by ultrasound together with 2D and 3D images, it was possible to provide the most accurate model for estimating salable meat yield. With respect to many others applications of ultrasound in food industry, ultrasonic inspection of meat quality has been developed to the stage of availability of commercial instruments. The use of ultrasound to measure muscle and fat depths for the initial screening of meat was proposed as an alternative method to X-ray computer tomography (CT) or MRI (Chi-Fishman *et al.*, 2004; Jones *et al.*, 2004). The CT measure of muscularity was positively associated with those performed by ultrasound. Compared with CT and MRI, ultrasound is a considerably less expensive and relatively more portable imaging technique. Ultrasound technology provides quantitative and qualitative information about mass features that may be linked to measures of muscles strength. There are two modes for ultrasonic imaging of biological tissue. One is the A-mode (amplitude modulation) and the other is the B-mode (brightness modulation). The first mode is one-dimensional and is used to measure depth of tissue, whereas the B-mode provides the characterization of biological tissue. Real-time ultrasound (RTU) technique is used as a special version of the B-mode technique in order to provide images of moving objects (Du and Sun, 2004).

An application of ultrasound that is becoming increasingly popular in the food industry is the determination of creaming and sedimentation profiles in emulsions and suspensions (Basaran *et al.*, 1998). Acoustic techniques can also assess nondestructively the texture of aerated food products such as crackers and wafers. Air cells, which are critical to consumer appreciation of baked product quality, are readily probed due to their inherent compressibility (Elmehdi *et al.*, 2003). Kulmyrzaev *et al.* (2000) developed an ultrasonic reflectance spectrometer to relate ultrasonic reflectance spectra to bubble characteristics of aerated foods. Experiments were carried out using foams with different bubble concentration and the results showed that ultrasonic reflectance spectrometry is sensitive to changes in bubble size and concentration of aerated foods.

Some of the simplest ultrasonic measurements involve the detection of the presence/absence of an object or its size from ultrasonic spectrum (Coupland and McClements, 2001). An ultrasonic wave incident on an ensemble of particles is scattered in an amount depending on size and concentration of the particles. As the ultrasonic parameters depend on the degree of the scattering, it can therefore be used to provide information about particle size.

Chow *et al.* (2004) reported dynamic video images of the influence of ultrasonic cavitation on the sonocrystallization of ice at a microscopic level. The ultrasonic device was used in combination both with an optical microscope and with an imaging system in order to observe the production of secondary ice nuclei under an alternating acoustic pressure.

In comparison with other techniques, the major advantages of ultrasound are that it is nondestructive, rapid, and can easily be adapted for on-line measurements.

Despite its desirable attributes, ultrasound is not without deficiencies. It can be applied to systems that are concentrated and optically opaque. One of its major disadvantages is that the presence of small gas bubbles in a sample can attenuate ultrasound and the signal from the bubbles may obscure those from other components. Another potential problem occurs when ultrasound is used to follow complex biochemical and physiological events. In this case, it is difficult to attribute specific mechanisms to the observed changes in velocity and attenuation. In addition, velocity may strongly depend on the temperature of the foods, therefore in real processing operations with gradient temperature, it is critical to evaluate the effects of such gradients on the ultrasonic velocity measurements (Coupland, 2004; Povey, 1997).

F. ATOMIC FORCE MICROSCOPY

In 1986, Binning *et al.* provided a remarkable solution to the impossibility of molecular or submolecular resolution by scanning a sharp stylus attached to a flexible cantilever across a sample surface. This invention was known as AFM. This instrument is a new example of scanning probe microscopy (SPM) techniques in which the interaction between tip and specimen is not represented by a current deriving from tunneling electrons but rather by force interaction. The principle is very similar to the way in which a record stylus plays a record, with the exception that the stylus is much smaller (a few micrometers) (Kirby *et al.*, 1995). In the AFM, the stylus is rigidly fixed onto an elastic cantilever. When the stylus is close to the sample, the repulsive forces determine the bending away of the cantilever from the surface. By monitoring the extent of the cantilever bending, any undulations in the sample can be recorded and detected by a laser beam, which is reflected into a photodetector (Morris, 2004). The conventional microscopes look at samples, while the AFM feels the details on the surface of the specimen. The sample is felt by scanning it with a sharp probe attached to a cantilever or spring (Kirby *et al.*, 1995). Most AFM cantilevers are microfabricated from silicon oxide, silicon nitride, or pure silicon by applying photolithographic techniques. The sample is applied to a solid substrate, such as mica or glass, and its roughness dictates the restriction in the use of this technique.

Mica is a cleavable aluminum silicate crystal, whereas glass is a rougher substrate useful for imaging larger structures. Biopolymer samples are generally applied onto cleaved mica and then air dried if a better resolution is to be realized. Solid samples can just be glued onto a metal plate before imaging. The force applied on the sample by the stylus is very important for determining the contrast. The movements of the samples in small 3D ranges are achieved by mean of piezoelectric devices. In general, there are two AFM-imaging methods, the contact mode and the noncontact mode. In the latter, the shearing forces exerted by the stylus scanning over the sample may be reduced or eliminated. In the contact mode, the atoms of the stylus are so close to the sample that they touch it. The most important contact mode of AFM imaging is that based on the constant repulsive forces on the cantilever, which are kept constant with a feedback circuit. When a variable force is exerted on the sample, the image is obtained by recording and amplifying the signal of the piezoelectric device. A particular imaging method, named "error signal mode" and suitable for emphasizing molecular structure of rough samples, involves the direct monitoring of the cantilever deflection as it senses features on the surface. In the contact mode, the forces on the sample are not always desirable because they can destroy the sample. This inconvenience can be solved by using the noncontact mode. In this case, the cantilever is bonded onto a small slab of piezoelectric material and then it is vibrated close to its resonant frequency. Images can be obtained in two distinct modes: true noncontact imaging mode, when the cantilever is vibrated so gently above the sample that it does not touch it, and tapping mode, when the cantilever is vibrated more strongly so that the stylus intermittently touches the sample. This last type of imaging attracts the interest of the users of this technique because it provides high resolution if compared to that obtained by performing the analysis by keeping both sample and cantilever immersed under a liquid and, furthermore, it does not require time for the instrument stabilization (Kirby *et al.*, 1995).

The AFM technique is easy to apply, the specimen can be imaged in air or liquid, the resolution is very high, and the sample preparation is much simpler if compared to those required by traditional microscopy.

AFM technique is able to provide information about the individual molecules of the material and the way in which size and shape of molecules affect their behavior in foods.

Biological nonconducting material can be easily imaged by AFM (Gunning *et al.*, 1996). AFM allowed the study of irregular polysaccharide structures and their function as suspending agents in foods (Kirby *et al.*, 1995). The polysaccharides were immersed in alcohol because the moisture present in the atmosphere can condense around the stylus or the surface of the samples, causing a poor quality image. By AFM in noncontact mode (tapping mode), Elofsson *et al.* (1997) characterized different whey protein preparations such as pure β-lactoglobulin standard whey protein concentrate and cold gelling whey protein concentrate. The samples were diluted at three different concentrations, dried into mica sheets, and imaged in the AFM microscope. 3D views and cross-section topography images of monolayer coverage of β-lactoglobulin standard whey protein concentrate and heat-modified whey protein concentrate were obtained to clearly distinguish the different states of protein aggregation at a submicrometer level. In food context, AFM has also been used to study polysaccharide networks such as starch granules and cell walls from fruit and vegetables (Kirby *et al.*, 1995). AFM allows the study of interfacial phenomena, such as bacterial boils and fouling and air–water or oil–water interfaces, which stabilize emulsions or food foams. The technique provides the resolution suitable to visualize these structures and to study the surfactant-induced destabilization of protein-stabilized foams or emulsion, but it cannot be directly used to study interfaces in foods (Morris *et al.*, 1999). Moreover, AFM was used to visualize the internal structure of starch granules without inducing the necessary contrast in the images (Ridout *et al.*, 2002). The images allow the examination of the possible mutations that affect starch structure and its functionality.

G. VIBRATIONAL MICROSPECTROSCOPY

Raman microspectroscopy results from coupling of an optical microscope to a Raman spectrometer. The high spatial resolution of the confocal Raman microspectrometry allows the characterization of the structure of food sample at a micrometer scale. The principle of this imaging technique is based on specific vibration bands as markers of Raman technique, which permit the reconstruction of spectral images by surface scanning on an area.

While an optical microscopy gives only a mapping of the whole mixture, the Raman microspectroscopy offers selective image contrast of each molecular component because it uses a fixed wave number characteristic of each component of the mixture (Huong, 1996). Components larger than 1 μm can be illuminated by the micro-Raman setup and their spectra can be recorded without interference. The Raman spectroscopy measurements are a function of vibrations of all bonds, geometries, distances, angles, and polarizability of the chemical bonds. For these reasons, Raman spectroscopy can differentiate the single bond from the double or triple ones, whereas the other microscopy techniques give information only about the nature of the bonded atoms.

Also the infrared microspectroscopy (IR) is a vibrational spectroscopy, but it presents some differences with respect to Raman spectroscopy and also provides different information. In infrared spectroscopy the sample is radiated with infrared light, whereas in Raman spectroscopy a monochromatic visible or near infrared light is used. In this way, the vibrational energy levels of the molecule are brought to a short-lived, high-energy collision state. The return to a lower energy state occurs by emission of a photon. Raman microspectroscopy is based on the detection of the vibrations of molecules whose polarizability changes, whereas IR spectroscopy detects vibrations of molecules whose electrical dipole moment changes (Thygesen *et al.*, 2003). The limiting spatial resolution is of the order of 1 μm × 1 μm in Raman microspectroscopy, whereas it is around 20 × 20 μm^2 in IR. Each food system shows characteristic absorbance bands for both Raman and IR microspectroscopy. The four major food chemical compounds (water, fat, protein, and carbohydrates) absorb in Raman and IR but with different intensities. For example, water presents very strong absorption in the IR but it is invisible in Raman spectroscopy because of the weak vibration of the O–H (Huong, 1996). FT-IR and Raman microspectroscopy may be combined with three different mapping techniques: point, line, and area. With point acquisition, several spectra are measured from different places in the sample. Line mapping defines a series of spectra along one dimension. Area-mapping technique uses two dimensions, providing a spectroscopic image that can be related to the corresponding visual image with an entire spectrum in each pixel. Raman technique is rapidly performed and does not require any destructive preparation of the sample, even if the sample could be destroyed due to the heating determined by the laser light (West, 1996). Compared with the infrared spectroscopy, it permits a better spatial resolution, an easier setting up, and in addition makes possible the focusing of a sample through a food-packaging material without exposing it to the atmosphere. Two important limitations of this technique are the signal-to-noise ratio, which can be very low if the sample fluoresces, and the fact that the surface of the sample cannot be planar to allow a correct evaluation of the repartition of each component (Huong, 1996). Raman and Infrared microspectroscopy may reveal useful information about food samples. Vibrational microspectroscopy was applied to a number of different problems related to food analysis to obtain information about microstructure and chemical composition. Samples of microscopic size can be directly analyzed in air, at ambient temperature and pressure, and under wet or dry conditions. By using confocal Raman microspectrometry, Piot *et al.* (2000) followed the evolution of protein content and structure during grain development of various wheat varieties. The technique is not only a powerful method to identify cereal components but it also gives information about the secondary structure and configuration of the proteins. The originality of the technique used in the work resides in the coupling between a Raman spectrometer and an optical microscope. The confocality was assured by a diaphragm located in the focal image plane of the sample, just before the input of the spectrograph. By using marker vibration bands, spectral images were generated on one or more particular components. The whiter the points in the images, the more intense the Raman scattering was.

IR could represent a complementary technique with respect to Raman spectroscopy, to better understand structure and molecular bond at a micrometer scale (Wetzel *et al.*, 2003). The synchrotron infrared microspectroscopy is superior to the same technique using a conventional global as a source because it is 1000 times brighter and highly directional and there is no thermal noise. Nevertheless, it is also more expensive. The vibrational microspectroscopy was applied to other different heterogeneous food systems providing information about the microstructure (Thygesen *et al.*, 2003). For example, high-quality spectra of starch granules in potatoes were acquired. Using Raman microspectroscopy it was possible to study the distribution of amygdalin in bitter almond cotyledons. IR microspectroscopy was used to study the nature of the blisters contained on bread crust or microstructure of high-lysine barley.

H. PHASE-CONTRAST MICROTOMOGRAPHY

Tomography is usually defined as the quantitative description of a slice of matter within an object. Several sources can be used, in particular X-rays sources, widely used in both the medical and industrial fields. The experimental implementation of tomography requires an X-ray source, a rotation stage, and a radioscopic detector.

A complete analysis is made by acquiring a number of radiographs (typically about 1000) of the same sample under different viewing angles (one orientation for each radiograph). A final computed reconstruction step is required to produce a 3D map of the linear attenuation coefficients in the material. This 3D map indirectly gives a picture of the structure density. In the X-ray-computed tomography, the X-ray source and detector are placed at the opposite sides of the sample. The spatial resolution of the attenuation map depends on the characteristics of both the detector and number of X-ray projections.

X-rays can be absorbed or scattered and the attenuation of the incident radiation can be expressed by the Beer's law:

$$I = I_0 \cdot \mathrm{e}^{-\int_l \mu(x) \cdot \mathrm{d}x}$$

where I and I_0 are the transmitted and incident radiation, respectively, μ is the linear attenuation coefficient (cm^{-1}), and l (cm) is the path of the radiation inside the sample. As a consequence, the obtained image is a map of the spatial distribution of the μ in which the brighter region corresponds to the higher level of attenuation if the detector used is a CCD camera. Differences in the linear attenuation coefficients within a material are responsible for the X-ray image contrast.

The main contrast formation in X-ray tomography is due to absorption contrast. This is often a limitation for imaging of low-Z or low-density materials. Synchrotron radiation (SR) sources are, essentially, large multidisciplinary research facilities supporting a broad research portfolio in physics, chemistry, biology, and engineering. These sources are based on high-energy electron accelerators producing electromagnetic radiation that covers a wide spectral range from the far infrared to hard X-rays. Compared with conventional laboratory sources, SR can deliver several orders of magnitude greater photon flux with a well-collimated beam and other properties that make them extremely powerful tools for a whole range of scientific and technological applications.

Advances in electron storage ring and the use of the so-called insertion devices (wigglers and undulators) have led to the development of third generation sources with another important characteristic: the small divergence of the beam as seen from the sample, due to the very small area of the electron beam that acts as the source of SR, combined with the increased distance between the source itself and the sample. These qualities of the X-ray beam, defining its spatial coherence, have been used to offer new opportunities in the field of X-ray imaging, such as phase-contrast and diffraction-enhanced imaging.

When X-rays interact with any kind of materials, *absorption* and *phase shifts* effects occur. Conventional X-ray radiography relies on the *absorption properties* of the sample. The image contrast is produced by a variation of density, a change in composition or thickness of the sample, and is based exclusively on the detection of an *amplitude variation of X-rays* transmitted through the sample itself. Information about the phase of X-rays is not considered. The main limitation of this technique is the poor intrinsic contrast in samples with low atomic number (i.e., the case of "soft matter") or in materials with low variation of absorption from point to point.

If X-ray beams have a high spatial coherence—as for third generation SR sources–contrast may be originated by the interference among parts of the wave front that have experienced different *phase shifts* through the sample (Fresnel diffraction). In the energy range of 10–25 keV, the phase shift contribution can be up to 1000 times greater than the absorption one and allows the detection of the phase effects even if the absorption contrast is low (Cloetens *et al.*, 1996; Snigerev *et al.*, 1995). Among the different techniques available for phase-sensitive imaging (Fitzgerald, 2000), the PHase contrast (PHC) microtomography setup is the same as that of absorption microtomography with the difference that the detector is positioned at a certain distance d from the sample. The choice of d depends on the size a of the feature to be identified, measured perpendicularly to the beam direction. In the edge detection regime ($d \ll a^2/\lambda$, where λ is the X-ray wavelength), images can be directly used to extract morphological information. Larger values of d lead to the holography regime ($d \approx a^2/\lambda$), and have not been used here. In the images, the Fresnel diffraction pattern appears superimposed to the absorption contrast and contributes strongly to enhance the visibility of the edges of the sample features.

The main limitation of tomographic setups based on conventional X-ray generators is obviously related to the lower flux in comparison with synchrotron radiation sources. As X-ray tubes generate a polychromatic spectrum, moreover, the different attenuation of photons as a function of their energy leads to a fast attenuation of the less energetic photons and, as a consequence, to an increase of the mean energy along the path of the X-rays. This effect, called "beam-hardening," generates different kinds of artifacts that must be taken into account during the reconstruction or, better, during the data acquisition (Kaftandjian *et al.*, 1996).

Conventional systems, on the other hand, have their own advantages. First, the access to such systems is much easier than to a synchrotron. Second, they can deliver X-rays with higher energies compared with typical SR sources, with evident advantages when bulky or high-Z samples are considered. Finally, thanks to the last generation of scanners based on micro- and nanofocus generators, the space-resolving power of these equipments has increased dramatically (Hirakimoto, 2001).

According to Evans (1995), differentiation of features within the materials is possible because μ at each point directly depends on the electron density of the material in that point (ρ_e), the atomic number (Z) of the chemical components of the materials in that point, and the energy of the incoming X-ray beam (I_0). In particular, the linear attenuation coefficient can be approximately considered as the sum of the Compton scatter and photoelectric contributions:

$$\mu = \rho_e \times \left(a + \frac{bZ^{3.8}}{I_0^{3.2}} \right)$$

where b is a constant (Vinegar and Wellington, 1987), a is the weakly energy-dependent Klein–Nishina coefficient (related to the angular distribution of the scattered photons as a function of the initial energy), ρ_e is given by $\rho_e = \rho \cdot (Z/A) \cdot N_{AV}$ (where ρ is the material density, Z and A are the atomic number and atomic weight, respectively, and N_{AV} is the Avogadro's number). The first term in the above equation represents the Compton scattering (direction change with loss of energy), which is predominant at X-ray energies above 100 keV, whereas the second term accounts for the photoelectric absorption (deposition of all energy in the matter), which is predominant below 100 keV (Vinegar and Wellington, 1987). Therefore, when various parts of the sample display contrast in density, these parts will be characterized by distinct values of the linear attenuation coefficient and differentiated regions contrasting brightness in the CT image.

Spatial distribution of the linear attenuation coefficients within material is based on the average linear attenuation coefficient along the projected line through the sample that can be calculated from the measured X-ray intensities as follows:

$$\bar{\mu} = \frac{\ln\left(\frac{I}{I_0}\right)}{-N \cdot \Delta x} I_0 \times e^{\mu \cdot x}$$

where $\bar{\mu}$ is the average attenuation coefficient in 1/cm and I/I_0 is the normalized X-ray intensity. The length of the projection is a constant approximated by the product of the step size (Δx) and the number of steps (N).

In the case of conventional sources, the ability to discriminate among materials with closely similar linear attenuation values (or bulk density) strongly depends on the accuracy of the μ_{voxel} value determination (Denison *et al.*, 1997). For each individual object voxel within a digital image, it is possible to compute a normalized attenuation coefficient known as *CT number* from the linear attenuation:

$$\text{CT number} = k \cdot \frac{(\mu_{\text{voxel}} - \mu_{\text{water}})}{\mu_{\text{water}}}$$

where μ_{voxel} is the linear X-ray absorption coefficient of a given matter expressed in 1/m, μ_{water} is the linear X-ray absorption coefficient of water in 1/m, and k is a constant value. In the case of $k = 1000$, the CT number is called "Hounsfield unit." It has an arbitrary scale in which a CT number of -1000 is attributed to air and a CT number of zero is attributed to water. This parameter allows the obtainment of relative measures of structural features.

The measurements obtained by a CT scanner results in a series of attenuation coefficients that are function of position and angle $\mu(t, \theta)$. Computer manipulation is required to convert these physical data into digital image arrays $g(x, y)$ in order to determine the distribution of the attenuation coefficients and hence the density distribution within the sample. There are many different algorithms developed to accomplish this task. One of most popular used today is called "filtered backprojection" algorithm (Hermann, 1980; Kak and Slaney, 1988) that is the base of the reconstruction software programs conventionally used to obtain 2D and 3D images. Figures from 1 to 5 show examples of both reconstructed 2D and rendered 3D images of porous and cellular food products obtained by SR X-ray absorption microtomography.

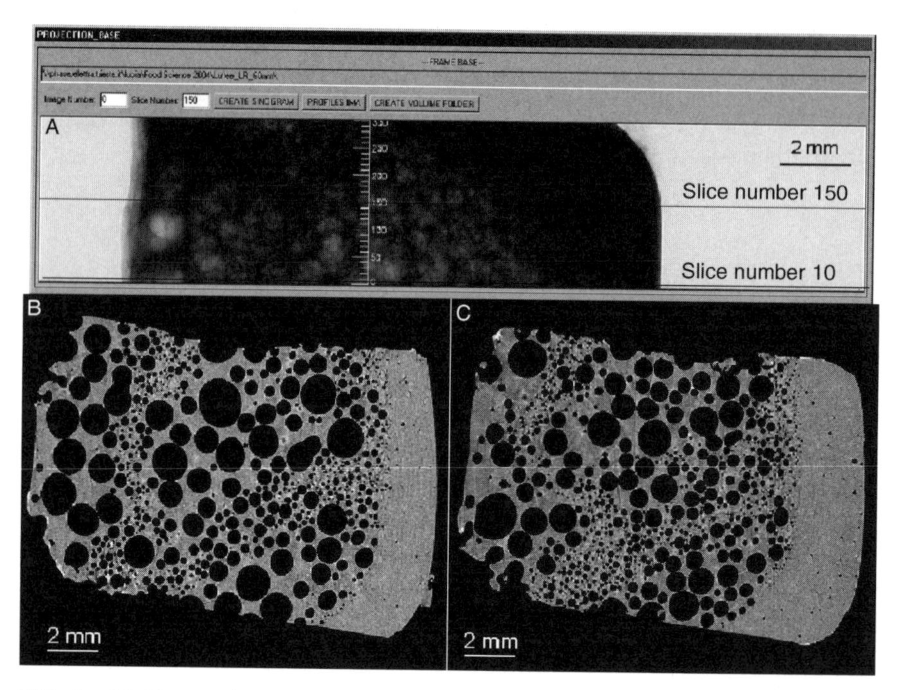

FIG. 1 3D X-ray microtomographic (XRM) images of a milk aerated chocolate bar. (A) X-ray absorption radiograph, (B) reconstructed slices corresponding to the slice number 10 within (A), (C) reconstructed slices corresponding to the slice number 150 within (A). The images show a closed cell structure at microscopic level (pixel size = 14 μm, $E = 13$ keV).

The attenuation along a line through the sample is calculated. This procedure simply assigns the mean attenuation coefficient to each point (or pixel) along that line. This backprojection is repeated for all angles. The attenuation coefficient for a particular point will be built up from all the projections passing through that point (or pixel). Nevertheless, the image under reconstruction is not continuous but composed of discrete pixels. The projection lines will not pass perfectly through the center of each pixel in their path and therefore a mathematical method is used to describe the projection line in terms of individual pixels within a matrix.

The reconstruction algorithms to be used for conventional sources are substantially different from the codes described in the previous section, due to the different acquisition geometry related to the conical shape of the beam (Burch and Lawrence, 1992; Feldkamp *et al.*, 1984).

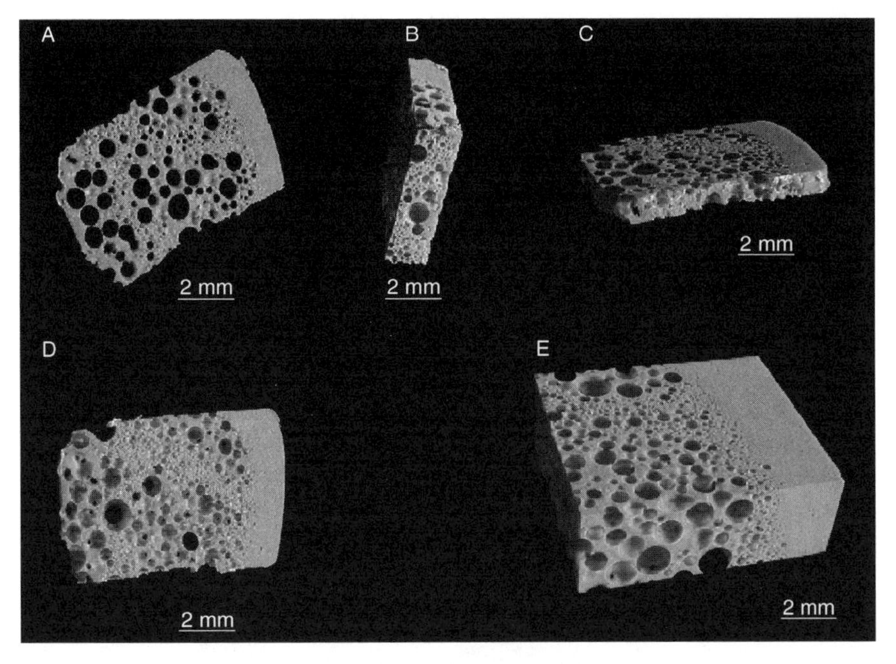

FIG. 2 3D XRM images of a milk aerated chocolate bar at middle level of sample. Rendered volumes numerically cut with (A) 20 slices, (B) 200 slices, (C) 100 slices, (D) 60 slices, and (E) 200 slices. The numerically rendered models reveal the foamed structure of sample from different angles of view (pixel size = 3.85 μm, $E = 13$ keV).

III. DATA PROCESSING

A. COMPUTERIZED SIMULATIONS OF X-RAY IMAGING TECHNIQUE

In order to use the X-rays in a wide range of application areas of agricultural and food industry, it is necessary to optimize the ability of the detection system in determining small spatial density and/or atomic number differences within a specific food substrate. For each specific application, there is generally a complex task to design the appropriate optimal parameter values such as the current and voltage of the X-ray tube, the geometric adjustment, the exposure time, and so on.

There is an increasing interest in the simulation codes of X-rays interactions with samples with the purpose to develop and optimize new imaging systems and to assess the influence of the various adjustable parameters in

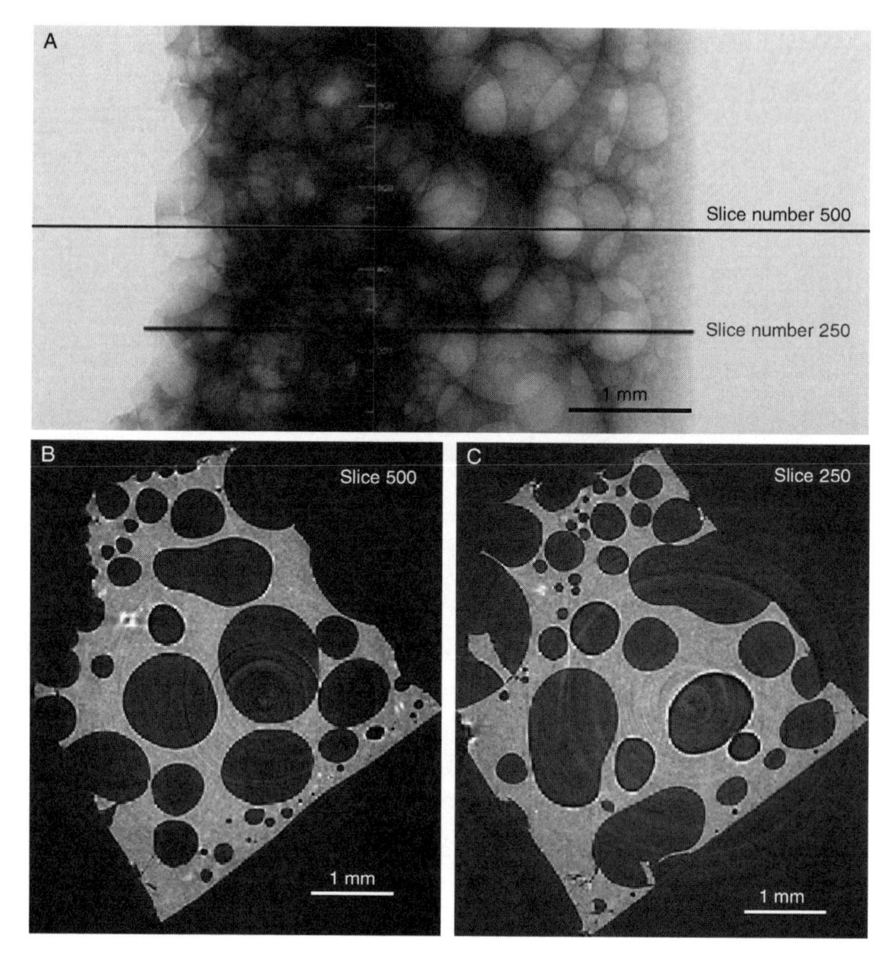

FIG. 3 Milk aerated chocolate bar. (A) X-ray absorption radiograph, (B) reconstructed 2D XRM images corresponding to the slice number 500 within (A), (C) reconstructed 2D XRM images corresponding to the slice number 250 within (A).

the image formation (Duvauchelle *et al.*, 2000; Zwiggelaar *et al.*, 1996). This approach allows to choose the most suitable components and to predict the future device performance by acting as a virtual experimental bench. Simulated images that can be obtained in little time and at low cost may enable the behavior of the whole imaging system to be investigated in complex situations (Inanc *et al.*, 1998).

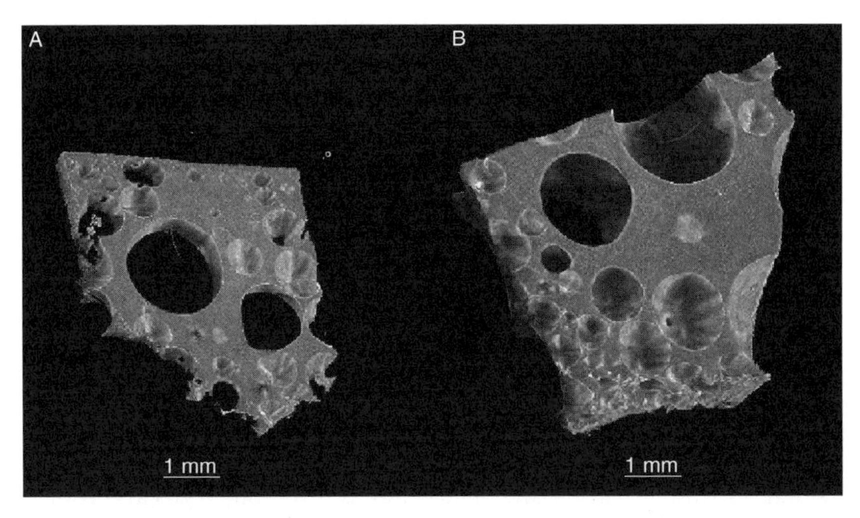

FIG. 4 3D XRM images of a milk aerated chocolate bar at middle level of sample revealing its foamed structure. (A) Rendered model numerically cut with 100 slices. (B) Rendered model numerically cut with 200 slices (pixel size = 3.85 μm, $E = 13$ keV).

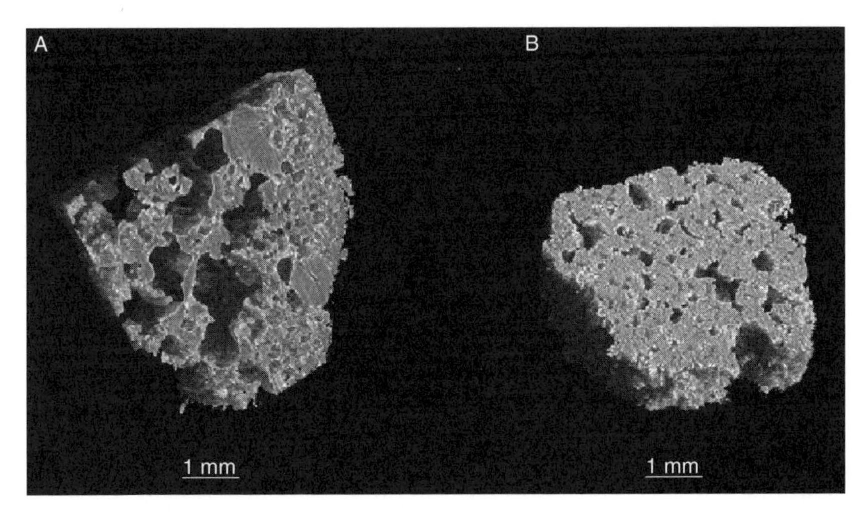

FIG. 5 3D XRM images of samples of commercial (A) and home made (B) crust pastry biscuits. Rendered volumes from (A) 300 and (B) 200 reconstructed slices (pixel size = 3.85 μm, $E = 13$ keV). It is worth noting a highly interconnected sponge-like structure: air cells are open and with highly irregular edges highlighting a coarse structure in the commercial sample and fine structure in the home made one.

Zwiggelaar *et al.* (1996) developed a specific computer code named XCOM/EGS4 to optimize the X-ray detection with respect to the following aspects: energy/dose ranges useful to detect variation in density and chemical composition, trade-off between signal-to-noise and intensity as a function of energy, and optimum energy range. XCOM/EGS4 code could be used to determine the optimal imaging conditions in more specific problems such as the foreign body detection or 3D distribution of chemical constituents. Possible application areas are the meat industry (control of the fat/meat ratio), potato industry (starch and sugar contents), and fruit quality assessment (sugar content, bruising, and problems such as hollow heart). Simulation of X-ray techniques using scattered photons could find applications in food processing to predict the structure development in baking process or the crystallization during cooling (Zwiggelaar *et al.*, 1996).

Another computer code was developed by Duvauchelle *et al.* (2000) to simulate the tomographic scan. Simulation is based on ray-tracing techniques and X-ray attenuation law. Automatic translations or rotations of the object can be performed to simulate tomographic image acquisition. Simulations can be carried out with monochromatic or polychromatic beam spectra. The simulation principle is completely deterministic and consequently the computed images present no photon noise. This computer code was designed to accept standard computer-aided drawing (CAD) files describing the sample geometry. The use of CAD models is a helpful tool to enable simulations with complex 3D samples since the geometry of every component of the imaging chain, from the source to the detector, can be defined. Geometric sharpness, for example, can be easily taken into account, even in complex configurations. Many software packages enable complex 3D objects to be drawn and CAD files to be generated in a short time, for example, with the stereolithographic (STL) or unstructured cell data (UCD) format. These files contain a list of nodes and meshes (triangular facets) that fit the object surface. The precision of this fit, which is linked to the size of the meshes, can be adjusted. The object may consist of different parts, possibly made of different materials, assumed to be homogeneous. The CAD model of each part can be independently handled. If a compound material is specified, the corresponding mass attenuation coefficient μ/ρ is calculated with the well-known formula

$$\frac{\mu}{\rho} = \sum_i \omega_i \left(\frac{\mu}{\rho}\right)_i$$

where the index i refers to each element of the compound and ω_i stands for the fraction by weight associated to the element i.

B. CT NUMBER AND FOOD QUALITY

Food quality inspection is extremely important in current agricultural production. X-ray CT technique shows a great potential for nondestructive, online sorting of immature from mature fruits (Brecht *et al.*, 1991; Yantarasri *et al.*, 2000). For example, X-ray-computed tomographic (CT) scanner was used to monitor and predict the internal quality changes in peaches during ripening (Barcelon *et al.*, 1999). As ripening time increased, the region surrounding the peach stone became drier and contained voids (dark region in the image) as showed by the low X-ray absorption and negative value of the CT number. The CT number was calculated on the radial and longitudinal cross-sections of the peach CT images. A histogram was produced reflecting the frequency of the CT number of each peach slice. Authors satisfactorily determined the relationships between the CT number and peach characteristics such as moisture, density, soluble solids, pH.

Yantarasri *et al.* (2000) observed significant differences between durian fruit of 50% maturity (CT number −70.59) and durian fruit of 80% maturity (CT number −45.18). The CT number computed from the image intensity showed an increasing trend with the increase in maturity according to a polynomial relationship with $R^2 = 0.99$. Brecht *et al.* (1991) reported similar results in tomato in which X-ray CT images showed intense signal in the gel tissue of fully mature tomatoes that appeared brighter than that of immature fruit. Suzuki *et al.* (1994) reported CT number referred to unripened papaya lower than that of the ripened ones.

X-rays have been used to detect structural discontinuities caused by voids and cracks and density variation. X-ray line scans were used to determine bruising, density, and water content of apples (Diener *et al.*, 1970; Tollner *et al.*, 1992), to select mature lettuce heads (Lenker and Adrian, 1971), and to detect hollow heart in potatoes and split pit in peaches (Han *et al.*, 1992). X-ray-computed tomography has also been used to detect density changes in tomatoes (Brecht *et al.*, 1991), seeds in oranges and pommels (Sarig *et al.*, 1992), defects in watermelons and cantaloupes (Tollner, 1993), internal changes of peaches, and stones in apricots (Zwiggelaar *et al.*, 1997). Tomographic images have been numerically analyzed to assess the quality of small food structures such as tomato seedlings (Van der Burg *et al.*, 1994), kernels (Schatzki and Fine, 1988), and walnuts (Crochon *et al.*, 1994). 3D numerical reconstructions by CT techniques have been used to obtain the density distribution and water content of apples. This technique has also been applied to determine injury by vapor heat treatment in papaya (Suzuki *et al.*, 1994), woolly breakdown in nectarines (Sonego *et al.*, 1994), internal defect disorder of durian, mangosteen, and mango (Yantarasri and Sornsrivichai, 1998; Yantarasri *et al.*, 1997), dry juice sac and granulation of tangerine fruits (Yantarasri and Sornsrivichai, 1998), and internal bruising and ripeness of pineapple (Yantarasri *et al.*, 1999).

Computed tomography offers a great potential for nondestructive studies of transport phenomena in foods. For example, accurate quantification of salt concentration in cured pork was made by computed tomography relating the CT values to the chemical analysis of salt (Vestergaard *et al.*, 2004).

C. CT NUMBER AND FOOD SAFETY

Patel *et al.* (1996) developed a strategy to process X-ray images with the purpose of detecting internal inhomogeneities in food products due to foreign objects including stone, metal, and glass. Food substrates in X-ray image show high textured appearance but contaminants or defects do not necessarily exhibit different textural properties. To overcome this problem, authors used a convolution-based method to detect anomalies within the texture of the food background. The convolution masks acted as matched filters for several types of textural variation found in the image. The convolution operation transformed the digital data from the spatial domain to the feature space domain allowing the highlighting of each local texture properties. This approach was able to recognize very small foreign objects or defects of 2×2 pixel size from the rest of the textured food substrate.

D. INTERNAL FEATURES AND MECHANICAL PROPERTIES

X-MT proved to be a very useful technique to image the 3D microstructure of cellular and porous food products (Falcone *et al.*, 2004a,b; Lim and Barigou, 2004; Schatzki and Fine, 1988; Tollner *et al.*, 1992; Van Dalen *et al.*, 2003; Van der Burg *et al.*, 1994). Effective numerical algorithms are developed to derive some important structure descriptors from digital images such as local density, volume fraction, specific surface (Falcone *et al.*, 2004b; Lim and Barigou, 2004; Maire *et al.*, 2003), spatial cell size distribution, cell wall thickness distribution (Lim and Barigou, 2004; Maire *et al.*, 2003), index of connectivity of the pores (Lim and Barigou, 2004), fractal dimension (Falcone *et al.*, 2004a), cell wall number, cell wall thickness, cell wall spacing, anisotropy degree of the solid phase, and main orientations of the cell walls in the space (Falcone *et al.*, 2004b). Approximate information on volume fraction, cell wall thickness, and cell size can be obtained using conventional 2D image analysis techniques (Whitworth and Alava, 1999), but information about anisotropy degree, predominant orientation of the structural elements, and index of connectivity requires the use of 3D data to obtain a correct quantification.

1. Volume fraction

Both 2D and 3D image analysis allows to get quantitative information on the volume fraction in foamed foods including air cell count, total cell area, relative cell area, and pore volume, but the 2D approach results in great limitations. Lim and Barigou (2004) extracted data concerning the above features from reconstructed CT images of foamed foods such as foamed chocolate bar, honeycomb chocolate bar, chocolate muffin, marshmallow, and strawberry mousse. They worked to provide a quantitative comparison between two commercial softwares, that is, the T-View software package (Skayscan, Belgium, 2003) used to analyze 3D images and the Leica Q-Win Pro image analysis software [Leica Microsystems (UK) Ltd., 2000] used to analyze 2D images. Both the software packages were run only on the central area of each image, about 42 mm^2, excluding the sample edges that had been affected by the physical cutting. In general, the T-View software allows to detect more air cells than the Leica Q-Win Pro image analyzer. This arises from the fact that, in its automatic object measurement mode, the image analysis of the Leica software did not include those parts of the cells situated on the borders of the measurement region, whereas the T-View software also included these incomplete cells. However, the bi-dimension cell parameters measured by the two methods are in close agreement, with a mean difference of 12% in relation to the cell size and less than 1% in the relative area. The smallest difference in the values of the relative area measured by the two methods is due to the fact that the automatic field measurement mode of the Leica software measures the whole cell areas detected in the region of interest in the same way as the T-View software. Falcone *et al.* (2004b) developed an algorithm able to perform the stereological analysis of the CT images and used it in order to evaluate the bread crumb porosity. The results were compared with those obtained by means of the ImageJ software package (version 1.29, National Institute of Health, USA, in the public domain from http://rsb.info.nih.gov/ij site at April 16, 2003). Authors found that the observed differences between the solid volume fractions as calculated by means of the two software packages were not statistically significant.

Advances in 3D image analysis can be attributed to Maire *et al.* (2003) who quantified the local volume fraction of bread crumb samples by means of the 3D image analysis. It is known that the gray level in each voxel of a tomographic image is proportional to the local density of the constituents so that it can be directly used to quantify the global and also the local value of the density in a foamed food. According to Maire *et al.* (2003), global density does not provide enough information to characterize cellular architecture of bread. For this purpose, density profile or local density variation can be more interesting. These authors investigated the local density variation in bread crumb by determining the histogram of the local density. For each voxel of the volume, they calculated an average gray level of the tomographic images considering the neighboring voxels in a predefined box of size $N \times N \times N$. The choice of the value of N was made according to the characteristics of the architecture of the bread crumb. N was taken larger than the mean cell size and smaller than the size of the volume studied. This approach allows to take into account all the incomplete cells within an image. N value was thus chosen by means of the following equation:

$$N = \frac{-\pi d^3}{6(1-\rho)} \sqrt{\sqrt{1/3}}$$

where d is the mean cell size and ρ is the relative density of bread crumb.

2. *Cell size distribution*

Thanks to the high resolution of the tomographic 3D images, it is possible to accurately calculate the cell size distribution (Lim and Barigou, 2004; Maire *et al.*, 2003). Conventional 2D image analysis allows to get a more approximate evaluation of the cell size distribution than 3D image analysis. To perform this calculation, it is necessary to isolate each cell considered as a cluster of connected voxels separated from the others (i.e., each cell does not have voxel connection with the other cells). Once each cell is isolated, standard morphological information can be obtained by using conventional image analysis techniques. Saltykov area analysis method or other techniques are usually applied to calculate 3D parameters on cross-section tomographic images. This approach gives good results only if a simple shape (spherical or convex) can be assumed for each cell (DeHoff and Rhines, 1972). This technique cannot be satisfactorily used for the analysis of sponge-like materials, such as the bread crumb, (Maire *et al.*, 2003) due to the number of the open gas cells that is greater than that of the closed ones. The same consideration can be made for the cellular-like foods whose resolution image setup is not enough to resolve the cell walls within internal structure or also when the segmentation process is not adequate to preserve the whole portion of the image that is related to the continuous solid phase.

The cell size distribution is very easy to get if the cells are perfectly closed. If the closed cells are connected, the cell size distribution can be determined according to two methods (Maire *et al.*, 2003). The first possibility is to try to close the cells within the digital image using morphological operations such as erosion/dilation. When this activity is done successfully, it is possible to extract some morphological 3D parameters of cells such as volume, surface, aspect ratio, and sphericity. The cell size distribution can be easily obtained since the size is calculated for each cell. The second possibility is to use morphological operations, for example, opening granulometry, to directly retrieve the cell size distribution from the image. This technique can be also applied on 3D images allowing, in this way, the measurement of the size distribution even if the cells are connected. The method is applied by placing virtual spheres having a different radius within digital arrays representing the solid microstructure. Obviously, it works well if the cells are equiaxed. Otherwise, only the smallest dimension of the cell is retrieved. This method is particularly interesting in the case of the bread crumb or of other sponge-like structures for which morphological closing would be completely impossible. Authors determined the normalized volume frequency of the cell size for bread crumb. Opening granulometry technique can be used not only to quantify a cell size distribution but also a wall thickness distribution. This technique, instead, does not supply information on the cell shape.

Another approach to evaluate the cell size distribution is the Saltykov area analysis of the CT images that is based on the stereological measurement of the 2D cell areas (Lim and Barigou, 2004). Authors determined the volume cell size frequency for foods such as foamed chocolate bar, honeycomb chocolate bar, chocolate muffin, marshmallow, and strawberry mousse. A foamed food can be considered as an idealized polydispersed system of suspended spherical or convex particles. A test plane, which cuts the system, will encounter circular sections of different sizes. The basic idea in the Saltykov area analysis is to work backward from this distribution of 2D circular sections to the real spatial size distribution of 3D particles. The circular sections may or may not coincide with the particle diameters, so the source of each section must be determined. For nonidealized particle systems, this method generally assumes that all particles are spherical and that the distribution of particle size can be represented by a discontinuous distribution. The Saltykov stereological technique provides a function based on the probability of random planes intersecting a sphere of diameter D to give sections of diameter d. Provided that the number of classes is not too small, the greatest inaccuracies are expected to arise from nonspherical particles. The assumption of sphericity, however, is usually found to be satisfactory for spheroidal and equiaxed particles.

According to Underwood (1970) and Xu and Pitot (2003), the general Saltykov working formula is the following:

$$(N_v)_j = \frac{1}{D_j}\begin{bmatrix} +1.6461(N_A)_j & -0.4561(N_A)_{j-1} & -0.1162(N_A)_{j-2} \\ -0.0415(N_A)_{j-3} & -0.0173(N_A)_{j-4} & -0.0079(N_A)_{j-5} \\ -0.0038(N_A)_{j-6} & -0.0018(N_A)_{j-7} & -0.0010(N_A)_{j-8} \\ -0.0003(N_A)_{j-9} & -0.0002(N_A)_{j-10} & -0.0002(N_A)_{j-11} \end{bmatrix}$$

where $(N_v)_j$ is the number of particles per unit of sample volume in the class interval *j*; D_j is the actual diameter of the particle in the class interval *j*; $(N_A)_j$ is the number of circular sections in the class interval *j* per unit of cross-section sample area; $(N_A)_{j-1}, (N_A)_{j-2} \cdots (N_A)_{j-11}$ are the number of circular sections in successively smaller size intervals $j-1, j-2, \ldots, j-11$, found per unit of cross-section area. In this formula, *j* has an integer value ≤ 12, that is, the method uses a maximum of 12 class intervals to construct the particle size distribution.

In order to specify the size of a circular section, instead of the absolute area A, the Saltykov method adopts the ratio A/A_{max}, where A_{max} is the maximum circular section area in the whole population. The first step in determining the cell size distribution is to measure all the 2D circular sections, identifying the largest cell section in the total cell section population found on the section planes. This maximum area can be used to divide the size distribution of the cell sections into 12 classes and to determine the total number of sections in each class.

Lim and Barigou (2004) used a scale factor to determine the class intervals for A/A_{max} based on a logarithmic scale of diameters with a factor of $10^{0.1}$ (=0.794). Consequently, a logarithmic scale with a factor $(10^{0.1})^2$(=0.631) is used by authors for the sectional areas. In their study, a total of 1293 cell sections were traced on 41 horizontal slice images with a spacing of 0.189 mm. The maximum section area in the whole population of the cell sections was found to be equal to 3.83 mm^2, giving a maximum section diameter of 2.209 mm. After grouping all the circular sections into 12 class intervals, the Saltykov working formula was used to convert this 2D information into a 3D bubble size distribution. Thus, the number of particles per unit of volume in the largest class interval $j = 1$ was given by

$$(N_v)_j = \frac{1}{D_j}\left[1.6461(N_A)_1\right] = \frac{1}{0.2209}[1.6461(2)] = 15$$

where D_j is expressed in cm, $(N_A)_j$ is expressed as bubble cm^2, and $(N_v)_j$ is expressed as bubble cm^3. For the smaller bubble size intervals, the procedure is similar except that for the subtraction of the number of those sections contributed by the spheres in the previous classes. Therefore, for $j = 2$, the calculation proceeded as follows:

$$\begin{aligned}(N_v)_j &= \frac{1}{D_j}\left[1.6461(N_A)_2 - 0.4561(N_A)_1\right]\\ &= \frac{1}{0.1755}[1.6461(6) - 0.4561(2)]\\ &= 51\end{aligned}$$

The calculations are then carried out for the remaining class intervals until all particle cross-sections have been accounted. It is worth noting that for each class interval, the value of D used in the calculation was that of the largest section in the interval. The accuracy of the stereological analysis was checked calculating the total volume of the air cells in each distribution: the maximum difference observed was about 4%. Authors found three apparent peaks in the 3D cell size distribution relative to the aerated chocolate. The first peak around 0.2 mm represented the population of the smallest bubbles interdispersed between the larger ones, the second peak around 0.4 mm represented the population of medium size cells near the top and bottom regions of the chocolate foam, the third peak around 1.4 mm represented the population of the largest cells contained within the bulky foam structure. Such a cell size distribution was best fitted by a three-parameter gamma distribution according to the following probability density function:

$$f(x) = \frac{(x - \vartheta)^{a-1}\left(e^{-(x-\vartheta)/b}\right)}{\Gamma(a)b^{a}}$$

where $a = 1.0015$ is the shape parameter, $b = 0.5795$ is the scale parameter, $\vartheta = 0.1381$ is the threshold parameter, and Γ is the gamma function. Concerning the honeycomb chocolate sample, the cell size distribution data were best fitted by a lognormal distribution according to the following probability density function:

$$f(x) = \frac{e^{\left[\log\left((x-\vartheta)-\zeta\right)^{2}/2\sigma^{2}\right]}}{(x - \vartheta)\sqrt{2\pi\sigma}}$$

where $\sigma = -1.8284$ is the location parameter, $\zeta = 1.2719$ is the scale parameter, and $\vartheta = 0.2432$ is the threshold parameter.

The cell size distribution in products, such as marshmallow, muffin, and mousse, was also fitted by a lognormal distribution.

3. *Cell wall thickness distribution*

The technique described in the previous section can be easily applied also in the analysis of the solid phase (Maire *et al.*, 2003). Closed cell foams can be generally described with edges and walls. An important morphological parameter is the solid fraction contained in the cell edges "ϕ" that is used in some mathematical models (Gibson and Ashby, 1997). A value of ϕ equal to 1 indicates that the foam presents open cells. In order to measure this parameter, it is very sensitive the resolution used during the image acquisition and also the segmentation performed on the digital volume. Sometimes, when the edges and the walls present two different values of thickness, a bimodal distribution is observed: the distribution showing the lowest values corresponds to the cell wall thickness distribution, whereas that showing the highest values corresponds to the edges thickness distribution. The case in which the deconvolution is feasible, the value of ϕ corresponds to the integral of the second peak. The latter represents the edges divided by the integral of the total curve.

4. Degree of anisotropy

Anisotropy primarily derives from the spatial arrangements of the structure elements or particulates and manifests itself in the directional dependence of both mechanical and transport properties. Anisotropy degree is a measure of the 3D structural symmetry, indicating the presence or absence of a preferential alignment of the structure elements along a particular direction. The anisotropy quantification primarily has importance in the modeling of the mechanical behavior of a cellular or porous structure (Odgaard, 1997). For many porous materials, the structural anisotropy is quantified by a "fabric" measure (Cowin, 1985; Kanatani, 1984) that represents the preferential orientations of the cell walls and the degree of dispersion of the solid phase around the main directions. The analysis of the anisotropy consists of a numerical approach based on the stereological calculus and also on the anisotropy tensor eigenanalysis of direction-dependent structural descriptors such as the mean intercept length (Harringan and Mann, 1984; Whitehouse, 1974), the star length distribution, and the star volume distribution (SVD) or areal pore size (Inglis and Pietruszczak, 2003).

Computerized algorithms were developed in order to perform the stereological calculation on natural materials such as soil, trabecular bone (Kuo and Carter, 1991; Odgaard, 1997; Odgaard *et al.*, 1997; Pietruszczak *et al.*, 1999; Whitehouse, 1974), and manufactured fabric materials such as sedimentary rocks and reinforced composites (Kanatani, 1985; Pietruszczak and Mroz, 2001). The basic idea is that, for each considered angle, many parallel lines are sent in order to cover the whole test volume and the mean intercept length is calculated as the average for all these lines. In the case of the isotropic objects, lines traversing the object at any angle will pass through a similar length of the solid phase. Anisotropy tensor eigenanalysis is then used to extract some numerical parameters that define orientation and anisotropy of these mean intercept length distributions. In the case of the orthotropic materials, a mathematical relationship exists between morphological features (volume fraction and fabric measures) and elastic properties (Young's modulus, shear modulus, and Poisson ratios) (Cowin, 1985). This is a general finding based on the experimental evidence that the anisotropy tensor main orientations correspond to the mechanical main orientations of these materials.

The bread crumb is an anisotropic material, as suggested by the fabric tensor value (Falcone *et al.*, 2004b). Authors proved that the volume fraction (in terms of porosity or density), which is a first measure of the bread architecture, cannot predict the compressive elastic modulus and a measure of the "structural anisotropy" must be performed. They used a CT scanner able to obtain high-contrast tomographic images with a nominal resolution of 14 μm along each edge of the voxels in the 3D arrays. The high performance of the CT system used allowed to successfully resolve the whole 3D map of the solid phase of the bread crumb. Authors used a numerical procedure able to perform an accurate characterization of the internal microarchitecture of the bread crumb (Falcone *et al.* 2004b). They obtained some direction-dependent measurements (basic quantities) by probing segmented 3D images with an appropriate volumetric array (3D testing sphere) of parallel test lines. The obtained directional data were used to derive a specific anisotropy descriptor named "mean intercept length vectors" (MILv). An ellipsoid was used as 3D distribution function for the visualization and interpretation of the MILv spatial distribution. A multivariable linear least square fitting technique was used to fit the ellipsoid function to the MILv data. Finally, a MIL tensor (or anisotropy tensor) based on the ellipsoidal coefficients was defined and calculated by means of the eigenanalysis. Eigenvalues and eigenvectors of the anisotropy tensor were used to obtain the summary numerical parameters (derivative quantities) defining orientation and anisotropy of the bread crumb. These numerical parameters are related to both anisotropy and favorite orientations of the cell walls. The anisotropy tensor eigenanalysis developed by Harringan and Mann (1984) leads to the following primary definition of the anisotropy degree (*DA*):

$$DA = 1 - \frac{\text{min eigenvalue}}{\text{max eigenvalue}}$$

A value of 0 corresponds to the total isotropy, whereas a value of 1 indicates the total anisotropy.

For some foods having a cellular structure, such as foamed chocolate bar, honeycomb chocolate bar, chocolate muffin, marshmallow, and strawberry mousse (Lim and Barigou, 2004) and for bread crumb (Falcone *et al.*, 2004b), the degree of anisotropy was directly determined from the 3D images. Results show that cellular food products have a concave cellular structure and a good degree of isotropy, whereas bread crumb has highly connected cells and preferential orientations of cell walls that demonstrate the orthotropic nature of this food product. Falcone *et al.* (2004b) proved that even if dissimilarities in the bread crumb macrostructure cannot be evident to naked eye, the microarchitecture has a crucial effect on the global mechanical behavior. In fact, two samples obtained from the same slice of a bread loaf showed different shapes of the stress–strain curve.

Falcone *et al.* (2004b) determined the structural descriptors applying an algorithm able to scan the 3D digital arrays by means of a 3D version of the directed secant method (Saltykov, 1958). They extracted parameters, such as local density (or porosity), anisotropy degree, and favorite orientation of the cell walls, from a 3D tomographic image. This approach is the adaptation of the "parallel plate model" proposed by Harringan and Mann (1984) for other porous materials. According to this model, the solid phase is made of cell walls that interconnect pores; the cell walls have parallel sides (some of them are open and others are closed) and the solid phase is unevenly distributed in both the faces and edges. The algorithm used includes the following six routines.

a. 1° routine—definition of the testing volume and of the unbiased test probe. Numerical procedure begins with a specification of the size of the image array and the center coordinate of the sampling domain, a testing sphere within digital data. Next, a set of points is defined on the equatorial plane of the testing sphere so that they generate a 3D array of isotropically spaced parallel test lines (in the following called 3D lineal probe or 3D test grid). A straight line (ω_i) normal to the equatorial plane defines the sampling direction within the digital volume. Each 3D image is scanned by rotating the 3D linear probe in a lot of randomly prescribed sampling directions (ω_i). The hemispherical orientations ω_i in the coordinate system with axes k_1, k_2, and k_3 are defined by the polar coordinate (θ, φ) or by the directional cosines (x, y, z). Using polar coordinates, the colatitude θ is the angle between ω_i-direction and k_3-axis and the longitude φ is the angle anticlockwise measured between the i-axis and the projection of the ω_i-direction onto the k_1–k_2 plane.

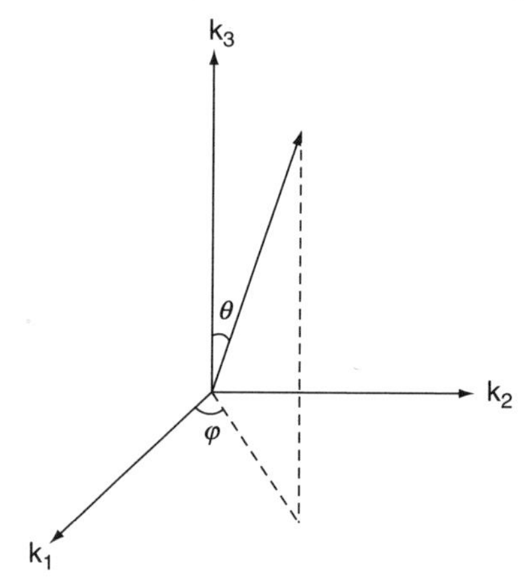

b. 2° routine-image segmentation. Based on the threshold value, the image is converted into a binary format and two basic quantities, that is, the number of voxels enclosed in the solid phase (having gray values smaller than the threshold value) and the number of voxels enclosed in the void phase (having gray values higher than the threshold one) were counted. Then the solid volume fraction δ (or density) was computed with the following formula:

$$\delta = \frac{S_v}{T_v}$$

where S_v is the number of voxels corresponding to the crumb within the testing sphere and T_v is the total number of voxels within the same volume. Porosity P (or void volume fraction) is derived as $P = 1 - \delta$.

Porosity or density are used as global structural descriptors.

c. 3° routine-voxel identification and intercept length measurement. By setting the starting position of a line segment on the boundary of the sampling domain, each test line within the testing sphere is systematically and incrementally scanned until the end point on the boundary is reached in order to determine the voxel position and the solid or void phase classification as a function of the prescribed sampling directions (ω_i). A "two-voxels method"-based subroutine is performed as a voxel identification procedure according to Simmons and Hipp (1997). Once the scan is complete, three basic directionally dependent quantities are computed. They include $\mathbf{N(\omega_i)}$ (the number of intercept lengths), $\mathbf{\Sigma I(\omega_i)}$ (the sum of intercept lengths arising from the intersection of the test lines and the phase of interest), and $\mathbf{L(\omega_i)}$ (the length of test lines intersecting the sampling volume). Since the sampling domain has a spherical volume, $\mathbf{L(\omega_i)}$ is constant for all the sampling directions and is computed as twice the sum of all the perpendicular distances from the equatorial plane to the edge of the 3D testing sphere. $\mathbf{N(\omega_i)}$ is an *orientation-dependent* parameter recorded along a test line when the binary value of the current voxel differs from the binary value of the previous voxel. In this stereological procedure "intercept length" was defined as an isolate line segment arising from the intersection of a line test with the boundary of the phase of interest (solid or void) and lying in this phase. The automatic storage process and intercept identification is run for all test lines under a given number of sampling directions ($\omega_I = 128$).

Based on these three basic quantities, the mean intercept length vector $\mathbf{MIL(\omega_i)}$ is defined as an anisotropy descriptor according to Harringan and Mann (1984). This parameter indicates the dispersion of the intercept length around the sampling direction ω_I and is conceptually defined as the total line length divided by the number of intersections along the ω_i direction:

$$MIL(\omega_i) = \frac{L(\omega_i)}{N(\omega_i)}$$

Generally, **MIL(ω)** can be related to the solid phase as well as to the void phase according to Whitehouse (1974):

$$MIL_p(\theta, \varphi) = V_v(\omega_i) \times 2 \times MIL(\omega_i)$$

where p is the phase of interest, solid (s) or void (v), and $V_v(\omega_i)$ is the lineal fraction of p in the direction ω_i. The reason for the factor 2 is that $[2 \times MIL]$ is the mean length of a solid + void intercept.

In the numerical procedure proposed by Falcone *et al.* (2004b), the mean intercept length is related to the solid phase and $MIL_s(\omega_i)$ and is expressed in polar coordinate as the following:

$$MIL_s(\theta, \varphi) = \delta(\omega_i) \times 2 \times MIL(\omega_i)$$

where $\delta(\omega_i)$ is the lineal fraction of the solid phase in the sampling direction ω_i.

Another way to calculate the solid-related $\mathbf{MIL(\omega_i)}$ vectors is the following:

$$MIL_s(\theta, \varphi) = 2^*[1 - P(\omega_i)] \times \frac{L(\omega_i)}{N(\omega_i)}$$

where $P(\omega_i)$ is the lineal fraction of the void phase in the sampling direction ω_i. Mean intercept length in the solid phase gives information about the distances between solid-void interfaces and the width of the cell walls as a function of the orientation and therefore of the 3D anisotropy. Since the $P(\omega_i)$ is an orientation-independent quantity, the effect of the anisotropy is only contained in the second term. The presence of $P(\omega_i)$ in the above-mentioned formula takes into account the fact that the mechanical properties of foamed materials apart from cellular anisotropy are strictly dependent on their porosity.

d. 4° routine—3D fitting of the MIL data distribution. The next step in the anisotropy analysis is the mathematical characterization of the directional data distribution, that is, the solid-related $\mathbf{MIL(\theta, \varphi)}$ vectors in the 3D space. This method is based on the 2D method of Whitehouse (1974) and Harringan and Mann (1984). It consists of the plot of the positions of the end points of the MIL vectors issuing from a common centre as a function of their spherical angles. In the 3D space, this creates a pincushion-like effect with lines going in all the directions at different lengths. The distribution function that better describes the 3D shape of the directional MIL data is represented by the equation of an ellipsoid:

$$A \cdot n_1^2 + B \cdot n_2^2 + C \cdot n_3^2 + D \cdot n_1 \cdot n_2 + E \cdot n_1 \cdot n_3 + F \cdot n_2 \cdot n_3 = \frac{1}{\mathrm{MIL}^2}$$

where MIL is the magnitude of the **MIL**(θ, φ) vectors; n_i are the direction cosines between **MIL(θ, φ)** vectors and the base vectors in an arbitrary coordinate system with k_1, k_2, and k_3 axis; $A \ldots F$ are the ellipsoid coefficients. For the purpose of fitting the ellipsoid equation to the MIL data, a multivariable linear least square fitting technique is performed by solving the following linear system:

$$A_{\omega_i} \cdot \vec{x}(6) = \vec{b}_{\omega_i}$$

where A is the matrix containing the six projection data n_i, that is, the direction cosines (x, y, z), of the solid-related directional MIL(θ, φ) data for each orientation ω_I; $\vec{x}(6)$ is the six-order column vector for the corresponding six ellipsoid coefficients and $\vec{b}$ is the column vector of $1/\mathrm{MIL}^2$.

For each sampling direction ω_i, the projection data x, y, and z of each MIL(θ, φ) vector are calculated by means of the following formulas:

$$\omega_i \begin{cases} x = \sin(\theta) \times \cos(\varphi) \\ y = \sin(\theta) \times \sin(\varphi) \\ z = \cos(\varphi) \end{cases}$$

then A matrix is defined as the following:

$$\begin{aligned} A_{\omega_i}(1) &= x \cdot x \\ A_{\omega_i}(2) &= y \cdot y \\ A_{\omega_i}(3) &= z \cdot z \\ A_{\omega_i}(4) &= x \cdot y \\ A_{\omega_i}(5) &= y \cdot z \\ A_{\omega_i}(6) &= x \cdot z \end{aligned}$$

and finally, $\vec{b}$ is calculated as the following:

$$b_{\omega_i} = \frac{1}{[MIL(\omega_i) \cdot MIL(\omega_i)]}$$

The solution is:

$$\vec{x}(6) = (At \times A)^{-1} \times At \times \vec{b}$$

where At is the transpose matrix of the A matrix and $\vec{x}(6)$ provides the six ellipsoid coefficients as external variables:

$$\vec{x}(1) = A$$
$$\vec{x}(2) = B$$
$$\vec{x}(3) = C$$
$$\vec{x}(4) = D$$
$$\vec{x}(5) = E$$
$$\vec{x}(6) = F$$

The goodness of the fit is evaluated on the base of the sum of the squares of the residuals and the correlation coefficient. A great number of randomly sampled orientations were required to obtain a set of fabric descriptors that could be subjected to the application of a least squares fitting procedure. In particular, Falcone *et al.* (2004b) worked by setting 200 μ as $\tau\eta\varepsilon$ test line spacing and 128 random sampling directions for the 3D test grid so that $0 \leq \varphi \leq 2\pi$ and $0 \leq \sin(\theta) \leq 1$.

e. 5° routine-anisotropy tensor eigenanalysis. The MIL ellipsoid is related to the average distribution of the samples in the space. This ellipsoid is characterized by three axes, which are orthogonal at right angles to each other. These axes, named prinMIL_1, prinMIL_2, and prinMIL_3, respectively, describe the longest orientation and the length and width (major and minor axes) of the ellipse section at right angles to the longest orientation. When the test lines are oriented along the principal favorite orientation of the cell walls, the magnitude of the MIL is maximum. When the test lines are perpendicularly oriented to this direction, the magnitude of the MIL is minimum. The algorithm performs the descriptive statistics for the MIL vectors by computing the mean MIL value and the standard deviation and the descriptive statistics for the ellipsoid coefficients by computing the goodness of fit.

Several numerical parameters defining the orientation and the anisotropy degree of the directional *MIL* distribution can be computed by solving a second-rank tensor according to Harringan and Mann (1984). The second-rank tensor is usually named "fabric tensor" since it gives information about the structural symmetry and the orientation of the fabric structure. In particular, by assuming the material orthotropy, the above ellipsoid equation can be equivalently obtained as the inner product of a second-rank tensor A_{ij} with two vectors x_i and x_j as the following:

$$x_i \cdot A_{ij} \cdot x_j = 1$$

or

$$\{x_1 x_2 x_3\} \cdot \begin{Bmatrix} A_{11} & A_{12} & A_{13} \\ A_{21} & A_{22} & A_{23} \\ A_{31} & A_{32} & A_{33} \end{Bmatrix} \cdot \begin{Bmatrix} x_1 \\ x_2 \\ x_3 \end{Bmatrix} = 1$$

where the A coefficients derive from the ellipsoid fit and define the MIL tensor and x_i are the direction cosines of the mean intercept lengths in the base coordinate system with axes k_1, k_2, and k_3. In particular, the MIL tensor is defined as the following:

$$A_{ij} = \begin{vmatrix} A & D/2 & C/2 \\ D/2 & B & F/2 \\ E/2 & F/2 & C \end{vmatrix}$$

The eigenanalysis of the MIL tensor is run via Jacobi method to calculate the main characteristics values, that is, *eigenvalues* ($\hat{e}_{0-8}$), and characteristic directions, that is, *eigenvectors* ($\vec{e}_{0-8}$).

The three main eigenvectors are associated with the main orientation of the MIL ellipsoid axes, giving information about the favorite directions of the porous structure in the space. On the other hand, the three major eigenvalues are functions of the amount of solid phase (oriented intercept lengths) along each of three main eigenvectors. As a consequence, the eigenvalues are related to the anisotropy degree, whereas the eigenvectors are related to the favorite orientation of the considered porous structure.

A common way to evaluate the anisotropy of porous structures requires that the three *principal* **MILs(θ, φ)** *vectors* are derived from the 3D distribution of the directional MILs**(θ, φ)** *data*. The inverse of the square root of the absolute value of the three main eigenvalues (A, B, and C) is defined as the magnitude of the *principal* **MIL(θ, φ)** *vectors* and supplies information about the dispersion degree of the solid phase around the *principal* **MIL(θ, φ)** *vectors*.

A single fabric measure is not sufficient for the distinction of the anisotropy classes. As a consequence, several asymmetry measures are evaluated by algebrically combining the normalized eigenvalues ($\hat{\tau}_i = \hat{e}_i/\Sigma(\hat{e}_i), \hat{\tau}_1 + \hat{\tau}_2 + \hat{\tau}_3 = 1$) or the principal MILs values.

An immediate visual interpretation of the symmetry degree via the relative magnitudes of the normalized eigenvalues may be provided through the geometric visualization of the MIL tensor (Westin *et al.*, 1999), since a second-order tensor may be represented as a composite object, which is a combination of fabric shapes such as a line, disk, and sphere. The line segment describes the major principal direction of the tensor and its length is proportional to the largest eigenvalue ($\hat{\tau}_1$). The disk describes the plane spanned by the eigenvectors that corresponds to the two largest eigenvalues ($\hat{\tau}_1$ and $\hat{\tau}_2$).

The sphere has the radius proportional to the smallest eigenvalue ($\hat{\tau}_3$). In the following is shown the above mentioned scheme:

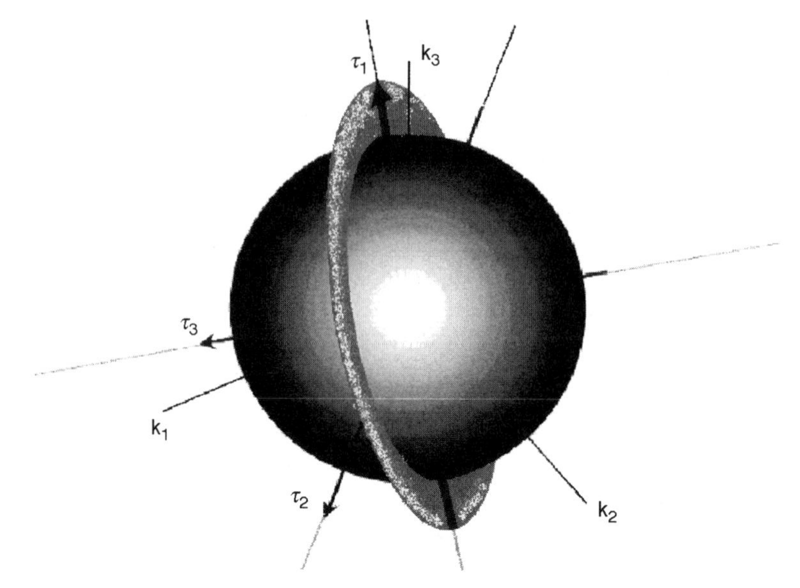

Here, the three component objects are plotted as an ellipsoid with semi-axes scaled to a maximum of 1. Thus, the length of the line segment is 1, the radius of the disk is $\hat{\tau}_2/\hat{\tau}_1$, whereas the radius of the sphere is $\hat{\tau}_3/\hat{\tau}_1$ (with $|\hat{\tau}_1| \geq |\hat{\tau}_2| \geq |\hat{\tau}_3|$).

The main orientations of the porous structure are determined by calculating both the polar coordinates, colatitude (θ) and longitude (φ), for the three main eigenvector directions with respect to the chosen coordinate system with axes k_1, k_2, and k_3:

$$O_{\tau_1}\{\vartheta\} = \tan^{-1}\left[\frac{(\vec{\tau}_6)}{\sqrt{(\vec{\tau}_0)^2 + (\vec{\tau}_3)^2}}\right]$$

$$O_{\tau_1}\{\varphi\} = \left\{\tan^{-1}\left[\frac{(\vec{\tau}_3)}{(\vec{\tau}_0)}\right]\right\}$$

$$O_{\tau_2}\{\vartheta\} = \tan^{-1}\left[\frac{(\vec{\tau}_7)}{\sqrt{(\vec{\tau}_1)^2 + (\vec{\tau}_4)^2}}\right]$$

$$O_{\tau_2}\{\varphi\} = \left\{\tan^{-1}\left[\frac{(\vec{\tau}_4)}{(\vec{\tau}_1)}\right]\right\}$$

$$O_{\tau_3}\{\vartheta\} = \tan^{-1}\left[\frac{(\vec{\tau}_8)}{\sqrt{(\vec{\tau}_2)^2 + (\vec{\tau}_5)^2}}\right]$$

$$O_{\tau_3}\{\varphi\} = \left\{\tan^{-1}\left[\frac{(\vec{\tau}_5)}{(\vec{\tau}_2)}\right]\right\}$$

The statistical independence of the eigenvalues at a 0.05 confidence level (α) is also tested. In this way, it is possible to distinguish among *orthotropic* structure (in which all the three eigenvalues differ each other), *transverse isotropic* structure (with two eigenvalues statistically equivalent), and *isotropic* structure (where all the three eigenvalues are equivalent).

Tables I and II report the morphological indices and the anisotropy measures obtained by Falcone *et al.* (2004b), respectively, from the stereological analysis and the eigenanalysis of the MIL tensor in order to characterize the inner crumb microstructure. By using these indices, it is possible to compare with high sensitivity the spatial arrangement of the cell walls in bread samples.

Further researches on the microstructure characterization from 3D tomographic images could be used in assessing the effect of the tomography scanning

TABLE I

ALGEBRIC EXPRESSIONS OF THE MORPHOLOGICAL FEATURES OF BREAD CRUMB SAMPLES AS OBTAINED BY THE 3D STEREOLOGICAL CALCULATION

	Symbol	Unit	Equation
Morphological descriptors	P	(*)	$P = 1 - \frac{S_v}{T_v}$
	W_N	mm^{-1}	$\frac{N(\omega_i)}{L}$
	W_Th	mm	$\delta \times \frac{L}{N(\omega_i)}$
	W_Sp	mm	$1 - \delta \times \frac{N(\omega_i)}{L}$
	SS	mm^2/mm^3	$\frac{2}{\delta} \times \frac{N(\omega_i)}{L}$

(*) nondimensional parameter.

TABLE II

ALGEBRAIC EXPRESSIONS OF THE ANISOTROPY MEASURES OF THE MIL TENSOR

Anisotropy degrees	$\text{PrinMIL}_{1,2,3}$	(*)	$\frac{1}{\sqrt{\lvert\hat{\tau}_i\rvert}}$
	Is_Ix	(*)	$\frac{\hat{\tau}_3}{\hat{\tau}_1}$
	El_Ix	(*)	$1-\frac{\hat{\tau}_2}{\hat{\tau}_1}$
	DA	(*)	$\frac{\hat{\tau}_1}{\hat{\tau}_3}$
	DA1	(*)	$\frac{\text{prinMIL}_1 - \text{prinMIL}_\text{m}}{\text{prinMIL}_\text{m}}$
	DA2	(*)	$\frac{\text{prinMIL}_2 - \text{prinMIL}_\text{m}}{\text{prinMIL}_\text{m}}$
	DA3	(*)	$\frac{\text{prinMIL}_3 - \text{prinMIL}_\text{m}}{\text{prinMIL}_\text{m}}$
	FA	(*)	$\sqrt{\frac{3}{2}}\cdot\frac{\sqrt{(\hat{\tau}_1-\hat{\tau}_m)^2+(\hat{\tau}_2-\tau_m)^2+(\hat{\tau}_3-\hat{\tau}_m)^2}}{\sqrt{\hat{\tau}_1^2+\hat{\tau}_2^2+\hat{\tau}_3^2}}$

resolution and the voxel size reconstructing resolution on the accuracy of the stereological measurements in foamed, sponge-like, and composite foods. A long-term perspective of the stereological analysis in food evaluation is the large-scale finite element analysis of the reconstructed 3D microstructure in order to evaluate or predict the material mechanical behavior.

5. *Index of connectivity*

Connectivity may be defined as the maximal number of particles (branches, cell walls, and so on) that may be cut without separating the structure. For several porous foods, such as foamed chocolate bar, honeycomb chocolate bar, chocolate muffin, marshmallow, and strawberry mousse, air cell connectivity was defined as a measure of the relative convexity or concavity of the total slid surface (Lim and Barigou, 2004). Concavity indicates connectivity, whereas convexity indicates isolated disconnected structures. Through the image analysis, authors compare the solid area and perimeter before and after an image dilation operation and calculate the index of connectivity as the following:

$$\text{Connectivity} = \frac{P_1 - P_2}{A_1 - A_2}$$

where P and A are solid area and perimeter, respectively, and subscripts 1 and 2 refer, respectively, to values measured before and after image dilation operation. Where solid connectedness results in enclosed air cell spaces, the dilation of solid surfaces will contract the perimeter. By contrast, open ends will have their perimeters expanded by surface dilation. As a consequence, a low index of connectivity indicates better-connected solid lattices, whereas a high value indicates a more disconnected solid structure. A concave structure will have a negative connectivity value, whereas a convex structure will have a positive index value.

Another method, applied to evaluate the connectivity of two-phase materials, is based on the Euler–Poincaré formula that establishes for a single solid, independently on its complexity, the following rule:

$$0.5 \times (V - E + F) + G = 1$$

where V, E, F, and G are, respectively, the number of the vertices, edges, faces, and genus that compose the surface mesh of that single solid (De Oliveira *et al.*, 2003). The genus represents the number of the handles present at the solid. In the Poincaré formula, the term $0.5 \times (V - E + F)$ is called Euler–Poincaré (EP) number that represents the key of all the determination of connectivity. This parameter is related to the connectivity and number of the connected components (particles, cells, and so on), furthermore, the connectivity is identical for the two phases. Connectivity is a scalar number and is not dependent on the orientation of the structure.

The Euler number may easily be calculated from the voxel-based data set of a 3D reconstruction. For example, the test object is fragmented into voxels, each of them representing a single solid itself. It is possible to compute the 3D morphological parameters of these solids by taking into account the following basic rule:

$$EP(A \cup B) = EP(A) + EP(B) - EP(A \cap B)$$

according to which the EP number of the union of the two sets is equal to the sum of their individual numbers minus the number of the intersections between these sets. The case in which a voxel is homeomorphic to the sphere, its EP number is 1 since their genus value is 0. When two voxels are put together side-by-side forming a new solid, it is necessary to subtract one face, that is, the contact face or interface. Then, the EP number of the new solid is 2–1 (2 of the two voxels and 1 of the one interface). The result is 1 that is in accordance with the EP number of the new solid (till homeomorphic to sphere, genus 0). Adding more and more voxels and subtracting the number of the interfaces, it is possible to calculate the EP number of the total solid. It is important to put in evidence that if a torus is formed, the EP number returns to 0 because the genus of a torus is 1 and the sum of the EP number with the genus value has to match 1, established by the Euler–Poincaré formula. A general consequence is that when the number of the handles increase in a solid object, their *EP* number becomes more and more negative.

IV. SUMMARY

Sensorial and mechanical characteristics of food products are strongly affected by the food structural organization that can be studied at molecular, microscopic, and macroscopic levels.

Since the microstructure affects food sensorial properties, foods having a similar microstructure also have a similar behavior. Studies on food microstructure can be performed by means of a large variety of techniques allowing the generation of data in the form of images. With the development of more powerful tools, such as the X-ray computed tomographic scanners, both 2D and 3D digital images of the food internal structure can be readily acquired with high resolution and contrast and without any sample preparation. These images can be processed by means of the fractal and stereological analysis in order to quantify a number of structural elements. Fractal analysis allows the investigation of the fractal geometry in both 2D and 3D digital images. Stereology, instead, allows the obtainment of 3D features from 2D images. These data-processing techniques represent new promising approaches to a full characterization of complex internal structures. Such advances in food evaluation open new horizons for the development of mathematical and computational models able to individuate the interactions between product microstructure and their mechanical properties.

REFERENCES

Aguilera, J.M. 2000. Microstructure and food product engineering. *Food Technol.* **54**(11), 56–65.

Aguilera, J.M. and Stanley, D.W. 1999. "Microstructural Principles of Food Processing and Engineering", 2nd Ed. Aspen Publisher, Inc., Gaitherburg, MD.

Allan-Wojtas, P.M., Forney, C.F., Carbyn, S.E., and Nicholas, K.U.K.G. 2001. Microstructural indicators of quality-related characteristics of blueberries—an integrated approach. *Lebensm.-Wiss. u.-Technol.* **34**, 23–32.

Autio, K. and Laurikainen, T. 1997. Relationships between flour/dough microstructure and dough handling and baking properties. *Rev. Trends Food Sci. Technol.* **8**, 181–185.

Autio, K. and Salmenkallio-Marttila, M. 2001. Light microscopic investigations of cereal grains, doughs and breads. *Lebensm.-Wiss. u.-Technol.* **34**, 18–22.

Bache, I.C. and Donald, A.M. 1998. The structure of the gluten network in dough: A study using environmental scanning electron microscopy. *J. Cereal Sci.* **28**, 127–133.

Barcelon, E.G., Tojo, S., and Watanabe, K. 1999. X-ray computed tomography for internal quality evaluation of peaches. *J. Agric. Eng. Res.* **73**, 323–330.

Basaran, T.K., Demetriades, K., and McClements, D.J. 1998. Ultrasonic imaging of gravitational separation in emulsions. *Physicochem. Eng. Aspects* **136**, 169–181.

Basset, O., Buquet, B., Abouelkaram, S., Delachartre, P., and Culioli, J. 2000. Application of texture image analysis for the classification of bovine meat. *Food Chem.* **69**, 437–445.

Benedito, J., Carcel, J.A., Sanjuan, N., and Mulet, A. 2000. Use of ultrasound to assess cheddar cheese characteristics. *Ultrasonics* **38**, 727–730.

Benn, D.I. 1994. Fabric shape and the interpretation of sedimentary fabric data. *J. Sediment. Res.* **A64** (4), 910–915.

Bertram, H.C., Whittaker, A.K., Shorthose, W.R., Andersen, H.J., and Kalsson, A.H. 2004. Water characteristics in cooked beef as influenced by ageing and high-pressure treatment—an NMR micro imaging study. *Meat Sci.* **66**, 301–306.

Binning, G., Quate, C.F., and Gerber, C.H. 1986. Atomic force microscopy. *Phys. Rev. Lett.* **56**, 930–933.

Bonny, J.M., Rouille, J., Della Valle, G., Devaux, M.F., Douliez, J.P., and Renou, J.P. 2004. Dynamic magnetic resonance microscopy of flour dough fermentation. *Magn. Reson. Imaging* **22**, 395–401.

Bows, J.R., Patrick, M.L., Nott, K.P., and Hall, L.D. 2001. Three-dimensional MRI mapping of minimum temperatures achieved in microwave and conventional food processing. *Int. J. Food Sci. Technol.* **36**, 243–252.

Brecht, J.K., Shewfelt, R.T., Garner, J.C., and Tollner, E.W. 1991. Using X-ray computed tomography to nondestructively determine maturity in green tomatoes. *Hort. Science* **26**, 45–47.

Brunello, N., McGauley, S.E., and Marangoni, A. 2003. Mechanical properties of cocoa butter in relation to its crystallization behavior and microstructure. *Lebensm.-Wiss. u.-Technol.* **36**, 525–532.

Buffa, M.N., Trujillo, A.J., Pavia, M., and Guamis, B. 2001. Changes in textural, microstructural, and colour characteristics during ripening of cheeses made from raw, pasteurized or high-pressure-treated goats' milk. *Int. Dairy J.* **11**, 927–934.

Burch, S.F. and Lawrence, P.F. 1992. Recent advances in computerized X-ray tomography using real-time radiography equipment. *Brit. J. NDT* **34**, 129–133.

Busk, H., Olsen, E.V., and Brondum, J. 1999. Determination of lean meat in pig carcasses with the Autofom classification system. *Meat Sci.* **52**, 307–314.

Chanamai, R. and McClements, D.J. 1999. Ultrasonic determination of chicken composition. *J. Agri. Food Chem.* **47**(11), 4686–4692.

Chi-Fishman, G., Hicks, J.E., Cintas, H.M., Sonies, B., and Gerber, L.H. 2004. Ultrasound imaging distinguishes between normal and weak muscle. *Arch. Phys. Med. Rehabil.* **85**, 980–986.

Chow, R., Blindt, R., Kamp, A., Grocutt, P., and Chivers, R. 2004. The microscopic visualization of the sonocrystallization of ice using a novel ultrasonic cold stage. *Ultrason. Sonochem.* **11**, 245–250.

Cloetens, P., Barrett, R., Baruchel, J., Guigay, J.P., and Schlenker, M. 1996. Phase objects in synchrotron radiation hard X-ray imaging. *J. Phys. D Appl. Phys.* **29**, 133–146.

Coupland, J.N. 2004. Low intensity ultrasound. *Food Res. Int.* **37**, 537–543.

Coupland, J.N. and McClements, D.J. 2001. Droplet size determination in food emulsions: Comparison of ultrasonic and light scattering methods. *J. Food Eng.* **50**(2), 117–120.

Cowin, S.C. 1985. The relationship between the elasticity tensor and the fabric tensor. *Mech. Mater.* **4**, 137–147.

Cremer, D.R. and Kaletunç, G. 2003. Fourier transform infrared microspectroscopic study of the chemical microstructure of corn and oat flour-based extrudates. *Carbohydr. Polym.* **52**, 53–65.

Crippen, K.L., Hamann, D.D, and Young, C.T. 1989. Effects of grind size, sucrose concentration and salt concentration on peanut butter texture. *J. Texture Stud.* **20**, 29–41.

Crochon, M., Guizard, C., Steinmetz, V., and Bellon, V. 1994. Walnut sorting according to their internal quality: Primary results. International Conference on Agricultural Engineering, Aug. 29–Sept. 1, Milano **2**, 901.

de Campo, L., Yaghmur, A., Garti, N., Leser, M.E., Folmer, B., and Glatter, O. 2004. Five-component food-grade microemulsions: Structural characterization by SANS. *J. Colloid Interface Sci.* **274**, 251–267.

De Oliveira, L.F., Lopes, R.T., de Jesus, E.F.O., and Braz, D. 2003. 3D X-ray tomography to evaluate volumetric objects. *Nucl. Instrum. Methods Phys. Res.* **505**, 573–576.

DeHoff, R.T. and Rhines, F.N. 1972. "Microscopie Quantitative". Masson et Cie eds, p. 404, Paris.

Del Nobile, M.A., Martoriello, T., Mocci, G., and La Notte, E. 2003. Modeling the starch retrogradation kinetic of durum wheat bread. *J. Food Eng.* **59**(2/3), 123–128.

Denison, C., Carslon, W.D., and Ketcham, R.A. 1997. Three-dimensional quantitative textural analysis of metamorphic rocks using high-resolution computed tomography: Part I. Methods and techniques. *J. Metamorphic Geol.* **15**, 29–44.

Diener, R.G., Mitchell, J.P., and Rhoten, M.L. 1970. Using an X-ray image scan to sort. *Bruised Apples Agri. Engr.* **51**(6), 356–361.

Ding, K. and Gunasekaran, S. 1998. Three-dimensional image reconstruction procedure for food afficrostructure evaluation. *Artif. Intell. Rev.* **12**, 245–262.

Du, C.J. and Sun, D.W. 2004. Recent developments in the applications of image processing techniques for food quality evaluation. *Trends Food Sci. Technol.* **15**, 230–249.

Duvauchelle, P., Freud, N., Kaftandjian, V., and Babot, D. 2000. A computer code to simulate X-ray imaging techniques. *Nucl. Instrum. Methods Phys. Res.* **170**, 245–258.

Elmehdi, H.M., Page, J.H., and Scanlon, M.G. 2003. Monitoring dough fermentation using acoustic waves. *Trans. Inst. Chem. Eng. Part C* **81**, 217–223.

Elofsson, C., Dejmek, P., Paulsson, M., and Burling, H. 1997. Atomic force microscopy studies on whey proteins. *Int. Dairy J.* **7**, 813–819.

Engelsen, S.B., Jensen, M.K., Pedersen, H.T., Nørgaard, L., and Munck, L. 2001. NMR-baking and multivariate prediction of instrumental texture parameters in bread. *J. Cereal Sci.* **33**, 59–69.

Evans, R.D. 1955. "The Atomic Nucleus". McGraw-Hill, New York, PA.

Falcone, P.M., Baiano, A., Zanini, F., Mancini, L., Tromba, G., Montanari, F., and Del Nobile, M. A. 2004a. A novel approach to the study of bread porous structure: Phase-contrast X-ray microtomography. *J. Food Sci.* **69**(1), 38–43.

Falcone, P.M., Baiano, A., Zanini, F., Mancini, L., Tromba, G., Dreossi, D., Montanari, F., Scuor, N., and Del Nobile, M.A. 2004b. Three-dimensional quantitative analysis of bread crumb by X-ray microtomography. *J. Food Sci.* (in press).

Fardet, A., Baldwin, P.M., Bertrand, D., Bouchet, B., Gallant, D.J., and Barry, J.L. 1998. Textural image analysis of pasta protein networks to determine influence of technological processes. *Cereal Chem.* **75**(5), 699–704.

Feldkamp, L.A., Davis, L.C., and Kress, I.W. 1984. Practical cone-beam algorithm. *Opt. Soc. Am.* **1**, 612–619.

Fitzgerald, R. 2000. Phase-sensitive X-ray imaging. *Phys. Today* **53**(7), 23–26.

Fortin, A., Tong, A.K.W., Robertson, W.M., Zawadski, S.M., Landry, S.J., Robinson, D.J., Liu, T., and Mockford, R.J. 2003. A novel approach to grading pork carcasses: Computer vision and ultrasound. *Meat Sci.* **63**, 451–462.

Gibson, L.J. and Ashby, M.F. 1997. "Cellular Solids: Structure and Properties", 2nd Ed. Cambridge University Press, Cambridge, UK.

Goodrum, J.W. and Elster, R.T. 1992. Machine vision for cracks detection in rotating eggs. *Trans. ASAE* **39**, 2319–2324.

Grenier, A., Lucas, T., Collewet, G., and Le Bail, A. 2003. Assessment by MRI of local porosity in dough during proving. Theoretical considerations and experimental validation using a spin-echo sequence. *Magn. Reson. Imaging* **21**, 1071–1086.

Gunasekaran, S. 1996. Computer vision technology for food quality assurance. *Trends Food Sci. Technol.* **7**, 245–256.

Gunning, A.P., Wilde, P.J., Clark, D.C., Morris, V.J., Parker, M.L., and Gunning, P.A. 1996. Atomic force microscopy of interfacial protein films. *J. Colloid Interface Sci.* **183**, 600–602.

Han, Y.J., Bowers, S.V., and Dodd, R.B. 1992. Nondestructive detection of split-pit peaches. *Trans. ASAE* **35**(6), 2063–2067.

Harringan, T.P. and Mann, R.W. 1984. Characterisation of microstructural anisotropy in orthotropic materials using a second rank tensor. *J. Mater. Sci.* **19**, 761–767.

Hermann, G.T. 1980. "Image Reconstruction from Projections", p. 316. New York, Academic Press.

Hermansson, A.M., Langton, M., and Lorén, N. 2000. New approaches to characterizing food microstructures. www.mrs.org/pubblications/bulletin, December, 30–36.

Hirakimoto, A. 2001. Microfocus X-ray computed tomography and it's industrial applications. *Anal. Sci.* **17**, 123–125.

Huong, P.V. 1996. New possibilities of Raman micro-spectroscopy. *Vibr. Spectrosc.* **11**, 17–28.

Inanc, F., Gray, J.N., Jensen, T., and Xu, J. 1998. Proceedings of the SPIE Conference on Physics of Medical Imaging. Vol. 3336, p. 830.

Inglis, D. and Pietruszczak, S. 2003. Characterization of anisotropy in porous media by means of linear intercept measurements. *Int. J. Solids Struct.* **40**, 1243–1264.

Inoue, S. 1986. "Video Microscopy". Plenum Press, New York.

Ishida, N., Takano, H., Naito, S., Isobe, S., Uemura, K., Haishi, T., Kose, K., Koizumi, M., and Kano, H. 2001. Architecture of baked breads depicted by a magnetic resonance imaging. *Magn. Reson. Imaging* **19**, 867–874.

Javanaud, C. 1998. Applications of ultrasound to food systems. *Ultrasonics* **26**, 117–123.

Jones, H.E., Lewis, R.M., Young, M.J., and Simm, G. 2004. Genetic parameters for carcass composition and muscularity in sheep measured by X-ray computer tomography, ultrasound and dissection. *Livestock Prod. Sci.* **90**, 167–179.

Kaftandjian, V., Peix, G., Babot, D., and Peyrin, F. 1996. High resolution X-ray computed tomography using a solid state linear detector. *J. X-Ray Sci. Technol.* **6**, 94–106.

Kak, A.C. and Slaney, M. 1988. "Principles of Computerized Tomographic Imaging". IEEE Press.

Kaláb, M. 1984. Artifacts in conventional scanning electron microscopy of some milk products. *Food Microstruct.* **3**(2), 95–111.

Kaláb, M., Allan-Wojtas, P., and Miller, S.S. 1995. Microscopy and other imaging technique in food structure analysis. *Trends Food Sci. Technol.* **6**(6), 177–186.

Kamman, P.W. 1970. Factors affecting the grain and texture of white bread. *Baker's Dig.* **44**(2), 34–38.

Kanatani, K. 1985. Measurement of crack distribution in a rock mass from observation of its surfaces. *Soil Found.* **25**(1), 77–83.

Kanatani, K.I. 1984. Distribution of directional data and fabric tensors. *Int. J. Eng. Sci.* **22**, 149–164.

Kerr, W.L., Kauten, R.J., McCarthy, M.J., and Reid, D.S. 1998. Monitoring the formation of ice during food freezing by magnetic resonance imaging. *Lebensm-Wiss. u.-Technol.* **31**, 215–220.

Kim, N.D., Amin, A., Wilson, D., Rouse, G., and Udpa, S. 1998. Ultrasound image texture analysis for characterizing intramuscular fat content of live beef cattle. *Ultrason. Imaging* **20**, 191–205.

Kirby, A.R., Gunning, A.P., and Morris, V.J. 1995. Atomic force microscopy in food research: A new technique comes in age. *Trends Food Sci. Technol.* **6**, 359–365.

Kokelaar, J.J., van Vliet, T., and Prins, A. 1996. Strain hardening properties and extensibility of flour and gluten doughs in relation to breadmaking performance. *J. Cereal Sci.* **24**(3), 199–214.

Kulmyrzaev, A., Cancelliere, C., and McClements, D.J. 2000. Characterization of aerated food using ultrasonic reflectance spectroscopy. *J. Food Eng.* **46**, 235–241.

Kuo, A.D. and Carter, D.R. 1991. Computational methods for analyzing the structure of cancellous bone in planar sections. *J. Orthop. Res.* **9**, 918–931.

Kuo, M.I., Anderson, M.E., and Gunasekaran, S. 2003. Determining effects of freezing on pasta filata and non-pasta filata mozzarella cheeses by nuclear magnetic resonance imaging. *J. Dairy Sci.* **86**, 2525–2536.

Langton, M., Aström, A., and Hermansson, A. 1996. Texture as a reflection of microstructure. *Food Qual. Pref.* **7**(3/4), 185–191.

Leemans, V., Magein, H., and Destain, M.F. 1998. Defects segmentation on "Golden Delicious" apples by using colour machine vision. *Comput. Electron. Agri.* **20**, 117–130.

Lenker, D.H. and Adrian, P.A. 1971. Use of X-ray for selecting mature lettuce heads. *Trans. Amer. Soc. Agr. Eng.* **84**, 491–500.

Li, J, Tan, J., and Shatadal, P. 2001. Classification of tough and tender beef by image texture analysis. *Meat Sci.* **57**, 341–346.

Lim, K.S. and Barigou, M. 2004. X-ray micro-tomography of cellular food products. *Food Res. Int.* **37**, 1001–1012.

Liu, Z., Chuah, C.S.L., and Scanlon, M.G. 2003. Compressive elastic modulus and its relationship to the structure of a hydrated starch foam. *Acta Materialia* **51**, 365–371.

Maire, E., Fazekas, A., Salvo, L., Dendievel, R., Youssef, S., Cloetens, P., and Letang, J.M. 2003. X-ray tomography applied to the characterization of cellular materials. Related finite element modeling problems. *Components Sci. Technol.* **63**, 2431–2443.

Martens, H.J. and Thybo, A.K. 2000. An integrated microstructural, sensory and instrumental approach to describe potato texture. *Lebensm.-Wiss. u.-Technol.* **33**, 471–482.

Martinez, O., Salmerón, J., Guillén, M.D., and Casas, C. 2004. Texture profile analysis of meat products treated with commercial liquid smoke flavourings. *Food Control* **15**, 457–461.

Martinez, I., Aursand, M., Erikson, U., Singstad, T.E., Veliyulin, E., and van der Zwaag, C. 2003. Destructive and non-destructive analytical techniques for authentication and composition analyses of foodstuffs. *Trends Food Sci. Technol.* **14**, 489–498.

Matusani, Y., Aoky, S., Abe, O., Hayashy, N., and Otomo, K. 2003. MR diffusion tensor imaging: Recent advance and new techniques for diffusion tensor visualization. *Eur. J. Radiol.* **46**, 53–66.

McCarthy, M.J. and McCarthy, K.L. 1996. Applications of magnetic resonance imaging to food research. *Magn. Reson. Imaging* **14**, 799–802.

McClements, D.J. 1995. Advances in the application of ultrasound in food analysis and processing. *Trends Food Sci. Technol.* **6**, 293–299.

McClements, D.J. and Povey, M.J.W. 1988. Comparison of pulsed NMR and ultrasonic velocity measurements for determining solid fat contents. *Int. J. Food Sci. Technol.* **23**, 159–170.

Miquel, M.E., Carli, S., Couzens, P.J., Wille, H.J., and Hall, L.D. 2001. Kinetics of the migration of lipids in composite chocolate measured by magnetic resonance imaging. *Food Res. Int.* **34**, 773–781.

Morris, V.J. 2004. Probing molecular interactions in foods. *Trends Food Sci. Technol.* **15**, 291–297.

Morris, V.J., Kirby, A.J., and Gunning, A.P. 1999. "Atomic Force Microscopy for Biologists". Imperial College Press, London.

Novaro, P., Colucci, F., Venora, G., and D'Egidio, M.G. 2001. Image analysis of whole grains: A non-invasive method to predict semolina yield in durum wheat. *Cereal Chem.* **78**(3), 217–221.

Odgaard, A. 1997. Three-dimensional methods for quantification of cancellous bone architecture. *Bone* **20**, 315–328.

Odgaard, A., Kabel, J., Rietbergen, B., Dalstra, M., and Huiskes, R. 1997. Fabric and elastic principal directions of cancellous bone are closely related. *J. Geomech.* **30**, 487–495.

Ozutsumi, K., Nade, T., Watanabe, H., Tsujimoto, K., Aoki, Y., and Aso, H. 1996. Non-destructive, ultrasonic evaluation of meat quality in live Japanese black steers from coloured images produced by a new ultrasonic scanner. *Meat Sci.* **43**, 61–69.

Patel, D., Davies, E.R., and Hannah, I. 1996. The use of convolution operators for detecting contaminants in food images. *Pattern Recognit.* **29**(6), 1019–1029.

Peyrin, F., Salome-Pateyron, M., Cloetens, P., Laval-Jeantet, A.M., Ritman, E., and Ruegsegger, P. 1998. Micro-CT examinations of trabecular bone samples at different resolutions: 14, 7 and 2 micron level. *Technol. Health Care* **6**, 391–401.

Peyrin, F., Salome-Pateyron, M., Nuzzo, S., Cloetens, P., Laval-Jeantet, A.M., and Baruchel, J. 2000. Perspectives in three-dimensional analysis of bone samples using synchrotron radiation microtomography. *Cell. Mol. Biol.* **46**(6), 1089–1102.

Pietruszczak, S. and Mroz, Z. 2001. On failure criteria for anisotropic cohesive-frictional materials. *Int. J. Numer. Anal. Methods Geomech.* **25**, 509–524.

Pietruszczak, S., Inglis, D., and Pande, G.N. 1999. A fabric-dependent fracture criterion for bone. *J. Biomech.* **32**, 1071–1079.

Piot, O., Autran, J.C., and Manfait, M. 2000. Spatial distribution of protein and phenolic constituents in wheat grain as probed by confocal raman microspectroscopy. *J. Cereal Sci.* **32**, 57–71.

Povey, M.J.W. 1997. "Ultrasonic Techniques for Fluids Characterization". Academic Press, San Diego.

Povey, M.J.W. and McClements, D.J. 1988. Ultrasonic in food engineering: Part I. Introduction and experimental methods. *J. Food Eng.* **8**, 217–245.

Quevedo, R., López, G.C., Aguilera, J.M., and Cadoche, L. 2002. Description of food surface and microstructural changes using fractal image texture analysis. *J. Food Eng.* **53**, 361–371.

Rao, A.R., Ramesh, N., Wu, F.Y., Mandville, J.R., and Kerstens, P.J.M. 1992. Algorithms for a fast confocal optical inspection system. *In* "Proceedings of the IEEE Workshop on Applications of Computer Vision", pp. 298–305.

Renou, J.P., Foucat, L., and Bonny, J.M. 2003. Magnetic resonance imaging studies of water interactions in meat. *Food Chem.* **82**, 35–39.

Ridout, M.J., Gunning, A.P., Wilson, R.H., Parker, M.L., and Morris, V.J. 2002. Using AFM to image the internal structure of starch granules. *Carbohydr. Polym.* **50**, 123–132.

Ritman, E.L., Borah, B., Dufresne, T.E., Phipps, R.J., Sacha, J.P., Jorgensen, S.M., and Turner, R.T. 2002. 3-D Synchrotron μCT allows unique insight of changes in bone quality. The American Society for Bone and Mineral Research Annual Meeting, San Antonio TX, September 20–24.

Ross, K.A., Pyrak-Nolte, L.J., and Campanella, O.H. 2004. The use of ultrasound and shear oscillatory tests to characterize the effect of mixing time on the rheological properties of dough. *Food Res. Int.* **37**, 567–577.

Ryan, T.M. and Ketcham, R.A. 2002. The three-dimensional structure of trabecular bone in the femoral head of strepsirrhine primates. *J. Hum. Evol.* **43**(1), 1–26.

Saltykov, S.A. 1958. "Stereometric Metallography", 2nd ed., p. 81. Metallurgizdat, Moscow.

Sarig, Y., Gayer, A., Briteman, B., Israeli, E., and Bendel, P. 1992. Nondestructive seed detection in citrus fruits. 7th International Citrus Congress, Acireale, Italy, International Society of Citriculture **3**, 1036–1039.

Scanlon, M.G. and Liu, Z.Q. 2003. Predicting the functional properties of bread crumb from its structure. *In* "Proceedings of the 4th AOCS Annual Meeting & Expo", May 4–7, 2003, Kansas City, Kansas, USA.

Schatzki, T.F. and Fine, T.A. 1988. Analysis of radiograms of wheat kernels for quality control. *Cereal Chem.* **65**, 233–239.

Severini, C., Baiano, A., Del Nobile, M.A., Mocci, G., and De Pilli, T. 2004. Effects of blanching on firmness of sliced potatoes. *Ital. J. Food Sci.* **16**(1), 31–34.

Simmons, C.A. and Hipp, J.A. 1997. Method-based differences in the automated analysis of three-dimensional morphology of trabecular bone. *J. Bone Miner. Res.* **12**(6), 942–947.

Snigerev, A., Snigireva, I., Kohn, V., Kuznetsov, S., and Schelokov, I. 1995. On the possibilities of X-ray phase-contrast microimaging by coherent high-energy synchrotron radiation. *Rev. Sci. Instrum.* **66**, 5486–5492.

So, J.D. and Wheaton, F.W. 1996. Computer vision applied to detection of oyster hinge lines. *Trans. ASAE* **39**, 1557–1566.

Sonego, L., Ben-Arie, R., Raynal, J., and Pech, J.C. 1994. Biochemical and physical evaluation of textural characteristics of nectarines exhibiting woolly breakdown. *Postharv. Biol. Technol.* (Amsterdam) **54**, 58–62.

Stanley, D.W. 1987. Food texture and microstructure. *In* "Food Texture" (H.R. Moskowitz, ed.). Marcel Dekker, Inc., New York.

Stokes, D.J. and Donald, M. 2000. *In situ* mechanical testing of dry and hydrated breadcrumb in the environmental scanning electron microscope (ESEM). *J. Mater. Sci.* **35**, 599–607.

Sun, D.W. and Brosnam, T. 2003. Pizza quality evaluation using computer vision—part 1—pizza base and sauce spread. *J. Food Eng.* **57**, 81–89.

Suzuki, K., Tajima, T., Takano, S., Asano, T., and Hasegawa, T. 1994. Nondestructive methods for identifying injury to vapor heat-treated papaya. *J. Food Sci.* **59**(4), 855–857, 875.

Takano, H., Ishida, N., Koizumi, M., and Kano, H. 2002. Imaging of the fermentation process of bread dough and the grain structure of baked breads by magnetic resonance imaging. *J. Food Sci.* **67**(1), 244–250.

Tao, Y., Heinemann, P.H., Vargheses, Z., Morrow, C.T., and Sommer, H.I. 1995. Machine vision for color inspection of potatoes and apples. *Trans. ASAE* **38**, 1555–1561.

Thorvaldsson, K., Stading, M., Nilsson, K., Kidman, S., and Langton, M. 1999. Rheology and structure of heat-treated pasta dough: Influence of water content and heating rate. *Lebensm.-Wiss. u.-Technol.* **32**, 154–161.

Thybo, A.K., Szczypinski, P.M., Karlsson, A.H., Donstrup, S., Stodkilde-Jorgensen, H.S., and Andersen, H.J. 2004. Prediction of sensory texture quality attributes of cooked potatoes by NMR-imaging (MRI) of row potatoes in combination with different image analysis methods. *J. Food Eng.* **61**, 91–100.

Thygesen, L.G., Løkke, M.M., Micklander, E., and Engelsen, S.B. 2003. Vibrational microspectroscopy of food. Raman vs. FT-IR. *Trends Food Sci. Technol.* **14**, 50–57.

Tollner, E.W. 1993. X-ray technology for detecting physical quality attributes in agricultural produce. *Postharv. News Infor.* **4**, N149–N155.

Tollner, E.W., Hung, Y.C., Upchurch, B.L., and Prussia, S.E. 1992. Relating X-ray absorption to density and water content in apples. *Trans. ASAE* **35**, 1921–1928.

Underwood, E. 1970. "Quantitative Stereology", pp. 25–103. Addison-Wesley Publishing Company, New York.

Utku, H. and Koksel, H. 1998. Use of statistical filters in the classification of wheats by image analysis. *J. Food Eng.* **36**, 385–394.

Van Dalen, G., Blonk, H., van Aalst, H., and Hendriks, C.L. 2003. 3D Imaging of foods using x-ray microtomography. *In* "G.I.T. Imaging & Microscopy", pp. 18–21. GIT VERLAG GmbH & Co. KG, Darmstadt, Germany, www.imaging-git.com.

Van der Burg, W.J., Aartse, J.W., van Zwol, R.A., Jalink, H., and Bino, R.J. 1994. Predicting tomato seedling morphology by X-ray analysis of seeds. *J. Am. Soc. Hortic. Sci.* **119**, 258–263.

van Duynhoven, J.P.M., van Kempen, G.M.P., van Sluis, R., Rieger, B., Weegels, P., van Vliet, L.J., and Nicolay, K. 2003. Quantitative assessment of gas cell development during the proofing of dough by magnetis resonance imaging and image analysis. *Cereal Chem.* **80**, 390–395.

Vestergaard, C., Risum, J., and Adler-Nissen, J. 2004. Quantification of salt concentrations in cured pork by computed tomography. *Meat Sci.* **68**, 107–113.

Vinegar, H.J. and Wellington, S.L. 1987. Tomographic imaging of three-phase flow experiments. *Rev. Sci. Instrum.* **58**(1), 96–107.

Wendin, K., Langton, M., Caous, L., and Hall, G. 2000. Dynamic analyses of sensory and microstructural properties of cream cheese. *Food Chem.* **71**, 363–378.

West, Y.D. 1996. Study of sample heating effects arising during laser Raman spectroscopy. *Int. J. Vibr. Spectrosc.* **1**, 5.

Westin, C.-F., Maier, S., Khidhir, B., Everett, P., Jolesz, F., and Kikinis, R. 1999. Image processing for diffusion tensor magnetic resonance imaging. *In* "Medical Image Computing and Computer-Assisted Intervention". Lecture Notes in Computer Science, pp. 441–452.

Wetzel, D.L., Srivarin, P., and Finney, J.R. 2003. Revealing protein infrared spectral detail in a heterogeneous matrix dominated by starch. *Vibr. Spectrosc.* **31**, 109–114.

Whitehouse, W.J. 1974. The quantitative morphology of anisotropic trabecular bone. *J. Microsc.* **101**(2), 153–168.

Whitworth, M.B. and Alava, J.M. 1999. The imaging and measurement of bubble in bread doughs. *In* "Bubbles in Foods" (G.M. Campbell, C. Webb, S.S. Pandiella, and K. Niranjan, eds), pp. 221–231. Eagan Press, St. Paul, MN.

Wildmoser, H., Scheiwiller, J., and Windhab, E.J. 2004. Impact of disperse microstructure on rheology and quality aspects of ice cream. *Lebensm.-Wiss. u.-Technol.* **37**, 881–891.

Wium, H., Pedersen, P.S., and Qvist, K.B. 2003. Effect of coagulation conditions on the microstructure and the large deformation properties of fat-free Feta cheese made from ultrafiltered milk. *Food Hydrocolloids* **17**, 287–296.

www.geog.ubc.ca/courses/geog470/notes/image_processing.

www.hei.org/research/depts/aemi/emt.htm.

www.nikon-instruments.jp/eng/tech/1-0-4.aspx.

www.plbio.kvl.dk/~als/confocal.htm.

www.ruf.rice.edu/~bioslabs/methods/microscopy/microscopy.html.

Xu, Y.-H. and Pitot, H.C. 2003. An improved stereologic method for three-dimensional estimation of particle size distribution from observations in two dimensions and its application. *Comput. Methods Programs Biomed.* **72**, 1–20.

Yada, R.Y., Harauz, G., Marcone, M.F., Beniac, D.R., and Ottensmeyer, F.P. 1995. Visions in the mist: The Zeitgeist of food protein imaging by electron microscopy. *Trends Food Sci. Technol.* **6**, 265–270.

Yan, Z.Y., McCarthy, M.J., Klemann, L., Otterburn, M.S., and Finley, J. 1996. NMR applications in complex food systems. *Magn. Reson. Imaging* **14**, 979–981.

Yantarasri, T., Kalayanamitra, K., Saranwong, S., and Sornsrivichai, J. 2000. Evaluation of the maturity index for durian fruit by various destructive and non-destructive techniques. Quality assurance in agricultural produce. *In* "ACIAR Proceedings" (G.I. Johnson, Le Van To, Nguyen Duy Duc, and M.C. Webb, eds), pp. 700–705.

Yantarasri, T. and Sornsrivichai, J. 1998. Internal quality evaluation of tangerines. *Acta Hortic.* **464**, 494.

Yantarasri, T., Sornsrivichai, J., and Chen, P. 1997. Nondestructive X-ray and NMR imaging for quality determination of mango fruit. Presented at the 5th International Symposium on Fruit, Nut and Vegetable Production Engineering, Davis, California.

Yantarasri, T., Sornsrivichai, J., and Kalayanamitra, K. 1999. Nondestructive techniques for quality evaluation of pineapple fruits. The Third International Pineapple Symposium, Pattaya, Thailand.

Zwiggelaar, R., Bull, C.R., Mooney, M.J., and Czarnes, S. 1997. The detection of 'soft' materials by selective energy X-ray transmission imaging and computer tomography. *J. Agric. Eng. Res.* **66**(3), 203–212.

Zwiggelaar, R., Bull, C.R., and Mooney, M.J. 1996. X-ray simulations for imaging applications in the agricultural and food industries. *J. Agric. Eng. Res.* **63**, 161–170.

ELECTRODIALYSIS APPLICATIONS IN THE FOOD INDUSTRY

MARCELLO FIDALEO AND MAURO MORESI

Department of Food Science and Technology, University of Tuscia
Via San Camillo de Lellis, 01100 Viterbo, Italy

ABBREVIATIONS

A_m	Overall membrane surface area ($= N_{cell}\ a_{mg}$)
A_k	Generic k-th empirical constant in Eq. 9
a_{ERS}	Exposed surface area of the electrodes (m^2)
a_i	Activity of species i (dimensionless)
a_k	Effective surface area involved in the ion flow pattern (m^2)

ISSN: 1043-4526
DOI: 10.1016/S1043-4526(06)51005-8

a_{me}	Effective membrane surface area (m^2)
a_{mg}	Geometrical membrane surface area (m^2)
C	Weight concentration (kg/m^3)
c	Molar concentration ($kmol/m^3$)
$c^{\pm}$	Anion or cation molar concentration in solution ($kmol/m^3$)
c_{Cl^-}	Molar concentration of chloride ions in the electrode channels ($kmol/m^3$)
D_B	Solute diffusion coefficient (m^2/s)
D_{B0}	Solute diffusion coefficient at infinite dilution (m^2/s)
d_e	Equivalent diameter of ED channel (m)
d_{eR}	Reference equivalent diameter of ED channel (m)
E	Overall potential drop across an ED stack (V)
E_T	Thermodynamic electrode potential (V)
E_T^0	Standard thermodynamic electrode potential at 25°C and unit activity (V)
E_D	Donnan potential difference across membranes (V)
E_d	Thermodynamic cell potential difference (V)
E_{el}	Electrode potential for anode and cathode processes (V)
E_j	Junction potential difference across the boundary layers (V)
E_{MP}	Potential drop across an ED membrane pack (V)
F	Faraday's constant (96,486 C/mol)
h	Channel interval (m)
I	Electric current intensity (A)
$J^{\pm}$	Ion flux in solution (mol $m^{-2}s^{-1}$)
J_B	Overall solute permeation flux (mol $m^{-2}s^{-1}$)
J_W	Overall water flux (mol $m^{-2}s^{-1}$)
J_{Wd}	Water flux due to osmosis (mol $m^{-2}s^{-1}$)
j	Electric current density (A/m^2)
k_m	Solute mass transfer coefficient (m/s)
L	Overall energy consumption in an ED stack (J)
L_B	Membrane constant for solute transport by diffusion (m/s)
L_p	Membrane hydraulic permeability (mol $m^{-2}s^{-1}bar^{-1}$)
L_W	Membrane constant for water transport by diffusion (m/s)
M_B	Solute molecular mass (g/mol)
m	Solute molality (mol/kg)
m_e	Theoretical amount of equivalents transported to an electrode (eq)
m_{F0}	Initial solute mass in the feed (kg)
N_{cell}	Overall number of cell pairs (dimensionless)
N	Number of electrons participating in the electrode reaction (dimensionless)
n_B	Solute mass (mol)

n_W	Water mass (mol)
Q	Electrical charge passed through a cell (C)
R	Overall electric resistance of the membrane stack (Ω)
Re	Reynolds number (defined in Table III; dimensionless)
R_{ERS}	Electric resistance of electrode rinsing solution (Ω)
R_f	Boundary layer electric resistance (Ω)
R_{fg}	Electric resistance of the fouling gel layer (Ω)
R_G	Gas-law constant (= 8.314 J mol^{-1}K^{-1})
R_K	Electric resistance of the bulk solution in the k-th compartment (Ω)
R_k	Electric resistance of the k-th electromembrane (Ω)
R_{MP}	Apparent membrane pack electric resistance (Ω)
r_k	Generic electromembrane surface resistance (Ωm^2)
s	Thickness of a generic electrolytic layer (m)
Sc	Schmidt number [$= \eta/(\rho D_B)$; dimensionless]
Sh	Sherwood number ($= k_m\, h/D_B$; dimensionless)
s_i	Stoichiometric coefficient of species i involved in the electrode reaction (dimensionless)
T_K	Absolute temperature (K)
$t^{\pm}$	Ion transport numbers in solution (dimensionless)
$t_m^{\pm}$	Ion transport number in a generic electromembrane (dimensionless)
t_s	Effective cation- ($t_c^+ - t_a^+$) or anion- ($t_a^- - t_c^-$) transport number (dimensionless)
t_W	Water transport number (dimensionless)
V_W	Molar volume of pure water (m^3/mol)
v_S	Superficial flow velocity (m/s)
W	ED channel width (m)
x	Generic spatial coordinate (m)
$z^{\pm}$	Ion charge number (dimensionless)

Greek Symbols

α_k	Empirical k-th constant in Eq. 17 (dimensionless)
α_m	Membrane perm-selectivity (dimensionless)
β_k	Generic k-th empirical constant in Eq. 4
$\gamma^{\pm}$	Mean molal activity coefficient (dimensionless)
Δc_B	Difference in solute concentrations in C and D compartments ($= c_{BC} - c_{BD}$) (mol/m^3)
Δl	Separation length between successive eddy promoters (m)
Δm_B	Solute mass transferred into compartment C (kg)
ΔP	Transmembrane pressure difference (MPa)
$\Delta \pi$	Osmotic pressure difference (MPa)

δ	Diffusion boundary layer thickness (m)
δ_{gel}	Gel layer thickness (m)
ε	Specific energy consumption (kWh/kg)
ε_S	Volume fraction of spacer (dimensionless)
ζ	Solute recovery efficiency (dimensionless)
η	Dynamic viscosity (Pa s)
η_r	Relative viscosity ($= \eta/\eta_W$; dimensionless)
Ψ_{ED}	Electrodialysis membrane-fouling index (Ω/A)
θ	Time (s)
θ'	Dummy variable of integration (s)
Λ	Equivalent conductance (S m^2/keq)
Λ_0	Equivalent conductance at infinite dilution (S m^2/keq)
$\lambda^{\pm}$	Equivalent ion conductance (S m^2/keq)
$\lambda_0^{\pm}$	Equivalent ion conductance at infinite dilution (S m^2/keq)
ρ	Density (kg/m^3)
σ	Reflection coefficient (dimensionless)
χ	Electric conductivity (S/m)
Ω	Total current efficiency (dimensionless)
Ω_C	Faraday efficiency (dimensionless)
Ω_L	Current leakage efficiency (dimensionless)

Subscripts

a	Referred to the anion-exchange membrane
B	Referred to solute
c	Referred to the cation-exchange membrane
C	Referred to the concentrating compartment
D	Referred to the diluting compartment
e	Effective
ERS	Referred to electrode-rinsing solution
g	Geometric
gel	Referred to the gel layer
lim	Limiting
m	Referred to the membrane surface
max	Referred to the theoretical maximum concentration
W	Referred to water

This paper reviews the most recent innovations in electrodialysis (ED) modules and/or processes that appear to affect the food and drinks industries in the short-medium term, together with their basic mass transport equations that might help ED unit design or optimization. Future perspectives for ED processing in the food sector are also outlined.

I. INTRODUCTION

Electrodialysis (ED) is a unit operation for the separation or concentration of ions in solutions based on their selective electromigration through semipermeable membranes under the influence of a potential gradient (Lacey and Loeb, 1972; Strathmann, 1992). Owing to their selectivity, ion-exchange membranes (IEM) allow transport of only cations (cation-exchange membranes) or anions (anion-exchange membranes) and thus can be used to concentrate, remove, or separate electrolytes.

Despite the first industrial ED application in the food sector dated back to 1960 and concerned the demineralization of cheese whey for use in baby foods, the history of ED is longer than that usually acknowledged (Shaposhnik and Kesore, 1997).

A technique combining dialysis and electrolysis was first proposed to de-ash sugar syrup by Maigrot and Sabates (1890). The archetype consisted of two carbon electrodes piloted by a dynamo and separated by a permanganate paper-based membrane. Once the sugar syrup had been poured into the central anodic compartment and the dynamo had been turned on, potassium, sodium, magnesium, and calcium cations tended to migrate into the cathodic compartment. By controlling the pH in this compartment via the so-called litmus paper, it was possible to prevent sparingly soluble hydroxides from precipitating. Thus, as the indicator turned blue, the ED process was stopped.

In 1939, Manegold and Kalauch assembled a three-compartment ED apparatus consisting of a permselective anion-exchange membrane and a cation-exchange one. It was, however, only in the early 1950s that the manufacture of selective membranes from ion exchangers allowed the multicompartment electrodialysers to be assembled (Shaposhnik and Kesore, 1997).

The development of such membranes in England and the United States was not an easy task, as finely outlined by Solt (1995). In those days, the combined efforts of the Netherlands National Research Organisation (TNO) and the South African Council for Scientific and Industrial Research resulted in the development of the ED process for demineralizing saline waters from mines. In the late 1950s, the Office of European Economic Cooperation, as well as in the 1960s the Institute for Arid Zone Research at Beersheva (Israel) and several Japanese manufacturers, contributed to further R&D in this sector (Lacey and Loeb, 1972; Solt, 1995).

The first scientific paper on ED was published in 1903 by Morse and Pierce (Shaposhnik and Kesore, 1997), but the quantitative theory of charged membranes was available only 50 years later (Teorell, 1953). It incorporated the electrostatic repulsion of co-ions from the fixed charges of the membrane, as predicted by the thermodynamical theory of membrane equilibrium developed by Donnan (1911).

The present largest area of application for ED is in the desalination of brackish water for the production of potable water (Audinos, 1992; Strathmann, 1992) and de-ashing of milk whey to obtain valuable raw materials for baby foods (Batchelder, 1987). However, ED applications are still in their infancy and ED-processing potentialities have not been completely exploited probably because of the high specific electromembrane costs or their short lifetime (this being not longer than 1 year, especially if the feed solution is fouling or ED separation plant has not been well designed or is not properly conducted).

The main aim of this paper was to review the present and potential ED applications that appear to be of particular interest for the food and drinks sector in the short-medium term, as well as the basic mass transport equations that might help ED unit design or optimization.

II. ED PRINCIPLES

A. BASIC CONCEPTS AND DEFINITIONS

For a better comprehension of the ED processes it is necessary to refresh a few basic concepts and definitions regarding the electrolytic cell and thermodynamic electrode potential, Faraday's laws, current efficiency, ion conduction, diffusivity, and transport numbers in solution.

An electrolytic cell is essentially composed of a pair of electrodes submerged into an electrolyte for conduction of ions and connected to a direct current (DC) generator via an external conductor to provide for continuity of the circuit. The electrode connected to the positive pole of the DC generator is called anode, while that linked to the negative one, cathode. The current flow in an electrolyte results from the movement of positive and negative ions and is assumed as positive when directed as the positive charges or opposite to the electrons in the external circuit. When the cell is not operating under conditions of standard concentration, the thermodynamic electrode (or cell) potential (E_T) can be estimated from the Nernst equation:

$$E_T = E_T^0 - \frac{R_G T_K}{nF} \ln\left(\prod_i a_i^{s_i}\right) \quad (1)$$

where E_T^0 is the standard potential at 25°C and unit activity, a_i is the activity of species i, s_i the corresponding stoichiometric coefficient, which is positive for products or negative for reactants, F the Faraday's constant (= 96,486 C mol^{-1}), R_G the universal gas constant (= 8.31 J mol^{-1}K^{-1}), T_K the absolute temperature, and n the overall number of electrons participating in the reaction (Prentice, 1991).

Thus, in a cell supplied with two graphite electrodes submerged into a brine solution, the following electrode reactions may be accomplished:

$$\text{Cathode: } 2H_3O^+ + 2e^- = H_2 \uparrow + 2H_2O$$

$$\text{Anode: } 2Cl^- = Cl_2 \uparrow + 2e^-$$

The evolution of H_2 and Cl_2 transforms the cathode into a hydrogen electrode and the anode into a chlorine one (Prentice, 1991), thus resulting in an electrochemical cell with a thermodynamic cell potential difference (E_d) at 25°C equal to:

$$E_d = 1.358 + 0.059 \quad \text{pH} - 0.059 \ \log c_{Cl^-} \tag{2}$$

where c_{Cl^-} is the molar concentration of chloride ions in the electrode channels. So, when using an electrode-rinsing solution at 0.5 kmol/m^3 and an initial pH value of about 6.5, the cell potential difference is initially equal to 1.76 V. As the H_3O^+ reduces at the cathode, its concentration reduces, thus increasing the pH value and cell potential difference (E_d).

According to Michael Faraday, the theoretical amount of equivalents (m_e) of the product transported to one or another electrode is directly proportional to the electric charge (Q) passed through the cell:

$$m_e = \frac{Q}{F} = \frac{\int_0^{\theta} I \ d\theta'}{F} \tag{3}$$

where θ is the time elapsed. In real electrolytic processes some of the charge is consumed in parasitic processes or other deviations are to be accounted for. Thus, the ratio between the effective and theoretical amounts of equivalents transported is a measure of the process efficiency, that is the so-called current or Faraday efficiency (Ω).

The specific conductivity (χ) is a measure of the mobility of ions in an electrolyte or electrons in a metallic conductor. Thus, χ is about 1 or 10^7 S/m for a 0.1 kmol/m^3 aqueous salt solution or for a metal such as iron, respectively. Such a difference in charge mobility makes the temperature dependence of χ [i.e.,$(1/\chi)\partial\chi/\partial T_K$] positive for ions of about 2.5% per K, but negative for metals and alloys of approximately an order of magnitude lower (Prentice, 1991).

In aqueous systems χ is proportional to the bulk electrolyte concentration (c_B) in dilute solutions, reaches a maximum, and then decreases due to ion association and viscosity effects. Friedrich Kohlrausch found useful to introduce a conductivity function not changing abruptly with c_B, that is, the so-called *equivalent conductance* (Λ), defined as the ratio between χ and c_B, the latter being expressed in equivalents per unit volume. Because of incomplete dissociation at higher concentrations and higher electrolyte viscosity, Λ decreases at higher c_B values. Empirically, it was observed a linear relationship between Λ and the square root of c_B, which is generally known as the Kohlrausch limiting law. As an example, Figure 1 shows its validity either for a strong electrolyte (i.e., NaCl) or for the sodium salts of a few weak monocarboxylic acids such as acetic and propionic acids. For $c_B > 0.1$ mol/m^3 Kohlrausch limiting law is generally regarded as inadequate to describe the variation of Λ with c_B (Robinson and Stokes, 2002) and it may be empirically expanded in powers of $\sqrt{c_B}$ as follows:

$$\Lambda = \frac{\chi}{c_B} = \Lambda_0 + \beta_1 c_B^{1/2} + \beta_2 c_B + \beta_3 c_B^{3/2} \quad (4)$$

where β_i is a generic coefficient and Λ_0 is the equivalent conductivity at infinite dilution.

Since the current results from the motion in opposite directions of anions and cations, Λ can be considered as the sum of two ionic conductivities:

$$\Lambda = \lambda^+ + \lambda^- \quad (5)$$

According to Kohlrausch's law of the Independent Migration of Ions the equivalent conductivity at infinite dilution of a cation ($\lambda_0{}^+$) or an anion ($\lambda_0{}^-$) depends only on the nature of the ion and properties of the medium, such as

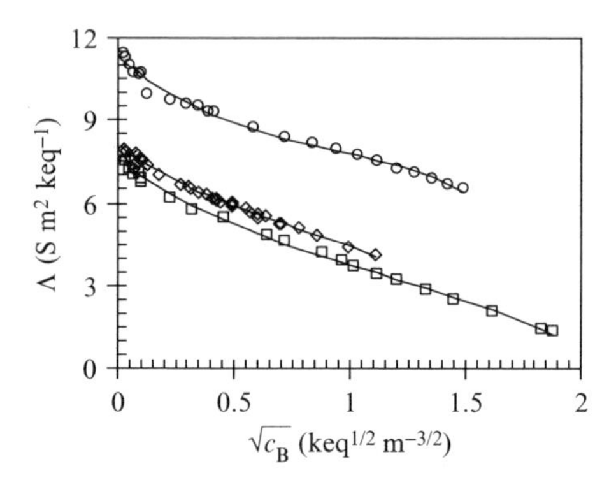

FIG. 1 Equivalent conductance (Λ) for sodium chloride (○), sodium acetate (◇), and sodium propionate (□) at 20°C against the square root of solute concentration ($\sqrt{c_B}$), as extracted from Fidaleo and Moresi (2005a,b, 2006). The continuous lines were calculated using Eq. 4 and the empirical parameters Λ_0 and β_i extracted from Fidaleo and Moresi (2005a,b, 2006).

TABLE I

EQUIVALENT CONDUCTIVITY AT INFINITE DILUTION OF SELECTED CATIONS ($\lambda_0{}^+$) AND ANIONS (λ_0^-) IN WATER AT 25°C[a]

Anion	λ_0^-(S m^2 keq^{-1})	Cation	λ_0^+ (S m^2 keq^{-1})
OH^-	19.76	H^+	34.98
Cl^-	7.63	Na^+	5.01
NO_3^-	7.14	K^+	7.35
SO_4^{2-}	8.00	NH_4^+	7.34
HCO_3^-	4.45	Mg^{2+}	5.31
CH_3COO^-	4.09	Ca^{2+}	5.95

[a]As extracted from Prentice (1991) and Reid *et al.* (1987).

temperature and viscosity, since each ion is moving in a medium where the ions are so far apart they do not interfere one another. Table I shows the equivalent conductivity at infinite dilution of a few ions at 25°C, as extracted from Prentice (1991) and Reid *et al.* (1987). Generally speaking, such a parameter is of the order of 5 S m^2/keq with notable exceptions for hydrogen and hydroxyl ions (Table I).

The transport (or transference) numbers ($t^{\pm}$) represent the fractions of current carried out by such specific ions in the absence of concentration gradients. In dilute solutions $t^{\pm}$ can be estimated by dividing the equivalent ion conductance ($\lambda_0^{\pm}$) by Λ_0. When concentration gradients are present, some of the current arises from diffusion and this affects the effective $t^{\pm}$ numbers. Generally, the transference numbers are weak functions of concentration and temperature and may be regarded as practically independent of salt concentration (Prentice, 1991; Robinson and Stokes, 2002).

As an example, by referring to Figure 1, it is possible to determine the limiting conductivity for NaCl by extrapolating the Λ versus $\sqrt{c_B}$ plot for c_B tending to zero ($\Lambda_0 = 11.33 \pm 0.09$ S m^2/keq) and to extract the limiting conductance for Na^+ from literature ($\lambda_0{}^+ = 4.495$ S m^2/keq at 20°C). Thus, the transport number for sodium ion (t^+) can be directly estimated as 0.397, while that for the chloride ion (t^-) is 0.603 since the sum of the transport numbers must equal one (Prentice, 1991; Robinson and Stokes, 2002). This means that Na^+ ions carry about 40% of the current, whereas the Cl^- ions the remaining 60%. Thus, the ion flux in solution resulting from the applied electric field, in the absence of concentration gradients, can be predicted as:

$$J^{\pm} = \frac{t^{\pm}}{F} j \tag{6}$$

where j is the electric current density.

The mechanism of ion transport is altered by the contribution of diffusion. To account for this effect, it is necessary to know the electrolyte diffusivity. The diffusion coefficient (D_{B0}) at infinite dilution can be estimated via the Nernst-Haskell equation (Reid *et al.*, 1987):

$$D_{B0} = \frac{R_G T_K}{F^2} \frac{2}{\frac{1}{\lambda_0^+} + \frac{1}{\lambda_0^-}} \tag{7}$$

In the concentration range regarding the ED processes, the effective diffusion coefficient (D_B) can be predicted via the Gordon relationship (Reid *et al.*, 1987), which accounts for the partial derivative of the natural logarithm of the mean molal activity coefficient ($\gamma^{\pm}$) with respect to molality (m) and solvent relative viscosity (η_r):

$$D_B = \frac{D_{B0}}{\eta_r} \left(1 + m \frac{\mathrm{dln}\ \gamma^{\pm}}{\mathrm{d}m} \right) \tag{8}$$

Figure 2 shows $\gamma^{\pm}$ for a few electrolytes as a function of m, as extracted from Robinson and Stokes (2002), as well as their calculated values using the Debye-Hückel model integrated with a linear function of m to extend its accuracy beyond the dilute region:

$$\ln \gamma^{\pm} = \frac{-A_1\sqrt{m}}{1 + A_2\sqrt{m}} + A_3 m \tag{9}$$

In this way, it was possible to correlate $\gamma^{\pm}$ data for the above-mentioned electrolyte solutions with molality up to 2.5–6 mol/kg (Figure 2). For the

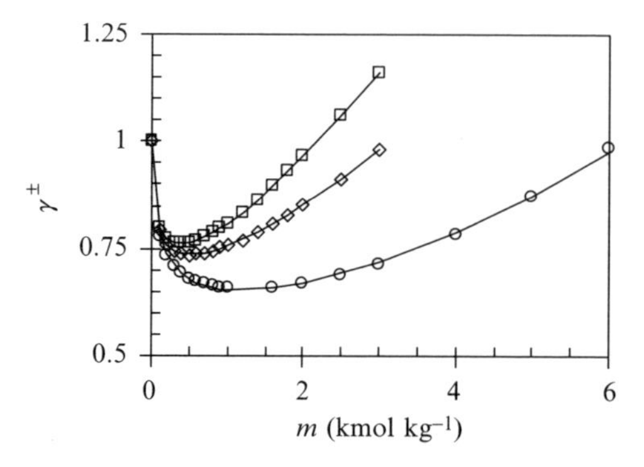

FIG. 2 Effect of solute molality (m) on the mean molal activity coefficient ($\gamma^{\pm}$) for sodium chloride (○), sodium acetate (◇), and sodium propionate (□) at 25°C, as extracted from Robinson and Stokes (2002). The continuous lines were calculated using Eq. 9 and the empirical coefficients A_i extracted from Fidaleo and Moresi (2005a,b, 2006).

NaCl solutions, when setting A_1, A_2, and A_3 equal to 0.979 $kg^{1/2}/mol^{1/2}$, 0.732 $kg^{1/2}/mol^{1/2}$, and 0.139 kg/mol, respectively (Fidaleo and Moresi, 2005a), the average experimental error was as little as 0.6%, about an order of magnitude lower than the average error (5%) pertaining to the well-known correlation by Bromley (1973).

B. ELECTROMEMBRANES

1. Monopolar membranes

The IEM used in ED are essentially sheets of ion-exchange resins (IER). Whereas IERs are generally weak and dimensionally unstable for they may adsorb different amounts of water depending on pH, electrolyte concentration, and temperature, the IEMs possess greater mechanical strength and flexibility as conferred by appropriate reinforcing materials. The commercially available IEMs can be subdivided into two major categories, either homogeneous or heterogeneous (Strathmann, 1992).

Homogeneous IEM are produced either by polymerization of functional monomers (e.g., polycondensation of phenol or phenol-sulphonic acid with formaldehyde) or by functionalization of a polymer film by sulphonation of a polystyrene film.

Heterogeneous IEMs are produced by melting and pressing of dry IERs with a granulated polymer or by dispersion of IERs in a solution or a melted polymer matrix. To present a low electrical resistance, such membranes have to contain more than 65% w/w of cross-linked ion-exchange particles, resulting in inadequate mechanical strength and dimensional instability. Thus, the heterogeneous membranes usually hold a higher electrical resistance and a more uneven distribution of fixed charges than the homogeneous ones.

Owing to the negatively (e.g., $-SO_3^-$, $-COO^-$, $-PO_3^{2-}$, $-HPO_2^-$) or positively (e.g., $-NH_3^+$, $-RNH_2^+$, $-R_3N^+$, $-R_2N^+$, where the R group generally coincides with the methyl one, $-CH_3$) charged groups chemically attached to the resin matrix, an IEM can be classified as a *cationic* membrane or an *anionic* one, respectively. It allows intrusion and exchange of counterions from an external source, as well as exclusion of the co-ions, that is the ions with a charge opposite or equal to its fixed charge, respectively (Figure 3). Such membranes are monopolar since they are permeable to a single type of ions only.

Whatever the IEM, counterions carry most of the electric current, the concentration of co-ions being relatively low within the membrane itself. Moreover, the fixed ions are in electroneutrality with mobile ions in the interstices of the membrane and repel the co-ions, this type of action being usually called Donnan exclusion.

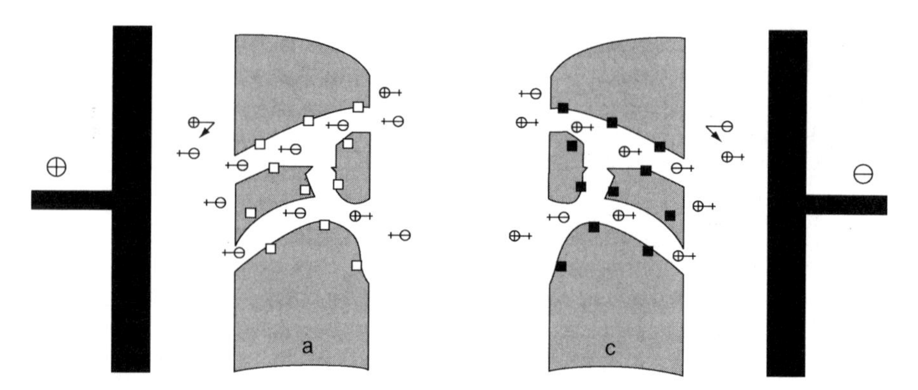

FIG. 3 Schematic diagram of the internal structure of anionic (a) and cationic (c) membranes and their operation under the influence of an electric field: ○, mobile anions (−) and cations (+); □, positive fixed charge; ■, negative fixed charge.

Ion-exchange polymers, such as polystyrene sulphonic acid, are water soluble, so cross-linking with divinylbenzene (DVB) is used to prevent dissolution of ion-permeable membranes. As the degree of cross-linking is increased, the membrane selectivity, stability, and electrical resistance increase. On the contrary, as the fixed-charge density is increased, there is a positive effect on membrane selectivity and conductivity, but a negative one on membrane swelling. Thus, a compromise among selectivity, electrical resistance, and dimensional stability has to be achieved by controlling appropriately cross-linking and fixed-charge densities.

Also the different ionic groups significantly affect the selectivity and electric resistance of IEMs. For instance, the sulphonic acid ($-SO_3^-$) or quaternary ammonium ($-R_3N^+$) group is completely dissociated over the entire pH range, while the primary ammonium group ($-NH_3^+$) is only weakly dissociated and the carboxylic one ($-COO^-$) is virtually undissociated for pH < 3 (Strathmann, 1992).

IEMs can be discriminated according to their mechanical and electrical properties, perm-selectivity, and chemical stability.

First, to distinguish an anion-exchange membrane from a cation-exchange one, it is sufficient to drip a few drops of 0.05% solution of methylene blue and methyl orange on the top of any of them and to note whether the color of the resulting stain tends to a golden yellow or a deep blue, respectively (Strathmann, 1992).

Second, an efficient use of such membranes asks for the absence of pinholes, this being easy to check via the following procedure (Strathmann, 1992):

A. Place a wet membrane sheet on a white absorbent paper sheet.
B. Spread a few drops of 0.2% (w/v) solution of methylene blue or erythrocin-B over the entire surface of the anionic or cationic membrane, respectively.
C. Remove the membrane and check whether the paper sheet has been stained.

Lack of dyed spots on the white paper assures the membrane is free of pinholes.

The membrane swelling capacity (or gel water content) is defined as the percentage ratio of the difference between the mass of a wet sample (m_W), equilibrated in deionized water for 2 days, and that of the same sample (m_D), dried at 75°C over phosphorous oxide (P_2O_5) under vacuum, by m_D (Strathmann, 1992).

The ion-exchange capacity (IEC) of a charged membrane can be determined as follows:

A. Equilibrate a sample of anionic or cationic membrane in 1 kmol/m^3 HCl or NaOH, respectively, for about 24 hours.
B. Rinse the sample with deionized water for about 24 hour.
C. Titrate the sample with 1 kmol/m^3 NaOH or HCl, respectively.
D. Dry the sample at 75°C over P_2O_5 under vacuum and weight it (m_D).
E. Refer the equivalents of alkali or acid used to m_D.

The membrane perm-selectivity (α_m) is defined as the ratio between the actual and theoretical transfer of counterions through any IEM. It can be simply determined as the percentage ratio between the experimental and theoretical Donnan potential differences as measured using a test system consisting of two cells provided with calomel electrodes and filled with well-mixed standardized aqueous solutions of KCl (at 0.1 and 0.5 kmol/m^3), kept at 25°C, and separated by the IEM sample under testing.

The area resistance (r_m) of a charged membrane is indirectly measured using a conductivity cell composed of two well-stirred chambers containing two electrodes submerged into an aqueous solution of NaCl (0.5 kmol/m^3). The electric conductivity of the cell is determined in presence or absence of a generic IEM sample in a bridge circuit using alternating current and both measures allow the membrane area resistance to be calculated using the second Ohm's law (Strathmann, 1992). This value is generally provided by the manufacturer (Table II) and is to be multiplied by 1.75 to estimate roughly the membrane surface resistance to DC (Davies and Brockman, 1972).

The long-term chemical stability of the electromembranes affects the economics of any ED application and is generally determined by assessing

TABLE II

MAIN CHARACTERISTICS[a] OF SOME REPRESENTATIVE COMMERCIALLY AVAILABLE ION-EXCHANGE MEMBRANES[b]

Membrane	Type	Main properties	Category	Thickness (mm)	Burst strength (bar)	IEC (meq/g)	Gel water (%)	r_m (Ω cm^2)	α_m (%)
ASAHI GLASS ENGINEERING Co. (Chiba-shi, Chiba, Japan)									
SELEMION (http://www.agc.co.jp/english/chemicals/ion-maku/selemion/sele.htm#anchor795659)									
CMV	Cation	Styrene	Homo	0.13–0.15	3–5	2.4	25	2.0–3.5	91
AMV	Anion	Butadiene	Homo	0.11–0.15	2–5	1.9	19	1.5–3.0	93
CMD	Cation		Homo	0.38–0.42	10–20	–	–	8–10	–
AMD	Anion		Homo	0.38–0.42	10–20	–	–	8–10	95
ASTOM Co. (Minato-Ku, Tokyo, Japan; http://www.astom-corp.jp)									
AMF-IPEX (Asahi Chemical Industry Co. Ltd., Chiyoda-ku, Tokyo, Japan; http://www.asahi-kasei.co.jp)									
C-60	Cation	Polyethylene-styrene	Hetero	0.3	3	1.6	35	5	92
A-60	Anion	Polyethylene-styrene	Hetero	0.3	3	2	22	7	93
ACIPEX (Asahi Chemical Industry Co. Ltd., Tokyo, Japan)									
K-101	Cation	Styrene/DVB		0.24	–	1.2	24	2.1	91
A-101	Anion	Styrene/DVB		0.21	–	1.4	31	2–3	45
NEOSEPTA® (Tokuyama Soda Co. Ltd., Nishi-Shimbashi, Minato-ku, Tokyo, Japan; htpp://www.tokuyama.co.jp)									
CMX	Cation			0.17–0.19	5–6	1.5–1.8	25–30	2.5–3.5	98
AMX	Anion			0.16–0.18	4.5–5.5	1.4–1.7	25–30	2.5–3.5	98
CMS	Cation	Univalent		0.14–0.17	3–4	2.0–2.5	35–45	1.5–2.5	98
ACS	Anion	Univalent		0.15–0.20	3–5	1.4–2.0	20–30	2.0–2.5	98
AFN	Anion	Antifouling		0.15–0.20	2–3.5	2.5–3.7	40–55	0.4–1.5	98
AXE 01	Anion	Antifouling 60°C		0.17	4.1	2	–	1.4	–
ELECTROPURE EXCELLION™ (Laguna Hills, CA, USA; http://www.electropure-inc.com/downloads/Excellion%20Specifications.pdf)									
I-100	Cation	100°C	Hetero	0.32–0.34	3.0–3.2	–	–	7.5–12.5	–
A-20	Anion	100°C	Hetero	0.32–0.34	2.8–3.0	–	–	5–10	–
DU PONT Co. (Fayetteville, NC, US; http://www.dupont.com/fuelcells/pdf/nae101.pdf)									
NAFION®									
N 117	Cation	Perfluorinated		0.2	–	0.9	16	1.5	–
N 901	Cation	Perfluorinated		0.4	–	1.1	5	3.8	96

IONICS Inc. (Watertown, MA, US; htpp://www.ionics.com)									
CR61 CZL183	Cation	Polystyrene	Homo	0.6	–	2.7	40	–	–
AR103 PZL183	Anion	Polystyrene	Homo	0.6	–	1.8	43	–	–
61CZL386	Cation			0.6	–	2.7	40	9	–
103QZL386	Anion			0.63	–	2.1	36	6	–
LINAN EURO-CHINA Co. (Linan City, China; http://www.Linanwindow.Com/Qianqiu/Membeng.Htm)									
–	Cation		Hetero	0.42	3.5	2	40–55	15	90
–	Anion		Hetero	0.42	3.5	1.8	30–45	15	88
MEMBRANES INTERNATIONAL Inc. (Glen Rock, NJ, USA; http://www.membranesinternational.com/tech.htm)									
CMI-7000S	Cation	90°C	Hetero	0.45	13.6	1.3	–	40	94
AMI-7001S	Anion	90°C	Hetero	0.45	13.6	1.0	–	85	90
PALL RAI Inc. (Hauppauge, NY, US; http://phychem.kjist.ac.kr/312.pdf)									
R-1010	Cation	Perfluorinated		0.05	–	1.2	20	0.4	94
R-1035	Anion	Perfluorinated		0.05	–	1.0	10	1.3	77
PCA GmbH (Heusweiler, D; www.pca-gmbh.com)									
PC-100D	Anion	Polyester reinforced		0.08–0.10	4–5	1.2	50	5	>93
PC-200D	Anion	Polyester reinforced		0.08–0.10	4–5	1.3	45	2	>93
PCA GmbH (Heusweiler, D; http://www.pca-gmbh.com)									
PC-100D	Anion	Polyester reinforced		0.08–0.10	4–5	1.2	50	5	>93
PC-200D	Anion	Polyester reinforced		0.08–0.10	4–5	1.3	45	2	>93
PERMUTIT Co. (USFilter Corp., Warrendale, PA, US; http://www.usfilter.com/water/Business+Centers/Industrial_Process_Water/Industrial_Process_Water_Products/pw_permutit.htm)									
PERMAPLEX									
C-20	Cation			0.8	–	3	30–40	–	98
A-20	Anion			0.8	–	2	30–40	–	98
RHONE-POULENC CHEMIE GmbH (Frankfurt, D; http://phychem.kjist.ac.kr/312.pdf)									
CRP	Cation			0.6	–	2.6	40	6.3	65
ARP	Anion			0.5	–	1.8	34	6.9	79
SYBRON CHEMICALS Inc. (Birmingham, NJ, US; htpp://www.ion-exchange.com/products/membranes/index.html)									
IONAC									
MC-3470	Cation	80°C	Hetero	0.38	10	1.5	35	10–25	96
MA-3475	Anion	80°C	Hetero	0.41	10	0.9	31	25–50	99

[a]Type; main properties; category; thickness, burst strength; ion-exchange capacity, IEC; gel water content; area resistance, r_m, at 25°C, 0.5 kmol $NaCl/m^3$; membrane perm-selectivity, α_m, at 1.0/0.5 $kmol/m^3$ KCl.

[b]As claimed by the manufacturers or extracted from Elmidaoui *et al.* (2002); Lacey (1972); Lee *et al.* (2002d); Strathmann (1992); and several online publications (http://phychem.kjist.ac.kr/312.pdf; http://phychem.kjist.ac.kr; http://www.pca-gmbh.com).

changes in the membrane mechanical or electrical properties after exposure to acidic or alkaline solutions, as well as oxidizing agents, for prefixed time intervals. For instance, by reducing the DVB content in the resin matrix, it was possible to improve the membrane chemical resistance, thus resulting in an extended membrane life. In the circumstances, the novel IEM labelled as AXE 01 (Eurodia Industrie, Wissous, F) can resist to irreversible organic fouling, since its soaking in aqueous solutions containing 1% (w/w) NaOH and 1% (w/w) NaCl at 60°C for 5 days makes the membrane swell so as to remove easily the organic molecules trapped inside the membrane itself with just a 10% or 20% reduction in IEC or burst strength (Elmidaoui *et al.*, 2002). Moreover, whereas almost all electromembranes operate at temperatures smaller than 40°C, this novel membrane can be operated up to 60°C, thus favoring mass transport while limiting microbial growth in the solutions undergoing ED treatment.

Membrane fouling due to adsorption of polyelectrolytes (such as humic acids, surfactants, and proteins) may severely reduce ion permeability, especially in the anion-exchange membranes. However, exhausted anion-exchange membranes used in the ED of molasses, whey, citric acid, or sodium dodecylbenzenesulfonate can be reactivated by circulating simultaneously an acidic solution in one compartment and an alkaline solution in the other one, both solutions at titres greater than 0.1 kmol/m^3 (Tokuyama Soda Co., 1983).

Table II shows the main characteristics of some representative commercially available IEMs as claimed by the manufacturers or extracted from Elmidaoui *et al.* (2002), Lacey (1972), Lee *et al.* (2002d), Strathmann (1992), and several online publications (http://phychem.kjist.ac.kr/312.pdf; http://phychem.kjist.ac.kr; http://www.pca-gmbh.com).

Finally, the commercial ED membranes made "tight" or "loose" have an average pore size of 1 or 10 nm, respectively. This gives rise to a smaller or a greater permeability toward co-ions. A typical ED membrane has usually a pore size of 1–2 nm with an IEC of 1.6–3.0 meq/g of dry mass of membrane.

a. ED stack construction. An ED system consists of the following items: one or more ED stacks, a DC generator, pumps, piping, tanks, and measuring devices for pressure, temperature, pH, electric conductivity, and flow rate.

The ED stack is the unit holding together anionic and cationic membranes assembled in parallel as in a filter press between two electrode-end blocks in such a manner that the stream undergoing ion depletion (i.e., the *diluate* or *diluting stream*) is kept separated from the other solution (*concentrate* or *concentrating stream*) undergoing ion enrichment. Figure 4 shows an exploded view of it.

The two electrode-end blocks contain the in- and outlet adapters and the electrical connections. They are pressed together by a steel frame to hold the

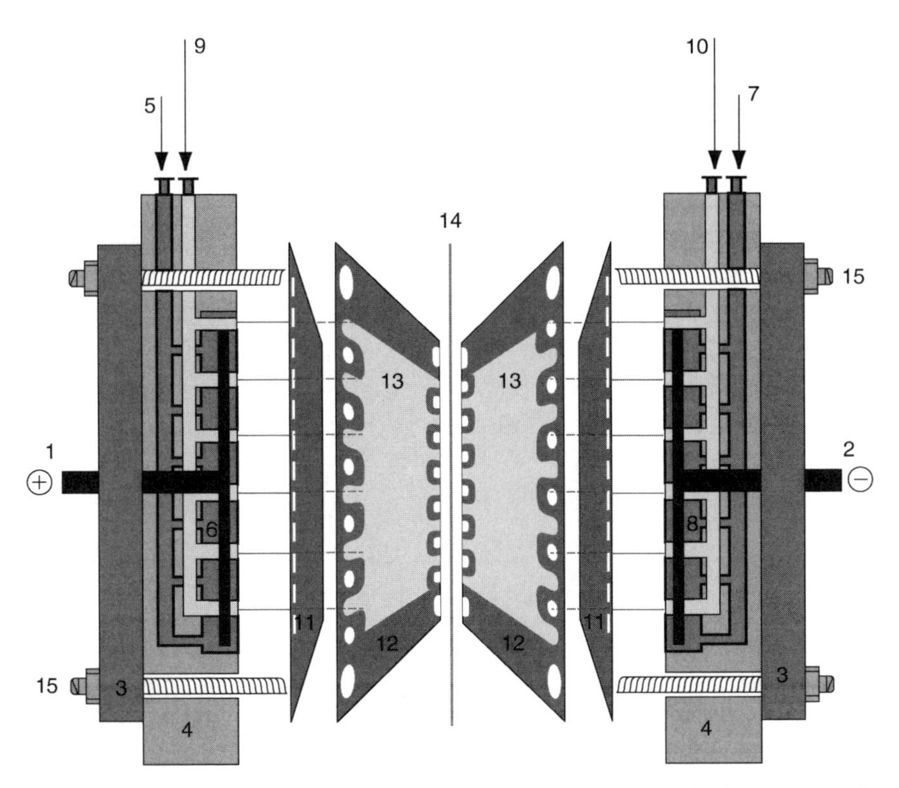

FIG. 4 Exploded view of an ED stack with indications of its main items: 1, anode; 2, cathode; 3, steel frame; 4, plastic end plate; 5, inlet anode compartment; 6, anode chamber; 7, inlet cathode compartment; 8, cathode chamber; 9, inlet-concentrating compartment; 10, inlet-diluting compartment; 11, cation-exchange membrane; 12, spacer-sealing frame; 13, spacer net; 14, anion-exchange membrane; 15, screws.

stack components together. The inside surfaces of the electrodes, generally consisting of carbon, stainless steel, or titanium, are recessed (Figure 4) so as to form an electrode-rinse compartment when an IEM or a neutral one is clamped in place.

The basic unit of an ED stack is the so-called *cell pair*, which consists of a couple of anionic and cationic membranes together with their adjacent compartments (Figure 5A). By applying a direct electric voltage to the electrodes, the ions in the solution flowing in a generic compartment tend to migrate toward the electrode with opposite charge. Cations freely flow through the cationic membranes, but they are retained by the anionic ones. On the contrary, anions can permeate across the anionic membranes, while they are rejected by the cationic ones. The overall result of this process is that

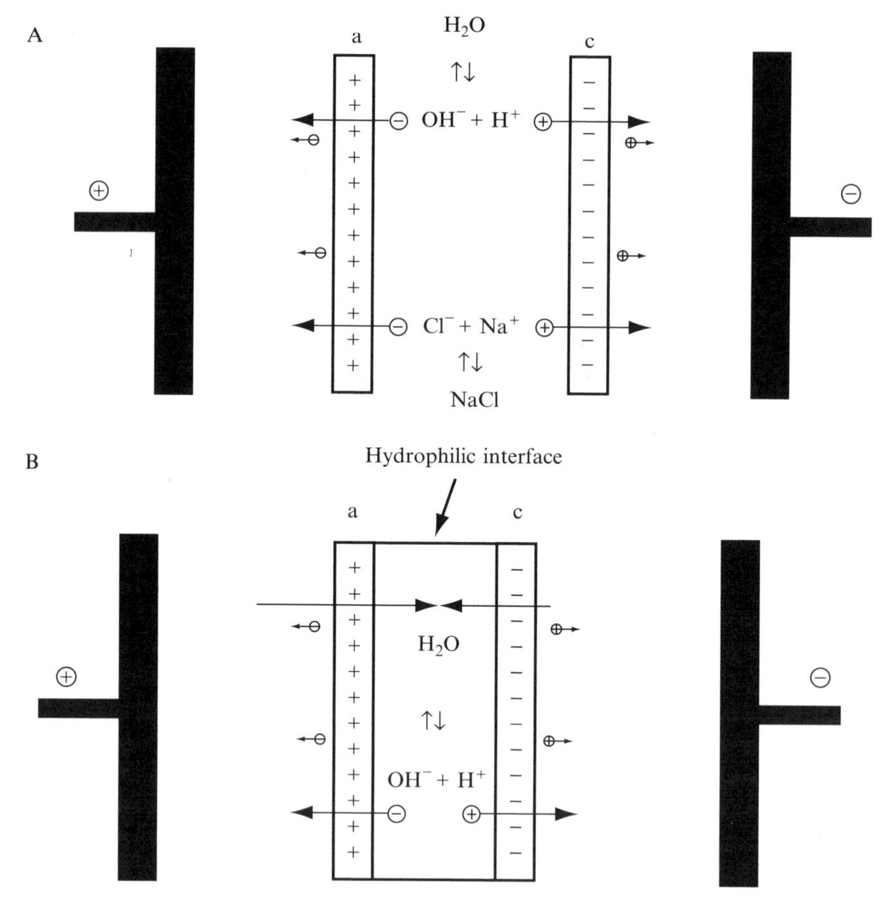

FIG. 5 Basic operating principles of an electrodialysis process using (A) a couple of monopolar (anionic, a, and cationic, c) membranes or (B) a bipolar membrane.

the electrolyte concentration increases in a series of alternate (e.g., *concentrating*) compartments and simultaneously decreases in the other adjacent (e.g., *diluting*) compartments.

The number of cell pairs can vary up to 1000 in commercial units. For instance, the stack EUR40 commercialized by Eurodia Industrie (Wissous, F) can assemble up to 1000 cells using various intermediate plates with and without electrodes, each cell pair having an effective membrane surface area of 0.4 m^2. Pilot-scale units, such as EUR2 or EUR6, may contain up to 10 or 80 cells with a surface area of 0.02 or 0.06 m^2, respectively (htpp://www.eurodia.com).

The usual thickness of commercial membranes ranges from 0.05 to 0.8 mm (Table II), while each channel gap is of a few millimetres. The thinner the membrane, the smaller the electric resistance and mechanical strength become, whereas the narrower the channels, the greater their conductance and pressure drops are.

ED stacks can be subdivided into two basic types: tortuous path or sheet flow (Figure 6). In the former, the solution is forced at superficial velocities (v_S) of 30–50 cm/s through a long (about 100 times the nominal membrane width) narrow channel with numerous 180° bends between the inlet and outlet ports of a compartment and a great number of cross-straps to support the membrane itself and promote turbulence in the flowing solution (Figure 6A). On the contrary, in the latter the solution approximately flows along the straight line linking the entrance and exit ports of any compartment (Figure 6B) at quite smaller v_S values (5–15 cm/s).

The differences in v_S, direction changes, and path length make the pressure drop (200–400 kPa) through a tortuous-path stack by far greater than that (20–70 kPa) in a sheet-flow one, as reported by Lacey (1972) and Strathmann (1992).

In tortuous-path stacks there is no need for spacer screens as thicker membranes, narrow channels, and plenty of cross-straps are used. On the contrary, in sheet-flow stacks spacers of different geometry and thickness are necessary to prevent membrane contact (that would result in burning through), as well as to induce turbulence in the flowing solution (Kuroda *et al.*, 1983). Spacers generally consist of a sealing frame and a net in the

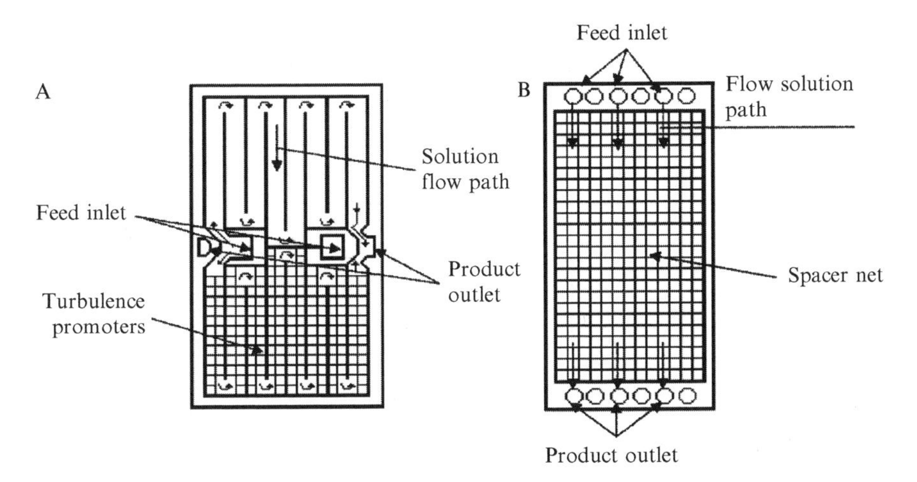

FIG. 6 Schematic diagrams of (A) tortuous-path and (B) sheet-flow ED spacer gaskets.

active area, which is filled with the electrolyte. Once the spacers have been stacked, the openings in the sealing frame result in tubes, which are arranged so as to build both the channel systems for the concentrating and diluting streams (Figure 6).

Whereas the tortuous path stacks are manufactured by Ionics Inc. (Watertown, MA, US) (http://www.ionics.com) only, the sheet-flow ones are produced by several companies, such as Astom Co. (Minato-Ku, Tokyo, Japan; http://www.astom-corp.jp), a jointed company of Tokuyama Co. (Shibuya-ku, Tokyo, Japan) and Asahi Chemical Industry Co. (Chiyoda-ku, Tokyo, Japan) founded in 1995; Asahi Glass Engineering Co. (Chiyoda-ku, Tokyo, Japan; http://www.agc.co.jp/english/chemicals/ion-maku/selemion/sele.htm#anchor795659); Du Pont Co. (Fayetteville, NC, US; http://www.dupont.com/fuelcells/pdf/nae101.pdf); Electropure Excellion™ (Laguna Hills, CA, US; http://www.electropure-inc.com/downloads/Excellion%20Specifications.pdf); Linan Euro-China Co. (Linan City, China; http://www.Linanwindow.Com/Qianqiu/Membeng.Htm); Membranes International Inc. (Glen Rock, NJ, US; http://www.membranesinternational.com/tech.htm); Pall RAI Inc. (Hauppauge, NY, US); PCA GmbH (Heusweiler, D; http://www.pca-gmbh.com); Permutit Co. (UK), that became a part of the USFilter Corp. (Warrendale, PA, US; http://www.usfilter.com/water/Business+Centers/Industrial_Process_Water/Industrial_Process_Water_Products/pw_permutit.htm) in 1993; Rhone-Poulenc Chemie GmbH (Frankfurt, D); and Sybron Chemicals Inc. (Birmingham, NJ, US; http://www.ion-exchange.com/products/membranes/index.html).

b. ED stack arrangement. Depending on the peculiar application, several arrangements of ED stacks are used. The simplest case, that is, the *batch desalination process*, is carried out by circulating the solution through the stack from a storage vessel and vice versa (Figure 7A) until the desired degree of ion depletion or enrichment is achieved, this being characterized by specific target values for the electric conductivity of the dilute and/or concentrate. As the electric conductivity of the dilute reduces, the voltage potential applied to the ED stack has to be increased to keep the current density constant, thus raising the overall electric energy consumption.

It is also possible to run an ED process in the *continuous* mode (Figure 7B) by letting simultaneously a fresh feed enter the ED unit and the final product flow out of it. This operation can be performed in one or more than one turn. The latter is generally referred to as the *feed-and-bleed* mode (Figure 7C) and needs a portion of the product to be blended with the raw inlet feed solution. This procedure is applied when the feed solute concentration fluctuates and a continuous flow of product is required or when the demineralization degree is low. To minimize the overall membrane

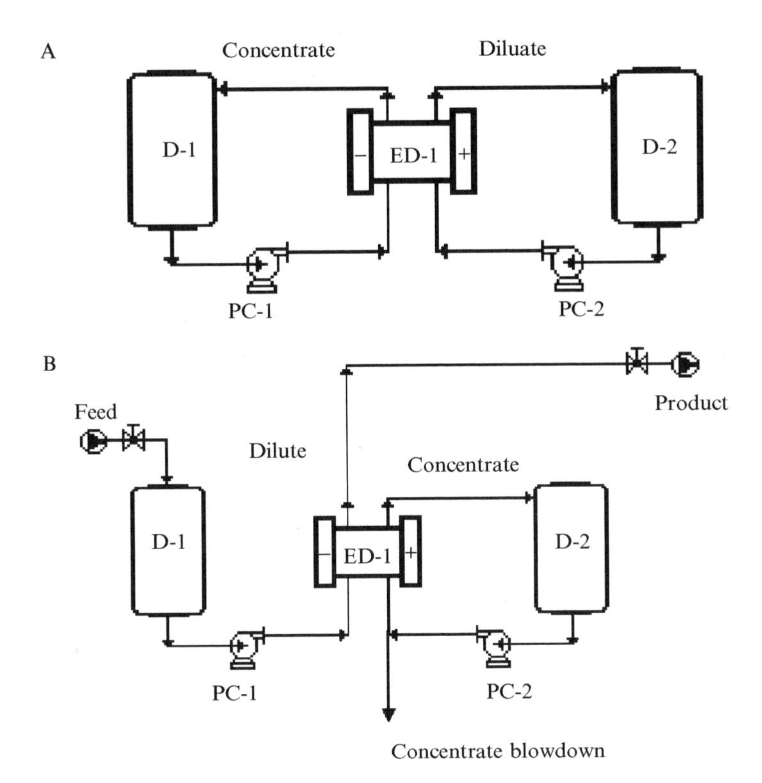

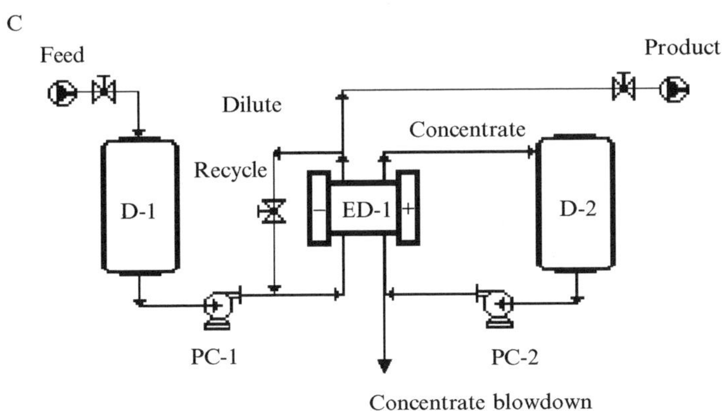

FIG. 7 Flow diagrams of the different operating modes for an ED unit: (A) batch process; (B) continuous single-passage process; and (C) feed-and-bleed process.

surface area installed, the de-ashing process can be fractionated into several feed-and-bleed stages, arranged in series and operated at progressively smaller current densities.

In the continuous mode, scaling of carbonates and hydroxides may be prevented by resorting to automatic acid addition to control the pH of the concentrate or to periodic or continuous concentrate blowdown.

Whatever the operating mode, an ED unit can work with constant polarity (that is in the conventional or unidirectional ED) or with reversed polarity. The latter was first proposed by Ionics Inc. (Watertown, MA, US) in the early 1970s (http://www.ionics.com/products/membrane/WaterDesalting/edr/edr2020.htm) as the *ED reversal process* (EDR) to solve one of the problems encountered in water desalination processes, that is, membrane fouling or scaling as due to the organic or inorganic substances present in brackish water. This process consists of inverting the polarities of the electrodes periodically (i.e., at time intervals varying from a few minutes to several hours), as well as the hydraulic flow streams. In the circumstances, fouling or scaling constituents that build up especially on the anionic membrane surfaces of the concentrating compartments in one cycle are removed or redissolved in the next cycle, when the concentrating compartments are reverted to the diluting ones. For instance, in the new Ionics EDR 2020 system the polarity of the DC power is reversed from two to four times per hour so as to allow automatic self-cleaning of membrane surfaces.

For more than 30 years, the EDR mode has been the process of choice in the desalination industry. It is today applied in other ED applications (e.g., whey demineralization) by other companies too (http://www.ameridia.com).

2. *Bipolar membranes*

At the end of the 1980s a new type of electromembrane (e.g., the bipolar membrane, bpm) started to be commercialized (Bazinet *et al.*, 1998). It consists of an anion- and a cation-exchange membrane that are bound together, either physically or chemically, and of a very thin hydrophilic layer (< 5 nm), where water molecules diffuse from the outside aqueous salt solutions (Figure 5B). In the presence of an electrical field, these molecules are dissociated into hydrogen (H^+) and hydroxyl (OH^-) ions (Mani, 1991). These ions can migrate out of that layer, provided that the bipolar membrane is correctly oriented (no current reversal being allowed in water splitting). With the anion-exchange side facing the anode and the cation-exchange side facing the cathode, the hydroxyl anions will flow across the anion-exchange layer and the hydrogen cations across the cation-exchange layer. In the circumstances, bipolar membranes allow an efficient generation and concentration (up to 10 kmol/m^3) of hydroxyl and hydrogen ions at their surfaces, thus resulting in the production of basic and acidic solutions, respectively.

The main advantages of bipolar membranes are no formation of gases at their surfaces or within the bipolar membranes themselves, a power consumption to dissociate water into O_2 and H_2 about half that used in electrolytic cells, a minimum formation of by-product or waste streams in the case of dilute (< 1 kmol/m^3) acids or bases, and reduced downstream purification steps.

The requirements for a bipolar membrane are low electrical resistance at high current density, high water dissociation rates, low co-ions transport rates, high ion selectivity, and good chemical and thermal stability in strong acids and bases. Low electrical resistance of the cationic and anionic layers of bpms can be obtained by using, for instance, sulphonic acid groups in the cation-exchange layer and quaternary amines in the anion-exchange layer of the bipolar membranes as fixed charges at high concentrations in the polymer matrix.

Tokuyama Soda Co. Ltd. (www.tokuyama.co.jp), the main shareholder of Eurodia Industries (http://www.eurodia.com), has developed a bipolar membrane type BP-1E, that is characterized by a thickness of 0.2–0.35 mm, a burst strength of 0.4–0.7 MPa, and water-splitting voltage and efficiency of 1.2–2.2 V and 98%, respectively (http://phychem.kjist.ac.kr/312.pdf). Other bpms with quite similar characteristics (http://www.hansei.com/filtra/ion-b2.htm; http://phychem.kjist.ac.kr/312.pdf) are manufactured by Aqualytics Inc. (Warren, NJ, USA), that has been acquired by FuMA-Tech GmbH (St. Ingbert, D; http://www.fumatech.com/).

a. ED stack arrangement. ED stacks using bpms can be arranged in two or three compartments and operated in a single- or multipassage continuous or batch mode as any conventional ED stack. The most common three-compartment configuration, obtained by interposing a bpm between an anionic (a) and a cationic (c) membrane, is used to split a salt solution into an acidic and a basic stream, each flowing out of the compartments limited by a bipolar and an anionic membrane and by a bipolar and a cationic membrane, respectively (Figure 8A). In this configuration, it is possible to arrange up to 200 cells in a single stack provided with a system of manifolds suitable to feed the three corresponding compartments in parallel, that is, acid, base, and dilute, besides the two electrode compartments.

By feeding a salt solution into the diluting compartments and water into both acid and base compartments, and by supplying a DC across the electrodes, it is possible to split a salt of an organic acid into the free acid and the corresponding hydroxide. In this process, only a small percentage (1–2%) of the overall power is consumed at the electrodes where small amounts of hydrogen and oxygen are generated.

In the two-compartment configuration, it is possible to couple a bpm with a cation- or an anion-exchange membrane only, thus minimizing both the overall investment and operating costs for the ED unit needs one less process loop, less membranes to install and replace and, consequently, less power to operate under constant-current density.

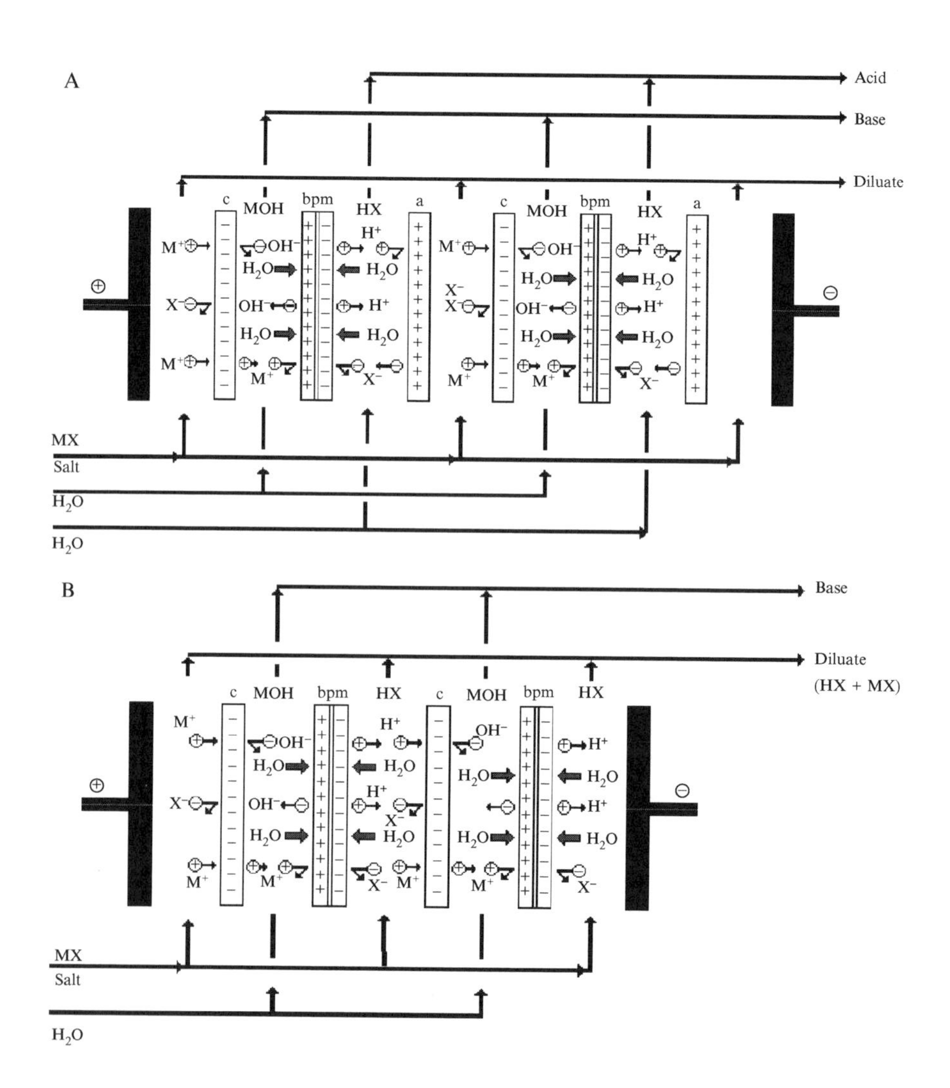

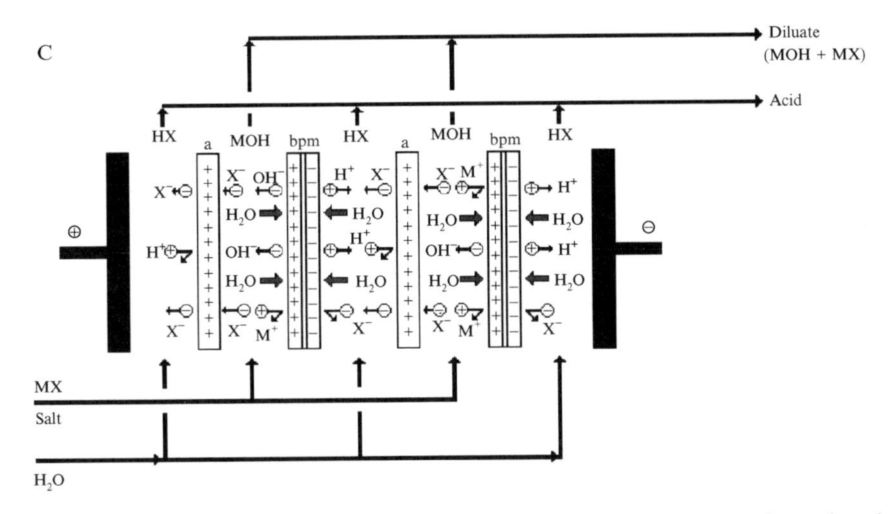

FIG. 8 Schematic diagrams of different arrangements for an ED unit equipped with bipolar membranes (bmp) coupled to anionic (a) and/or cationic (c) membranes: (A) three-compartment configuration; (B) two-compartment configuration using bipolar and cationic membranes; (C) two-compartment configuration using bipolar and anionic membranes.

The two-compartment cells made of bipolar and cation- or anion-exchange membranes (Figure 8B and C) are useful to convert a salt of a weak acid and a strong base (such as the sodium salts of acetic, lactic, and formic acids, or sodium glycinate) or a weak base and a strong acid (such as ammonium chloride or sulfate) into a concentrated stream containing the corresponding strong base (Figure 8B) or acid (Figure 8C) and a diluting stream containing the residual fraction of the salt (e.g., 1–2% w/w) together with its corresponding free weak acid (Figure 8B) or base (Figure 8C). In these cases, it is impractical to use the three-compartment cell owing to poor electric conductivity of the corresponding weakly dissociated acid or base.

Over the last 15 years, about 2500 m^2 of bpms have been installed throughout the world to produce specialty and fine chemicals, such as amino and organic acids (Gillery *et al.*, 2002). A successful operation of an ED unit using bpms depends on several factors, such as a feed multivalent cation content not greater than 1–5 g/m^3 to avoid their precipitation over the cation-exchange membranes as they react with hydroxyls in the stack, a salt concentration greater than 1 $kmol/m^3$ to operate at high current density and reduce the membrane area required, an oxidizing compound-free feed, and a maximum operating temperature smaller than 40°C (Gillery *et al.*, 2002).

C. MASS TRANSFER IN AN ED STACK

Despite there is not yet enough knowledge on the exact phenomena taking place in the electromembranes, the ion flux in a generic electromembrane can, for the sake of simplicity, be described by means of an expression analogous to Eq. 6, where the ion transport number in solution ($t^{\pm}$) is replaced with that in the membrane phase ($t_{\mathrm{m}}^{\pm}$). Owing to the charged groups chemically attached to their support matrix, anionic or cationic membranes allow intrusion and exchange of negative or positive ions from an external source, as well as exclusion of the cations or anions, respectively (Figure 5). Thus, the transport number for anions in anion-exchange membranes would tend to 1, while that for cations to zero and vice versa for a cation-exchange membrane.

The anionic or cationic membrane perm-selectivity (α_{m}) for the corresponding counterions can be expressed as the ratio by the differences between the real and theoretical transport numbers in the membrane and solution phases:

$$\alpha_{\mathrm{m}} = \frac{t_{\mathrm{m}}^{\pm} - t^{\pm}}{1 - t^{\pm}} \tag{10}$$

When the counterion transport number in the membrane is the same as in the electrolytic solution, the perm-selectivity tends to zero (Strathmann, 1992).

As a result of the different ionic mobility in the solution and membrane phase, some species are rejected and, therefore, accumulate at one side of a semipermeable barrier, producing a concentration gradient and the so-called phenomenon of *concentration polarization* (CP). For instance, in the case of the flux of Cl^- in an ED stack, the transport numbers in the anionic membrane (t_{a}^{-}) and bulk solution (t^{-}) are about unitary and 0.6, respectively. This means that the capacity of the anionic membrane to convey Cl^- is greater than that of the concentrated solution, thus making the side of the anionic membrane facing the concentrated stream richer in Cl^-. Similarly, at that side there is also accumulation of Na^+ since the transport number for Na^+ in solution ($t^{+} \approx 0.4$) is greater than that in the anionic membrane ($t_{\mathrm{a}}^{+} \approx 0$). On the other side of the anionic membrane facing the diluting stream, the flux of Cl^- is greater in the membrane than in the bulk solution, thus making the membrane surface depleted of Cl^-.

Figure 9 shows a schematic diagram of an ED unit with the simplified concentration profiles in an ED cell pair.

In agreement with the great majority of the authors involved in ED stack design and optimization (Bailly *et al.*, 2001; Boniardi *et al.*, 1996, 1997; Fidaleo and Moresi, 2005a; Ibanez *et al.*, 2004; Lee *et al.*, 1998, 2002d; Nikonenko *et al.*, 2002, 2003; Yen and Cheryan, 1993), the overall ion flux

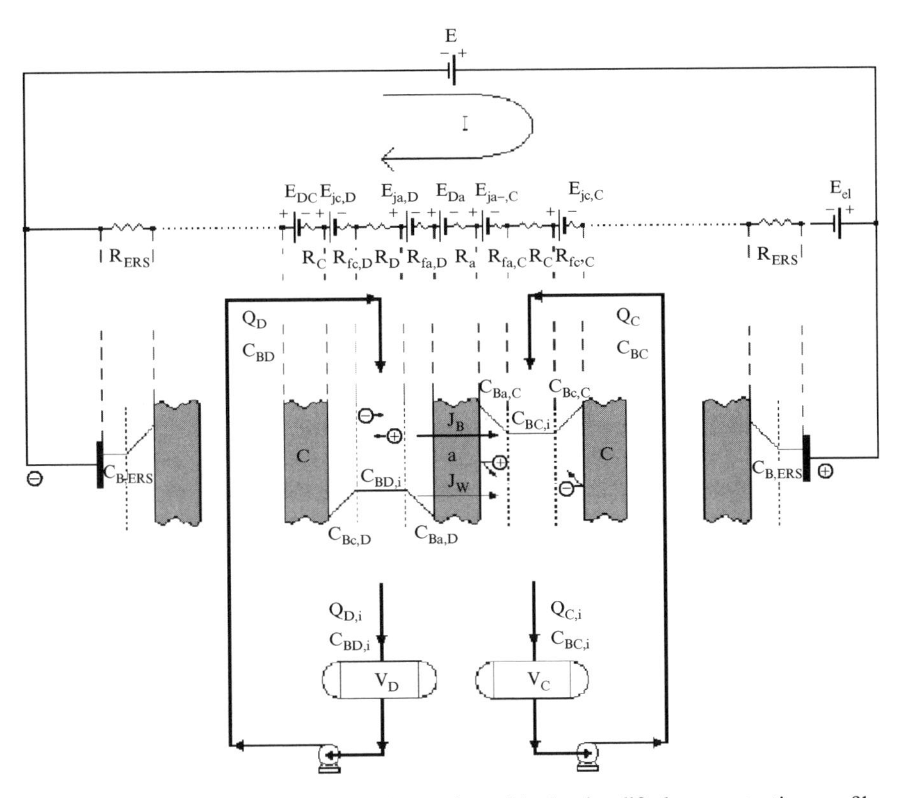

FIG. 9 Flow diagram for an ED unit together with the simplified concentration profiles in an ED cell pair and the analogous electrical circuit. All symbols are defined under the heading "Abbreviations".

in an ED cell pair was described using the Nernst–Planck (NP) equation that contains two terms expressing the contribution of diffusion and electromigration in the ionic transport (Buck, 1984) and can be regarded as a reduced form of the Maxwell-Stefan (MS) equation in the case of dilute systems (Krishna, 1987; Krishna and Wesselingh, 1997). In this way, only one diffusion coefficient per ionic species in each phase is necessary and this simplicity allows such equation to be easily coupled to other equations describing hydrodynamic conditions, water transport through the electromembranes, chemical reactions in the solutions and membrane, boundary, and other conditions. Even if such oversimplification should restrict the applications of NP equation for salt concentrations below 0.1 $kmol/m^3$ (Krishna, 1987), NP-based models were used to describe the ED desalination performance up to feed sodium chloride concentrations of 0.5 $kmol/m^3$

(Lee *et al.*, 2002d) or up to 1.7 kmol/m^3 (Fidaleo and Moresi, 2005a) provided that the model parameters were estimated using existing correlations or determined by independent experiments, respectively.

For the ED cell pair sketched in Figure 9, the overall flux of solute will be:

$$J_B = \frac{t_s}{F} j - L_B \Delta c_B \tag{11}$$

where L_B is the membrane constant for solute transport by diffusion, t_s is the effective cation ($= t_c^+ - t_a^+$) or anion ($= t_a^- - t_c^-$) transport number, while Δc_B ($= c_{BC} - c_{BD}$) is the difference in solute concentrations in C and D compartments, provided that the polarization effect is negligible. The negative sign at the right hand side of Eq. 11 accounts for the reverse direction of solute diffusion with respect to ion electromigration.

In the same ED cell pair, the overall water transport through the electromembranes from the dilute stream to the concentrate one can be expressed by accounting for electroosmosis (i.e., the migration of water molecules associated with ions, this being proportional to j) and osmosis phenomena:

$$J_W = \frac{t_W}{F} j + J_{Wd} \tag{12}$$

In accordance with the Spiegler-Kedem model (Krishna and Wesselingh, 1997), J_{Wd} is proportional to the net pressure difference across the membranes. Since the pressure difference (ΔP) may be regarded as negligible, J_{Wd} is mainly controlled by the corresponding instantaneous osmotic pressure difference ($\Delta\pi$), this being proportional to the difference in solute concentration across the membranes for a great number of solutes (Lo Presti and Moresi, 2000):

$$J_{Wd} = L_p(\sigma\Delta\pi - \Delta P) \approx L_p\sigma\Delta\pi \approx L_W \Delta c_B \tag{13}$$

where L_W is the membrane constant for water transport by diffusion.

1. *Limiting current concept*

With reference to Figure 9, at the boundary layers adjacent each electromembrane in the diluting or concentrating compartment it is possible to establish the following ion mass balance by accounting for the ion transport corresponding to NP equation both in solution and membrane phases:

$$J^\pm = \frac{t^\pm}{z^\pm} \frac{j}{F} - D_B \frac{dc^\pm}{dx} = \frac{t_m^\pm}{z^\pm} \frac{j}{F} \tag{14}$$

Upon integration over the generic boundary layer of thickness δ, in the case of a monovalent electrolyte, it follows:

$$j = \frac{k_m F(c_{Bk} - c_{Bmk})}{t_m^{\pm} - t^{\pm}} \tag{15}$$

where k_m is the mass transfer coefficient ($= D_B/\delta$), c_{Bk} and c_{Bmk} are the solute molar concentrations in the bulk solution in compartment k and at the generic membrane surface.

The so-called *limiting current density* is the first value at which current density is diffusion limited (Cowan and Brown, 1959), that is, the value at which the electrolyte concentration at any membrane surface falls to zero:

$$j_{lim,c} = \frac{F c_{BD} k_m}{(t_c^+ - t^+)}; \quad j_{lim,a} = \frac{F c_{BD} k_m}{(t_a^- - t^-)} \tag{16}$$

where c_{BD} is the bulk solute concentration in the dilute solution.

Equation 16 assumes that a concentration gradient exists only in the direction perpendicular to the membrane surface, that is, the flow pattern in the cell is turbulent in such a way that vertical concentration gradients are eliminated, as observed using a tracer dye by Krol *et al.* (1999).

As shown by Eq. 16, $j_{lim,k}$ should vary linearly with c_{BD} and increase as k_m increases.

2. *Mass transfer coefficient*

As shown in Table III, several authors (Fidaleo and Moresi, 2005a; Kraaijeveld *et al.*, 1995; Kuroda *et al.*, 1983; Sonin and Isaacson, 1974) established power function relationships between the Sherwood number (*Sh*) and the Reynolds (*Re*) and Schmidt (*Sc*) numbers in ED cells equipped with different eddy promoters, even if different definitions of the equivalent diameter were used to calculate the Reynolds number.

In a great number of papers dealing with the design of ED stacks, and especially in the recent and comprehensive paper by Lee *et al.* (2002d), the solute mass transfer coefficient (k_m) is expressed as a nonlinear function of the superficial flow velocity (v_S):

$$k_m = \alpha_0 v_S^{\alpha_1} \tag{17}$$

where α_0 and α_1 are empirical coefficients depending on the cell and spacer construction with α_1 ranging from 0.5 to 1.

One of the main disadvantages of the above-mentioned relationship is that it does not account for the variation of density (ρ), viscosity (η), and solute diffusivity (D_B) with c_B. On the contrary, the use of dimensionless *Sh*, *Re*, and *Sc* numbers has been traditionally regarded as the most accurate mode to deal with such variations. The empirical dimensionless correlations reported in Table III, except for the one established by Kraaijeveld *et al.* (1995),

TABLE III

EMPIRICAL CORRELATION FOR PREDICTING THE SOLUTE MASS TRANSFER COEFFICIENT (k_m) IN ED CELLS WITH EDDY PROMOTERS

Correlation	*Re* Definition	*Re* Range	d_e (mm)	References
$Sh = 0.43\mathrm{Re}^{1/2}Sc^{1/3}\sqrt{\frac{d_e}{d_{eR}}}$	$Re = \frac{\rho v_S d_e}{\eta(1-\varepsilon_S)}$	50–70	$d_e = 3.7–4.2$	Kuroda *et al.*, 1983
	$d_e = \frac{2hW(1-\varepsilon_S)}{h+W}$		$d_{eR} = 1.58$	
$Sh = 1.9Re^{1/2}Sc^{1/3}\sqrt{\frac{h}{\Delta l}}$	$Re = \frac{\rho v_S h}{\eta}$	300–2000	–	Sonin and Isaacson, 1974
$Sh = 8.09Re^{0.13}Sc^{1/3}\sqrt{\frac{h}{\Delta l}}$	$Re = \frac{\rho v_S h}{\eta}$	100–300	–	Kraaijeveld *et al.*, 1995
$Sh = (0.53 \pm 0.01)Re^{1/2}Sc^{1/3}$	$Re = \frac{\rho v_S h}{\eta}$	10–25	–	Fidaleo and Moresi, 2005a

show that k_m is proportional to the square root of v_S, this being in agreement with the average experimental value (0.47 ± 0.05) obtained by Lee *et al.* (2002d).

For instance, by using the correlations previously developed as referred to an ED cell, equipped with AMV or CMV membranes (Table II) and 0.4-mm thick spacers, it was possible to estimate that, as NaCl concentration increased from 0 to 2 kmol/m^3, D_B was of the order of 1.2×10^{-9} m^2/s, the *Sh* was practically constant and equal to about 22, and k_m ranged from 53 to 77 μm/s (Fidaleo and Moresi, 2005a).

D. OVERALL POTENTIAL DROP ACROSS AN ED STACK

Anytime a generic boundary layer or semipermeable barrier separates two phases or zones at different electrolyte concentrations a junction (E_j) or a Donnan (E_D) potential is established, the value of which can be estimated in accordance with Boniardi *et al.* (1996), Prentice (1991), and Vetter (1967).

By referring to Figure 9, in any cell pair the presence of an anionic and a cationic membrane gives rise to four boundary layers and thus to four junction potentials differences ($E_{ja,k}$ and $E_{jc,k}$) and two Donnan potential differences (E_{Da} and E_{Dc}), their mathematical expressions being given in Table IV.

In the equivalent electrical circuit of the ED stack sketched in Figure 9, such potentials were regarded as a series of batteries the polarities of which were *a priori* assumed as coincident with those of the external DC generator (E), except for the DC generator representing the thermodynamic and overpotential of electrodes (E_{el}). For instance, since $t_a^- > t_a^+$ and $c_{Ba,D}$ is generally smaller than $c_{Ba,C}$, E_{Da} is negative. Thus, it should be represented as a DC generator with inverted polarities with respect to that sketched in Figure 9. By contrast, in the case of $t^- > t^+$, since $c_{Ba,C}$ is generally greater than c_{BC}, $E_{ja,C}$ is positive, thus being equivalent to a DC generator with the same polarities of the external DC generator, that is, as the corresponding DC generator shown in Figure 9.

To estimate E_j and E_D, it is necessary to estimate the electrolyte concentrations at the membrane surfaces. On the assumption that the boundary layers established are invariant, such concentrations can be calculated on the basis of the bulk concentrations, current density, transport number of ions, and mass transfer coefficient (k_m), via Eq. 15.

By applying the second Kirchhoff's law to the equivalent electrical circuit of the ED stack (Figure 9), the overall potential drop across an ED stack can be written as:

$$E - E_{el} + (E_j + E_D)N_{cell} = RI \tag{18}$$

TABLE IV

MATHEMATICAL EXPRESSIONS OF THE JUNCTIONS ($E_{ja,k}$ AND $E_{jc,k}$) AND DONNAN (E_{Da} AND E_{Dc}) POTENTIAL DIFFERENCES FOR A GENERIC ANIONIC OR CATIONIC MEMBRANE; RESISTANCES OF THE BOUNDARY LAYERS ADJACENT TO THE ANIONIC ($R_{fa,k}$) AND CATIONIC ($R_{fc,k}$) MEMBRANES, OF BULK SOLUTIONS (R_k) IN D AND C COMPARTMENTS, AND ELECTRODE-RINSING SOLUTION (R_{ERS})[a]

Parameter	C compartment	D compartment
$E_{ja,k}$	$\frac{R_G T_K}{F}(t^- - t^+)\ln\left(\frac{c_{Ba,C}}{c_{BC}}\right)$	$\frac{R_G T_K}{F}(t^- - t^+)\ln\left(\frac{c_{BD}}{c_{Ba,D}}\right)$
$E_{jc,k}$	$\frac{R_G T_K}{F}(t^- - t^+)\ln\left(\frac{c_{BC}}{c_{Bc,C}}\right)$	$\frac{R_G T_K}{F}(t^- - t^+)\ln\left(\frac{c_{Bc,D}}{c_{BD}}\right)$
E_{Da}	$\frac{R_G T_K}{F}(t_a^- - t_a^+)\ln\left(\frac{c_{Ba,D}}{c_{Ba,C}}\right)$	
E_{Dc}	$\frac{R_G T_K}{F}(t_c^+ - t_c^-)\ln\left(\frac{c_{Bc,D}}{c_{Bc,C}}\right)$	
$R_{fc,k}$	$\frac{2D_B F}{ja_{me}\Lambda_0(t_c^+ - t^+)}\ln\left[\left(\frac{\Lambda_0 + \beta_1\sqrt{c_{BC}}}{\Lambda_0 + \beta_1\sqrt{c_{Bc,C}}}\right)\left(\frac{\sqrt{c_{Bc,C}}}{\sqrt{c_{BC}}}\right)\right]$	$\frac{2D_B F}{ja_{me}\Lambda_0(t_c^+ - t^+)}\ln\left[\left(\frac{\Lambda_0 + \beta_1\sqrt{c_{Bc,D}}}{\Lambda_0 + \beta_1\sqrt{c_{BD}}}\right)\left(\frac{\sqrt{c_{BD}}}{\sqrt{c_{Bc,D}}}\right)\right]$
$R_{fa,k}$	$\frac{2D_B F}{ja_{me}\Lambda_0(t_a^- - t^-)}\ln\left[\left(\frac{\Lambda_0 + \beta_1\sqrt{c_{BC}}}{\Lambda_0 + \beta_1\sqrt{c_{Ba,C}}}\right)\left(\frac{\sqrt{c_{Ba,C}}}{\sqrt{c_{BC}}}\right)\right]$	$\frac{2D_B F}{ja_{me}\Lambda_0(t_a^- - t^-)}\ln\left[\left(\frac{\Lambda_0 + \beta_1\sqrt{c_{Ba,D}}}{\Lambda_0 + \beta_1\sqrt{c_{BD}}}\right)\left(\frac{\sqrt{c_{BD}}}{\sqrt{c_{Ba,D}}}\right)\right]$
R_k	$\frac{h - 2\delta}{a_{me}c_{BC}\Lambda(c_{BC})}$	$\frac{h - 2\delta}{a_{me}c_{BD}\Lambda(c_{BD})}$
R_{ERS}	$\frac{h_{ERS}}{a_{ERS}c_{B,ERS}\Lambda(c_{B,ERS})}$	

[a] As extracted from Fidaleo and Moresi (2005a).

where I is the current flowing through the ED device; E_{el} the electrode potentials for anode and cathode processes; R the overall resistance of the membranes, bulk solutions, boundary layers, and electrode-rinsing solutions; E_j and E_D are the overall junction and Donnan potential differences across the boundary layers and membranes pertaining to any cell, respectively; and N_{cell} the overall number of cells, each one being composed of a couple of anionic and cationic membranes.

The overall electric resistance of the ED stack can be expressed as follows:

$$R = (R_c + R_{fc,D} + R_D + R_{fa,D} + R_a + R_{fa,C} + R_C + R_{fc,C})N_{cell} + 2R_{ERS} \tag{19}$$

where $R_{fa,k}$ and $R_{fc,k}$ are the resistances of the boundary layers adjacent to the anionic and cationic membranes in the generic k-th compartment, R_c and R_a, R_C and R_D, and R_{ERS} are the resistances of the cationic and anionic membranes, the bulk solution in C and D compartments, and electrode-rinsing solution, respectively.

Except the membrane resistances, any other generic k-th ohmic resistance can be estimated by applying the second Ohm's law:

$$R_k = \int_0^s \frac{dx}{\chi a_k} \tag{20}$$

where a_k is the effective membrane or electrode surface area involved in the ion flow pattern, while χ and s are the electric conductivity and thickness of the electrolyte solution involved.

Table IV also reports the mathematical expressions useful to estimate such resistances, where the electric conductivity of the solutions involved was expressed in terms of equivalent conductance. For the sake of simplicity, the resistances of the boundary layers adjacent to any membrane were estimated by expressing the equivalent conductance as a linear function of the square root of solute concentration, that is, by neglecting the contribution of the empirical coefficients β_2 and β_3 of Eq. 4.

When recirculating the same electrolytic solution through the concentrating and diluting compartments of an ED cell while increasing the voltage applied by the external DC generator, it is possible to establish a characteristic current–voltage relationship. For $E < E_d$ negligible current flows through the cell, while for $E > E_d$ the current starts to increase with E, thus exhibiting a typical three-region pattern. In the first region (the so-called *ohmic region*) a linear relationship is observed. As E increases, the solute concentration at the membrane surface reduces. As c_B falls to zero, a first deviation in the linear trend of the E versus I plot is observed. Any further increase in E gives rise to a smaller increase in I, thus leading to the

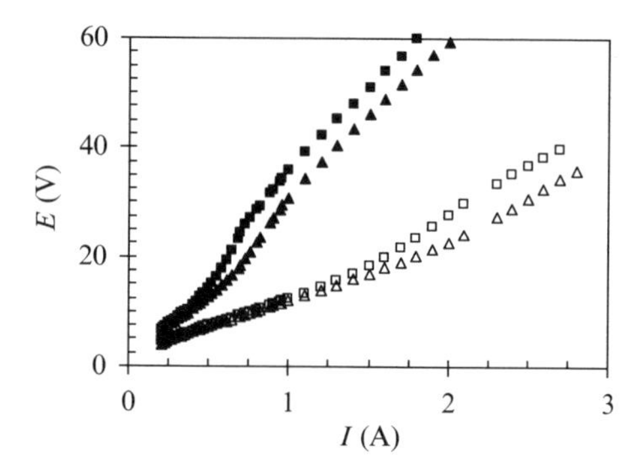

FIG. 10 Main results of typical limiting current tests performed using a membrane pack composed of 19 CMV membranes (Table II) under two levels of NaCl concentration ($c_B = 9$ mol/m^3, closed symbols; $c_B = 28$ mol/m^3, open symbols) and superficial velocity v_s (□, ■: 3.4 cm/s; △, ▲: 5.9 cm/s), as derived from Fidaleo and Moresi (2005a): voltage (E) vs. current intensity (I).

second region (sometimes called *plateau region*) (Krol *et al.*, 1999). By continuing to increase E, I tends to increase again, thus allowing the presence of the third over-limiting region to be revealed. In conventional ED stacks, polarization results in the following phenomena: as the limiting current is exceeded, the apparent resistance of the cell rises sharply, the pH of the dilute falls, while that of the concentrate increases, the Coulomb efficiency falls. Wen *et al.* (1996) demonstrated that the main mechanism responsible for all these phenomena is a large fall in the net electrical motive force (emf), mainly due to a great increase in the back emf, due to the membrane concentration potential, while the increase in the real ohmic resistance of the stack was minimal as the contribution of water dissociation was negligible, ranging from 1% (Wen *et al.*, 1996) to 3% (Krol *et al.*, 1999).

Both the first (ohmic) and second (polarization controlled) regions can be noted in Figure 10, reporting the results of typical limiting current measurements performed on an ED stack composed of only 19 cationic membranes (CMV type, see Table II) and model solutions containing 9 and 28 mol of NaCl per m^3 for superficial velocities (v_s) ranging from 3.4 to 5.9 cm/s (Fidaleo and Moresi, 2005a).

E. OVERALL ED PERFORMANCE INDICATORS

The major factors that determine the performance of an ED process are the solute recovery efficiency (ζ), total current efficiency (Ω), the specific energy consumption (ε), and the maximum solute weight concentration achievable in the concentrate ($C_{BC,max}$).

The solute recovery efficiency (ζ) is defined as the ratio between the solute mass (Δm_B) transported into the C stream or tank and the initial solute mass (m_{F0}) in the feed:

$$\zeta = \frac{\Delta m_B}{m_{F0}} \tag{21}$$

It can vary up to 95% depending on the application of choice and influences the overall energy consumption.

The total current efficiency (Ω) can be expressed as the product of two efficiencies:

$$\Omega = \Omega_C \Omega_L \tag{22}$$

where Ω_C is the Coulomb's efficiency and Ω_L is the current leakage efficiency. Whereas the former accounts for the imperfection of the electromembranes used, the latter for the inefficiency resulting from current leakage through pipes and manifolds. When the contribution of diffusion is negligible, Ω_C coincides with t_s. Since the co-ion transport numbers (t_a^+, t_c^-) in commercial electromembranes usually vary in the range 0.01–0.05, Ω_C is generally greater than 0.9. For instance, in ED stack equipped with AMV and CMV membranes (Table II) and treating aqueous solutions containing 12–37 kg of NaCl per m^3, t_s was 97 $\pm$ 2% in the constant-current region (Fidaleo and Moresi, 2005a). Thus, for well-designed laboratory-scale equipment and dilute solutions Ω is greater than 90%, while for commercial units it may reduce to about 50% (López Leiva, 1988).

The overall energy consumption (L) in an ED stack, as that sketched in Figure 9, can be calculated as:

$$L = \int_0^{\theta} E(\theta')I(\theta')\, d\theta' \tag{23}$$

where θ is the overall process time. When the operation is performed in the constant-current and -tension regions, L reduces to:

$$L = R_{MP} I^2 \theta \tag{24}$$

where R_{MP} is the overall apparent electric resistance of the ED stack.

The overall energy consumption per kilogram of solute recovered ($\varepsilon = L/\Delta m_B$) also accounts for the contributions of the thermodynamic potential and overpotential of electrodes ($E_{el}\, I$) and ohmic resistance of the electrode-rinsing solution ($R_{ERS}\, I^2$). However, such contributions are negligible in the industrial-scale ED stacks, that may be composed of 500–1000 cells. By neglecting such contributions, the specific energy consumption (ε) was found to increase from 0.18 (Fidaleo and Moresi, 2005a) to 0.22 (Fidaleo and Moresi, 2004) kWh/kg for a sodium chloride or lactate recovery yield (ζ) of 90% at 1 A and 20°C.

The water transport number (t_W) accounts for water transport associated to ion electromigration and thus controls the maximum solute weight concentration theoretically achievable in the concentrate. By dividing the instantaneous solute mass transferred into the C compartment by the actual volume accumulated into tank C for θ tending to infinite, the following can be obtained:

$$C_{BC,max} = \frac{t_s M_B}{t_w V_W} \tag{25}$$

where V_W (= 0.018 m^3/kmol) is the molar volume of pure water and M_B is the molecular mass of the generic solute. The effective t_W value depends on the electromembrane used and was found to vary in the range 2–10 in the case of water desalination (Lee *et al.*, 2002a) and 14–26 in the case of sodium acetate or lactate recovery (Boniardi *et al.*, 1996; Chukwu and Cheryan, 1999; Fidaleo and Moresi, 2004, 2005b; Yen and Cheryan, 1993). Thus, such a large variation in t_W results in quite different $C_{BC,max}$ values.

F. FOULING AND SCALING OF ELECTROMEMBRANES

Membrane fouling or scaling is generally due to the organic or inorganic substances that are present in the solution to be electrodialytically treated. These may deposit onto the membrane surface or inside the membrane, thus resulting in a decline in the solute permeation flux and an increase in the electrical resistance and energy consumption of the ED stack. To restore the original membrane performance, membrane cleaning is carried out from time to time. Whether the fouling is irreversible, membranes are to be replaced. Both cases affect the ED process economics, membrane cleaning and replacement accounting up to 47% of the overall water desalting costs (Grebenyuk *et al.*, 1998).

When internal fouling takes place, the transport of ions through the electromembrane is hindered by slow-moving organic molecules in the membrane and/or by adsorption or precipitation of organic molecules in the membrane restricting the free-flow area of the membrane. External fouling occurs when a layer of precipitated or adsorbed organic molecules on the membrane surface adds an additional transport resistance to the intrinsic resistance of the membrane.

The precipitation mechanism is governed by the solubility of the foulant, while the adsorption is affected by electrostatic and hydrophobic interactions between the foulant and the membrane surface. The molecular size and the pH of the solution affect the solubility of organic acids. The larger the molecules the lower their solubility and mobility in the membrane become, this improving their adsorption affinity to the membrane. Consequently, fouling cannot be predicted simply by measuring the concentration of the organic matter present in the feed (Lindstrand *et al.*, 2000). First of all, the foulants are to be identified and their solubility has to be taken into account when predicting the degree of fouling.

Numerous foulants are known. It is worth citing some anionic surfactants such as sodium octanoate and sodium dodecylbenzene sulfonate (Lindstrand *et al.*, 2000), sulfonated lignin (Watkins and Pfromm, 1999), sodium humate (Korngold *et al.*, 1970; Lee *et al.*, 2002a), starch, gelatin, egg albumin (De Korosy *et al.*, 1970), grape must (Audinos, 1989), and milk whey (Lonergan *et al.*, 1982).

Table V classifies the main types of foulants together with their charge and methods used to prevent their fouling aptitude.

Among the numerous approaches studied so far to minimize such phenomena in ED, it is worth citing pretreatment of the feed solution by coagulation (De Korosy *et al.*, 1970) or microfiltration (MF) or ultrafiltration membrane processing (Ferrarini, 2001; Lewandowski *et al.*, 1999; Pinacci *et al.*, 2004), turbulence in the compartments, optimization of the process conditions, as well as modification of the membrane properties (Grebenyuk *et al.*, 1998). However, all these methods are partially effective and hydraulic or chemical cleaning-in-place (CIP) is still needed today, thus

TABLE V

MAIN FOULANT CATEGORIES, TYPES, AND CHARGE, AS WELL AS ANTIFOULING METHODS[a]

Foulant category	Type	Electric charge	Tendency to	Fouling prevention method
Scale	$CaCO_3$, $CaSO_4 \cdot 2H_2O$, $Ca(OH)_2$, etc.	Neutral	Precipitate on membrane surface	pH adjustment, EDTA or CA addition; lower concentration ratio
Colloids	SiO_2, $Fe(OH)_3$, $Al(OH)_3$, etc.	Negative	Agglomerate on membrane surface	MF or UF feed pretreatment; higher feed flow rate; lower concentration ratio
Organics	Polysaccharides, proteins, polyelectrolytes, humate, surfactans, etc.	Negative	Attach to membrane surface	AC, MF, or UF feed pretreatment; alkali rinsing, EDR process

[a]As extracted from Lee *et al.* (2002b).

AC, activated carbon; CA, citric acid; EDR, electrodialysis reversal; EDTA, ethylenediaminetetraacetic acid; MF, microfiltration; UF, ultrafiltration.

asking for additional chemicals or instruments that expand the investment and operating costs.

Many fouling indices have been defined to describe quantitatively fouling phenomena, such as the silt density index (SDI) and the modified fouling index (MFI) in membrane processing (Schippers and Verdouw, 1980). Actually, MFI was intended to be an aid in predicting the rate of fouling of reverse osmosis (RO) membranes by a specific type of water. It can be easily determined by plotting the ratio between the filtration time and filtrate volume as a function of total filtrate volume. In this way, three different regions can be generally identified in which blocking filtration, cake or gel filtration without or with compression consecutively occur. The linear section of the curve shows the contribution of the cake or gel filtration without compression and its slope represents the MFI, which is a measure of the fouling tendency of the feed under study (Schippers and Verdouw, 1980).

Such an index may be useful in the case of uncharged foulants, but it should be insufficient to assess the rate of fouling in ED membranes. To this end, another fouling index for ED was proposed by Lee *et al.*, 2002b. In presence of an electric field as that shown in Figure 9, the negatively charged foulants tend to migrate toward the anion-exchange membranes. Being rejected by the perm-selectivity of the anionic membranes, these foulants accumulate on the membrane surface owing to the concentration polarization phenomenon. As the foulant concentration at the membrane surface exceeds its saturation or gel concentration, the foulant precipitates, thus forming a fouling gel layer attached to the membrane surface. As the operation proceeds, such a layer becomes thicker and more compact, thus hindering the anions migration across the anionic membranes. Moreover, in accordance with the second Ohm's law the electrical resistance of the fouling gel layer (R_{fg}) will progressively augment as its thickness (δ_{gel}) increases, thus asking for a greater overall potential drop (E) across the ED stack of concern to keep the electric current constant through the process. Both the gel concentration (c_{gel}) and electric conductivity (χ_{gel}) vary with the size, shape, and chemical structure of the foulant, as well as the hydrophilicity of the membranes (Lee *et al.*, 2002b). Thus, the overall electric resistance (R) of the ED stack, as expressed by Eq. 19, has to include the additional term R_{fg}, which is proportional to the electrical charge Q $(= I\,\theta)$ passed through the ED stack (Lee *et al.*, 2002b). In the circumstances Eq. 18 can be expressed as

$$E - E_{el} + (E_j + E_D)N_{cell} = (R + R_{fg})I \qquad (26)$$

Under the constant electric current operating condition, all the terms in Eq. 26 except R_{fg} are independent of time for the continuous mode, this being also approximately valid at the beginning of ED processes operating in the batch mode. Thus, by dividing both members of Eq. 26 by the square of I, the following can be obtained:

$$\frac{E(\theta)}{I^2} \approx \text{const} + \Psi_{ED}\theta \tag{27}$$

where Ψ_{ED} is the slope of the plot of the $E(\theta)/I^2$ *versus* time (θ), this being defined as the *ED membrane-fouling index* by Lee *et al.* (2002b). The greater such an index the greater the fouling potential for the foulant under testing will be.

Grebenyuk *et al.* (1998) suggested another method to assess the membrane stability against fouling. It was based on the ratio between the overall energy consumption (L) and Coulomb's efficiency (Ω_C) during a real ED process to that observed in ideally nonfouled (K_f) or modified (K_s) membranes.

Watkins and Pfromm (1999) suggested a further method based on capacitance spectroscopy in the range 10^2–10^5 Hz to identify and tracking membrane fouling with organic macromolecules, such as sulfonated lignin, in real times.

Lindstrand *et al.* (2000) investigated the effect of different organic foulants, such as octanoic acid, sodium octonate, and sodium dodecylbenzene sulfonate, on anionic and cationic membranes by monitoring the increase in the membrane resistance (R_m) with time.

Finally, use of square (Lee *et al.*, 2002b) or half (Lee *et al.*, 2003) wave-pulsed electric fields of given frequency was found to be effective to mitigate anionic membrane fouling in presence of sodium humate.

By using a few surfactants, such as oligourethanes with three polar groups (i.e., the so-called trianchor compounds, A-3) or the disodium salt (NB-8) of the α,ω-oligooxipropylene-bis(o-urethane-2.4,2.6-tolueneurylbenzensulphonic acid), at different concentrations, it was possible to minimize membrane fouling in virtue of the formation of a surface protective layer with a charge opposite to that of the base membrane, which was capable of rejecting antipolar ions. However, such modifying substances are to be harmless and safely linked to the membrane surface to avoid any leaching in the food solution undergoing treatment. Moreover, the capability of such layers to generate H^+ and OH^- ions is to be accounted for. For instance, the competing transfer of such ions may reduce the current efficiency (Ω_C), while an increase in the alkalinity of the concentrating stream may lead to the formation of lowly soluble hydroxides, as well as calcium and/or magnesium carbonates, which deposit on the membrane surface (Grebenyuk *et al.*, 1998). Such a deposit was also detected on the cationic membrane side in contact with the alkaline solution during skim milk electro-acidification (Bazinet *et al.*, 2000a).

In conclusion, the anionic membranes appear to be more prone to be fouled by organic matter (De Korosy *et al.*, 1970; Grebenyuk *et al.*, 1998; Lee *et al.*, 2002a, 2003; Lindstrand *et al.*, 2000), while the cationic ones to be scaled by inorganic matter (Atamanenko *et al.*, 2004; Bazinet *et al.*, 2000; Lindstrand *et al.*, 2000). Thus, it is actually impossible to recommend any general procedure to moderate fouling and scaling in ED processing and there is a need for additional research on this issue so that the process can be optimized taking scaling and fouling into account.

III. ED APPLICATIONS

The present ED industry has experienced a steady growth rate of about 15% since 15 years (Srikanth, 2004). The most important industrial ED application is still the production of potable water from brackish water. However, other applications either in the semiconductor industry for the production of ultrapure, that is, completely deionized water without the chemical regenerations of IERs or in the food industry (i.e., whey demineralization, tartaric stabilization of wine, fruit juice deacidification, and molasses desalting) are gaining increasing importance with large-scale industrial installations.

Table VI shows a synopsis of the main industrial ED applications in the food sector.

A few selected examples will be briefly pointed out in the following sections.

TABLE VI

MAIN APPLICATIONS OF THE ED IN THE FOOD SECTOR

Application	Example
Fractionation	Brackish water desalination
	Nitrate removal from drinking water
	NaCl removal from amino acid solutions
	Cheese whey demineralization
	Desalting of protein hydrolysates (i.e., soy sauce), sugar solutions, molasses, and polysaccharide dispersions
	Deacidification of fruit juices
	Tartaric wine stabilization
	Flavor recover from pickle brines
Concentration	Edible table salt production from seawater
	Salts of organic acids from exhausted fermentation media
	Amino acids from protein hydrolysates
Splitting	Conversion of salts into their corresponding free acids and bases

A. BRACKISH AND SEAWATER DESALINATION

Depending on the total dissolved solid content (TDS, expressed in ppm or g/m^3), water may be classified as (http://www.tcn.zaq.ne.jp/membrane/english/DesalE.htm):

- Fresh Water: <1000 TDS
- Brackish: 1000–5000 TDS
- Highly Brackish: 5000–15,000 TDS
- Saline: 15,000–30,000 TDS
- Sea Water: 30,000–40,000 TDS
- Brine: 40,000–300,000 TDS

A variety of desalting technologies has been developed over the last 40 years. Based on their commercial success, they can be classified into major (viz., multistage flash distillation, MSFD; multiple-effect distillation, MED; vapor compression, VC; ED; RO) and minor (i.e., freezing, membrane distillation; solar humidification) processes.

In the inventory compiled for the International Desalination Association (IDA) by Wangnick (1998) the total capacity of installed desalination plants worldwide was estimated as equal to 22.7×10^6 m^3/day, almost half of which was used in the Middle East and North Africa (http://www.membranes-amta.org/media/pdf/reliable.pdf). About 24% of the world's capacity was concentrated in Saudi Arabia, most of which derived from distillation processes. About 16% was produced in the United States by resorting to RO-treated brackish water. MSFD and RO processes made up to about 86% of the total capacity, while the remaining 14% consisted of MED, ED, and VC processes, all the minor processes amounting to less than 1%. Despite the larger number of RO plants, MSFD plants have a higher total production capacity than any other process.

Table VII lists the global production of desalinated water, by process and plant capacity, as extracted from Wangnick (1998) and web site (http://www.desware.net/desa3.aspx).

Figure 11 shows a simplified schematic of a modern ED brackish water desalination plant. Similar to RO, ED plants require water pretreatment to maximize their performance. Coarse and 5- to 10-μm cartridge filtration prevents large particles from plugging the narrow channels of ED stacks. If the raw water Fe^{2+} or Mn^{2+} ion content is greater than 0.3 or 0.1 g/m^3, respectively, an oxidation step followed by filtration is needed to remove the insoluble Fe^{3+} or Mn^{4+} species formed. Raw water must also be accurately filtered to reduce turbidity and colloidal substances (silica), thus minimizing fouling. Chlorination is generally used to control biological fouling (as due to microorganisms, algae, and so on). Dechlorination via activated carbon

TABLE VII

SUMMARY OF WORLDWIDE DESALINATION CAPACITY TILL 1998, SPLIT BY PLANT TYPE AND PROCESS CAPACITY IN THE RANGE 100–60,000 m^3/day[a]

Desalting process	Percentage (%)	Capacity ($\times 10^{-6}$ m^3/day)	No. of plants
Multistage flash	44.4	10.02	1244
Reverse osmosis	39.1	8.83	7851
Multiple effect	4.1	0.92	682
Electrodialysis	5.6	1.27	1470
Vapor compression	4.3	0.97	903
Membrane softening	2.0	0.45	101
Hybrid	0.2	0.05	62
Others	0.3	0.06	120
Overall	100.0	22.57	12,433

[a]Wangnick (1998).

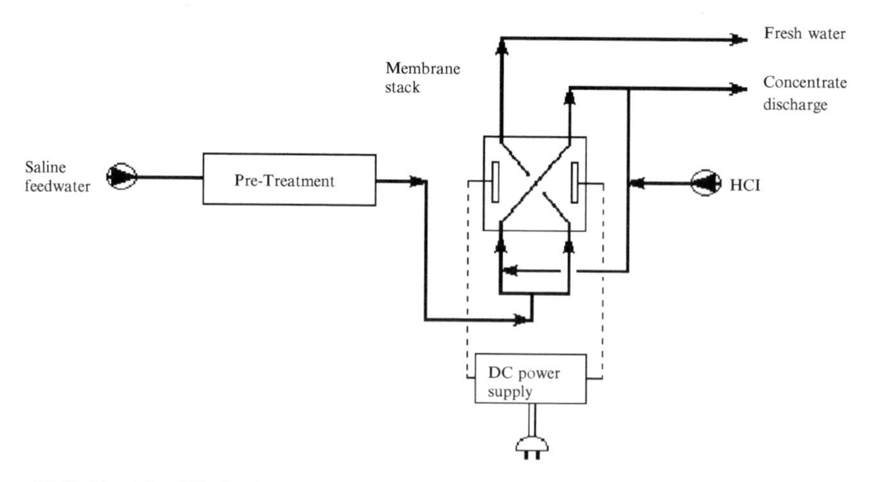

FIG. 11 Simplified schematic of a modern ED brackish water desalination plant.

filters or sodium bisulphite addition is required before water is fed into RO or ED modules, since both membranes are respectively more or less damaged by dissolved free chlorine. Well water with H_2S content greater than 0.3 g/m^3 is to be aerated and/or chlorinated to avoid fouling from elementary sulfur that results from H_2S oxidation in water. Scaling in the concentrating compartments by $CaCO_3$, $CaSO_4$, $BaSO_4$, and $SrSO_4$ decreases stack efficiency and is circumvented by adding an acid (i.e., HCl) or an inhibitor (i.e., sodium hexametaphosphate) into the brine stream or by using the EDR process. The latter is largely more complex than the ED conventional system owing to the automatic valves in the external piping that reverse the feed stream to the brine compartment and the product stream to the diluting one. Moreover, periodic acid cleaning is still required. By reversing the electrode polarity no more than four times per hour and using a 2-minute purge, it is possible to limit production losses to less than 13%.

In the case of a feed TDS ranging from 1.5 to 2 kg/m^3, the energy consumption is of the order of 1.6–2.6 kWh/m^3 of treated water. In the circumstances, there is no need to remove organic matter, colloids, and SiO_2 from raw water, while treated water is just chlorinated for disinfection. On the contrary, degasification and alkaline injection for pH adjustment are used with RO systems. Also, caustic neutralization is needed before RO retentate is discharged to the environment.

The main bottlenecks of RO processes are thus water pretreatment and post-treatment owing to higher RO membrane sensitivity to different organic and inorganic impurities, smaller water recovery (65–75% vs 80–90%), and shorter membrane lifespan (5–7 years against 7–10 years) with respect to ED membranes (Pilat, 2001). With water salinity up to 5000 ppm, ED is generally regarded as the most economic desalination process, while at TDS values higher than 12 kg/m^3 RO is more profitable than conventional ED units as far as power consumption is accounted for (Pilat, 2001). However, the overall energy consumed to desalt seawater using a cascade of two electrodialyzers, equipped with Asahi Glass CMV and AMV membranes (Table II) at current densities of 300–600 A/m^2 in the 1st stage and 300 A/m^2 in the second one, was estimated as equal to 6.6–8.7 kWh/m^3 of treated water at TDS of 0.45 kg/m^3, thus involving a desalination cost of US$1.05 per m^3 (Turek, 2003a).

Today, drinking water supply systems are still lacking in many countries of Asia, Africa, and South America and this makes the market of seawater desalination extremely larger than that of brackish water one. To this end, quite useful portable ED units are currently manufactured by EIKOS Co. (Almaty, Kazakhstan; http://www.eikos.ru/pageng/Eindex.htm), that can be simply connected to the kitchen tap through a flexible tube and, once switched on, can provide individual houses, hospitals, restaurants, and cottages with 5–130 dm^3/h of potable water (Pilat, 2001).

There is also interest in water desalination by solar-powered ED processes (AlMadani, 2003), especially in many Arabic countries where solar energy is available from 10 to 13 hours of sunshine per day during winter- and summertime, respectively. By using a small-scale commercial ED stack, equipped with 24 cell pairs, arranged in four hydraulic stages, and powered by photovoltaic cells, it was possible to desalt laboratory or groundwater of medium salinity at different degrees by varying the operating temperature and product flow rate in the ranges 10–40 °C and 8–47 dm^3/h, respectively.

ED is also applied to remove nitrates in drinking water. Owing to the intensive utilization of fertilizers in EU and US agriculture over the last decades, the concentration of nitrates in surface and underground water has steadily increased up to exceed largely the maximum EU (and WHO) norm of 50 g/m^3 for use as drinking water. Among the available competitive techniques (biological processes, BP; IER; and RO) for nitrate removal, the EDR system has the main advantages of high selectivity and water recovery yields of 96–98%, while it suffers from the same disadvantage of IER and RO of asking for further treatment to dispose the concentrate off. On the contrary, BPs allow nitrates to be directly degraded to molecular nitrogen, even if such a process is lengthy and needs bioreactors of quite large volumes.

Several EDR plants with capacity ranging from 120 to 3500 m^3/day have been operating since the early 1990s in the EU. Use of monovalent-selective anion-exchange membranes allowed a selective removal of nitrates over other anions such as chlorides, bicarbonates, and sulfates. Table VIII shows the average composition, pH, and electric conductivity of raw and treated waters, as resulted from about 5-week continuous operation in an EDR pilot unit installed in Vouillé (France) (http://www.ameridia.com/html/ele.html). The raw water was filtrated using a 10-μm cartridge filter and fed to the pilot unit by the well pump at 13°C and a flow rate of 11.5 m^3/h. Automatic polarity and hydraulic inversion was carried out three times per hour to avoid any eventual membrane fouling. Moreover, about 0.6 m^3/h of feed water was used to dilute the concentrated loop. In this way, it was possible to achieve a 69% nitrate removal yield with a 95% water recovery and a power consumption of approximately 0.1 kWh/m^3 of treated water.

TABLE VIII

AVERAGE PERFORMANCE OF A 50-m^2 EDR PILOT UNIT INSTALLED IN VOUILLÉ (FRANCE) THROUGHOUT A 5-WEEK CONTINUOUS OPERATION TO DESALT DRINKING WATER[a]

Parameter	Unit	Raw water	Treated water	Removal efficiency (%)
Ca^{2+}	g/m^3	108	61	44
Mg^{2+}	g/m^3	3	2	35
Na^+	g/m^3	13	12	8
K^+	g/m^3	1	1	0
HCO_3^-	g/m^3	250	171	32
Cl^-	g/m^3	25	12	52
SO_4^{2-}	g/m^3	18	17	6
NO_3^-	g/m^3	75	23	69
pH	–	7.9	7.7	3
Total dissolved solids	g/m^3	492	298	39
Electric conductivity	S/m	55.4	34.3	40

[a]http://www.ameridia.com/html/ele.html

The waste stream from any desalination process contains the salts removed from the saline feed to produce the fresh water product, as well as some of the chemicals added and corrosion by-products. Thus, its disposal may cause environmental problems. If the desalting plant is located near the sea, its disposal in the sea may be a minor problem provided that environmental changes related to the receiving waters for the accumulation of added constituents, dissolved oxygen levels, and different water temperatures are taken into account. If the desalting plant is sited inland, away from a natural saltwater body, waste disposal may involve dilution, injection into a saline aquifer, evaporation, or transport by pipeline to a suitable disposal point. Thus, the overall operating costs for a desalination plant depend on its capacity, type, and location, feed water, labor, energy, financing, concentrate disposal, and plant reliability. In general, the cost of desalted seawater is about three to five times greater than that of desalted brackish water when using the same plant size.

B. TABLE SALT PRODUCTION

The production of table salt from seawater, by the use of ED to concentrate sodium chloride up to 200 kg/m^3 prior to evaporation and salt crystallization, has achieved a certain commercial importance, especially in Japan and Kuwait, even if it seems to be highly subsidized (Strathmann, 1992). The key to the success of this technology has been the low-cost, highly conductive membranes with a preferred permeability of monovalent ions. This allowed chloride ions to be cumulated in the concentrated stream, while Ca^{2+} and Mg^{2+} ions and sulfates were quite totally rejected in the diluting stream.

More recently, Turek (2003b) suggested to desalt further such a diluate to produce potable water and reduce the overall ED-operating costs. Owing to the difficulty of such an operation, being the salinity of partially desalinated water as high as 22.9 kg/m^3 of TDS, such a stream was treated by using another EDR stage equipped with normal grade CMV, AMV Asahi Glass membranes (Table II), thus avoiding gypsum crystallization in the concentrated stream and yielding 90% recovery. By assuming the salt sell price at US$30 per metric ton (Mg), the potable water cost was estimated as US$0.44 per m^3.

C. DAIRY INDUSTRY

In the dairy industry the high content of minerals in cow milk (Table IX) restricts the commercial utilization of its main by-products, that is, whey and ultrafiltration permeates. The discovery that desalted whey could be used in baby food production as an economic alternative to the more expensive skim

TABLE IX

AVERAGE pH, BIOLOGICAL OXYGEN DEMAND (BOD_5), AND GROSS COMPOSITION OF COW MILK AND SWEET OR ACID WHEY[a]

Component	Cow milk	Sweet whey	Acid whey	Unit
pH	6.9	<6	<4	–
BOD_5	–	31	35	kg/m^3
Water	87.13	93.6	93.5	% w/w
Dry matter	12.87	6.4	6.5	% w/w
Lactose	4.9	4.9–5.0	4.48–4.76	% w/w
Lactic acid	–	0.03–0.04	0.42–0.49	% w/w
Raw protein	3.37	0.84–1.10	0.84–1.10	% w/w
Fat	3.9	0.06–0.07	0.03–0.04	% w/w
Ash	0.7	0.49–0.56	0.67–0.81	% w/w

[a]Moresi (1981), Zall (1992).

milk converted whey, a troublesome effluent to dispose of, into a value-added raw product. Thus, whey de-ashing has become a prerequisite to produce infant formulae closely resembling human milk, as well as to lactose crystallization or hydrolysis (Hoppe and Higgins, 1992).

Whey is a yellow–green liquid which is the major by-product of cheese making, the weight ratio of whey to cheese ranging from 9 to 11. It easily acidifies and can be distinguished into sweet or acid whey. Table IX shows its average composition (expressed as weight fraction), pH, and biological oxygen demand (BOD_5). It contains 4.5–5.0% lactose, 0.04% nitrogen components and approximately 0.5% ash. Its economical use is influenced by the high water content (94–95%). Owing to its BOD_5 content varying from 30 to 45 kg/m^3, whey cannot be directly discharged as wastewater into sewage.

The first commercial whey-processing ED plant was commissioned in 1961 (Stribley, 1963), while the first European plant was bespoke in Holland in 1963 (Batchelder, 1987). They consisted of a typical monopolar two-compartment ED process equipped with tortuous-path or sheet-flow spacer gaskets (Figure 6) and operated in the batch or continuous mode (Figure 7) provided that a 90–95% or 60–75% decrease in ash is to be respectively achieved (Ahlgren, 1972; Batchelder, 1987). According to Iaconelli (1973), the output capacity of an ED plant doubles or quadruplicates if the demineralization degree is reduced from 90% to 75% or 50%, respectively. Also the electric power consumption depends on the de-ashing level being not greater than 1 kWh/kg of ash removed if the demineralization degree is ranging from 50% to 75% (Ahlgren, 1972).

The ion removal medium is generally a sodium chloride solution, while the ideal electrode-rinsing solution contains sodium sulfate rather than sodium chloride to avoid chlorine evolution at the anode compartment (Shaffer and Mintz, 1966). Lifelong stainless steel or graphite electrodes are generally used, these being designed to operate also in the current reversal mode (EDR) (http://www.ameridia.com/html/dry.html).

The optimal performance of whey de-ashing is determined by many factors depending on feedstock composition and de-ashing degree.

By increasing whey concentration up to 15–20% TDS (w/v), the increase in electric conductivity enhances the Faraday efficiency with no counter effect resulting from the concomitant increase in solution viscosity (de Boer and Robbertsen, 1983; Johnson *et al.*, 1976). The optimal temperature is a compromise between de-ashing rates, which increase in the range of 30–40 °C, and bacteriological control, which is ideal at about 10 °C (de Boer and Robbertsen, 1983). Any increase in temperature beyond 40 °C was found to be ineffective because of protein agglomeration. This gives rise to particulate deposition onto the membrane surfaces, thus enhancing the polarization layer and overall stack resistance, and finally can lead to channel blocking and eventually to electrical short circuit.

During whey de-ashing by ED, multivalent ions (viz., PO_4^{-3} e Ca^{+2}) are generally removed after the monovalent ones (i.e., Cl^-, K^+, Na^+) have been essentially removed (Ahlgren, 1972; de Boer and Robbertsen, 1983; Hoppe and Higgins, 1992). For instance, conventional ED of an ultrafiltered (UF) permeate of *Grana* cheese whey allowed the ash content to be reduced from 0.42% to 0.05% (w/w), thus involving practically 100% reduction in the Na^+, K^+, and Cl^- ion contents, about 90% decrease in PO_4^{3-} anion level, and just as little as 25% decline in Ca^{2+} and Mg^{2+} cation concentrations (Mucchetti and Taglietti, 1993). To enhance removal of the latter, whey may be decalcified by using weak-acid IER in the sodium form (de Boer and Robbertsen, 1983) or acidified to a pH of about 4.6 (D'Souza *et al.*, 1973; de Boer and Robbertsen, 1983). Such a pH value, being quite coincident with the protein isoelectric point, was found to be optimal both for calcium solubility and for the removal rate of calcium salts, cation and anion transport being unhindered by charged protein groups (D'Souza *et al.*, 1973; de Boer and Robbertsen, 1983).

Demineralization of UF whey retentate as compared to UF whey permeate is generally slowed down by salt and/or proteins that build up onto and in the membranes (Pérez *et al.*, 1994). Finally, in the case of skim milk the aim of the ED process is not only to reduce the overall ash content, but also to increase the calcium/phosphate ratio to about 0.77 in skim milk powder used in infant formula (Batchelder, 1987), this goal being much easily achievable by replacing the conventional membranes with selective ones (Andrés *et al.*, 1995).

Membrane scaling and fouling are critical points in whey demineralization by ED. Whereas scaling is due to precipitation of $CaCO_3$, $CaSO_4$, $Ca_3(PO_3)_2$ on the brine side of the cationic membranes and can be easily removed by a normal acid cleaning, organic fouling is due to precipitation of charged protein fractions, as well as amino acids, onto the surfaces or pores of the anionic membranes, but it is poorly removed by normal CIP techniques, thus causing the continuous increase in the stack overall resistance and decline in process performance.

To minimize membrane fouling it was suggested to replace the anionic membranes with neutral cellulose-based membranes, such as the dialysis membranes (Ahlgren, 1972), thus leading to the so-called *transport-depletion ED* (TDED) process, as schematically shown in Figure 12. In this way, the only anion transport is limited, whereas cations can pass through any membrane and cumulate into the concentrating compartments. Even if the current efficiency of the process is approximately halved, the savings in membrane investment and replacement costs, as well as in CIP procedures, resulted in overall processing costs of the same order of magnitude of those associated to conventional ED, where the current efficiency initially approaching 100% progressively reduces to less than 50% owing to protein

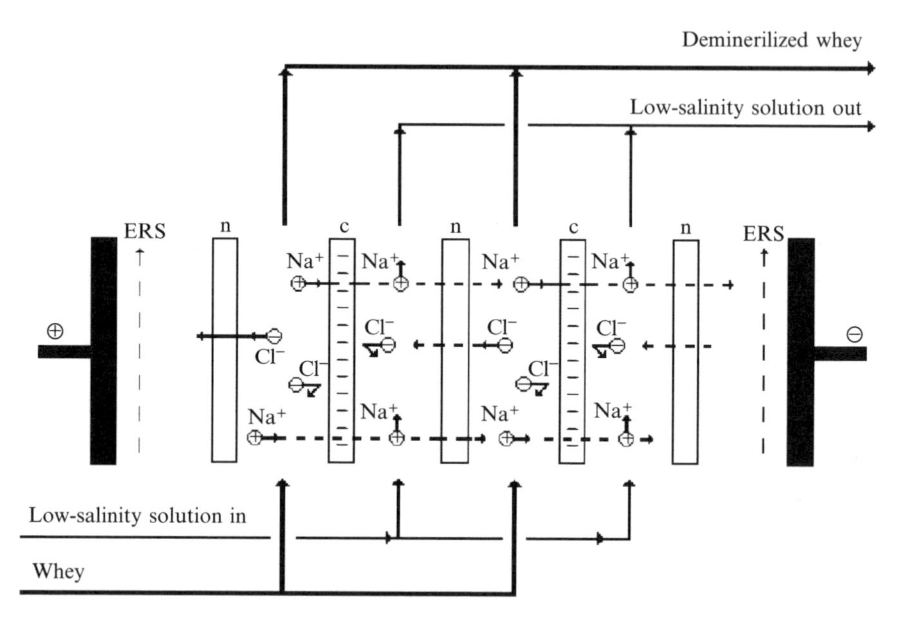

FIG. 12 Schematic layout of a transport depletion ED stack used to demineralize whey: c, cationic membrane; ERS, electrode-rinsing solution; n, neutral membrane.

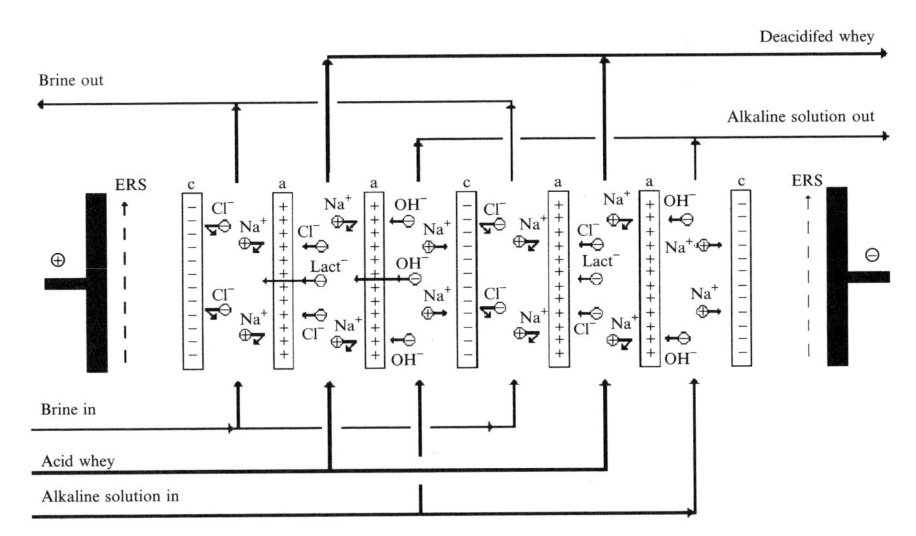

FIG. 13 Schematic layout of a three-compartment ED stack used to desalt acid whey: a, anionic membrane; c, cationic membrane; ERS, electrode-rinsing solution.

deposition onto and within anionic membranes (Ahlgren, 1972). Nevertheless, TDED process seems to have been overcome by the EDR process, which is today the process of choice in whey demineralization (http://www.ameridia.com; http://www.ionics.com).

When dealing with acid whey, and lactate removal is a priority task, it is possible to resort to a three-compartment configuration (Figure 13), obtained by assembling a series of two anionic membranes and a single cationic one (Williams and Kline, 1980). By feeding the compartments limited by two anionic membranes with acid whey and the other two adjacent compartments with a brine solution and an alkaline one, respectively, it is possible to remove selectively lactate anions from the product and replace them with hydroxyl ions. This procedure is also suggested to reduce the acidity of several acidic fruit juices without any chemical addition (Section III.E).

Besides ED, a number of other commercial processes using ion exchange (IE) or nanofiltration (NF) has been proposed for whey demineralization (Greiter *et al.*, 2004; Hoppe and Higgins, 1992). However, it is still impossible to assess definitively which is the optimal process for whey de-ashing, its capital and operating costs depending on several financial, operational, and geographical factors.

As an example, by referring to an IE or ED de-ashing unit with an overall capacity of 45 m^3/day of nanofiltrated whey (three times concentrated and partially desalted), Greiter *et al.* (2002) estimated that the cumulative energy

TABLE X

WHEY DEMINERALIZATION USING AN IE OR ED UNIT WITH AN OVERALL CAPACITY OF 45 m^3/DAY OF NANOFILTRATED WHEY: COMPARISON OF WASTEWATER FORMATION AND ENERGY DEMAND[a]

Alternative technology			
Parameter	IE	ED	Unit
Demineralization level	99	90	%
Wastewater formation			
Volume	3.7	1.25	m^3
Ash content	36.3	8.1	kg
Organic load (COD)	26.0	8.4	kg/m^3
Energy demand for			
Pumping	0.15	4.2	kWh/m^3
Production of regenerants	25.3	–	kWh/m^3
Electrodialysis desalting	–	5.4	kWh/m^3
Reduction of the organic load	9.8	3.2	kWh/m^3

[a]Greiter *et al.* (2002).
Note: IE, ion exchange; ED, electrodialysis.

demand for pumping, producing the IE regenerants, and reducing the organic charge of wastewaters in the case of the IE unit was about three times greater than that in the ED one (Table X). Even if their corresponding overall de-ashing rates were set to 99% or 90%, respectively, the volume, ash content, and organic load of IE effluents were from three to four times greater than those of ED ones (Table X). Despite this clearly proved the greater sustainability of ED with respect to IE, additional aspects, such as regenerant transportation, energy demand for resin production, membrane specific costs, and so on, as well as IE or ED process improvement, are to be accounted for. To this end, it is worth citing the ammonium bicarbonate or SMR process (*Svenska Mejeriernas Riksforening* or Swedish Dairy Association) as an example of how the use of a single regenerant (NH_4HCO_3) can reduce regenerant consumption by recycling, as well as associated waste disposal problems and costs (Hoppe and Higgins, 1992; Jönsson and Olsson, 1981). See also the comments on different ED processes reported earlier.

Today, improvement and economic evaluation of such competing technologies seem to identify the optimum process for whey demineralization as a hybrid process combining EDR, IE, and NF depending on the target de-ashing level (Table XI) (http://www.ameridia.com/html/dry.html). For instance, a single NF unit is regarded as inappropriate for demineralization rates greater than 35–40% because of high loss of lactose and divalent ions, but useful as a pre- or post-treatment step. This was also verified in the case

TABLE XI

SINGLE OR HYBRID PROCESS, BASED ON ED, IE, AND/OR NF, RECOMMENDED BY EURODIA/AMERIDIA TO DEMINERALIZE RAW OR PRECONCENTRATED WHEY AS A FUNCTION OF DIFFERENT DE-ASHING LEVELS[a]

	Feed total solid content (% w/w)	
De-ashing level (%)	6	18–24
30	NF	ED
50–70	IE + NF	ED
90	ED + IE + NF	IE + ED

[a]http://www.ameridia.com/html/dry.html

Note: ED, electrodialysis; IE, ion exchange; NF, nanofiltration.

of lactic acid recovery, its average recovery yield being of the order of 61% for ED against 53% for NF. Even a single IE unit is regarded as uneconomical because of high volume of effluents and heavy pollution load (Greiter *et al.*, 2002). These findings are also confirmed in Table XII, where the estimated specific overall costs to manufacture 90%-de-ashed whey powder are compared by referring to an industrial plant with a daily capacity of 400 m^3 of raw whey at 6.3% TDS using a single or hybrid process (http://www.ameridia.com/html/dry.html).

To separate the major protein constituents of whey, such as β-lactoglobulin (β-LG) and α-lactoalbumin (α-LA), Bazinet *et al.* (2004) recommended the use of bipolar membrane electroacidification (BMEA) by resorting to a three-compartment ED stack consisting of a bipolar membrane and two cationic membranes in parallel. Recycling of 5–20% (w/w) whey protein isolate (WPI) solutions in the compartments edging the cationic side of any bipolar membrane, where the H^+ ions are generated, allowed the pH to be lowered from about 6.9 to 4.6. Further centrifugation of the resulting acidic solution enabled about 53% of the initial β-LG to be recovered, thus yielding a residue, containing about 97% of β-LG and 2.7% of α-LA, and an α-LA-enriched supernatant (Bazinet *et al.*, 2004). This procedure proved not only much simpler, but also more effective than the numerous isolation methods available in the literature and based on ion-exchange chromatography, metaphosphate complex precipitation, heat/acid separation, and ion depletion at low pH. For instance, the procedure developed by Slack *et al.* (1986) yielded β-LG-enriched fractions containing just 33% of the original acid whey proteins or 17% of the original sweet whey proteins by submitting whey to concentration by UF; partial demineralization by ED or diafiltration, upon adjustment of the initial and final pH values to 4.65; and finally to centrifugation.

TABLE XII

EFFECT OF A SINGLE OR HYBRID PROCESS, BASED ON ED, IE, AND/OR NF, ON THE OVERALL OPERATING COSTS PER kg OF DM OF 90%-DEMINERALIZED WHEY POWDER, AS ESTIMATED BY EURODIA/AMERIDIA[a] BY REFERRING TO AN INDUSTRIAL PLANT WITH AN INPUT CAPACITY OF 400 m^3/day OF RAW WHEY AT 6.3% TDS[a]

Hybrid process	Demineralization operating costs (€/kg DM)
IE	0.37
IE + ED	0.18
IE + ED + IE	0.06
NF + ED + IE + NF	0.05

[a]http://www.ameridia.com/html/dry.html
Note: ED, electrodialysis; IE, ion exchange; NF, nanofiltration; DM, dry matter.

The BMEA procedure was also used to produce high-purity bovine milk casein (Bazinet *et al.*, 1999a) and soybean isolates (Bazinet *et al.*, 1997; 1999b) or to fractionate soybean 11S and 7S fractions (Bazinet *et al.*, 2000b).

Further information on such a novel process, as well as the preparation of acid caseinates, electrochemical coagulation of milk proteins or electro-reduction of disulfide bonds of milk proteins, using conventional or bipolar ED stacks, is given by Bazinet (2004). It is however worthy underlining that all these processes do not seem to have been used on an industrial scale yet.

D. WINE INDUSTRY

One of the main concerns in wine shipping is to avoid tartrate crystal precipitation in bottle. These crystals may be confused with fragments of broken glasses, sugar crystals, or chemical residues, thus inducing wine bottle refusal in the general consumer, especially in the United States and Japan. Moreover, in the case of sparkling wines, such crystals may cause excessive loss of product when the bottle is opened. Hence, there is a great deal of interest to improve wine stability, as well as to resort to reliable stabilization procedures.

Grapes naturally contain quite large levels of tartaric acid (H_2T) and potassium ions of the order of 1–3 and 0.8–1.5 kg/m^3, respectively. Tartaric acid is a weak dicarboxylic one that dissociates into tartrate and bitartrate forms as follows:

$$H_2T \leftrightarrows HT^- + H^+$$

$$HT \leftrightarrows T^{2-} + H^{+}$$

It can precipitate as potassium hydrogen tartrate (KHT) or as calcium tartrate (CaT), the latter being practically insoluble in aqueous solutions. Their equilibrium solubility varies with temperature, pH, and alcohol content, while the presence of a few wine components, such as polysaccharides and mannoproteins, may hinder spontaneous nucleation even if the solution is supersaturated. From Figure 14 that shows the equilibrium tartaric acid-dissociated fractions *versus* pH and ethanol volumetric fraction (Berta, 1993; Usseglio-Tomasset and Bosia, 1978), it can be seen that in the typical pH range (3–4) of wines KHT is predominant. As temperature is reduced from 20 to 0°C, KHT solubility in water or in a 12% (v/v) hydro-alcoholic solution reduces from 5.11 to 2.45 kg/m^3 or from 2.75 to 1.1 kg/m^3, respectively (Berta, 1993). Each of these data also varies with pH and reaches a minimum at the pH value associated with the maximum concentration of the hydrogen tartrate anions. For the above-mentioned solutions, the solubility minimum shifts from pH 3.57 to pH 3.73 as the ethanol content increases from 0 to 12% (v/v) (Berta, 1993).

The conventional tartaric stabilization techniques applied in the wine industry are based on two opposite principles. One is aimed at heightening HT^- and T^{2-} precipitation by reducing wine temperature and leads to the so-called cold stabilization technique. To accelerate nucleation, wines are seeded with exogenous KHT crystals, cooled and kept at −4°C for 4–8 days

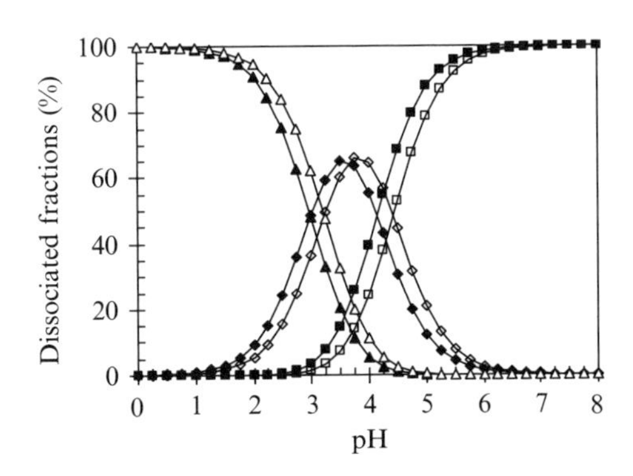

FIG. 14 Effect of pH on the equilibrium tartaric acid dissociated fractions (◆, ◇, [HT^-]/[T]; ■, □, [T^{2-}]/[T]; ▲, △, [H_2T]/[T]) at two different ethanol volumetric fractions (0% v/v, closed symbols; 18% v/v, open symbols), as estimated according to Berta (1993) and Usseglio-Tomasset and Bosia (1978): H_2T, free tartaric acid concentration; HT^-, hydrogen tartrate concentration; T^{2-}, tartrate concentration; T, overall tartaric acid concentration.

to favor crystal growth. The other principle is directed to impair the crystallization process, especially when KHT content is closely exceeding the equilibrium one, by enriching wine with some antiprecipitation additives, such as metatartaric acid, carboxycellulose or yeast mannoproteins.

Despite its quite general application, the refrigeration technique is not only expensive and time-consuming, but also not fully reliable, as KHT crystals can still precipitate during bottle storage and transportation. It may even affect negatively the wine sensory properties as a result of the preliminary wine clarification and crystal removal steps. Thus, the use of IER (Hernàndez and Mínguez, 1997; Mourgues, 1993) and ED has been alternatively proposed to remove K^+ cations. Whereas the former is not allowed in the EU for it alters the ionic balance of wine, as well as its taste, the latter is recognized as a good manufacturing practice by the International Wine Office (OIV) and is approved for commercial use by the EU regulatory no. 2053/97.

Although early ED tests for tartrate stabilization of wines had been performed by Paronetto (1941), more systematic experiments were carried out in the 1970s (Audinos *et al.*, 1979; Paronetto *et al.*, 1977; Wucherpfennig and Krueger, 1975) and led to the automatic method and device for tartaric stabilization of wines developed by Escudier *et al.* (1995) at the French National Agronomic Research Institute (INRA) in cooperation with Ameridia (Moutounet *et al.*, 1997).

Not only does ED allow removal of KHT and tartaric acid in almost the same way observed in the conventional cold-stabilization process, but it also permits a certain degree of reduction in lactic and malic acids, as well as Mg^{2+}, Ca^{2+}, and Na^+ ions, this however being within the limits imposed by the EU regulatory no. 2053/97 (Bach *et al.*, 1999). On the contrary, removal of catechins, leucoanthocyanins, and anthocyanins in ED-stabilized wines was significantly lower than in cold-stabilized ones (Riponi *et al.*, 1992). As an example, Figure 15 shows the percentage removal yields for the main anions and cations present in two typical white and red Portuguese wines as a function of the corresponding percentage reduction in wine electric conductivity (Gonçalves *et al.*, 2003). It can be noted that the removal of calcium is crucial to achieve tartrate stability in terms of both KHT and calcium tartrate.

A number of sensory tests have so far revealed no difference in the organoleptic properties of ED- and cold-stabilized wines (Bach *et al.*, 1999; Cameira dos Santos *et al.*, 2000; Gonçalves *et al.*, 2003; Paronetto *et al.*, 1977; Riponi *et al.*, 1992; Wucherpfennig and Krueger, 1975).

ED treatment may be tailored for any type of wine by referring to a specific stability test. A sample of the raw wine under testing is to be filtered,

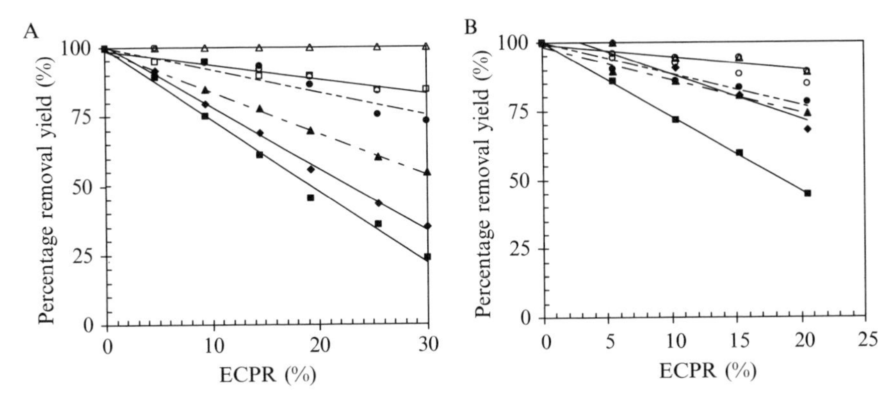

FIG. 15 Percentage removal yields for the main anions (○, tartaric acid; □, malic acid; △, lactic acid) and cations (▲, K^+; ●, Na^+; ■, Ca^{2+}; ◆, Mg^{2+}) present in two typical white (A) and red (B) Portuguese wines as a function of the corresponding percentage reduction in electric conductivity (ECPR), as extracted from Gonçalves *et al.* (2003).

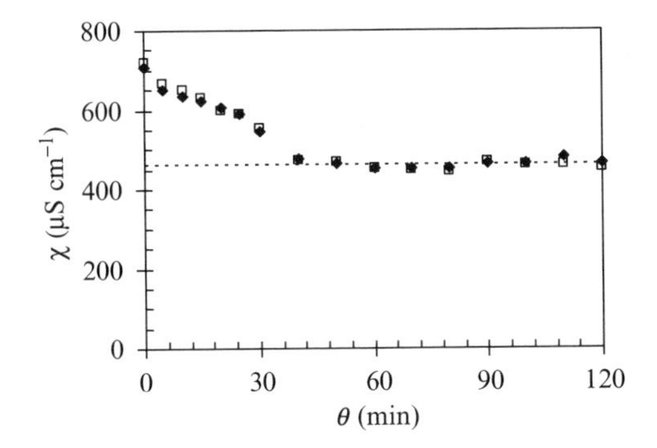

FIG. 16 Stability tests performed on two 200-cm^3 samples, seeded with fine crystals of KHT (4 kg/m^3) and chilled at −4°C, of San Flaviano Est! Est!! Est!!! white wine, manufactured at Montefiascone (Italy) during the 2000 vintage: electric conductivity (χ) vs. time.

seeded with fine crystals of KHT (4 kg/m^3), and then chilled at −4°C, while monitoring the time course of its electric conductivity (χ).

Figure 16 shows the typical evolution of such a test being performed on two 200-cm^3 samples of San Flaviano Est! Est!! Est!!! white wine, manufactured at Montefiascone (Italy) during the 2000 vintage. In less than 1 hour, it was possible to assess the equilibrium χ value (460 ± 9 μS/cm) or how much its initial χ value (721 ± 28 μS/cm) had to be reduced to avoid tartrate precipitation after bottling.

Based on the INRA experience (Escudier *et al.*, 1995), the average drop in χ to stabilize a wine is in the range 5–20% for red wines, but it can exceed 30% for white and young wines.

The ED-treated wines generally result to be completely stable once KHT and CaT have been selectively removed. Their basic characteristics (i.e., pH, acidity, sugar content, alcohol level), as well as taste, bouquet, and color, are practically unaltered, while their ethanol content, pH, and volatile acidity are reduced by less than 0.1% (v/v), 0.25 pH units, and 0.09 kg/m^3 (expressed as equivalent H_2SO_4), respectively (htpp://www.ameridia.com/html/wn.html).

Tartaric stabilization of wines can be performed using a conventional monopolar ED stack operating in the batch mode (Figure 5A). Anionic membranes are more liable to fouling than cationic ones, especially when dealing with red wines, but periodic (once per day) in-site rinsing and cleaning of the ED stack can minimize fouling and extend their life time.

After being microfiltered with no needs of filter aids and/or chemical additives, the raw wine is recirculated between the ED stack and a tank, equipped with level gauges and a conductivity meter. As soon as its conductivity has reached the predetermined equilibrium χ value, it is automatically discharged and replaced with another lot.

The brine solution may be enriched with NaCl or KHT to increase its conductivity and acidified to the same pH of the wine under treatment. It is recirculated between the concentrating compartments of the ED stack and another tank equipped with conductivity and pH automatic controls so as to avoid precipitation of potassium bitartrate onto the membranes by diluting the brine with deionized water and/or discharging more or less aliquots of the brine itself when its conductivity reaches 70–80% of its saturation value (Gonçalves *et al.*, 2003). In this way, a waste effluent in the range 10–20% of the wine volume treated is to be disposed of or used to recover KHT crystals (Nasr-Allah and Audinos, 1994).

For this treatment the typical permeation flux is of the order of 100 dm^{3}m^{-2}h^{-1} of wine treated, while the overall electric energy consumption ranges from 0.5 to 1.0 kWh/m^3, this being about 10 times lower than the energy needed for the refrigeration procedure. For an ED unit treating about 40,000 m^3 of wines per year it was estimated an overall investment of about US$400,000, and operating costs of circa cUS$0.6 per litre of wine treated. For much smaller capacities up to 4500 m^3/year, the ED unit can be installed on a truck and offered on rent at less than cUS$10 per bottle. By the year 2003, Eurodia/Ameridia claimed to have supplied over 40 plants in France, Italy, Spain, Australia, and United States with an overall capacity of 3 × 10^5 m^3/year (htpp://www.ameridia.com/html/wn.html).

Even in this case, the use of a hybrid process combining NF, IE, and ED appears to improve the economics and performance of the tartaric stabilization of wines. For instance, Ferrarini (2001) proposed to split raw wine into a retentate and permeate by NF. The permeate, being richer in minerals, was processed by using in sequence cationic and anionic exchange resins and ED to reduce its potassium, calcium, and tartrate ion contents. By recombining the de-ashed permeate with the NF retentate, Ferrarini (2001) asserted to obtain a stabilized wine retaining almost all the flavor and aroma compounds originally present in raw wine.

E. FRUIT JUICE INDUSTRY

In this industry ED processing might be applied in three main areas, that is, de-acidification, desalting, and enzymatic-browning inhibition.

Some acidic fruit juices are not fully appreciated by consumers and are to be de-acidified by sugar or alkali addition. Whereas the use of $CaCO_3$ is not recommended since the release of CO_2 induces foaming and poor pH control, that of $Ca(OH)_2$ may cause some precipitation problems in the final product. Such procedures may be limited by legislation and, even if they have so far conferred no sensory dislike, may involve chemophobic reactions in the general consumer, who is averse to any chemical addition in natural products (Vera *et al.*, 2003).

Use of conventional ED stacks composed of alternating cationic and anionic membranes was suggested to reduce the acidity of several fruit juices, such as grape, orange, lemon, pineapple (Adhikary *et al.*, 1983), and *ume* (pickled Japanese apricot) (Ono *et al.*, 1992) juices without any chemical addition. Preliminary clarification of fruit juices via ultrafiltration was found to minimize membrane fouling. In the case of UF mandarin orange juice, ED processing allowed its total acidity to be reduced by 30% with no detectable color change, almost constant ascorbic acid, amino-nitrogen compound, free sugar, and flavonoid contents, and a slight decrease in the pH (Kang and Rhee, 2002). Appropriate UF modules were also used to decolorize date juices before their demineralization via ED and concentration to yield liquid sugar (Lewandowski *et al.*, 1999).

To de-acidify orange juice it was also suggested using an ED stack composed of anionic membranes only and two compartments. In this way, only the anions, mainly citrate, were removed from the juice and replaced by the hydroxyl ions supplied by the KOH solution flowing in the adjacent compartments. Periodic reversal of polarity helped removal of the colloidal material deposited onto the membrane surfaces. Voss (1986) suggested two alternative ED processes. One was a three-compartment stack, obtained by arranging a cationic membrane and two anionic ones in sequence

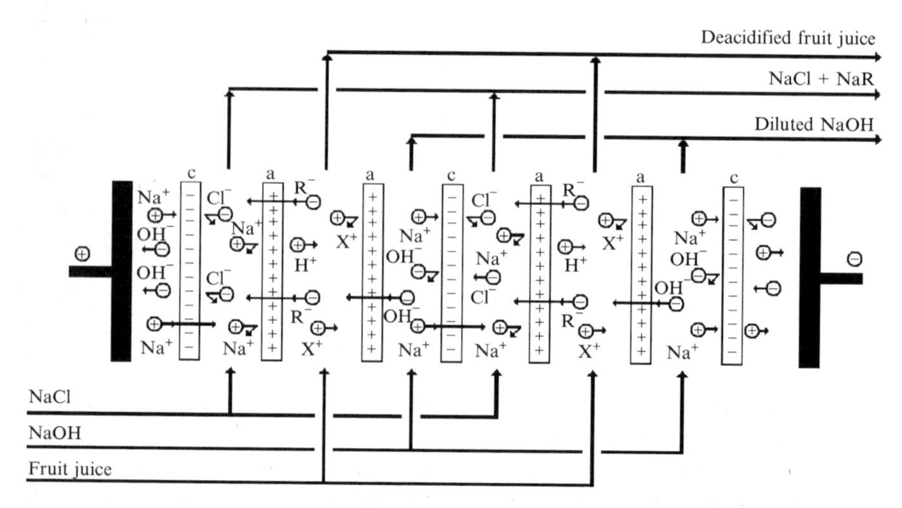

FIG. 17 Schematic layout of a three-compartment ED stack for the deacidification of fruit juices: a, anionic membrane; c, cationic membrane; R^-, generic anion; X^+, generic cation.

(Figure 17), as that used to remove selectively lactic acids from acid whey (Figure 13). The other was a two-compartment stack, made of a bipolar membrane and an anion-exchange one (Figure 8C), to attain a de-acidified juice and a concentrated stream containing free citric acid.

More recently, such processes were tested to reduce the acidity of clarified passion fruit (*Passiflora edulis v. flavicarpa*) juices from pH 2.9 to 4.0 in comparison with other conventional processes, such as calcium citrate precipitation as resulting from $CaCO_3$ or $Ca(OH)_2$ addition, or removal via weakly basic IER (Calle *et al.*, 2002; Vera *et al.*, 2003). Whatever the process tested, the physicochemical and sensory properties of the de-acidified juices were quite similar. In spite of the fact that their sodium concentration was higher when using any of the above-mentioned ED processes, the two-compartment stack using bipolar and anionic membranes (Figure 8C) was regarded as optimal, since no chemical consumption was needed and a valuable solution rich in citric acid (89% purity) was recovered (Vera *et al.*, 2003).

In the production of pickles, salted fruits are generally leached or flushed with water to reduce their sodium content. The waste stream may be split into a sugar-rich solution and a brine via ED. Whereas the latter is recycled to the brining process, the former is reused to return the natural sugar and color to the fruits undergoing leaching (Elankovan, 1996). By coupling ED and adsorption on weakly basic resins, the seasoning solution used to flavor salted ume may be selectively deprived of NaCl and organic acids, thus allowing its remaining sugars and aminoacids to be repeatedly reused (Takatsuji *et al.*, 1999).

Finally, a process was patented to minimize the nitrate nitrogen content in vegetable juices by ED, once concentrated to 20–40°Brix (Sumimura *et al.*, 2004).

There is an increasing consumer demand for cloudy apple juices owing to their greater sensory and nutritional quality. However, such juices are extremely sensitive to taste and color modification for the polyphenol oxidases (PPO) bound to suspended matter catalyze the oxidation of phenolic compounds to o-quinones, which readily polymerize into dark-colored pigments. By acidifying the juice to pH 2.0 (Zemel *et al.*, 1990), it is possible to affect irreversibly the PPO tertiary structure by inducing electrostatic repulsion between acids and positively charged amino groups and thus inhibit PPO activity. To avoid the dilution effect due to subsequent addition of acid (HCl) and base (NaOH), Tronc *et al.* (1998) suggested using ED stacks equipped with bipolar membranes to reduce temporarily the pH of cloudy apple juice to 2.0 and then to readjust its pH to the initial value (3.5). Of the two-compartment configurations, that coupling a bpm with an anion-exchange membrane (Figure 18) was found to be faster and more effective to accelerate the acidification and inhibit enzymic browning. By flowing 0.1 kmol/m^3 HCl between the anionic side of bpms and any anionic membrane, the Cl^- ions can permeate through the anionic membranes and cumulate in the juice, thus promoting the retention of the hydrogen ions generated by the cationic side of bpms. As referring to Figure 18 and reversing the juice and HCl compartments, it was possible to replace the Cl^- ions accumulated in the acidified juice with the hydroxyl ions generated by the anionic side of bpms, thus allowing the pH of the juice to return to its initial value. In this way, it was noted a slight reactivation of PPO activity, but browning inhibition was complete and irreversible. The treatment enhanced the color of cloudy apple juice during storage with a limited effect on its chemical composition and organoleptic quality (Quoc *et al.*, 2000).

F. SUGAR INDUSTRY

Beet or cane molasses are the main by-products in the sugar-manufacturing process. Despite an average sugar content of 50% (w/w), further sugar recovery is hampered by the presence of impurities ranging from suspended materials to inorganic salts and color substances. A great deal of emphasis has been put on the alkali metal cations, namely Na^+, K^+, Ca^{2+}, and Mg^{2+}, these being regarded as melassigenic ions (Elmidaoui *et al.*, 2002, 2004). To get rid of multivalent cations, molasses are generally integrated with lime or

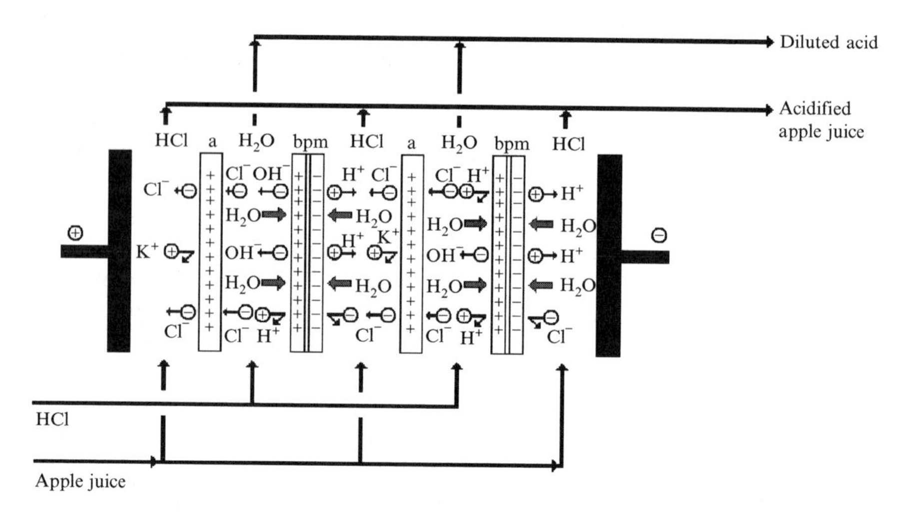

FIG. 18 Schematic layout of a two-compartment configuration using bipolar (bmp) and anionic (a) membranes proposed to inhibit enzymatic browning of cloudy apple juice (Tronc *et al.*, 1998) via preliminary acidification to pH 2.0 and subsequent de-acidification to pH 3.5 by reverting flow of HCl and apple juice streams.

$CaCl_2$ and heated, thus causing precipitation of $CaSO_4$ and $MgSO_4$,, which are then separated by centrifugation. Several alternative processes, such as IER, synthetic adsorbents, coagulants, membrane filtration, and ED, have been proposed so far.

For instance, the sugarcane juice might be clarified via filtration (Thampy *et al.*, 1999) or MF (Pinacci *et al.*, 2004) to minimize fouling phenomena, concentrated to about 30°Brix, desalted by 50–80% using conventional ED, further concentrated, and then crystallized, thus obtaining crystals of uniform size and molasses brown in color, but palatable (Thampy *et al.*, 1999).

The main bottlenecks of ED application in the sugar industry were both the short membrane life, especially for the anion-exchange membranes, and the high viscosity of cane- or beet-sugar syrups, the maximum working temperature for the electromembranes being generally less than 40°C.

By replacing the conventional anionic membranes with the novel ones type Neosepta® AXE 01 (Eurodia Industrie, Wissous, F; see Table II) as coupled to the conventional cationic membranes type CMX-Sb either in a laboratory-scale stack with an overall membrane surface area (A_m) of 0.2 m^2 (http://www.ameridia.com/html/fss.html) or in a pilot-scale one with A_m = 2.5 m^2 (Elmidaoui *et al.*, 2002), it was possible to overcome the earlier shortcomings, as shown in Table XIII. Further scaling-up in an industrial

TABLE XIII

MAIN RESULTS OF BEET SUGAR SYRUP DESALTING IN LABORATORY- AND PILOT-SCALE ED STACKS EQUIPPED WITH CMX-Sb AND AXE 01 MEMBRANES[a]

Beet sugar syrups	Laboratory plant		Pilot plant	Unit
Sucrose titre	30	50	55	°Brix
Temperature	50	50	60	°C
Current density	110	120	–	A/m^2
Voltage/cell	1	1	1.3	V
Conductivity decrease	70	70	75	%
Total cations removal yield	65	55	–	%
Ca^{2+} removal yield	60	35	65	%
Mg^{2+} removal yield	50	30	–	%
K^+ removal yield	75	75	86	%
Na^+ removal yield	25	45	75	%
Demineralization flux	3.5	4.5	–	eq $h^{-1}m^{-2}$
Current efficiency	90	90	–	%

[a]As extracted from Lutin (2000) and Elmidaoui *et al.* (2002), respectively.

plant consisting of two stacks EUR40 in series with capacity of 20 m^3/h yielded a demineralization degree of 60%, sugar loss less than 0.5%, a specific electric energy consumption of 1.1 kWh/m^3 for ion transport and pumping, and small consumption of chemicals for CIP of the order of 0.02 kg HCl/m^3 and 0.06 kg $NaOH/m^3$ (Lutin *et al.*, 2002).

Despite this, the European sugar industry has been so far extremely reluctant to replace the rather environmentally polluting traditional technology with such a novel procedure. Moreover, even in this case a hybrid process combining MF and ED appears to be effective to minimize organic fouling, operate up to 60 °C, reduce waste formation and pollution load, improve sugar yields, reduce the volume of molasses, and save capital and operating costs.

Finally, the main by-product of molasses fermentation and distillation, that is, vinasse, where ash represents from 26% to 34% of dry matter, has to be demineralized to expand its usage in feed preparation. To this end, vinasse can be concentrated from about 6° to 20° Brix and then submitted to conventional ED processing (Milewski and Lewicki, 1988). For instance, at an 80% demineralization level, the K^+ content was completely removed, while the Na^+, Ca^{2+}, or Mg^{2+} content was decreased by about 52%, 40%, or 19%, respectively. De-ashed vinasse exhibited no change in the betaine content, while the fraction of total amino acids in the dry matter increased. The specific electric energy consumed per kg of ash removed was found to range from 10 to as high as 40 kWh/kg, while it was about 5 kWh/kg in the case of potassium removal from grape juice (Wucherpfennig, 1975).

G. FERMENTATION INDUSTRY

A sector where the application of ED is potentially interesting is that of the fermentation industry, especially when the main product of the microbial metabolism is an electrolyte. This may exert an inhibitory effect on cell growth and/or metabolite production in either its free or dissociated form. Alternatively, it may be dissolved in media rich of impurities that are to be removed via numerous and expensive purification steps.

By resorting to the so-called membrane recycle bioreactors (MBR) (Bubbico *et al.*, 1997; Enzminger and Asenjo, 1986), continuous recycling of the culture broth through crossflow MF modules allows removal of the inhibiting metabolites, this helping to maximize cell density in the bioreactor, as well as bioproduct formation rate. Further ED treatment of MF permeates gives rise to two streams, a diluted one to be recycled back into the bioreactor, and a concentrated one to be supplementary refined.

Downstream processing may consist of several operations such as liming to precipitate the metabolite as the calcium salt, washing of the precipitate with water to remove soluble impurities, acidification using strong acids to convert the salt in its free acid form. The acidic liquor may be demineralized using IER, decolorized using active carbons, concentrated under vacuum, and finally crystallized.

Since the use of ED simplifies such a complex sequence of recovery techniques, ED is generally regarded as an environment-friendly alternative to the conventional bioproduct recovery processes.

Table XIV reviews a series of potential applications of ED to recover some microbial metabolites from the respective culture broths together with the microorganisms used and the main bottlenecks of conventional production processes, as well as their main references. For the most potentially relevant applications, more details are discussed later.

1. Acetic acid

Acetic acid (CH_3COOH) is a bulk commodity chemical with a world production of about 3.1×10^6 Mg/year, a demand increasing at a rate of +2.6% per year and a market price of US\$0.44–0.47 per kg (Anon., 2001a). It is obtained primarily by the Monsanto or methanol carbonylation process, in which carbon monoxide reacts with methanol under the influence of a rhodium complex catalyst at 180°C and pressures of 30–40 bar, and secondarily by the oxidation of ethanol (Backus *et al.*, 2003). The acetic fermentation route is limited to the food market and leads to vinegar production from several raw materials (e.g., apples, malt, grapes, grain, wines, and so on).

Despite only very large plants with capacities of (225–500) $\times 10^3$ Mg/year are economically feasible (Backus *et al.*, 2003), there are strong market, economic, and energy benefits to develop novel fermentation processes to produce acetic acid in scalable, regional-sized plants (Office of Industrial Technologies, 2003) for the current high cost of methanol and the fact that concentrating acetic acid production in large-size plants makes contract selling prices greatly affected by costs of transportation and distribution networks.

Because of the higher yield coefficient for acetic acid on glucose and lower energy requirements of the anaerobic acidogenesis of glucose by *Clostridia* with respect to aerobic route using *Acetobacter aceti* (Cheryan *et al.*, 1997; Ghose and Bhadra, 1985), the anaerobic process is presently regarded as the method of choice to produce acetic acid as a chemical feedstock.

Figure 19 shows the flowchart of the integrated process for production and purification of fermentation-derived acetic acid, which was recommended by the Office of Industrial Technologies of the US Department of Energy (2003) as a viable pathway for commercialization of regional smaller scale acetic acid production plants. Of course, the economic profitability of such a process relies on appropriate mutants of acetogenic bacteria, as well as efficient membrane modules, capable, on one side, of maximizing the concentration of free acid in the exhausted fermentation broth and, by the other side, of minimizing acid recovery costs. To this end, novel pervaporation membranes with high ammonia permeability might enable pervaporation-assisted thermal cracking of ammonium acetate, thus allowing the aqueous solutions of ammonium hydroxide or free acetic acid to be recycled into the bioreactor as a nitrogen source or further purified, respectively (Office of Industrial Technologies, 2003). Moreover, the ED unit in Figure 19 would help not only to recover acetates from the generally dilute broths in which they are produced in ionized form (Chukwu and Cheryan, 1999), but also to keep a low acidic level and control the pH in the fermenting broth throughout the course of the fermentation (Nomura *et al.*, 1988, 1994). In both cases, the broth is to be microfiltered and then fed to the ED unit to be separated into an enriched salt product (concentrate) and a salt-depleted broth (dilute). Since the residual sugars and nutrients are retained in the dilute, its recycling into the bioreactor would enable material utilization and microbial acetate productivity to be maximized. Simultaneous recovery of acetic acid in the concentrate would facilitate its further separation and purification.

Even in the aerobic fermentation by *A. aceti*, the use of a combined automatic system (to remove continuously the acetic acid from the fermentation broth and keep the pH in the fermenter about constant by regulating the DC potential difference applied to the ED stack and ethanol concentration in

TABLE XIV

POTENTIAL APPLICATIONS OF ED TO RECOVER MICROBIAL METABOLITES FROM FERMENTATION MEDIA

Microbial metabolite	Microorganism used	Typical operating limits	Main references
Acetic acid	*Acetobacter aceti* *Clostridium thermoaceticum*	Product inhibition	Chukwu and Cheryan, 1999; Fidaleo and Moresi, 2005b; Nomura *et al.*, 1988, 1994
Acetic and propionic acid	*Propionibacterium shermanii*	Product inhibition	Zhang *et al.*, 1993
Citric acid	*Aspergillus niger* *Yarrowia lipolytica*	Complex recovery, environmental impact	Datta and Bergemamm, 1996; Ling *et al.*, 2002; Mancini *et al.*, 1995; Moresi and Sappino, 1998, 2000; Novalic *et al.*, 1995, 1996; Sappino *et al.*, 1996
Gluconic acid	*Aspergillus niger*	Complex recovery	Novalic *et al.*, 1997
Itaconic acid	*Aspergillus terreus*	Complex recovery	Kobayashi, 1967, 1978; Moresi and Sappino, 2000; Nakagawa *et al.*, 1975
2-Keto-L-gulonic acid	–	–	Oka *et al.*, 1998

Lactic acid	*Lactobacillus delbruekii*, *Lactococcus lactis*	Product inhibition, complex recovery	Boniardi *et al.*, 1996; Boyaval *et al.*, 1987; Fidaleo and Moresi, 2004; Gillery *et al.*, 2002; Habova *et al.*, 2001, 2004; Hongo *et al.*, 1986; Ishizaki *et al.*, 1990; Kim and Moon, 2001; Lee *et al.*, 1998; Madzingaidzo *et al.*, 2002; Moresi and Sappino, 2000; Nomura *et al.*, 1987, 1988, 1991; Siebold *et al.*, 1995; Vonktaveesuk *et al.*, 1994; Xuemei *et al.*, 1999; Yamamoto *et al.*, 1993; Yen and Cheryan, 1993
Lysine and other amino acids	*Brevibacterium flavum*, *Corynebacterium glutamicum*	Product inhibition	Grib *et al.*, 2000; Kikuchi *et al.*, 1995; Lee *et al.*, 2002b, 2003; Nomura *et al.*, 1987
Malic acid	–	Complex recovery	Belafi-Bako *et al.*, 2004; Sridhar, 1987
Propionic acid	*Propionibacterium acidipropionici*	Product inhibition	Boyaval and Corre, 1995; Boyaval *et al.*, 1993; Fidaleo and Moresi, 2006; Weier *et al.*, 1992
Pyruvic acid	*Escherichia coli*	Product inhibition	Zelic *et al.*, 2004;
Succinic acid	–	–	Glassner and Datta; 1992

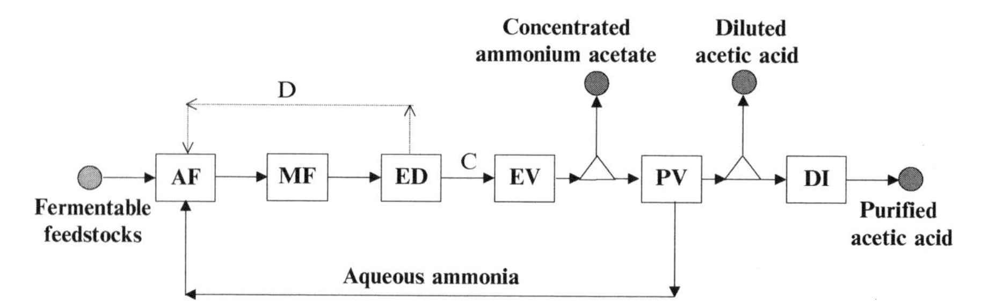

FIG. 19 Schematic flow diagram for the production and purification of fermentation-derived acetic acid, as modified from Office of Industrial Technologies (2003). Unit operation identification items: AF, anaerobic fermentation; DI, distillation; ED, electrodialysis; EV, evaporation; MF, microfiltration; PV, pervaporation-assisted thermal cracking.

the fermenting medium) resulted in acetate productivities (2.13 g/h) about 1.35 times greater than those (1.58 g/h) obtained in fermentation trials without any pH control and practically no effect on the yield coefficient for acetic acid on ethanol (1.13–1.18 g/g) (Nomura *et al.*, 1988).

The same combined fermenter-ED system was used to improve simultaneous excretion of acetic and propionic acids by *Propionobacterium shermanii* (Zhang *et al.*, 1993), their enhancement factors with respect to the conventional fermentation unit being of the order of 1.3 and 1.4, respectively.

2. *Citric acid*

The global market of citric acid ($C_6H_8O_7$) is about 1.4×10^6 Mg equivalent to US$5.3 billion and represents circa 70% of all food acidulants utilized in the field of foods and pharmaceuticals (http://www.foodproductiondaily.com/news/ng.asp?id=63089). It is mainly produced by high citric acid-yielding strains of *Aspergillus niger* by submerged-culture fermentation of production media based on molasses or other glucose/sucrose sources. In spite of the interest in citric acid production by yeast grown submerged in sugar- or hydrocarbon-based media to overcome the main disadvantages of the traditional mould fermentation (i.e., high sensitivity to trace metals and low production rates), no yeast-based process is known to be operating worldwide. In fact, among the fermentation plants using the Japanese industrial know-how (e.g., Takeda Chemical Industries) that constructed at Southport (North Carolina, USA) was shut down in 1982 after 3–4 year operation, while the Liquichimica plant (Saline, Italy) never came into operation (Moresi and Parente, 1999).

Once mycelia have been separated via continuous filtration from exhausted production media, citric acid may be recovered by using three different methods, such as *direct crystallization* upon concentration of the filtered liquor, *precipitation* as calcium citrate tetrahydrate, or *liquid extraction*. Since molasses are extremely rich in impurities, direct crystallization cannot be applied unless very refined raw materials, such as sucrose syrups or crystals, are used. The precipitation process (that is based on subsequent addition of sulfuric acid and lime to clarified fermentation broths) is used by the great majority of world citric acid manufacturers, including Archer Daniels Midland Co. (ADM) in the United States. Liquid extraction with mixtures of trilaurylamine, n-octanol, and C_{10} or C_{11} isoparaffin was used by Pfizer Inc. in Europe and Bayer Co. (formerly Haarmann & Reimer Co., subsidiary of Miles) in the Dayton (OH, USA) and Eikhart (IN, USA) plants only (Moresi and Parente, 1999), even if such plants might have been shut down in 1998.

To overcome such a drawback, that is, the formation and disposal of enormous amounts of liquid effluents (their Chemical Oxygen Demand being about 20 kg/m^3) and solid by-products (i.e., about 0.15 kg of dried mycelium and 2 kg of $CaSO_4.2H_2O$ per kg of monohydrated citric acid), several process alternatives have been so far suggested to minimize the overall environmental impact of this process (Moresi and Parente, 1999).

Citric acid recovery by ED was early proposed by Voss in 1986 (Ling *et al.*, 2002). The specific energy consumption (ε) depended on the electromembranes used and was about 1.7–2 or 0.3–0.9 kWh/kg of citrate recovered in the case of bipolar (Novalic *et al.*, 1996) or monopolar (Datta and Bergemann, 1996; Ling *et al.*, 2002; Mancini *et al.*, 1995; Novalic *et al.*, 1995; Sappino *et al.*, 1996) membranes, respectively.

ED appears to be an inefficient method to recover free citric acid because of its low electric conductivity (Novalic *et al.*, 1995). As it is converted into the monovalent (at pH ca. 3), divalent (at pH ca. 5), or trivalent (at pH about 7) citrate anion, there is a significant increase in the electric conductivity (χ), the latter increasing from 0.95 to 2.18 and to 3.9 S/m, respectively, in the case of an aqueous solution containing 50 kg/m^3 of citric acid equivalent (Moresi and Sappino, 1998). By increasing the pH from 3 to 7, ε reduced about eight times, the solute flux (J_B) practically doubled, while the overall water transport (J_W) increased 3–4 times. The latter partly counterbalanced the greater effectiveness of the electrodialytic concentration of citric acid at pH 7 with respect to that at pH 3. Table XV presents a summary of the effect of current density (j) on the main performance indicators of the electrodialytic recovery of the monovalent, divalent, or trivalent ionic fractions of citric acid (Moresi and Sappino, 1998). All the mean values or empirical correlations of the earlier indicators were useful to evaluate the economic feasibility of this separation technique (Moresi and Sappino, 2000).

TABLE XV

ELECTRODIALYTIC BATCH RECYCLE RECOVERY OF A FEW SODIUM SALTS OF LACTIC, ITACONIC, AND CITRIC ACIDS FROM AQUEOUS SOLUTIONS AT DIFFERENT pH VALUES AND 33°C[a,b]

	Mean values or empiric correlations					
	Citrate			Lactate	Itaconate	
Indicator	pH = 3	pH = 5	pH = 7	pH = 5	pH = 6	Unit
ζ	0.90 ± 0.08	0.90 ± 0.04	0.97 ± 0.02	1.06 ± 0.03	1.02 ± 0.04	dimensionless
Ω	0.09 ± 0.02	0.41 ± 0.03	0.50 ± 0.01	0.62 ± 0.07	0.61 ± 0.05	dimensionless
ε	$0.33\,j$	$0.064\,j$	$0.040\,j$	$0.0306\,j^{0.48}$	$0.00016\,j^{1.65}$	kWh/kg
J_B	$0.0084\,j$	$0.018\,j$	$0.016\,j$	$0.0026\,j$	$0.0020\,j$	kg m^{-2} h^{-1}
J_W	0.176 ± 0.004	$0.050\,j$	$0.050\,j$	$0.0083\,j$	$0.0067\,j$	dm^3 m^{-2} h^{-1}

[a]Mean values or empirical correlations between each performance indicator and current density (j), expressed in A/m^2.
[b]Moresi and Sappino (1998, 2000).
Note: Solute recovery efficiency, ζ; Faraday efficiency, Ω; specific energy consumption, ε; solute flux, J_B; and average water flux, J_W.

The effectiveness of this ED process increases with temperature, being presently limited by the maximum operating temperature (35°C) of the AMV and CMV electromembranes used (Table II).

Figure 20 shows a schematic of a novel membrane-integrated process for citric acid production from glucose syrups by *Yarrowia lypolitica* ATCC 20346, based on prolonged fed-batch fermentation carried out in a stirred bioreactor coupled to a MF unit equipped with tubular ceramic membranes, and disodium citrate recovery from MF permeates by ED (Moresi, 1995).

A combined system consisting of an anion exchanger and a two-compartment ED stack using bipolar and cationic membranes (Figure 8B) was patented by Morita *et al.* (1996) to recover citric acid and NaOH with reduced or nil consumption of reagents (H_2SO_4, lime) and no formation of residues ($CaSO_4$). After centrifugation to remove mycelia, the culture broth was fed to a column containing a weakly basic anion-exchange resin (Diaion WA-30; Mitsubishi Chem. Corp., Tokyo) to adsorb citric acid. By washing sequentially the column with equal volumes of demineralized water and aqueous NaOH (1 kmol/m^3), it was possible to regenerate the column, as well as to recover the sodium salts of citric acid by collecting the fractions at pH values ranging from 3.0 to 6.0. Such fractions were fed into the acid compartments of the aforementioned ED stack arrangement, thus resulting in an overall citric acid recovery yield of 96%. Further decolorization by activated charcoal, concentration, and crystallisation ensued citric acid crystals of 99.5% purity (Morita *et al.*, 1996).

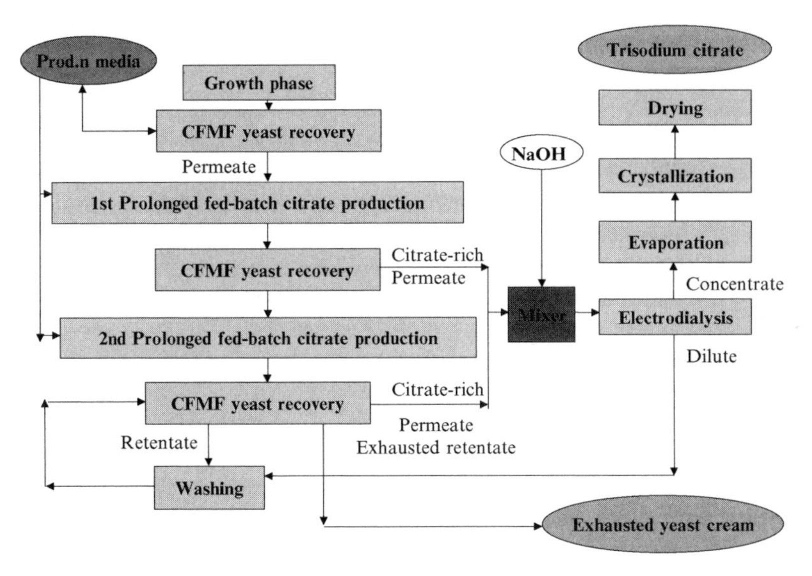

FIG. 20 Flow sheet of a novel integrated membrane process for citric acid production from glucose syrups by *Yarrowia lipolytica*, as proposed by Moresi (1995).

3. *Itaconic acid*

Itaconic acid (HOOC–CH=CH–COOH, $C_4H_4O_4$) is used as monomer or comonomer for plastics, resins, synthetic fibres, and elastomers (Milson and Meers, 1985) and it is produced by submerged culture fermentation with *Aspergillus terreus* in a medium containing molasses as the sugar source at 32–40°C and pH of 1.8–4.0 under 0.25–0.5 volumes of air per volume of medium per minute for 48–72 hour (Milson and Meers, 1985; Roehr and Kubicek, 1996). Use of ED was suggested to enhance the mycelial itaconate productivity in wood waste- (Kobayashi, 1967, 1978; Nakagawa *et al.*, 1975), pretreated beetjuice-, or molasses- (Nakagawa *et al.*, 1991) based media. More recently, Moresi and Sappino (2000) assessed the effect of j on the main performance indicators (Table XV) and specific recovery costs as a function of plant size.

4. *Lactic acid*

Lactic acid (CH_3–CHOH–COOH) is commonly used as a food additive for flavor and preservation. It is also converted into a polylactide polymer, which represents one of the first commercial applications of biodegradable polymers. It is produced by chemical synthesis or fermentation using lactococci, mesophilic lactobacilli (*Lactobacillus casei* subsp. *casei, Lb. amylophilus*), or thermophilic streptococci (*Streptococcus thermophilus*), these strains producing pure L-(+)lactic acid, the only enantiomer metabolized by humans (Moresi and Parente, 1999).

The global capacity of lactic acid is around 250,000 Mg/year and about 80% of the product is produced with the liquor resulting from corn wet milling (Anon, 2001b), since the use of the product chemically synthesized is restricted. China has more than 70 lactic acid producers with a total capacity of around 30,000 Mg/year (Anon, 2001b).

Lactic acid fermentation is hampered by two main bottlenecks (Yen and Cheryan, 1993). Undissociated lactic acid acts as a noncompetitive inhibitor for growth and lactic acid production by diffusing through the membrane and decreasing intracellular pH (Yen *et al.*, 1991). Thus, the pH is controlled at 5–6.5 by automatic addition of NaOH, Na_2CO_3, or NH_4OH. By referring to lactic acid production from yeast-extract enriched glucose-based media by *Lactobacillus delbruckii* (Yen *et al.*, 1991), the specific cell growth and lactic acid production rates were found to be practically independent of undissociated lactic acid concentration for concentrations lower than 0.12 kg/m^3. On the contrary for greater concentration values, both rates exhibited an almost linear decrease.

Batch fermentations result in high product concentration (120–150 kg/m^3) but in low productivity (2 kg m^{-3} h^{-1}). Dramatic improvements in productivity (20–80 kg m^{-3} h^{-1} in laboratory- and pilot-plant experiments) were obtained by using cell recycle via MF or immobilized cells, at the expense of lactate concentration in the effluent (usually lower than 50 kg/m^3).

Lactic acid recovery from fermentation broths, as well as its refining, is difficult owing to the high solubility of its salts, and significantly affects its downstream processing costs. The traditional process involves precipitation of calcium lactate and regeneration of lactic acid by addition of sulfuric acid followed by further purification steps (IE and decolorization). This gives rise to about 1 kg of $CaSO_4 . 2H_2O$ per kg of lactic acid, thus resulting in enormous amounts of crude calcium sulfate (circa 0.36×10^6 Mg/year) that are to be dumped as industrial wastes (U.S. Department of Energy, 1999). Alternative processes are solvent extraction (Yabannavar and Wang, 1991), adsorption (Kaufman *et al.*, 1994), and direct distillation (Cockrem and Johnson, 1993).

Combined use of lactate fermentation and ED separation has been proposed to overcome the main drawbacks of this fermentation process, that is, low microbial acidic productivity and expensive downstream processing of lactic acid fermentation broths (Boyaval *et al.*, 1987; Hongo *et al.*, 1986;

Ishizaki *et al.*, 1990; Kim and Moon., 2001; Lee *et al.*, 1998; Madzingaidzo *et al.*, 2002; Nomura *et al.*, 1987b, 1991).

As an example, Nomura *et al.* (1991) made use of a combined system (consisting of a fermenter, F, a MF module, and an ED stack), to recover first the microbial biomass from the culture broth by MF and second the lactate from the resulting MF permeate by ED. By recycling the cell-rich MF retentate, as well as the residual sugar- and nutrient-rich ED diluate, back into the bioreactor, it was possible to maximize material utilization and microbial lactate productivity. Then, simultaneous recovery of the lactic acid in the ED concentrate would also relieve the environmental impact of the conventional lactic acid recovery process. In a previous study carried out without the MF module (Ishizaki *et al.*, 1990), the experimental increase in microbial lactic acid production rate was less than expected as a consequence of cell damage and/or adhesion to the anionic membranes as the whole culture broth was recirculated through the ED stack. As referring to the lactate fermentation from glucose by *Lactococcus lactis* IO-1 in the combined F-MF-ED system, after 38 hour cell concentration (2.89 kg/m^3) and living cell count (4.8 $\times$ 10^{15} cell/m^3) were, respectively, circa 1.8 and 1.6 times greater than those observed during a conventional fermentation, thus involving greater glucose consumption and lactate production rates.

The performance of the F-MF-ED system was also assessed in the case of lactate fermentation from xylose (Nomura *et al.*, 1988). Starting with 50 kg of xylose per m^3, the conventional or combined system allowed full exhaustion of the carbon source after 60 or 32 hours, respectively. By further increasing the initial xylose concentration to 80 kg/m^3, both systems resulted in less (50 kg/m^3) or more (75 kg/m^3) consumption of xylose, respectively. Moreover, the simultaneous removal of microbial metabolites (lactate and acetate) via ED increased both lactate production and xylose consumption rates.

Simultaneous L-lactic acid fermentation (by *Rhizopus oryzae* immobilized in calcium alginate beads) and separation was carried out using a three-phase fluidized-bed bioreactor as a fermenter (F), an external electrodialyzer as a separator, and a pump to recycle the fermentation broth between the bioreactor and the separator. In this way, the experimental specific lactate productivity and yield practically coincided with those obtained in the $CaCO_3$-buffered fermentation process (Xuemei *et al.*, 1999), thus confirming the capability of the combined system to alleviate product inhibition without any addition of alkali or alkali salts. It was also shown that the adoption of ED-F for the production of inoculum reduced variability in inoculum quality, thus shortening the length of the lag phase of L-lactate production practically to zero as compared to that observed using an inoculum conventionally cultured for 24 hour (Yamamoto *et al.*, 1993). In the case of lactic acid production from glucose by *L. lactis* IO-1, about a double increase in lactate production rate was achieved using periodic ED in place of a combined F-MF-ED system (Vonktaveesuk *et al.*, 1994).

A two-stage ED process was proposed to recover lactic acid from model solutions and from fermentation media (Habova *et al.*, 2001, 2004). The broth was ultrafiltrated, decolorized by flowing through a granulated active charcoal-filled column, concentrated more than 2.5 times to 110–175 kg of sodium lactate per m^3 via ED using monopolar membranes, and finally separated into two streams rich in sodium hydroxide and free lactic acid (ca. 150–160 kg/m^3), respectively, using bipolar membranes Neosepta® (Tokuyama Corp., Japan) with an overall energy consumption of about 1.6 kWh/kg of lactate transported. A similar procedure was tested by Madzingaidzo *et al.* (2002), thus yielding almost congruent results in terms of average energy consumption in the 1st (0.6 kWh/kg) and second (0.6–1.0 kWh/kg) stage and lactate (15% w/w) or free lactic acid (16% w/v) concentration in the concentrated stream flowing out of them, respectively. In the optimal operating conditions current efficiency in the monopolar and bipolar ED was about 90%. Moreover, during the electrodialytic purification significant reduction in color and minerals in the free lactic acid solution was obtained, while the glucose and acetic acid contents were reduced to less than 1 kg/m^3.

On the contrary, a more efficient splitting of sodium lactate into lactic acid (0.96 kg/kg) and sodium hydroxide (0.93 kg/kg) was obtained using a single step three-compartment bipolar ED process, as that shown in Figure 6A, and no pretreatment of the fermentation broth (Kim and Moon, 2001).

The performance of the ED recovery of sodium lactate from model solutions was mathematically described by several authors (Boniardi *et al.*, 1996; Fidaleo and Moresi, 2004; Yen and Cheryan, 1993). It was found that the permeation flux of lactate was independent of the presence of lactose and glucose in culture medium and that the fluxes of both sugars were practically negligible (Yen and Cheryan, 1993). By using the performance indicators reported in Table XV, Moresi and Sappino (2000) estimated that the ED recovery of sodium lactate was less expensive than that of disodium itaconate or trisodium citrate.

Although promising prospects for lactate recovery by monopolar (Moresi and Sappino, 2000) or bipolar (Siebold *et al.*, 1995) ED have been presented, a resistance to embrace this technology still persists and just a fermentation industry in France seems to have adopted such a technology (Gillery *et al.*, 2002).

5. *Malic acid*

L-Malic acid (HOOC–CH_2–CHOH–COOH) for use in the pharmaceutical industry is manufactured by conversion of fumaric acid by the intracellular enzyme fumarase produced by various microorganisms. The excess fumaric acid is easily separated by crystallization after concentration of the mother solution. Further addition of lime allows malic acid to be separated as calcium malate within a bioreactor crystallizer system. By adding diluted sulfuric or oxalic acid, the salt is split into free malic acid and calcium sulfate or oxalate, the latter being removed by filtration (Mourgues *et al.*, 1997).

To simplify such a complex procedure, Sridhar (1987) suggested to make use of a single six-compartment ED stack, as schematically shown in Figure 21.

Fumaric acid crystals are dissolved in water and fed to compartment Z2, where hydrogen ions (H^+) are attracted by the cathode, pass through the cationic membrane, and accumulate into the cathodic compartment (Z1). On the contrary, fumarate ions (Fum^{2-}), direct toward the anode, permeate through the anionic membrane, and accumulate into the compartment Z3, where they are neutralized by the ammonium ions (${NH_4}^+$) arriving from compartment Z4. Malate ions (Mal^{2-}) migrate into compartment Z5, where they recombine with the hydrogen ions (H^+) coming from the anodic compartment (Z6), thus yielding the final product (malic acid) of this process. The ammonium fumarate is circulated through the bioreactor (R) to be enzymatically converted into ammonium malate. The electrode compartments are washed using acidic solutions and continuously refilled with water, the electrolysis of which generates gaseous hydrogen (H_2) or oxygen (O_2) at the cathode or anode, respectively.

In this way, it would be possible to convert electrodialytically fumaric acid into ammonium fumarate. This in turn may be enzymatically transformed into ammonium malate, which might finally be ED freed into malic acid with no reagent consumption and by-product formation, and minimum product loss.

This procedure was recently confirmed by Belafi-Bako *et al.* (2004).

6. *Propionic acid*

Propionic acid (CH_3CH_2COOH) is produced using mainly the oxo process (about 200,000 Mg), which involves reacting ethylene and carbon monoxide to produce propionaldehyde, to be further oxidized in the presence of cobalt or manganese ions at 40–50°C (Anon., 2002). About 45% of the overall consumption of propionic acid is used as such or as ammonium propionate

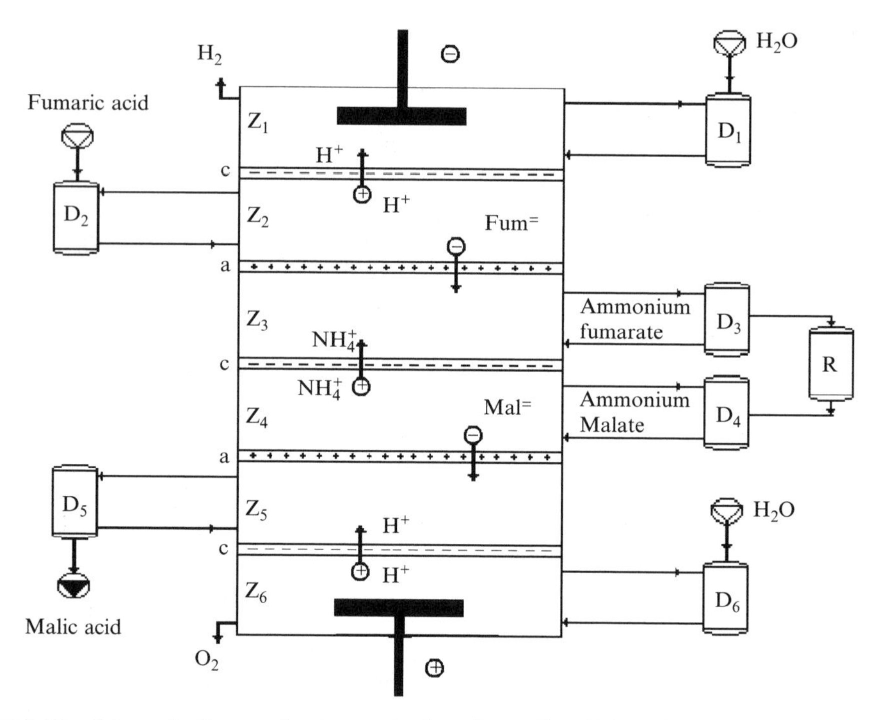

FIG. 21 Schematic diagram for the production of L-malic acid from fumaric acid using a six-compartment (Z_1–Z_6) ED stack composed of anionic (a) and cationic (c) membranes, as extracted from Sridhar (1987). As ammonium fumarate is formed, it is enzymatically converted into ammonium malate in an external bioreactor (R). The combined system is also provided with a series of storage tanks for the acidic cathode-(D1) and anode-(D6) rinsing solutions, raw materials (D2), ammonium fumarate (D3), ammonium malate (D4), and final product (D5).

to prevent mould in animal feeds; 21% is converted into sodium and calcium propionates to preserve baked goods and cheeses; 19% is used to produce diethyl ketone (DEK), that is, an intermediate in the manufacture of herbicides, such as pendimethalin; circa 11% is converted into cellulose acetate-propionate plastics (as moulding compounds for toothbrush handles, other brush handles, and eyeglass frames), and the remaining 4% into pharmaceuticals and propionate esters for solvents, flavors, and fragrances.

Propionic acid might be also produced by fermentation of *Propionibacterium acidipropionici*, thus providing an attractive alternative that can also meet consumer's demand for natural preservatives used in the food industry. However, conventional batch propionic acid fermentation suffers from low productivities (< 1 kg $m^{-3}h^{-1}$), low product concentrations (< 50 kg/m^3), low-propionic acid yield (0.548 g/g of glucose) due to the formation of acetic acid (0.222 g/g) as a main by-product and slow microbial growth as strongly inhibited by propionic acid and acidic pH (Boyaval and Corre, 1995; Moresi and Parente, 1999).

A great deal of research work has been carried out to enhance bioreactor productivity (2–14 kg $m^{-3}h^{-1}$) using cell recycle via membrane processing (Boyaval and Corre, 1987; Boyaval *et al.*, 1994) and recovering propionic acid by monopolar or bipolar ED (Boyaval *et al.*, 1993; Weier *et al.*, 1992; Zhang *et al.*, 1993).

To avoid the cosynthesis of acetate by *P. thoenii*, it was suggested to replace conventional carbon sources with glycerol, thus obtaining a theoretical yield of 0.804 g of propionic acid per g of glycerol and no acetic acid (Boyaval and Corre, 1995; Himmi *et al.*, 2000).

In the circumstances, it seems to be possible to readdress industrial manufacturers toward the fermentation route by resorting to an integrated process involving appropriate mutants of *Propionibacterium* spp., glycerol-based media, as well as efficient membrane modules capable of keeping the concentrations of free acid or bacteria in the culture medium to the minimum or maximum level, respectively. To this end, an ED unit would help not only to recover sodium propionate from the generally dilute broths in which it is produced in ionized form (Weier *et al.*, 1992), but also to keep a low acidic level and control the pH in the fermenting broth throughout the course of the fermentation (Zhang *et al.*, 1993). Generally, the broth is to be microfiltered and then fed to the ED unit to be separated into an enriched salt product (concentrate) and a salt-depleted broth (dilute). Since the residual sugars and nutrients are retained in the dilute, its recycling into the bioreactor would enable material utilization and microbial propionate productivity to be maximized. Use of bipolar membranes would also allow the aqueous solutions of sodium hydroxide or free propionic acid to be recycled into the bioreactor for the pH control or further purified, respectively (Boyaval *et al.*, 1993).

Boyaval *et al.* (1993), Weier *et al.* (1992), and Zhang *et al.* (1993) attempted to optimize the electrodialytic recovery of propionate from model or real solutions by resorting to a few performance indicators (i.e., current efficiency, solute recovery yield, solute and water fluxes). Despite such parameters were found to be dependent on the feed solutions used, that is, model sodium propionate and/or sodium acetate media and real fermentation broths, Weier *et al.* (1992) were able to estimate the membrane surface area required to guarantee low propionate levels (0.10–0.32 kmol/m^3) during the fermentation as a function of the fermenter size.

7. *Pyruvic acid*

Fed-batch production of pyruvic acid [$CH_3COCOOH$] from an engineered strain (*Escherichia coli* YYC202) was optimized by resorting to ED to prevent potential product inhibition in the bioreactor (Zelic *et al.*, 2004). In this way, by continuous separation of pyruvate from the fermentation medium, high values of the pyruvate-to-glucose molar yield (1.78 mol/mol), volumetric productivity (145 kg $m^{-3}day^{-1}$), and pyruvate concentration (79 kg/m^3) were achieved by the repeated fed-batch mode.

8. *Succinic acid*

A two-stage ED process was also proposed to recover succinic acid [$HOOC(CH_2)_2COOH$] from sugar- and triptophane-based fermentation media (Glassner and Datta, 1992). The broth was previously concentrated via ED using monopolar membranes and then separated into sodium hydroxide- and free succinic acid-rich streams using bipolar membranes. Further removal of sodium cations and sulfate anions was achieved using weakly acid and -basic IER.

9. *Lysine and other amino acids*

Lysine is an essential bibasic amino acid to support growth in children and general well-being in adults. It is also quite an important ingredient in feed mixes, especially for pigs and chickens. It is mainly produced by submerged fermentation using overproducing mutants of *Corynebacterium glutamicum* that are able to excrete up to 100 kg of lysine per m^3 at the end of fermentation, and commercialized as L-lysine hydrochloride. The process is generally carried out by intermittent feeding of the carbon source to prevent sugar inhibition on cell growth and lysine production rates.

Ajinomoto and BASF are leading lysine manufacturers. For instance, Ajinomoto is expected to expand its global lysine capacity to 300,000 metric tons by 2005 (http://static.highbeam.com/c/chemicalweek/january262000/ajinomotoraiseslysineproductionworldwide/).

Nomura *et al.* (1987a) attempted to minimize product inhibitory effect on the aspartate kinase step in lysine biosynthesis and enhance L-lysine production from *Brevibacterium flavum* QL-5 using a combined ED-F system. However, lysine production was not statistically different from that obtained in diffusion dialysis fermentation and about 20% greater than that achieved during conventional fermentation, thus making practically ineffective such a use of ED.

Desalting of lysine-rich fermentation broths by monopolar ED appears to be much more rewarding. To mitigate anionic membranes sensitivity to fouling, Lee *et al.* (2002b, 2003) resorted to pulsed electric field with the half-wave power and suggested the use of pulse power as an effective CIP method during ED of fermentation media.

A typical three-compartment ED stack using bipolar, cationic, and anionic membranes (Figure 8A) was patented by Mani (2000) to produce lysine hydrochloride. By feeding the salt compartments with a salt (e.g., NaCl, KCl, or LiCl) solution, the base and acid ones with water and a lysine-rich solution, respectively, the salt can be split into an alkali and hydrochloric acid. The latter reacts with lysine, thus yielding the amino hydrochloride.

Kikuchi *et al.* (1995) made use of an ED stack composed of Selemion-CMV and -AMV electromembranes (Table II) to separate almost completely a mixture of amino acids, that is, glutamic acid, methionine, and lysine. In this way, while methionine was not affected by the voltage applied, glutamic acid or lysine was found be transported across the anion- or cation-exchange membranes, respectively.

The recovery of phenylalanine from an industrial process stream was carried out using a conventional ED stack (Grib *et al.*, 2000). A preliminary soaking of such membranes in a bovine serum albumin solution for 2 hour allowed the reduction of phenylalanine loss to less than 5% and achievement of an average current efficiency of 98%. Such a process was also successful in removing 98% of Na_2SO_4 and $(NH_4)_2SO_4$.

H. OTHER FOOD INDUSTRIES

A few novel ED applications in the food sector have been reported in the literature and are generally aimed at desalting some food extracts.

For instance, the acidity of steam-stripped coffee extracts was reduced by about 50% using ED against a stream of 0.3% KOH at a current density of about 108 A/m^2 at 38 °C (Husaini, 1982).

The mussel cooking juice was desalted using two alternative technologies, that is, ED and diafiltration, thus resulting in about 80% or 77% demineralization degree with no or significant loss of flavor, respectively (Cros *et al.*, 2003). Alternatively, the centrifuged juice was previously desalted by ED and then concentrated by RO. Whereas the permeate might be disposed of without further depuration treatments, the retentate presented a sensory profile slightly different from that of unprocessed mussel cooking juice, but with its native characteristic aroma, thus making it useful as mussel flavor extract for human or pet food industries (Cros *et al.*, 2004).

ED was also suggested to deacidify liquorice extracts, containing 4–20% glycyrrhizic acid and 10–40% dry matter, or to desalt the glycyrrhizin-free juice recovered after acidification (Bozzi *et al.*, 1997).

Multistage ED stacks were also proposed to separate racemic mixtures of D,L-tryptophan in combination with a chiral selector (i.e., α-cyclodextrin). Despite the low selectivity (1.12) of such selector, the enantiomeric excess difference was found to range from 14% to 99% using ED stacks composed of 20 to 250 compartments, respectively (van der Ent *et al.*, 2002).

Genders and Hartsough (1999) patented an electrochemical method to recover ascorbic acid from an ascorbate salt and an inorganic salt (NaCl) without cogeneration of a waste salt stream and maintaining a high electric conductivity. The ascorbate-rich feed is dissociated into ascorbate anions and Na^+ cations under the influence of an electric field, while water splitting occurs in virtue of the bipolar membranes. As shown in Figure 8B, the hydrogen ions combine with the ascorbate anion to form ascorbic acid, while Na^+ cations migrate through the cationic membrane into the base compartment to combine with the hydroxyl ion to form the coproduct base (e.g., NaOH). The ascorbic acid is then crystallized and recovered.

IV. MATHEMATICAL MODELING OF AN ED DEVICE

To design or optimize an ED process several parameters are to be taken into account, namely stack construction and spacer configuration, operation mode, membrane perm-selectivity, feed and product concentration, flow velocities, current density and voltage applied to the electrodes, recovery rates, and so on.

The MS equation represents the simplest mathematical tool for linking the flux of a generic species through the membrane with its interfacial concentrations at the membrane left and right sides, as well as with the external electrical voltage applied to the ED electrodes (Krishna and Wesselingh, 1997). To overcome the main problem in the application of the MS mass transfer model to ED processes, that is, the large number of species diffusivities in the free solution and membrane phase (van der Stegen *et al.*, 1999), NP relationship is largely used to describe diffusion and electromigration contributions to ion transport in IEM (Bailly *et al.*, 2001; Boniardi *et al.*, 1996, 1997; Fidaleo and Moresi, 2004, 2005a; Ibanez *et al.*, 2004; Lee *et al.*, 1998, 2002d; Nikonenko *et al.*, 2002, 2003; Yen and Cheryan, 1993).

The basic mathematical model consists of water and solute mass balances in the concentrating and diluting tanks that are to be coupled with the solute—Eq. 11—and water—Eqs 12 and 13—mass transfer equations and voltage equation—Eq. 18—for the ED loop concerned.

Assuming perfect mixing in each compartment of the membrane pack and reservoir of the ED unit shown in Figure 9, the solute concentration in any of them is uniform and equal to that of the outlet stream. Therefore, by assuming pseudo-steady state conditions in any compartment, the differential solute and water mass balances in the diluted (D) and concentrated (C) reservoirs can be written as follows:

$$\frac{d(n_{BC})}{d\theta} = -\frac{d(n_{BD})}{d\theta} = J_B a_{me} N_{cell} \tag{28}$$

$$\frac{d(n_{WC})}{d\theta} = -\frac{d(n_{WD})}{d\theta} = J_W a_{me} N_{cell} \tag{29}$$

where n_{Bk} and n_{Wk} are the instantaneous solute and water mass in the k-th reservoir; θ is the process time; N_{cell} the overall number of cell pairs; and a_{me} the effective membrane surface area as viewed by the electrodes themselves.

Thus, any ED unit design or optimization exercise relies on quite a great number of engineering parameters, such as ion transport numbers in solution (t^+ and t^-) and electromembranes (t_a^-, t_c^+); effective solute (t_s) and water (t_W) transport numbers; solute (L_B) and water (L_W) transport rates by diffusion; effective membrane surface area (a_{me}); membrane surface resistances (r_a, r_c); solute mass transfer coefficient (k_m); and limiting current density (j_{lim}). These can be determined by independent experiments, tabulated data, or existing correlations.

In previous work (Fidaleo and Moresi, 2005a) a five-step sequential procedure was set up to assess such parameters by independent experiments. It mainly consisted of the following independent tests:

1. Zero-current leaching, osmosis, and dialysis tests to determine L_B and L_W.
2. Electroosmosis tests to estimate t_W.
3. Desalination tests to establish t_s.
4. Current-voltage tests to determine the limiting current intensity (I_{lim}), ion transport numbers (t_a^-, t_c^+), and surface resistances (r_a, r_c) in anionic and cationic membranes, as well as solute mass transfer coefficient (k_m).
5. Validation tests to assess the accuracy of all the parameters given earlier.

In this way, it was possible to extend the range of application of the NP equation from NaCl concentrations smaller than 0.1 (Krishna, 1987) or 0.5 (Lee *et al.*, 2002d) kmol/m^3 to about 1.7 kmol/m^3 (Fidaleo and Moresi, 2005a). Further details were reported elsewhere (Fidaleo and Moresi, 2005a).

As a result of a few parameter sensitivity analyses carried out in the case of the ED recovery of some sodium salts of hydrogen chloride (Fidaleo and Moresi, 2005a), lactic (Fidaleo and Moresi, 2004), acetic (Fidaleo and

Moresi, 2005b), or propionic (Fidaleo and Moresi, 2006) acids from model solutions, it is worth pointing out the following:

1. The solute and water transport numbers in the electromembranes reported in the literature may be accurate enough to predict the solute concentrations in C and D tanks and may need no extra experimental trials.
2. The contribution of solute diffusion (L_B) at zero current is usually negligible, that is, within the experimental error deviation band, with respect to that of electromigration. On the contrary, that of solvent diffusion (L_W) increases with the solute concentration difference at the membrane sides, especially at low current densities. It should be taken into account to reproduce accurately the experimental solute concentrations in the concentrate, especially at the end of ED recovery processes performed at low current densities (Fidaleo and Moresi, 2005a).
3. The solute mass transfer coefficient (k_m) in ED stacks approximately varies with the square root of the liquid superficial velocity (v_S) in agreement with the correlations reported in Table III, even if they can differ from those predicted within a ±30% deviation band because of the different cell and spacer configuration used.
4. As far as the overall potential drop (E) is concerned, the contribution of solute polarization, namely the electric resistance (R_f) and junction potential difference (E_j) across any boundary layer, may be neglected. On the contrary, the Donnan potential difference (E_D) in any cell pair, which behaves as a DC generator with inverted polarities with respect to those of the external DC generator (Figure 9), has to be accounted for as the solute concentration difference at both sides of the anionic and cationic membranes increases:

$$E_D = 2t_s \frac{R_G T_K}{F} \ln\left(\frac{c_{BD}}{c_{BC}}\right) \tag{30}$$

5. The ohmic resistances of the bulk solutions in the concentrating (C), diluting (D), or ERS compartment can be estimated via the second Ohm's law, as reported in Table IV.
6. The effective surface resistances of the anionic (r_a) and cationic (r_c) membranes are generally greater than those provided by the manufacturer (Table II), these being measured in a bridge circuit using alternating current. As a rule of thumb, such values are to be multiplied by 1.75 to express approximately the membrane surface resistance to DC (Davies and Brockman, 1972). Moreover, the membrane surface resistances may be regarded as

independent of the solute concentration (Lee *et al.*, 2002d). Both statements were confirmed by Fidaleo and Moresi (2005a).

7. Knowledge of the effective membrane surface area (a_{me}) is of paramount importance for a safe transfer of the data collected in a laboratory- or pilot-scale plant into an industrial-scale one. In previous work (Fidaleo and Moresi, 2004, 2005a,b), a_{me} was found to be about two-thirds of the geometrical membrane surface area (a_{mg}) and just 10% greater than the exposed surface area of electrodes (a_{ERS}). The latter may in principle be different from the former owing to the fact that the ion flow pattern starts in the direction orthogonal to the electrode surfaces and tends to diverge to maximize membrane area utilization and consequently minimize the electrical resistance of solutions fed in C and D compartments (Davies and Brockman, 1972). This is apparently in line with the empirical rule of overdesigning the geometrical membrane surface area by a correction factor of the order of 70% to account for the shadow effect of the spacer (Lee *et al.*, 2002d), since such an effect is nonlinearly related to the spacer-strand thickness and density (Turek, 2002). Thus, it is highly recommended to assess the effective surface area to assess the real limiting current density in the ED stack examined.

As an example of the application of the aforementioned sequence, Table XVI lists the main engineering parameters necessary to design or optimize ED stacks equipped with AMV and CMV electromembranes (Table II) and committed to the recovery of the sodium salts of some weak monocarboxylic acids of microbial origin (i.e., acetic, propionic, and lactic acid) and of a strong inorganic acid (i.e., chloride acid), as estimated by Fidaleo and Moresi (2005a,b, 2006).

As the salt molecular mass (M_B) increased from 58 to 112 Da, the transport number for Na^+ in the corresponding solution tended to increase from 0.4 to 0.6 for the progressively smaller equivalent conductance at infinite dilution (λ_0^-) of acetate, propionate, and lactate ions with respect to that of Cl^-. Nevertheless, the current within the electromembranes was almost exclusively carried by the counterions.

The effective solute (t_s) transport number, ranged from 93% to 98%, even if reduced to 88% for sodium lactate. The water transport number (t_W) increased from 9.3 to 15.6 and correspondently the maximum salt weight concentration in the concentrated stream ($C_{BC,max}$) ranged from 286 to 350 kg/m^3. Finally, while the surface resistance (r_c) of the cation-exchange membranes was found to be about constant ($5 \pm 2\ \Omega\ cm^2$), r_a tended to increase with M_B. However, the specific electric energy consumption (ε) slightly increased from 0.19 to 0.22 kWh/kg of salt recovered.

TABLE XVI

MAIN ENGINEERING PARAMETERS CHARACTERIZING THE ED RECOVERY OF NaCl, Na-A, Na-P, AND Na-L FROM MODEL SOLUTIONS[a,b]

	M_B				r_c	r_a			ε	$C_{BC,max}$
Salt	Da	t^+	t_s	t_w	Ω cm^2	Ω cm^2	t_c^+	t_a^-	kWh/kg	kg/m^3
NaCl	58.4	0.40	0.969 ± 0.002	9.3 ± 0.1	6 ± 1	6 ± 1	0.97 ± 0.01	1.00 ± 0.01	0.19	338
Na-A	82.0	0.56	0.931 ± 0.003	14.8 ± 0.1	6 ± 2	12 ± 2	0.93 ± 0.01	1.00 ± 0.01	0.21	286
Na-P	96.1	0.59	0.982 ± 0.002	15.23 ± 0.04	2 ± 1	25 ± 3	0.95 ± 0.04	1.03 ± 0.03	0.20	344
Na-L	112.1	0.60	0.876 ± 0.002	15.60 ± 0.05	5.5	16.4	0.92 ± 0.01	0.96 ± 0.01	0.22	350

[a] As extracted from Fidaleo and Moresi (2004, 2005a,b, 2006).

[b] Effect of the salt molecular mass (M_B) on the cation transport number (t^+) in the corresponding solution; effective solute (t_s) and water (t_W) transport numbers; surface resistances (r_c, r_a) of, and counterion transport numbers (t_c^+, t_a^-) in cation- and anion-exchange membranes; specific electric energy consumption (ε) in the case of 90% salt recovery at 1 A, and maximum solute concentration theoretically achievable in the concentrating stream ($C_{BC,max}$).

Note: NaCl, sodium chloride; Na-A, acetate; Na-P, propionate; Na-L, lactate.

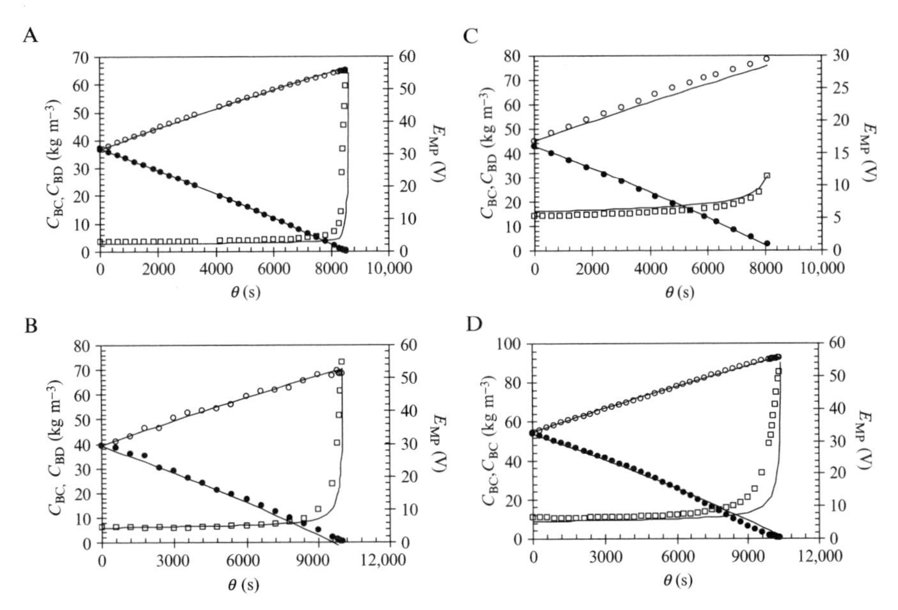

FIG. 22 Batch recovery of sodium chloride (A), acetate (B), propionate (C), and lactate (D) at 20°C, $v_s = 3$ cm/s, $I = 1$ A using 10 AMV and 9 CMV membranes (Table II): solute weight concentrations (C_{BC} and C_{BD}) in C (○) and D (●) tanks, and voltage (E_{MP}: □) applied to the membrane pack vs. time (θ). The continuous lines were calculated via Eqs 26, 27, and 18.

By integrating Eqs 28 and 29 and using Eq. 18 together with the independent parameters listed in Table XVI, it is possible to simulate the time course of the main variables involved in the batch electrodialytic recovery of the four sodium salts examined.

As an example, Figure 22 shows quite a satisfactory agreement between the experimental and calculated values of C_{BC}, C_{BD}, and E_{MP} against time (θ) for the laboratory-scale electrodialyzer used by Fidaleo and Moresi (2004, 2005a,b, 2006).

V. PRESENT PROBLEMS AND FUTURE PERSPECTIVES

Despite the ED industry has experienced a steady growth rate of about 15% since the late 15 years (Srikanth, 2004) and ED potentialities are numerous, its applications are still too marginally extended to the food industry. What are the reasons for such a scarce diffusion?

By tradition, the technical progress of the food industry has generally proceeded quite slowly with 20–30-year delay with respect to that of the chemical and pharmaceutical industries (Cantarelli, 1987). For instance, it is worth noting that falling film evaporators started to be used in the Italian citrus industry by the early 1970s in spite of the fact that they had been industrially manufactured in 1947 by *Majonnier Brothers Co.* on the behalf of *Florida Citrus Canners Cooperative* (Varsel, 1980) as a follow-up of the under vacuum concentration techniques developed during the World War II to concentrate highly thermosensitive materials, such as penicillin.

To counteract the typical misoneism of the world food industry, it is necessary to resort to appropriate scaling-up exercises in pilot- or industrial-plant scale to assess precisely the membrane process performance and reliability, as well as its economical feasibility. There are, however, a number of problems that have undoubtedly limited growth in ED membrane sales, like membrane-fouling problems, design considerations, cleanability, investment and membrane replacement costs, and competing technologies, such as NF and IER.

Membrane fouling and scaling are caused by soluble and insoluble impurities present in the feed such as insoluble salts, organic matter, colloidal substances, and microorganisms. Whereas the anionic membranes appear to be more liable to be fouled by organic matter, the cationic ones tend to be scaled by inorganic matter. To minimize such problems, as well as stack plugging, feed pretreatment via MF, UF, NF, or IER may be useful. Other approaches, such as the modification of the membrane properties (Grebenyuk *et al.*, 1998) or use of pulsed electric fields (Lee *et al.*, 2002b, 2003), appear to be expedient to alleviate anionic membrane fouling. In the great majority of cases, cationic membrane scaling was found to be almost reversible and kept under control by pH adjustment, EDTA, or citric acid addition. Thus, it is still lacking a well-defined procedure to moderate fouling and scaling in ED processing and there is a need for additional research on this issue so that the process can be optimized taking scaling and fouling into account.

The maximum cell voltage, which varies in the range of 0.8–1.5 V/cell under the current density recommended by the manufacturers, tends to increase with time as the charged groups in the electromembranes vanish with use as a result of their chemicophysical reactions with the feed contaminants. Beyond such potential difference limits, it is generally advisable to replace the membranes to limit the overall electric power consumption.

ED membranes can be generally cleaned with dilute acidic and alkaline solutions, as well as enzyme solutions.

When chemical cleaning is inefficient, current reversal is the method of choice.

In the case of clarified feeds and low current densities, membrane lifetime can be as long as seven or 10 years for brackish water desalination or drinking water nitrate removal, respectively. However, if the feed solution is fouling or scaling or the ED separation plant has not been well designed or is not properly conducted, membrane lifetime is no longer than a year.

The incomplete comprehension of mass transfer mechanisms in ED membrane systems is in all probability responsible for the difficult design of industrial plants and for their limited diffusion. For instance, in the food biotechnology sector ED applications are still in their infancy since quite a limited number of the novel processes studied so far in laboratory- and pilot-scales and reviewed here have been converted into industrial realities yet, except for the recovery of the sodium salt of unspecified organic acid from clarified fermentation broths, as well as amino and organic acids (Gillery *et al.*, 2002).

This unequivocally means that ED-processing potentialities have not been completely exploited and much more is needed to account for all the key parameters (i.e., current density, cell voltage, current efficiency, solute concentration in the diluting and concentrating streams) optimizing ED performance.

The electric potential difference is the driving force of the ED process since it determines the flux of ions across the membranes, which corresponds to a current density. The greater the current density, the smaller the membrane surface area required becomes. This minimizes the investment and maintenance costs, but boosts the electric power consumption per kg of solute recovered. Moreover, by increasing j the faster ion transfer through the membranes enhances the polarization concentration effects at the boundary layers. As soon as the solute concentration at the membrane surface is zero, the limiting current density (j_{lim}) is achieved. Since the Donnan potential difference (E_D) in any cell pair tends to infinite, this would result in quite a great increase in cell voltage. As j exceeds j_{lim}, the pH of the dilute or concentrate falls or increases, respectively, and Ω reduces. Even if it is recommended to keep j smaller than 70–80% of j_{lim}, for any given application it is essential to assess the effect of solute concentration and feed superficial velocity on this critical parameter by performing a series of current–voltage tests.

To maximize the overall current efficiency, all the parasitic phenomena occurring in the ED stack are to be minimized by getting rid of imperfect gaskets and membranes with pinholes leading to liquid leakage and dripping. Also the manifolds are to be made of nonconductive materials to limit shunt or stray currents running in the nonactive cell area. Thus, an optimal stack design feature is needed together with low values of cell voltage and ratio of conductivity in the concentrate and dilute loops. However, in the case of very soluble solutes, this ratio can be as high as 20, a much higher value than that achieved in RO units, which explains why ED is used to produce table salt from seawater in Korea and Japan. The minimum solute concentration in the diluting stream is generally associated to an electric conductivity (χ) of about 0.05 S/m to limit the ohmic resistance of the diluting compartments.

The maximum temperature in the ED stacks used for the food industry was generally 35–40 °C, but novel anionic membranes are claimed to allow operation up to 60 °C, this being useful for viscous and low-χ products such as sugar syrups or molasses.

Despite many of the ED membrane applications might be regarded as mature, being established in the 1970s, they are still relatively new owing to the fact that no membrane-based process is today an established procedure for given production processes. This is true even in the case of brackish- and sea-water desalination, the ED technology being competed by MSFD and RO depending on the TDS content of brackish or sea water and plant capacity (Table VII).

Moreover, the economics of such a separation technique is *a priori* expected to be controlled by electric energy costs, but this is not always true and should be verified case by case. For instance, when dealing with the batch electrodialytic recovery of trisodium citrate at pH 7, disodium itaconate at pH 6, or sodium lactate at pH 5 from aqueous solutions, it was found that, in the optimal operating conditions, the contribution of depreciation and general plant maintenance and annual membrane replacement ranged from 60% to 66% of the overall operating costs per unit mass of solute recovered, that of electric energy was about one third for citrate and itaconate recovery, and just 20% for lactate recovery while the remaining 8–14% was pertaining to labor costs (Moresi and Sappino, 2000). In this way, it was first shown that sodium lactate ED recovery was by far much less expensive than that of trisodium citrate or disodium itaconate. Second, it was possible to evaluate how any enhancement in ion mobility through the ED membranes or extension in their service life would have minimized the overall membrane surface area to be installed and thus reduced plant investment and maintenance costs. Third, the ED recovery of these sodium salts was mainly controlled by the so-called investment-related operating costs rather than the electric energy costs.

So, any attempt to minimize the specific electromembrane costs, especially now that the membrane market is open to new Far East manufacturers (Table II), is expected to reduce significantly the investment and membrane replacement costs.

This will be a prerequisite for promoting, in the short-medium term period, further interest toward ED since it might avoid the great pollution problems caused by the disposal of the enormous amounts of calcium sulfate produced by the traditional citrate or lactate recovery process most largely used in the industrial scale or simplify the complex operations actually used to purify for instance itaconic acid crystals.

In conclusion, a greater knowledge of the effect of the key controlling parameters of this powerful separation technique, as well as improvement in membrane life time of the currently available commercial electromembranes and reduction in their costs, would ensure further growth beyond desalination and salt production and foster ED applications in the food sector, as well as in the chemical, pharmaceutical, and municipal effluent treatment areas. This will of course need extensive R&D studies and will highly likely result in hybrid processes combining ED to other separation techniques, such as NF, IE, and so on, so as to shorten present downstream and refining procedures.

REFERENCES

Adhikary, S.K., Harkare, W.P., Govindan, K.P., and Nanjundaswamy, A.M. 1983. Deacidification of fruit juices by electrodialysis. *Indian J. Technol.* **21**, 120–123.

Ahlgren, R.M. 1972. Electromembrane processing of cheese whey. *In* "Industrial Processing with Membranes" (R.E. Lacey and S. Loeb, eds), pp. 57–69. Wiley-Interscience, John Wiley & Sons, New York.

AlMadani, H.M.N. 2003. Water desalination by solar powered electrodialysis process. *Renew. Energ.* **28**, 1915–1924.

Andrés, L.J., Riera, F.A., and Alvarez, R. 1995. Skimmed milk demineralization by electrodialysis: Conventional versus selective membranes. *J. Food Eng.* **26**, 57–66.

Anon. 2001a. Acetic acid. Chemical Market Reporter, February, 26th (online publication, http://www.findarticles.com/cf_dls/m0FVP/9_259/71324310/p2/article.jhtml?term).

Anon. 2001b. Lactic acid depending on export (market report) China Chemical Reporter, November, 26th (online publication at http://www.highbeam.com/library/doc0.asp?DOCID=1G1:80637162#=5).

Anon. 2002. Propionic acid. Chemical Market Reporter. April, 1st (online public-ation, http://www.findarticles.com/p/articles/mi_hb4250/is_200204/ai_n12947354).

Atamanenko, I., Kryvoruchko, A., and Yurlova, L. 2004. Study of the scaling process on membranes. *Desalination* **167**, 327–334.

Audinos, R. 1989. Fouling of ion-selective membranes during electrodialysis of grape must. *J. Membr. Sci.* **41**, 115–126.

Audinos, R. 1992. Liquid waste concentration by electrodialysis. *In* "Separation and Purification Technology" (N.N. Li and J.M. Calo, eds), pp. 229–301. Marcel Dekker, New York.

Audinos, R., Roson, J.P., and Jouret, C. 1979. Application of electrodialysis to the elimination of certain grape juice and wine components. *Connaissance de la Vigne et du Vin* **13**, 229–239.

Bach, H.-P., Scholten, G., and Friedrich, G. 1999. Tartar stabilization with electrodialysis in comparison to the contact process. *Wein-Wissenschaft* **54**, 143–156.

Backus, J., Fabiilli, M., Sanchez, D., and Wong, E. 2003. Acetic acid production via carbonylation of methanol: Technical and economical feasibility study, Vol. I, Fugacitech, Inc., Ann Arbor, Michigan, April, 4 (online publication, http://www-personal.engin.umich.edu/~mfabiill/Report%20rev06.doc).

Bailly, M., Roux-de Balmann, H., Aimar, P., Lutin, F., and Cheryan, M. 2001. Production processes of fermented organic acids targeted around membrane operations: Design of the concentration step by conventional electrodialysis. *J. Membr. Sci.* **191**, 129–142.

Batchelder, B.T. 1987. Electrodialysis applications in whey processing. *Bull. Int. Dairy Fed.* **212**, 84–90.

Bazinet, L. 2004. Electrodialytic phenomena and their applications in the dairy industry: A review. *Crit. Rev. Food Sci. Nutr.* **44**, 525–544.

Bazinet, L., Lamarche, F., Labrecque, R., Toupin, R., Boulet, M., and Ippersiel, D. 1997. Electro-acidification of soybean proteins for the production of isolate. *Food Technol.* **51**(9), 52–56, 58, 60.

Bazinet, L., Lamarche, F., and Ippersiel, D. 1998. Bipolar-membrane electrodialysis: Applications of electrodialysis in the food industry. *Trends in Food Sci. Technol.* **9**, 107–113.

Bazinet, L., Lamarche, F., Ippersiel, D., and Amiot, J. 1999a. Bipolar membrane electro-acidification to produce bovine milk casein isolate. *J. Agric. Food Chem.* **47**, 5291–5296.

Bazinet, L., Lamarche, F., and Ippersiel, D. 1999b. Ionic balance: A closer look at the K^+ migrated and H^+ generated during bipolar membrane electro-acidification of soybean proteins. *J. Membr. Sci.* **154**, 61–71.

Bazinet, L., Ippersiel, D., Montpetit, D., Mahdavi, B., Amiot, J, and Lamarche, F. 2000a. Effect of membrane permselectivity on the fouling of cationic membranes during skim milk electroacidification. *J. Membr. Sci.* **174**, 97–110.

Bazinet, L., Ippersiel, D., Labrecque, R., and Lamarche, F. 2000b. Effect of temperature on the separation of soybean 11S and 7S protein fractions during bipolar membrane electroacidification. *Biotechnol. Prog.* **16**, 292–295.

Bazinet, L., Ippersiel, D., and Mahdavi, B. 2004. Fractionation of whey proteins by bipolar membrane electroacidification. *Innovat. Food Sci. Emerging Technol.* **5**, 17–25.

Belafi-Bako, K., Nemestothy, N., and Gubicza, L. 2004. A study on applications of membrane techniques in bioconversion of fumaric acid to *L*-malic acid. *Desalination* **162**, 301–306.

Berta, P. 1993. La misura della stabilità tartarica dei vini. *Vignevini* **11**, 21–46.

Boniardi, N., Rota, R., Nano, G., and Mazza, B. 1996. Analysis of the sodium lactate concentration process by electrodialysis. *Separ. Technol.* **6**, 43–54.

Boniardi, N., Rota, R., Nano, G., and Mazza, B. 1997. Lactic acid production by electrodialysis. Part II: Modelling. *J. Appl. Electrochem.* **27**, 125–133.

Boyaval, P. and Corre, C. 1987. Continuous fermentation of sweet whey permeate for propionic acid production in a CSTR with UF recycle. *Biotechnol. Lett.* **9**(11), 801–806.

Boyaval, P. and Corre, C. 1995. Production of propionic acid. *Lait* **75**, 453–461.

Boyaval, P., Corre, C., and Terre, S. 1987. Continuous lactic acid fermentation with concentrated product recovery by ultrafiltration and electrodialysis. *Biotechnol. Lett.* **9**(3), 207–212.

Boyaval, P., Seta, J., and Gavach, C. 1993. Concentrated propionic acid production by electrodialysis. *Enzyme Microb. Technol.* **15**, 683–686.

Boyaval, P., Corre, C., and Madec, M.-N. 1994. Propionic acid production in a membrane bioreactor. *Enzyme Microb. Technol.* **16**, 883–886.

Bozzi, R., Gavach, C., and Petitbois, D. 1997. Method for processing liquorice extracts. *PCT Int. Appl.*, WO 9,700,264, January, 3rd, 1997.

Bromley, L.A. 1973. Thermodynamic properties of strong electrolytes in aqueous solutions. *AIChEJ* **19**, 313–320.

Bubbico, R., Lo Presti, S., and Moresi, M. 1997. Repeated batch citrate production by *Yarrowia lipolytica* in a membrane recycle bioreactor. *In* "Engineering & Food at ICEF7. Part I" (R. Jowitt, ed.), pp. B21–B24. Sheffield Academic Press, Sheffield, UK.

Buck, R.P. 1984. Kinetics of bulk and interfacial ionic motion: Microscopic bases and limits of the Nernst-Planck equation applied to membrane systems. *J. Membr. Sci.* **17**, 1–62.

Calle, E.V., Ruales, J., Dornier, M., Sandeaux, J., Sandeaux, R., and Pourcelly, G. 2002. Deacidification of the clarified passion fruit juice (*P. edulis f. flavicarpa*). *Desalination* **149**(1–3), 357–361.

Cameira dos Santos, P.J, Pereira, O.M., Gonçalves, F., Tomás Simões, J., and De Pinho, M.N. 2000. Ensaios de estabilização tartárica em vinhos portugueses: Estudo comparativo da electrodiálise e de um método tradicional. *Ciência Téc. Vitiv.* **15**, 95–108.

Cantarelli, C. 1987. Ricerca e formazione nel campo delle biotecnologie alimentari. *Industrie Alimentari* **26**, 333–349.

Cheryan, M., Parek, S., Shah, M., and Witjitra, K. 1997. Production of acetic acid by *Clostridium thermoaceticum*. *Adv. Appl. Microbiol.* **43**, 1–33.

Chukwu, U.N. and Cheryan, M. 1999. Electrodialysis of acetate fermentation broths. *Appl. Biochem. Biotech.* **77–79**, 485–499.

Cockrem, M.C.M. and Johnson, P.D. 1993. Recovery of lactate and lactic acid from fermentation broth. USP no. 5,210,296, May, 11th, 1993.

Cowan, D.A. and Brown, J.H. 1959. Effect of turbulence on limiting current in electrodialysis cells. *Ind. Eng. Chem.* **51**, 1445–1448.

Cros, S., Vandanjon, L., Jaouen, P., and Bourseau, P. 2003. Desalination by electrodialysis or diafiltration of juice from boiling of mussels: Consequences on the aromatic profile. *Recents Progres en Genie des Procedes* **90** (9e Congres de la SFGP, 2003), 87–94.

Cros, S., Lignot, B., Razafintsalama, C., Jaouen, P., and Bourseau, P. 2004. Electrodialysis desalination and reverse osmosis concentration of an industrial mussel cooking juice: Process impact on pollution reduction and on aroma quality. *J. Food Sci.* **69**, C435–C442.

D'Souza, S.V., Lund, D.B., and Amundson, C.H. 1973. Demineralization of untreated cottage cheese whey by electrodialysis. *J. Food Sci.* **38**, 519–523.

Datta, R. and Bergemamm, E.P. 1996. Process for producing of citric acid and monovalent citrate salts. USP no. 5,532,148, July, 2, 1996.

Davies, T.A. and Brockman, G.F. 1972. Physiochemical aspects of electromembrane processes. *In* "Industrial Processing with Membranes" (R.E. Lacey and S. Loeb, eds), pp. 21–37. Wiley-Interscience, John Wiley & Sons, New York.

de Boer, R. and Robbertsen, T. 1983. Electrodialysis and ion exchange processes: The case of milk whey. *In* "Progress in Food Engineering" (C. Cantarelli and C. Peri, eds), pp. 393–403. Forster-Verlag A.G., Switzerland.

De Korosy, F., Suszer, A., Korngold, E., Taboch, M.F., Flitman, M., Bandel, E., and Rahav, R. 1970. Membrane fouling and studies on new electrodialysis membranes. *US Office Saline Water, Res. Develop. Progr. Rep.* no. 605.

Donnan, F.G. 1911. Theorie der Membrangleichgewichte und membranpotentiale bei Vorhandsein von nicht dialysirenden Elektrolyten ein Beitrag zur physikalisch-chemischen Physiologie. *Z. Elektrochem. Angewandte Phys. Chem.* **17**, 572–581.

Elankovan, P. 1996. Processing food by desalting electrodialysis during leaching. *USP* no. 5,525,365, June, 11th, 1996.

Elmidaoui, A., Lutin, F., Chay, L., Taky, M., Tahaikt, M., and Alaoui Hafidi, My R. 2002. Removal of melassigenic ions for beet sugar syrup by electrodialysis using a new anion-exchange membrane. *Desalination* **148**, 143–148.

Elmidaoui, A., Chay, L., Tahaikt, M., Menkouchi Sahli, M.A., Taky, M., Tiyal, F., Khalidi, A., and Alaoui Hafidi, My R. 2004. Demineralization of beet sugar syrup, juice and molasses using an electrodialysis pilot plant to reduce melassigenic ions. *Desalination* **165**, 435.

Enzminger, J.D. and Asenjo, J.A. 1986. Use of cell recycle in the aerobic fermentative production of citric acid by yeast. *Biotechnol.* **8**, 7–12.

Escudier, J.-L., Saint-Pierre, B., Batlle, J.-L., and Moutounet, M. 1995. Automatic method and device for tartaric stabilization of wines. *PCT Int. Appl.*, WO 9,506,110, March, 2nd, 1995.

Ferrarini, R. 2001. A method for tartaric stabilization, in particular for wine, and apparatus for its implementation. *Eur. Pat. Appl.* EP no. 1,146,115, October, 17th, 2001.

Fidaleo, M. and Moresi, M. 2004. Modelling the electrodialytic recovery of sodium lactate. *Biotechnol. Appl. Biochem.* **40**, 123–131.

Fidaleo, M. and Moresi, M. 2005a. Optimal strategy to model the electrodialytic recovery of a strong electrolyte. *J. Membr. Sci.* **260**, 90–111.

Fidaleo, M. and Moresi, M. 2005b. Modelling of Sodium Acetate Recovery from Aqueous solutions by electrodialysis. *Biotechnol. Bioeng.* **91**, 556–568.

Fidaleo, M. and Moresi, M. 2006. Assessment of the main engineering parameters controlling the electrodialytic recovery of sodium propionate from aqueous solutions. *J. Food Eng.* **76**, 218–231.

Genders, J.D. and Hartsough, D.M. 1999. Electrochemical method for recovery of ascorbic acid from ascorbate salt without formation of waste salt stream. *PCT Int. Appl.*, WO 9,900,178, January, 7th, 1999.

Ghose, T.K. and Bhadra, A. 1985. Acetic acid. *In* "Comprehensive Biotechnology 3" (M. Moo-Young, ed.), pp. 701–729. Pergamon Press, New York.

Gillery, B., Bailly, M., and Bar, D. 2002. Bipolar membrane electrodialysis: The time has finally come. *In* "Proceedings of 16th International Forum on Applied Electrochemistry; Cleaner Technology—Challenges and Solutions." Amelia Island Plantation (FL, USA) November 10–14 (online publication htpp://ameridia.con.html/ebc.html).

Glassner, D.A. and Datta, R. 1992. Process for the production and purification of succinic acid. USP no. 5,143,834, September, 1st, 1992.

Gonçalves, F., Fernandes, C., Cameira dos Santos, P., and De Pinho, M.N. 2003. Wine tartaric stabilization by electrodialysis and its assessment by the saturation temperature. *J. Food Eng.* **59**, 229–235.

Grebenyuk, V.D., Chebotareva, R.D., Peters, S., and Linkov, V. 1998. Surface modification of anion-exchange electrodialysis membranes to enhance anti-fouling characteristics. *Desalination* **115**, 313–329.

Greiter, M., Novalin, S., Wendland, M., Kulbe, K.-D., and Fischer, J. 2002. Desalination of whey by electrodialysis and ion exchange resins: Analysis of both processes with regard to sustainability by calculating their cumulative energy demand. *J. Memb. Sci.* **210**, 91–102.

Greiter, M., Novalin, S., and Wendland, M. 2004. Development and state of the art of whey desalination. *Ernaehrung (Vienna, Austria)* **28**, 150–156.

Grib, H., Belhocine, D., Lounici, H., Pauss, A., and Mameri, N. 2000. Desalting of phenylalanine solutions by electrodialysis with ion-exchange membranes. *J. Appl. Electrochem.* **30**(2), 259–262.

Habova, V., Melzoch, K., Rychtera, M., Pribyl, L., and Mejta, V. 2001. Application of electrodialysis for lactic acid recovery. *Czech J. Food Sci.* **19**, 73–80.

Habova, V., Melzoch, K., Rychtera, M., and Sekavova, B. 2004. Electrodialysis as a useful technique for lactic acid separation from a model solution and a fermentation broth. *Desalination* **162**, 361–372.

Hernàndez, P. and Mínguez, S. 1997. Uso de resinas de intercambio iònico en enologia. Estabilización tartárica. *Revue Française d'Oenologie* **162**, 32–35.

Himmi, E.H., Bories, A., Boussaid, A., and Hassani, L. 2000. Propionic acid fermentation of glycerol and glucose by *Propionibacterium acidipropionici* and *Propionibacterium freudenreichii* ssp. *shermanii*. *Appl. Microbiol. Biotechnol.* **53**, 435–440.

Hongo, M., Nomura, Y., and Iwahara, M. 1986. Novel methods of lactic acid production by electrodialysis fermentation. *Appl. Environ. Microbiol.* **52**, 314–319.

Hoppe, G.K. and Higgins, J.J. 1992. Demineralization. *In* "Whey and Lactose Processing" (J.G. Zadow, ed.), pp. 91–131. Elsevier Applied Science, London.

Husaini, S.A. 1982. Electrodialysis of food products. *Eur. Pat. Appl.* EP no. 49,497, April, 14th, 1982.

Iaconelli, W.B. 1973. The use of electrodialysis in the food industry. *In* "IFT Annual Meeting," June 10–13, Miami Beach, USA.

Ibanez, J.P., Aracena, A., Ipinza, J., and Cifuentes, L. 2004. Modeling for copper transport within the boundary layer in an electrodialysis cell. *Revista de Metalurgia* (Madrid, Spain) **40**, 83–88.

Ishizaki, A., Nomura, Y., and Iwahara, M. 1990. Built-in electrodialysis batch culture, a new approach to release of end product inhibition. *J. Ferment. Bioeng.* **70**, 108–113.

Johnson, K.T., Hill, C.G., Jr., and Amundson, C.H. 1976. Electrodialysis of raw whey and whey fractionated by reverse osmosis and ultrafiltration. *J. Food Sci.* **41**, 770–777.

Jönsson, H.B. and Olsson, L.-E. 1981. The SMR process—a new ion exchange process to demineralize cheese whey. *Milchwissenschaft* **36**, 482–485.

Kang, Y.J. and Rhee, K.C. 2002. Deacidification of mandarin orange juice by electrodialysis combined with ultrafiltration. *Nutraceuticals and Food* **7**, 411–416.

Kaufman, E.N., Cooper, S.P., and Davison, B.H. 1994. Screening of resins for use in a biparticle fluidized-bed bioreactor for the continuous fermentation and separation of lactic acid. *Appl. Biochem. Biotech.* **45–46**, 545–554.

Kikuchi, K., Gotoh, T., Takashashi, H., Higashino, S., and Dranoff, J.S. 1995. Separation of amino acids by electrodialysis with ion-exchange membranes. *J. Chem. Eng. Jpn.* **28**, 103–109.

Kim, Y.H. and Moon, S.-H. 2001. Lactic acid recovery from fermentation broth using one-stage electrodialysis. *J. Chem. Technol. Biotechnol.* **76**, 169–178.

Kobayashi, T. 1967. Itaconic acid fermentation. *Proc. Biochem.* **2**(9), 61–65.

Kobayashi, T. 1978. Production of itaconic acid from wood waste. *Proc. Biochem.* **13**, 15–22.

Korngold, E., De Korosy, F., Rahav, R., and Taboch, M.F. 1970. Fouling of anion selective membranes in electrodialysis. *Desalination* **8**(2), 195–220.

Kraaijeveld, G., Sumberova, V., Kuindersma, S., and Wesselingh, H. 1995. Modelling electrodialysis using the Maxwell-Stefan description. *Chem. Eng. J.* **57**, 163–176.

Krishna, R. 1987. Diffusion in multicomponent electrolyte systems. *Chem. Eng. J.* **35**, 19–24.

Krishna, R. and Wesselingh, J.A. 1997. The Maxwell-Stefan approach to mass transfer. *Chem. Eng. Sci.* **52**, 861–911.

Krol, J.J., Wessling, M., and Strathmann, H. 1999. Concentration polarization with monopolar ion exchange membranes: Current-voltage curves and water dissociation. *J. Membr. Sci.* **162**, 145–154.

Kuroda, O., Takahasi, S., and Nomura, M. 1983. Characteristics of flow and mass transfer rate in electrodialyzer compartment including spacer. *Desalination* **46**, 225–232.

Lacey, R.E. 1972. Basis of electromembrane processes. *In* "Industrial Processing with Membranes" (R.E. Lacey and S. Loeb, eds), pp. 3–20. Wiley-Interscience, John Wiley & Sons, New York.

Lacey, R.E. and Loeb, S. 1972. "Industrial Processing with Membranes", pp. 21–106. Wiley-Interscience, John Wiley & Sons, New York.

Lee, E.G., Moon, S.-H., Chang, Y.-K., Yoo, I.-K., and Chang, H.N. 1998. Lactic acid recovery using two-stage electrodialysis and its modelling. *J. Membr. Sci.* **145**, 53–66.

Lee, H.-J., Choi, J.-H., Cho, J., and Moon, S.-H. 2002a. Characterization of anion exchange membranes fouled with humate during electrodialysis. *J. Membr. Sci.* **203**, 115–126.

Lee, H.-J., Moon, S.-H., and Tsai, S.-P. 2002b. Effects of pulsed electric fields on membrane fouling in electrodialysis of NaCl solution containing humate. *Sep. Purif. Technol.* **27**, 89–95.

Lee, H.-J., Oh, S.J., and Moon, S.-H. 2002c. Removal of hardness in fermentation broth by electrodialysis. *J. Chem. Technol. Biotechnol.* **77**(9), 1005–1012.

Lee, H.-J., Sarfert, F., Strathmann, H., and Moon, S.-H. 2002d. Designing of an electrodialysis desalination plant. *Desalination* **142**, 267–286.

Lee, H.-J., Oh, S.-J., and Moon, S.-H. 2003. Recovery of ammonium sulfate from fermentation waste by electrodialysis. *Water Res.* **37**, 1091–1099.

Lewandowski, R., Zghal, S., Lameloise, M.L., and Reynes, M. 1999. Purification of date juice for liquid sugar production. *Int. Sugar J.* **101**, 125–130.

Lindstrand, V., Sundström, G., and Jönsson, A.-S. 2000. Fouling of electrodialysis membranes by organic molecules. *Desalination* **128**, 91–102.

Ling, L.-P., Leow, H.-F., and Sarmidi, M.R. 2002. Citric acid concentration by electrodialysis: Ion and water transport modelling. *J. Membr. Sci.* **199**, 59–67.

Lo Presti, S. and Moresi, M. 2000. Recovery of selected microbial metabolites from model solutions by reverse osmosis. *J. Membr. Sci.* **174**, 243–253.

Lonergan, D.A., Fennemma, O., and Amundson, C.H. 1982. Use of Electrodialysis to improve the protein stability of frozen skim milks and milk concentrates. *J. Food Sci.* **47**, 1429–1434.

López Leiva, M.H. 1988. The use of electrodialysis in food processing. Part 1: Some theoretical concepts. *Lebensm. Wiss. Technol.* **21**, 119–125.

Lutin, F. 2000. Electrodialysis in the sugar industry as a purification technology. *In* "Proceedings of the Sugar Processing Research Conference", pp. 73–78. Porto, Portugal.

Lutin, F., Bailly, M., and Bar, D. 2002. Process improvements with innovative technologies in the starch and sugar industries. *Desalination* **148**, 121–124.

Madzingaidzo, L., Danner, H., and Braun, R. 2002. Process development and optimisation of lactic acid purification using electrodialysis. *J. Biotechnol.* **96**, 22–239.

Maigrot, E. and Sabates, J. 1890. Apparat zur läuterung von zuckersäften mittels elektrizität. *Germ. Pat.* no. 50,443.

Mancini, M., Moresi, M., and Sappino, F. 1995. Sodium citrate concentration by electrodialysis. *In* "Abstracts of the 2nd Italian Conference on Chemical and Process Engineering (ICheaP-2)", pp. 894–897. Florence, Italy, May, 15–17, 1995, AIDIC, Milano.

Mani, K.N. 1991. Electrodialysis water splitting technology. *J. Membr. Sci.* **58**, 117–138.

Mani, K.N. 2000. A process for simultaneous production of amino acid hydrochloride and caustic via electrodialytic water splitting. *Eur. Pat. Appl.*, EP no. 1,016,651, July, 5th, 2000.

Milewski, J.A. and Lewicki, P.P. 1988. Demineralisation of vinasse by electrodialysis. *J. Food Eng.* **7**, 177–196.

Milson, P.E. and Meers, J.L. 1985. Gluconic and itaconic acid. *In* "Comprehensive Biotechnology" (H.W. Blanch, S. Drew, and D.I. Wang, eds), Vol. 3, pp. 681–700. Pergamon Press Ltd., Oxford.

Moresi, M. 1981. Produzione di bioproteine da siero di latte. I) Analisi dei processi. *La Chimica e l'Industria* **63**, 593–603.

Moresi, M. 1995. Produzioni alternative di ingredienti alimentari. *In* "Atti del Convegno su La Ricerca Biotecnologica al Servizio del Consumatore attraverso l'Industria Alimentare", pp. 99–115. Bologna, Italy, November, 20–21, 1995, CNR-RAISA, Rome.

Moresi, M. and Parente, E. 1999. Production of organic acids. *In* "Encyclopedia of Food Microbiology" (R.K. Robinson, C.A. Batt, and P.D. Patel, eds), pp. 705–717. Academic Press, New York.

Moresi, M. and Sappino, F. 1998. Effect of some operating variables on citrate recovery from model solutions by electrodialysis. *Biotechnol. Bioeng.* **59**, 344–350.

Moresi, M. and Sappino, F. 2000. Electrodialytic recovery of some fermentation products from model solutions: Techno-economic feasibility study. *J. Membr. Sci.* **164**, 129–140.

Morita, M., Sato, M., Kono, S., Hanada, F., Matsunaga, Y., and Kobayashi, T. 1996. Method for separation and purification of polybasic organic acids and its apparatus. *Jpn. Kokai Tokkyo Koho* JP no. 08,325,191, December, 10th, 1996.

Mourgues, J. 1993. Use of ion-exchange resins. *Rev. des Oenologues* **69**, 51–54.

Mourgues, J., Robert, L., and Hanine, H. 1997. Extraction and purification of *D,L*-malic acid, produced by chemical synthesis, *L*-malic acid produced by microbiological synthesis or susceptible to be recovered during the manufacturing of food products. *Industries Alimentaires et Agricoles* **114**, 379–384.

Moutounet, M., Saint-Pierre, B., Batlle, J.L., and Escudier, J.L. 1997. Tartrate stabilization: Principle and description of the procedure. *Revue Française d'Oenologie* **162**, 15–17.

Mucchetti, G. and Taglietti, P. 1993. Demineralization of whey and ultrafiltration permeate by electrodialysis. *Scienza e Tecnica Lattiero-Casearia* **44**, 51–62.

Nakagawa, M., Nakamura, I., and Kobayashi, T. 1975. Product separation from fermented liquors. V. Process for concentrating and purifying itaconic acid from a fermented liquor by electrodialysis. *Hakko Kogaku Zasshi* **53**(5), 286–293.

Nakagawa, M., Ishibashi, K., and Hironaka, K. 1991. Itaconic acid fermentation with pretreated beet thick juice and molasses by *Aspergillus terreus* K 26. *Obihiro Chikusan Daigaku Gakujutsu Kenkyu Hokoku, Dai-1-bu* **17**(2), 123–127.

Nasr-Allah, A. and Audinos, R. 1994. A novel electromembrane process for recovery of tartaric acid and of an alkaline solution from waste tartrates. *In* "Actes du Colloque of the Congres International sur le Traitement des Effluents Vinicoles", pp. 199–202. Narbonne and Epernay (F), June 20–24, 1994.

Nikonenko, V., Zabolotsky, V., Larchet, C., Auclair, B., and Pourcelly, G. 2002. Mathematical description of ion transport in membrane systems. *Desalination* **147**, 369–374.

Nikonenko, V., Lebedev, K., Manzanares, J.A., and Pourcelly, G. 2003. Modeling the transport of carbonic acid anions through anion-exchange membranes. *Electrochim. Acta* **48**, 3639–3650.

Nomura, Y., Iwahara, M., and Hongo, M. 1987a. Application of electrodialysis fermentation to *L*-lysine fermentation. *Nippon Nogei Kagaku Kaishi* **61**, 1293–1295.

Nomura, Y., Iwahara, M., and Hongo, M. 1987b. Lactic acid production by electrodialysis fermentation using immobilized growing cells. *Biotechnol. Bioengng.* **30**, 788–793.

Nomura, Y., Iwahara, M., and Hongo, M. 1988. Acetic acid production by an electrodialysis fermentation method with a computerized control system. *Appl. Environ. Microb.* **54**, 137–142.

Nomura, Y., Yamamoto, K., and Ishizaki, A. 1991. Factors affecting lactic acid production rate in the built-in electrodialysis fermentation an approach to high speed batch culture. *J. Ferment. Bioeng.* **71**, 450–452.

Nomura, Y., Iwahara, M., and Hongo, M. 1994. Production of acetic acid by *Clostridium thermoaceticum* in electrodialysis culture using a fermenter equipped with an electrodialyzer. *World J. Microb. Biotechnol.* **10**, 427–432.

Novalic, S., Jagschitz, F., Okwor, J., and Kulbe, K.D. 1995. Behaviour of citric acid during electrodialysis. *J. Membr. Sci.* **108**, 201–205.

Novalic, S., Okwor, J., and Klaus, D.K. 1996. The characteristic of citric acid separation using electrodialysis with bipolar membranes. *Desalination* **105**, 277–282.

Novalic, S., Kongbangkerd, T., and Kulbe, K.D. 1997. Separation of gluconate with conventional and bipolar electrodialysis. *Desalination* **114**, 45–50.

Office of Industrial Technologies. 2003. Production and separation of fermentation-derived acetic acid. *In* "Energy Efficiency and Renewable Energy," US Department of Energy, Washington, DC, (online publication, March, 2003, http://www.oit.doe.gov/chemicals/factsheets/acetic_acid.pdf).

Oka, M., Yoneto, K., and Yamaguchi, T. 1998. Process for producing 2-keto-L-gulonic acid using electrodialysis. USP no. 5,747,306, May, 5th, 1998.

Ono, T., Teramoto, T., and Sawada, M. 1992. Electrodialysis of fruit juices for reduction of acidity. *Jpn. Kokai Tokkyo Koho*, JP no. 04,349,874, December, 4th, 1992.

Paronetto, L. 1941. The application of electrodialysis to wines. *Annuar. Staz. Sper. Viticolt. Enol. Conegliano* **10**, 123–149.

Paronetto, L., Paronetto, L., and Braido, A. 1977. Some tests on tartrate stabilization of musts and wines by electrodialysis. *Vignevini* **4**, 9–15.

Pérez, A., Andrés, L.J., Alvarez, R., Coca, J., and Hill, C.G. 1994. Electrodialysis of whey permeates and retentates obtained by ultrafiltration. *J. Food Process Eng.* **17**, 177–190.

Pilat, B. 2001. Practice of water desalination by electrodialysis. *Desalination* **139**, 385–392.

Pinacci, P., Radaelli, M., Bottino, A., and Capannelli, G. 2004. Molasses purification by integrated membrane processes. *Filtration* (Coalville, United Kingdom), **4**(2), 119–122.

Prentice, G. 1991. "Electrochemical Engineering Principles". Prentice-Hall International, Englewood Cliffs, New Jersey, USA.

Quoc, A.L., Lamarche, F., and Makhlouf, J. 2000. Acceleration of pH variation in cloudy apple juice using electrodialysis with bipolar membranes. *J. Agric. Food Chem.* **48**, 2160–2166.

Reid, R.C., Prausnitz, J.M., and Poling, B.E. 1987. "The Properties of Gases and Liquids", 4th Ed., pp. 620–624. McGraw-Hill Book Co., New York.

Riponi, C., Nauleau, F., Amati, A., Arfelli, G., and Castellari, M. 1992. Electrodialysis. 2. Tartrate stabilization of wines by electrodialysis. *Revue Française d'Oenologie* **137**, 59–63.

Robinson, R.A. and Stokes, R.H. 2002. "Electrolyte Solutions", 2nd Revised Ed., pp. 143–154. Dover Publications, Mineola.

Roehr, M. and Kubicek, C.P. 1996. Further organic acids. *In* "Biotechnology, Products of Primary Metabolism" (H.J. Rehm and G. Reed, eds), 2nd Ed., Vol. 6, pp. 364–379. VCH, Verlagsgesellschaft MDH, Weinheim.

Sappino, F., Mancini, M., and Moresi, M. 1996. Recovery of sodium citrate from aqueous solutions by electrodialysis. *It. J. Food Sci.* **8**, 239–250.

Schippers, J.C. and Verdouw, J. 1980. The modified fouling index, a method of determining the fouling characteristics of water. *Desalination* **32**, 137–148.

Shaffer, L.H. and Mintz, M.S. 1966. Electrodialysis. *In* "Principles of Desalination" (K.S. Spiegler, ed.), pp. 189–199. Academic Press, New York.

Shaposhnik, V.A. and Kesore, K. 1997. An early history of electrodialysis with permselective membranes. *J. Membr. Sci.* **136**, 35–39.

Siebold, M., Frieling, V.P, Joppien, R., Rindfleisch, D., Schügerl, K., and Röper, H. 1995. Comparison of the production of lactic acid by three different lactobacilli and its recovery by extraction and electrodialysis. *Process Biochem.* **30**, 81–95.

Slack, A.W., Amundson, C.H., and Hill, C.G., Jr. 1986. Production of enriched β-lactoglobulin and α-lactalbumin whey protein fractions. *J. Food Process. Pres.* **10**, 19–30.

Solt, G. 1995. Early days in electrodialysis. *Desalination* **100**, 15–19.

Sonin, A.A. and Isaacson, M.S. 1974. Optimization of flow design in forced flow electrochemical systems with special application to electrodialysis. *Ind. Eng. Chem. Process Des. Develop.* **13**, 241–248.

Sridhar, S. 1987. Method for recovery of L-malic acid. *Ger. Offen*, DE 3,542,861, June, 11th, 1987.

Srikanth, G. 2004. Membrane separation processes—technology and business opportunities. *In* "News and Views, Technology Information, Forecasting & Assessment Council". (online publication:http://www.tifac.org.in/news/memb.htm).

Strathmann, H. 1992. Electrodialysis. *In* "Membrane Handbook" (W.S.W. Ho and K.K. Sirkar, eds), Chapter 5, pp. 217–262. Chapman & Hall, New York.

Stribley, R.C. 1963. Electrodialysis first food use. *Food Process* **24**, 49–51.

Sumimura, K., Ushijima, H., Hayakawa, K., and Ishiguro, Y. 2004. Removal of nitrate nitrogen from vegetable juice by concentration followed by electrodialysis. *Eur. Pat. Appl.,* EP no. 1,466,534, October, 13th, 2004.

Takatsuji, W., Nakauchi, M., and Yoshida, H. 1999. Removal of salt and organic acids from solution used to season salted Japanese apricots (Ume) by electrodialysis, precipitation and adsorption. *J. Biosci. Bioeng.* **88**, 348–351.

Teorell, T. 1953. Transport processes and electrical phenomena in ionic membranes. *Prog. Biophysics* **3**, 305–369.

Thampy, S.K., Narayanan, P.K., Trivedi, G.S., Gohil, D.K., and Indushekhar, V.K. 1999. Demineralization of sugar cane juice by electrodialysis. *Int. Sugar J.* **101**, 365–366.

Tokuyama Soda Co., Ltd., Japan. 1983. Reactivation of spent anion exchange membrane used in food processing. *Jpn. Kokai Tokkyo Koho* JP no. 58,122,006, July, 20th, 1983.

Tronc, J.S., Lamarche, F., and Makhlouf, J. 1998. Effect of pH variation by electrodialysis on the inhibition of enzymic browning in cloudy apple juice. *J. Agric.Food Chem.* **46**, 829–833.

Turek, M. 2002. Optimization of electrodialytic desalination in diluted solutions. *Desalination* **153**, 383–387.

Turek, M. 2003a. Cost effective electrodialytic seawater desalination. *Desalination* **153**, 371–376.

Turek, M. 2003b. Dual-purpose desalination-salt production electrodialysis. *Desalination* **153**, 377–381.

U.S. Department of Energy. 1999. Advanced electro-deionisation technology for product purification, waste recovery and water recycling. Office of Industrial technologies (http://www.oit.doe.gov).

Usseglio-Tomasset, L. and Bosia, P.D. 1978. Determinazione delle costanti di dissociazione dei principali acidi del vino in soluzioni idroalcooliche di interesse enologico. *Riv. Vitic. Enol. Conegliano* **31**, 380–405.

van der Ent, E.M., van Hee, P., Keurentjes, J.T.F., van't Riet, K., and van der Padt, A. 2002. Multistage electrodialysis for large-scale separation of racemic mixtures. *J. Membr. Sci.* **204**, 173–184.

van der Stegen, J.H.G., van der Veen, A.J., Weerdenburg, H., Hogendoorn, J.A., and Versteeg, G.F. 1999. Application of the Maxwell-Stefan theory to the transport in ion-selective membranes used in the choralkali electrolysis process. *Chem. Eng. Sci.* **54**, 2501–2511.

Varsel, C. 1980. Citrus processing as related to quality and nutrition. *In* "Citrus Nutrition and Quality" (S. Nagy and J.A. Attaway, eds), pp. 225–271. ACS Symposium Series no. 143, American Chemical Society, Washington DC.

Vera, E., Ruales, J., Dornier, M., Sandeaux, J., Sandeaux, R., and Pourcelly, G. 2003. Deacidification of clarified passion fruit juice using different configurations of electrodialysis. *J. Chem. Technol. Biotechnol.* **78**, 918–925.

Vetter, K.J. 1967. "Electrochemical Kinetics". Academic Press, New York.

Vonktaveesuk, P., Tonokawa, M., and Ishizaki, Y. 1994. Stimulation of the rate of *L*-lactate fermentation using *Lactococcus lactis* IO-1 by periodic electrodialysis. *J. Ferment. Bioeng.* **77**, 508–512.

Voss, H. 1986. Deacidification of citric acid solutions by electrodialysis. *J. Membr. Sci.* **27**, 165–171.

Wangnick, K. 1998. IDA Worldwide Desalting Plants Inventory Report No. 15. Produced by Wangnick Consulting for International Desalination Association.

Watkins, E.J. and Pfromm, P.H. 1999. Capacitance spectroscopy to characterize organic fouling of electrodialysis membranes. *J. Membr. Sci.* **162**, 213–218.

Weier, A.J., Glatz, B.A., and Glatz, C.E. 1992. Recovery of propionic and acetic acids from fermentation broths by electrodialysis. *Biotechnol. Prog.* **8**, 479–485.

Wen, T., Solt, G.S., and Gao, D.W. 1996. Electrical resistance and Coulomb efficiency of electrodialysis (ED) apparatus in polarization. *J. Membr. Sci.* **114**, 255–262.

Williams, A.W. and Kline, H.A. 1980. Electrodialysis of acid whey. USP no. 4,227,981, October, 14th, 1980.

Wucherpfennig, K. 1975. The inhibition of tartrate precipitation in grape juice concentrate by means of electrodialysis. International Federation of Fruit Juice Producers. Scientific-Technical Commission Report 13, pp. 73–117.

Wucherpfennig, K. and Krueger, R. 1975. Stabilization of grape juice and wine against tartar by means of electrodialysis. *In* "Proceedings of the International Symposium on Separation Processes Membr. Ion-Exch. Freeze-Conc. Food Ind.", A.P.R.I.A., Paris.

Xuemei, L., Jianping, L., Mo'e, L., and Peilin, C. 1999. L-lactic acid production using immobilized *Rhizopus oryzae* in a three-phase fluidized-bed with simultaneous product separation by electrodialysis. *Bioprocess Eng.* **20**, 231–237.

Yabannavar, V.M. and Wang, D.I.C. 1991. Extractive fermentation for lactic acid production. *Biotechnol. Bioeng.* **37**, 1095–1100.

Yamamoto, K., Ishizaki, A., and Stanbury, P.F. 1993. Reduction in the length of the lag phase of *L*-lactate fermentation by the use of inocula from electrodialysis seed cultures. *J. Ferment. Bioeng.* **76**, 151–152.

Yen, P.L.-H., Bajpai, R.K., and Iannotti, E.L. 1991. An improved kinetic model for lactic acid fermentation. *J. Ferment. Bioeng.* **71**(1), 75–77.

Yen, Y.-H. and Cheryan, M. 1993. Electrodialysis of model lactic acid solutions. *J. Food Eng.* **20**, 267–282.

Zall, R.R. 1992. Sources and composition of whey and permeate. *In* "Whey and Lactose Processing" (J.G. Zadow, ed.), Chapter 1, pp. 1–72. Elsevier Applied Science, London.

Zelic, B., Gostovic, S., Vuorilehto, K., Vasic-Racki, D., and Takors, R. 2004. Process strategies to enhance pyruvate production with recombinant *Escherichia coli*: From repetitive fed-batch to *in situ* product recovery with fully integrated electrodialysis. *Biotechnol. Bioeng.* **85**, 638–646.

Zemel, G.P., Sims, C.A., Marshall, M.R., and Balaban, M. 1990. Low pH inactivation of polyphenol oxidase in apple juice. *J. Food Sci.* **55**, 562–565.

Zhang, S.T., Matsuoka, H., and Toda, K. 1993. Production and recovery of propionic and acetic acids in electrodialysis culture of *Propionibacterium shermanii*. *J. Ferment. Bioeng.* **75**, 276–282.

INDEX